DRUG METABOLISM
IN DRUG DESIGN
AND DEVELOPMENT

The Wiley Bicentennial—Knowledge for Generations

$\mathcal{E}$ach generation has its unique needs and aspirations. When Charles Wiley first opened his small printing shop in lower Manhattan in 1807, it was a generation of boundless potential searching for an identity. And we were there, helping to define a new American literary tradition. Over half a century later, in the midst of the Second Industrial Revolution, it was a generation focused on building the future. Once again, we were there, supplying the critical scientific, technical, and engineering knowledge that helped frame the world. Throughout the 20th Century, and into the new millennium, nations began to reach out beyond their own borders and a new international community was born. Wiley was there, expanding its operations around the world to enable a global exchange of ideas, opinions, and know-how.

For 200 years, Wiley has been an integral part of each generation's journey, enabling the flow of information and understanding necessary to meet their needs and fulfill their aspirations. Today, bold new technologies are changing the way we live and learn. Wiley will be there, providing you the must-have knowledge you need to imagine new worlds, new possibilities, and new opportunities.

Generations come and go, but you can always count on Wiley to provide you the knowledge you need, when and where you need it!

WILLIAM J. PESCE
PRESIDENT AND CHIEF EXECUTIVE OFFICER

PETER BOOTH WILEY
CHAIRMAN OF THE BOARD

DRUG METABOLISM IN DRUG DESIGN AND DEVELOPMENT

Basic Concepts and Practice

EDITED BY

DONGLU ZHANG

MINGSHE ZHU

W. GRIFFITH HUMPHREYS

WILEY-INTERSCIENCE
A JOHN WILEY & SONS, INC., PUBLICATION

Published by John Wiley & Sons, Inc., Hoboken, New Jersey
Published simultaneously in Canada

For general information on our other products and services or for technical support, please contact our Customer Care Department within the United States at (800) 762-2974, outside the United States at (317) 572-3993 or fax (317) 572-4002.

Wiley also publishes its books in a variety of electronic formats. Some content that appears in print may not be available in electronic formats. For more information about Wiley products, visit our web site at www.wiley.com.

Wiley Bicentennial Logo: Richard J. Pacifico

Library of Congress-in-Publication Data:

Drug metabolism in drug design and development : basic concepts and practice / edited by Donglu Zhang, Mingshe Zhu, and W. Griffith Humphreys.
    p. ; cm.
  Includes bibliographical references.
  ISBN 978-0-471-73313-3 (cloth)
1. Drugs–Metabolism. 2. Drugs–Design. 3. Drug development.
  I. Zhang, Donglu. II. Zhu, Mingshe. III. Humphreys, W. Griffith.
[DNLM: 1. Pharmaceutical Preparations–metabolism. 2. Drug Design. 3. Drug Evaluation–methods. 4. Pharmacokinetics. QV 38 D79367 2007]
  RM301.55.D79 2007
  615'.7–dc22

                  2007017661

Printed in the United States of America
10 9 8 7 6 5 4 3 2 1

CONTENTS

12 Introduction to NMR and Its Application in Metabolite Structure Determination

*Xiaohua Huang, Robert Powers, Adrienne Tymiak, Robert Espina,
and Vikram Roongta*

PREFACE

Information on the metabolism and disposition of candidate drugs has become a critical part of all aspects of the drug discovery and development process. This comprehensive involvement of drug metabolism information has been brought about by a desire for quality design at an early stage, sometimes referred to as designing good "developability" characteristics, and then to work proactively with clinical and safety organizations to impact the design of the various development programs. This desire is driven by the need to reduce attrition rates as a means to effectively lower the cost of drug development.

Drug metabolism information in the early stages of discovery can help guide medicinal chemistry efforts toward optimization of preclinical safety and efficacy properties. This approach can be made even more effective with the active involvement of other disciplines such as pharmaceutics and toxicology. Candidates can be optimized by examining a variety of parameters beyond potency and efficacy. During the development stages drug metabolism information can help guide drug–drug interaction and special population clinical studies. Metabolism information is also critical for designing toxicology studies to that ensure the safety of metabolites is adequately tested and can also be a key part of addressing whether toxicology found in animals is likely to translate to humans.

Drug metabolism, as practiced in the pharmaceutical industry, is a multidisciplinary field that requires knowledge of analytical technologies, expertise in mechanistic and kinetic enzymology, organic reaction mechanism, pharmacokinetic analysis, animal physiology, basic chemical toxicology, preclinical pharmacology, and molecular biology. Scientists entering the field from academia often receive coursework in many of the above areas, but have usually focused the bulk of their research efforts on only one of above mentioned fields. It often requires a number of years of practice for a new scientist to gain a comprehensive understanding of all the disciplines necessary to apply drug metabolism knowledge effectively to the drug discovery and development processes.

This book offers background information as well as practical descriptions of what happens during the drug design and development process. Emphasis will be placed on issues such as what data are needed, what experiments and analytical methods are typically employed, and how to interpret and apply data. The chapters of this book will highlight facts, detailed experimental designs, applications, and limitations of techniques.

The book was not intended to be a collection of individual reviews, rather a coherent integration of all relevant background information as well as detail of the experimental strategies and processes necessary for drug metabolism research during drug design and development. Authors aimed at providing a balanced, comprehensive perspective on their subject matter and were encouraged to include a full range of experimental approaches. The book contains four parts that should serve to integrate the entire process: Part I, Basic Concepts of Drug Metabolism; Part II, Role of Drug Metabolism in Pharmaceutical Industry; Part III, Analytical Techniques in Drug Metabolism; Part IV, Common Experimental Approaches and Protocols. This structure should provide a valuable resource to researchers seeking to broaden their knowledge of drug metabolism science as practiced in the modern pharmaceutical industry.

Donglu Zhang
Mingshe Zhu
Griff Humphreys

CONTRIBUTORS

Upendra Argikar, Metabolism and Pharmacokinetics Global Discovery Chemistry, Novartis Institutes for BioMedical Research, Inc., 250 Massachusetts Avenue, Cambridge, MA 02139 upendra.argikar@novartis.com

Suresh K. Balani, DMPK/NCDS, Millennium Pharmaceuticals, Inc., 40 Landsdowne St., Cambridge, MA 02139 suresh.balani@mpi.com

Swapan K. Chowdhury, Department of Drug Metabolism and Pharmacokinetics, Schering-Plough Research Institute, 2015 Gal-loping Hill Road, Kenilworth, NJ 07033 swapan.chowdhury@spcorp.com

S. Nilgun Comezoglu, Bristol-Myers Squibb Co., P.O. Box 4000 Mailstop: LVL F13-02, Princeton, NJ 08543 s.nilgun.comezoglu@bms.com

Renke Dai, Preclinical Research Department, Guangzhou Institute of Biomedicine and Health, Chinese Academy of Sciences, Guangzhou Science Park, China 510663 dai_renke@qibh.ac.cn

Robert Espina, Wyeth Research, 500 Arcola Road, Collegeville, PA 19426 espinaj@wyeth.com

Jinping Gan, Bristol-Myers Squibb Co., P.O. Box 4000 Mailstop: LVL MS 17-2.12, Princeton, NJ 08543 Jinping.gan@bms.com

Liang-Shang (Lawrence) Gan, Drug Metabolism and Pharmacokinetics Drug Discovery Support, Boehringer-Ingelheim Pharmaceuticals, Inc., 900 Ridgebury Road, Rigdefield, CT 06877 gan@rdg.boehringer-ingelheim.com

Scott J. Grossman, Bristol-Myers Squibb Co., P.O. Box 4000 Mailstop: LVL D2-497, Princeton, NJ 08543 scott.grossman@bms.com

F. Peter Guengerich, Vanderbilt University School of Medicine, Center in Molecular Toxicology, 638 Robinson Research Building, 23rd & Pierce Avenues, Nashville, TN 37232-0146 f.guengerich@vanderbilt.edu

Thomas M. Guenther, Department of Pharmacology, College of Medicine, University of Illinois at Chicago, 835 South Wolcott Avenue, Chicago, IL 60612 tmg@uic.edu

Steven Hansel, Pfizer, Global Department of Pharmacodynamics, Dynamics, and Metabolism, 2800 Plymouth Road, Ann Arbor, Ml 48105 Steven.hansel@pfizer.com

Xiaohua Huang, Bristol Myers Squibb Co., 5 Research Parkway, Wallingford, CT 06492 xiaohua.huang@bms.com

W. Griffith Humphreys, Bristol-Myers Squibb, P.O. Box 4000 Mailstop: LVL F13-04, Princeton, NJ 08543 william.humphreys@bms.com

Susan Hurst, Pfizer, Global Department of Pharmacodynamics, Dynamics, and Metabolism, 2800 Plymouth Road, Ann Arbor, Ml 48105 susan.hurst@pfizer.com

Ramaswamy Iyer, Britstol-Myers Squibb Co., P.O. Box 4000 Mailstop: LVL F13-07, Princeton, NJ 08543 Ramaswamy.iyer@bms.com

Laurence S. Kaminsky, Laboratory of Human Toxicology and Molecular Epidemiology, Wadsworth Center, P.O. Box 509, Albany, NY 12201 kaminsky@wadsworth.org

Gang Luo, Metabolism and Pharmacokinetics, Bristol-Myers Squibb Co., 311 Pennington Rockyhill Road, Pennington, NJ 08534 gang.luo@bms.com

Shuguang Ma, Amgen, Inc., PKDM, One Amgen Center Dr, Thousand Oaks, CA 91320 sma@amgen.com

Swati Nagar, Assistant Professor, Department of Pharmaceutical Sciences, School of Pharmacy, Temple University, Philadelphia, PA 19140 swati.nagar@temple.edu

Robert Powers, University of Nebraska-Lincoln, Department of Chemistry, 722 Hamilton Hall, Lincoln, NE 68588 rpowers3@unl.edu

Rory Remmel, University of Minnesota, Dept of Medicinal Chemistry, 308 Harvard St SE, Minneapolis, MN 55455 remme001@tc.umn.edu

A. David Rodrigues, Bristol-Myers Squibb Co., P.O. Box 4000 Mailstop: LVL F14-04, Princeton, NJ 08543 david.rodrigues@bms.com

Vikram Roongta, Bristol-Myers Squibb Co., P.O. Box 4000, Princeton, NJ 08543 vikram.roongta@bms.com

Magang Shou, Amgen, Inc., Department of Pharmacokinetics and Drug Motabolism, One Amgen Center Drive, Thousand Oaks, CA 91320-1799 mshou@amgen.com

Michael W. Sinz, Bristol-Myers Squibb Co., 5 Research Parkway, Mailstop: WFD 3AB-525, Wallingford, CT 06492 michael.sinz@bms.com

Adrienne Tymiak, Bristol-Myers Squibb Co., P.O. Box 4000, Princeton, NJ 08543 Adrienne.tymiak@bms.com

Timothy S. Tracy, Department of Experimental and Clinical Pharmacology, College of Pharmacy, University of Minnesota, 7-115B Weaver-Densford Hall, 308 Harvard St. SE, Minneapolis, MN 55455 tracy017@umn.edu

Xiaoxiong "Jim" Wei, Office of Clinical Pharmacology, Food and Drug Administration, WO Bldg 21, Room 4660, 10903 New Hampshire Ave, Silver Spring, MD 20993 Xiaoxiong.wei@fda.hhs.gov

J. Andrew Williams, Pfizer, Global Department of Pharmacodynamics, Dynamics, and Metabolism, 2800 Plymouth Road, Ann Arbor, MI 48105 James Williams2@pfizer.com

Cindy Q. Xia, DMPK/NCDS, Millennium Pharaceuticals, Inc., 40 Landsdowne St., Cambridge, MA 02139 Cindy.Xia@mpi.com

Johnny J. Yang, DMPK/NCDS, Millennium Pharmaceuticals, Inc., 40 Landsdowne St., Cambridge, MA 02139 johnny.yang@mpi.com

Donglu Zhang, Bristol-Myers Squibb Co., P.O. Box 4000 Mailstop: LVL F13-09, Princeton, NJ 08543 donglu.zhang@bms.com

Hongjian Zhang, Bristol-Myers Squibb Co., P.O. Box 4000 Mailstop: LVL F13-07, Princeton, NJ 08543 hongjian.zhang@bms.com

Zhi-Yi Zhang, Drug Disposition Department, Eisal Research Institute, 4 Corporate Drive, Andover, MA 01810 zhi-yi zhang@eri.eisai.com

Weiping Zhao, Bristol-Myers Squibb Co., P.O. Box 4000 Mailstop: LVL F13-01, Princeton, NJ 08543 Weiping.zhao@bms.com

Zhoupeng Zhang, Merck & Co. Inc., 126 E. Lincoln Avenue, Rahway, NJ 07065 zhoupeng.zhang@merck.com

Mingshe Zhu, Bristol-Myers Squibb Co., P.O. Box 4000 Mailstop: LVL F13-01, Princeton, NJ 08543 mingshe.zhu@bms.com

PART I

BASIC CONCEPTS OF DRUG METABOLISM

1

OVERVIEW: DRUG METABOLISM IN THE MODERN PHARMACEUTICAL INDUSTRY

Scott J. Grossman

1.1 INTRODUCTION

It is interesting to contrast contemporary pharmaceutical biotransformation with that practiced by R.T. Williams. The fundamental objectives are virtually unchanged, to characterize the disposition of a drug in animals. In addition, then and now the routes of excretion and overall molecular transformation are still, arguably, the most important aspects of the discipline. However, in the intervening years the scope of technological advancement, scientific breadth of knowledge, and range of impact has expanded in a manner that could not have been foreseen. This chapter will give an overview of biotransformation as it is practiced in the pharmaceutical industry today.

The role of any pharmaceutical biotransformation scientist is to characterize the disposition of a drug to relate this to overall safety and efficacy. The range of information needed to characterize overall disposition is so broad that it is unlikely any single scientist will accomplish the entire *characterization* alone. However, it is critically important that the entire disposition process is thoroughly understood, and then intelligently integrated with other pertinent aspects of the drug's behavior. The history of contemporary pharmaceutical industry is replete with examples of how the lack of fundamental scientific knowledge (e.g., mechanism and effects of enzyme induction), appreciation

Drug Metabolism in Drug Design and Development,   Edited by Donglu Zhang, Mingshe Zhu and W. Griffith Humphreys
Copyright © 2008 John Wiley & Sons, Inc.

of known metabolic effects (e.g., metabolic activation to toxic reactive metabolites), or incomplete integration of existing information (e.g., drug–drug interactions) led to drastically adverse outcomes. It could be argued that proper integration of information is both more difficult and important than the process of collecting the data itself. Thus, the challenge to the scientist today is to be able to comprehend decades of scientific knowledge, master an array of sophisticated technology, and integrate a diverse range of information to form a sound understanding of a drug's ultimate clinical behavior.

1.2 TECHNOLOGY

There is now an awe-inspiring array of technology available to aid the study of drug disposition. Consider that what once may have taken Williams nearly 6 months to accomplish, might only take about 20 min for a contemporary biotransformation scientist. This modern armamentarium has done much to integrate the power of biotransformation into pharmaceutical discovery and development. However, this tremendous evolution in technology presents its own set of dilemmas.

Taking full advantage of any technology requires an understanding of the technology itself. Fortunately, software and hardware engineering have greatly simplified common use of very sophisticated technologies. The LC/MS/MS instrument today is as common as the HPLC diode array UV instrument 15 years ago. This easy accessibility was greatly facilitated through robust instrument design and great software engineering.

Increasingly, the dilemma is not so much instrument access, as it is a thoughtful choice of exactly what experimental approaches and technology should be chosen to answer the question at hand. The biotransformation scientist is obliged to stay aware of technological innovations of all sorts, including instrumentation. However, the ultimate challenge should always be how to answer the most critical questions in the soundest way. True mastery of technology allows the scientific approach to follow naturally. The temptation to throw technological "sleights of hand" at a problem is often hard to resist.

Every technology has its inherent limits. Often, the specificity that enables prodigious sensitivity can also be a powerful filter of other important information. A rigorous biotransformation scientist is able to stand back and thoughtfully interrogate the strength of her own conclusions, including the technological blind spots of the approach. With thoughtful consideration, complementary technology may be applied judiciously to either flesh out a previous area of ambiguity or address the question from an entirely different perspective. In either case, scientific credibility is served well.

1.3 BREADTH OF SCIENCE

1.3.1 Chemistry

Biotransformation is fundamentally a chemical process. Likewise, the most frequently employed and valuable studies make heavy use of analytical and bioorganic chemistry. Over time, the underlying technology has become sufficiently complex that subspecialization in individual analytical techniques is common. For example, nuclear magnetic resonance spectroscopy (NMR) is invaluable for many unambiguous metabolite structural assignments. In most pharmaceutical companies, NMR specialists are employed to completely master the various facets of the technology. In many cases, these scientists will create sophisticated coupling and decoupling sequences to provide highly specific structural information. Often, their training also makes them most qualified to interpret all forms of NMR spectroscopic data. However, the "complete" biotransformation scientist will, at a minimum, know how to employ NMR spectroscopy to advance their structural understanding of a metabolite. Increasingly, the use of heteronuclear decoupling experiments is considered almost routine in the art.

Furthermore, biotransformation scientists are often fully capable of interpreting the spectra to deduce structure and are also able to recognize when such spectra still leave absolute structural assignments tentative. When one then considers the broader range of additional spectroscopic and chromatographic techniques employed in biotransformation studies, one soon recognizes the degree of technical sophistication required to be an effective biotransformation scientist.

Often, the definitive elucidation of a molecule's metabolic pathway is considered the ultimate goal of biotransformation studies. Proper application of analytical techniques, for the most part, will often be sufficient to achieve this goal. However, as often as it is "good enough" to simply define *what* has happened to a molecule, there are probably twice as many instances where it is also important to understand *how* these changes happened. The best biotransformation scientists are usually good "electron pushers." That is, their knowledge of bioorganic chemistry allows them to understand the mechanism of the molecular rearrangements taking place in each biotransformation process. They are able to both rationalize most biotransformations in a mechanistic sense and recognize when a proposed metabolite structure seems untenable. It is not uncommon to encounter a set of spectroscopic data that seems quite inconsistent with the parent molecule. In these cases, the fundamental principles of bioorganic chemistry are employed to rationalize putative structures that would be consistent with the data.

Increasingly, the roles of medicinal chemists and biotransformation scientists intersect in the discipline of bioorganic chemistry. Frequently, they share a mutual interest in decreasing metabolic liability through structural modification as well as avoiding creation of reactive metabolites

through informed molecular design. Fortunately, their common understanding of bioorganic chemistry also greatly facilitates the intelligent
redesign of structures to mitigate these liabilities. At its best, this requires
the best of both disciplines and each scientist can develop a deeper
fundamental understanding of the other's craft.

1.3.2 Enzymology and Molecular Biology

Although each of these disciplines could be discussed separately, for the
contemporary biotransformation scientist these areas are intimately inter-
twined. Since biotransformations are enzyme mediated, complete understanding of xenobiotic disposition is only achieved when one also considers the
role and impact of the individual enzymes involved.

Enzymological techniques allow the study of individual enzymatic reactions
as well as the role of individual enzymes in complex systems. Each of the
questions "What happens?" "What enzymes contribute?" "How does it
happen?" will require separate techniques. It is not unusual to ask and answer
these questions in a very short period of time. This obviously requires a certain
degree of breadth, versatility, and flexibility along with a fundamentally strong
understanding of the literature.

Cells and subcellular fractions from humans and many preclinical species
are readily available. These reagents make it possible to make interspecies
extrapolations easily. At one time, a major reason cited for early drug attrition
was pharmacokinetic failure, attributable to the difficulty in extrapolating
pharmacokinetic behavior from animals to humans. In this author's experience, unexpected pharmacokinetic performance in humans is now a rare event.
In addition, it is now commonplace to obtain very mechanistic information
revealing the probability of observing quite specific molecular events (e.g.,
toxicity) in humans (Mutlib et al., 2000).

While the availability of trans-species enzyme systems has had a major
impact, advances in molecular biology have also enabled the query of
increasingly sophisticated questions. Molecular biological methods have made
it possible to clone and express enzymes to study reactions at a molecular level.
This has improved our ability to study enzyme reactions at a fine molecular
level, to discern the contributions of individual enzymes in complex systems,
and even to employ them as "bioreactors" to generate small quantities of
metabolite standards.

The basis for many metabolizing enzyme polymorphisms is becoming
better understood, allowing one to anticipate potential interindividual
disposition differences. Molecular biological techniques have defined the
basis for polymorphisms and have described the distribution of the variants
in a population. It is now quite easy to discern whether a drug may
behave differently in one individual compared to another and to even exclude
anticipated poor responders from trials in a controlled fashion (Murphy
et al., 2000).

The means by which enzyme systems are regulated are now being appreciated and studied in a mechanistic fashion. Tools available today make it possible to screen against enzyme inducers as well as inhibitors in a relatively inexpensive, well-defined fashion.

1.4 IMPACT OF DRUG METABOLISM ON EFFICACY AND SAFETY

Even in the simplest case, a drug that is injected intravenously and excreted completely unchanged in the urine, there are likely important implications to the human risk/benefit evaluation. Is the excretion so fast that efficacy is compromised? Will the dose need to be adjusted in patients with compromised renal function? Will high drug concentrations in the urinary tract lead to important safety concerns? The biotransformation scientist who only asks "What is happening?" without "What could it mean?" is missing an opportunity to play a larger role in making important decisions. In fact, it can be argued that the biotransformation scientist is perhaps *best* suited to raise these concerns and is neglecting a critical aspect of their profession by not leading these discussions.

1.4.1 Efficacy

At the earliest stages of drug discovery, an important transition must be made from the screening well to the functioning cell. Even at this stage, there are often significant hurdles related to biotransformation. At every step along the way to higher levels of biological organization, biotransformation inevitably imposes further challenges to the goal of therapeutic efficacy. Understandably then, significant time and resource in drug discovery is spent optimizing a molecule's disposition properties. Perhaps it is more precise to say that much effort is put into the overall process of molecular optimization to yield a molecule with acceptable disposition properties. This distinction, though subtle, is critically important. For once a molecule is made, its properties are cast and its biological fate cannot be changed. Thus, it is critically important in drug discovery to get the optimization done right.

Few would argue that molecular optimization to achieve adequate pharmacokinetic properties is a high priority in early discovery. Practically speaking, much of this work could be accomplished with little biotransformation insight. By using *in vitro* and/or *in vivo* models, a chemistry team may certainly achieve the necessary degree of optimization. However, even when the optimization comes as a result of a well-developed *sense* of SAR, one recognizes that substantial amounts of intuition and good fortune were also necessary. Luck is fleeting and intuition has its limits. This is particularly true when there is little baseline data and the problem is complex. Thus, a purely empirical pharmacokinetic approach is not likely to be the most efficient path for success.

Pharmacokinetic optimization can be greatly aided with further biotrans-formation information. A limited disposition study can be extremely useful. Simply looking for intact drug in urine and bile, one may be able to discern significant clearance by renal or biliary excretion. Neither of these disposition routes are normally modeled by high throughput *in vitro* clearance assays. One may quickly learn that information from *in vitro* screening is not likely to have the desired benefits. Unfortunately, given current state of knowledge of transporter ligand affinity, screening in these instances is likely to remain a largely "black box" screening effort with *in vivo* models.

A simple study designed to identify biotransformation "hot spots" is frequently invaluable during pharmacokinetic optimization. Samples from either *in vitro* or *in vivo* studies analyzed by HPLC with parallel UV/MS detectors can often quickly identify those aspects of a chemotype most susceptible to metabolism. Now the challenge becomes an exercise of molecular modifications, informed by knowledge of the area of the molecule needing attention.

1.4.2 Safety

By definition, xenobiotic metabolism considers how an organism disposes of a foreign chemical. It is the study of what the body does to the drug. Whether intentional or unintentional, these xenobiotics often have physiological effects. Thus, a major role for biotransformation is to understand how metabolic processes terminate or limit desired physiological effects (efficacy) as well as how other processes may lead to unintended consequences (toxicity).

A drug's duration of action, its intensity of action, and interindividual variability in responsiveness are frequently related to its disposition properties. For drugs with a narrow therapeutic index, these sources of variability can and do lead to adverse effects and may significantly limit the full therapeutic usefulness of the product. Likewise, drug–drug interactions also lead to unintended effects. As an inhibitor or inducer of enzymes involved in the disposition of other co-medications the drug may cause exacerbated pharmacological effects (inhibitors) or therapeutic lapses (inducers). Again, drugs of this nature may have severely restricted use, depending on the therapeutic utility and the co-medication environment in which they would be used. Thus, without even considering how a drug is metabolized, safety can be affected.

Dr. James Gillette, the Millers, their coworkers and colleagues, and generations after them have documented how molecular biotransformation leads to toxicity (Brodie et al., 1971; Miller and Miller, 1955). Molecular activation (or biological reactive intermediates) is one of the most intensively studied aspects of both drug metabolism and toxicology. Thousands of publications have documented the breadth of reactions leading to reactive metabolites, and thousands of others have shown the breadth of impact throughout the body and among all species. Consequently, there is a well-developed basis for anticipating structural features that may predispose a molecule to form reactive metabolites. Once discovered, reactive metabolites can often be avoided or minimized by judicious molecular redesign. In fact, both biotransformation scientists and medicinal chemists are obligated to know this area. This knowledge facilitates design of molecules without known liabilities, or at least guides the incorporation of certain worrisome features in a way that can be carefully evaluated.

Perhaps because of the well-developed literature linking biotransformation and toxicity, there seems to be a widely held perception that "most toxicity is due to metabolism." This author does not subscribe to that thesis and will not discuss it further here (Grossman, 2006). However, xenobiotic-induced toxicity is a substantial issue to be dealt with. By most accounts, toxicity is the single most common cause for drug attrition. It is inconceivable that a contemporary pharmaceutical biotransformation scientist will not be involved in toxicity-related investigations in their career, and probably will be involved many times. However, it is human nature to view everything as a nail if you are a hammer. The most tempting course of action for a biotransformation scientist is to "start with the molecule" and posit putative reactive metabolites that could give rise to the observed effects. An alternate approach is to "start with the lesion" and query the pathophysiological drivers that give rise to observed effects. This would include the consideration of unanticipated interactions of the parent molecule or its stable metabolites with any of the 40,000 gene products expressed in the affected organism. Either approach, applied with prudence and substantial good fortune, can yield the answer. With maturity and discipline, the biotransformation scientist learns to dissect toxicology issues through the scientific method, proposing hypotheses and carefully designing experiments to eliminate false hypotheses in a definitive fashion.

1.5 REGULATORY IMPACT AND IP POSITION

Understandably, the majority of biotransformation scientists would not readily consider intellectual property a fundamental aspect of their work. Likewise, the regulatory impact of findings (or their lack of study) is infrequently thought of in the normal course of work. However, it is inarguable that if a drug is not registered there will be no derived therapeutic benefit. Without therapeutic benefit, the entire discovery and development endeavor is simply a very, very expensive set of experiments.

Registration and marketing approval of a new chemical entity (NCE) is increasingly difficult. In short, the era of the contemporary pharmaceutical industry has provided a long list of "lessons learned" from the accumulation of good and bad experiences. The inevitable evolution and improvements derived from this have benefited society with more efficacious and better tolerated drugs. As a consequence, the formalization of these "lessons learned" has created a seemingly incomprehensible list of points to consider, document, and evaluate in the risk/benefit decision-making process. Global regulatory bodies are charged with promulgating these various considerations through guidances and regulations. Further, it is their responsibility to act as an advocate for the consumer, wherein a consistent set of practices will be applied to form a judgment regarding benefit versus risk, devoid of bias from financial considerations. Within this arena, the complete body of a drug's study must be summarized intelligently such that all pertinent facts are properly accounted for and put into perspective.

If one now considers this final full evaluation as the ultimate basis for a drug's approval, it becomes clear that one must take additional steps to evaluate the data one produces. In short, one really needs to ask "I know I like what the data say, but will someone else accept the data as readily as I did and see it in the same light?" One must ensure that the proper scope of data has been collected (to meet regulatory standards). The data must be collected and documented in a proper fashion (to meet GLP requirements) and attain a proper degree of precision and accuracy in the samples studied (to weigh the value of derived conclusions). Information must be integrated and described in a fashion that tells a compelling story, and importantly, it must be looked at from many different perspectives to ensure that there are no holes or flaws that undermine important decisions. In the end, none of this is done without consistently good science. But one must also recognize that substantial good judgment and perspective is necessary that comes only through experience.

If the drug application is successful, the pharmaceutical sponsor must create a "label" that guides the physician's prescribing practice. The statements and claims in a label are negotiated between the pharmaceutical sponsor and regulatory agencies. Ultimately, the label is the physician's most accessible source of information describing the drug's various properties and how these properties may affect the drug's impact on patients. As mentioned previously, knowing what the body does to the drug can often be as valuable as knowing what the drug does to the body. A well-written label should give a concise understanding of the drug's disposition properties, particularly when dispositional characteristics may result in widely varying responses among patients. Given the frequency of drug registrations in any individual's career, the creation of the product label may be the ultimate pièce de résistance for the biotransformation scientist. It is likely their last chance to leave an indelible mark of their work, in this austere summary of many years of investigation.

It should not be overlooked either that label claims are often the basis of marketing advantages. It is quite common that a drug's dispositional characteristics provide benefit to the patient relative to other agents in the class. Without doubt, this differentiation may play a prominent role in subsequent marketing efforts. Thus, there is a financial imperative that differentiating features be perceived properly, as well as the ethical obligation that they are described accurately.

The cumulative cost of bringing a drug to market is staggering. Without some assurance that these costs will be recovered through eventual sales, it is economically untenable to take on the high risks of failure attendant with drug development. This fiscal assurance is provided, in part, through various patents assigned during discovery and development. Patents covering composition of matter are among the most valuable for a novel drug, as this allows the patent holder exclusive rights for a reasonable period of time. Eventually, much of the

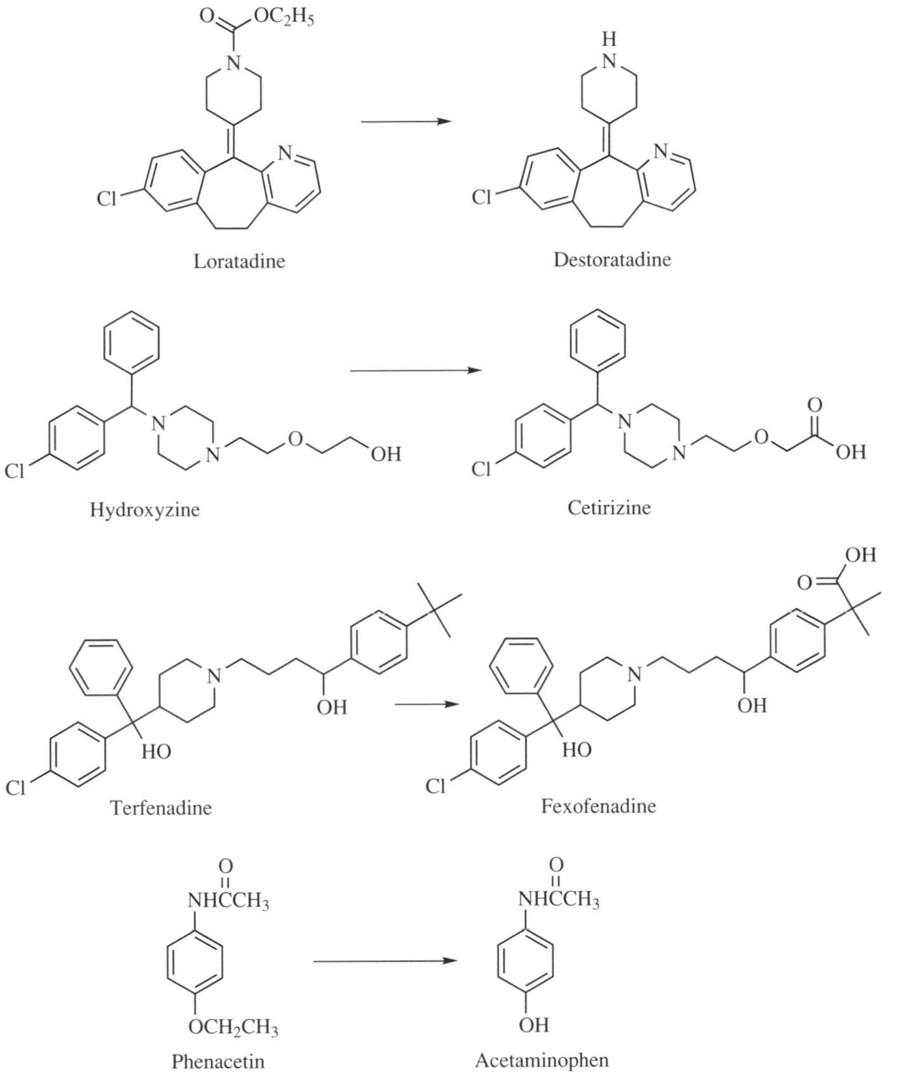

FIGURE 1.1 Examples of biotransformation reactions leading to a preferred marketed drug (Fura et al., 2004).

information learned from the drug's discovery and development is shared in the public domain. A carefully crafted patent protects the patent owner from having this substantial investment from being misappropriated.

The astute biotransformation scientist will readily recognize that many disposition-related events can affect the propriety of the original intellectual property. Absorption by itself can be a fairly complicated matter. In several cases, a better understanding of the processes affecting absorption has led to better versions of the original discoveries (e.g., prodrugs). Perhaps the most complex aspect of disposition relevant to this topic is biotransformation itself. The range of transformations can range from reasonably subtle chiral inversions to fairly dramatic intramolecular rearrangements. Along this continuum are a range of molecular conformations that may either retain much of the parent molecule's original attributes, or in many cases be either the actual entity responsible for activity or an improved version of the original. Therefore, it is important to contemplate these events and carefully characterize the range of molecular species derived. The incredible biosynthetic diversity of metabolism enzymes can infer important new structures never envisioned in the original patent claim. If not considered, it is quite possible, or even inevitable, that someone else will recognize this opportunity. Indeed, many prominent drugs are metabolism-related variants marketed by another company (Fig. 1.1).

1.6 SUMMARY

Drug disposition is an incredibly broad discipline. The range of skills and knowledge required to master this discipline is easy to underappreciate. While somewhat daunting in its own right, mastery of the various skills needed to acquire data is the easiest part of the process. The mark of a scientist will always be creating knowledge from data. For most scientists, the most satisfying aspect of their profession is the creation of highly impactful knowledge. This requires the foresight to ask the right questions, collect data in the right way, correctly integrating all of the information, and having the ability to apply this newfound knowledge in a meaningful way. Moreover, the astute scientist recognizes significant unexpected findings and has the perspective to see new and important questions raised in the process.

REFERENCES

Brodie BB, Reid WD, Cho AK, Sipes G, Krishna G, Gillette JR. Possible mechanism of liver necrosis caused by aromatic organic compounds. Proc Natl Acad Sci USA 1971;68(1):160–164.

Fura A, Shu YZ, Zhu M, Hanson RL, Roongta V, Humphreys WG. Discovering drugs through biological transformation: role of pharmacologically active metabolites in drug discovery. J Med Chem 2004;47(18):4339–4351.

Grossman, SJ. Are most toxicities caused by reactive metabolites? In: Molecular Toxicology, Josephy PD, Mannervik AB, editors. New York: Oxford University Press; 2006. p 373–374.

Miller EC, Miller JA. Biochemical investigations on hepatic carcinogenesis. J Natl Cancer Inst 1955;15(5 Suppl):1571–1590.

Murphy MP, Beaman ME, Clark LS, Cayouette M, Benson L, Morris DM, Polli JW. Prospective CYP2D6 genotyping as an exclusion criterion for enrollment of a phase III clinical trial. Pharmacogenetics 2000;10(7):583–590.

Mutlib AE, Gerson RJ, Meunier PC, Haley PJ, Chen H, Gan LS, Davies MH, Gemzik B, Christ DD, Krahn DF, Markwalder JA, Seitz SP, Robertson RT, Miwa GT. The species-dependent metabolism of efavirenz produces a nephrotoxic glutathione conjugate in rats. Toxicol Appl Pharmacol 2000;169(1):102–113.

2

OXIDATIVE, REDUCTIVE, AND HYDROLYTIC METABOLISM OF DRUGS

F. Peter Guengerich

2.1 INTRODUCTION

Oxidation, reduction, and hydrolysis reactions are common in the metabolism of drugs. In this chapter, each of these three types of chemistry will be treated briefly, with further division regarding individual enzymes. The analysis is brief. For more information about these enzymes, an older reference (Guengerich, 1997) covers each (plus the conjugation enzymes); in some cases more recent monographs are available, for example, for P450 (Ortiz de Montellano, 2005).

2.2 NOMENCLATURE AND TERMINOLOGY

Almost all of the enzymes to be discussed here exist in multigene families of varying complexity. These enzymes will be treated in a rather generic manner, with only reference to specific enzymes as examples or to specific reactions. With almost all of the enzymes under consideration, the individual genes and proteins are organized into specific families using a system of letters and numbers. The use of the terms "isozymes" and "isoforms" is no longer encouraged in that these are simply distinct "enzymes" that are organized into large gene families. Isozymes, by definition, catalyze the same reaction. With most of the systems under consideration here, the emphasis is on selectivity toward different reactions.

Drug Metabolism in Drug Design and Development, Edited by Donglu Zhang, Mingshe Zhu and W. Griffith Humphreys
Copyright © 2008 John Wiley & Sons, Inc.

Another issue is that the use of the terms Phase I, Phase II, and Phase III is discouraged, for a number of reasons (Josephy et al., 2005). First, this older terminology implies a temporal relationship, which is definitely not the case. Hydrolytic reactions have often been grouped with oxidation/reduction in Phase I but have more chemistry in common with conjugation (Phase II). Grouping reactions in these "phase" groups has no real usefulness, is confusing, and has been eliminated in the author's teaching and writing in this area.

The concept has often been taught that the drug metabolism process is one of making compounds more polar to facilitate excretion. This view has some general validity but many exceptions exist.

2.3   GENERAL FEATURES OF THE ENZYMES

As already mentioned, these enzymes are usually found in multigene families that can be complex (e.g., cytochrome P450 (P450)). The selectivity of the individual substrates and reactions is influenced by the nature of the protein (and ultimately, its amino acid sequence). This selectivity is a very significant issue in drug development, although it will be mentioned only in some of the examples used here. In some cases, the selectivity may be quite strict among the individual forms of an enzyme, for example, only P450 2D6 catalyzes debrisoquine 4-hydroxylation (Distlerath et al., 1985; Guengerich, 2005). However, in other cases, several enzymes within a group or even enzymes from different groups can catalyze the same reaction, even using different chemistry. For instance, several P450s and flavin-containing monooxygenase (FMO) can form certain *N*-oxides (Seto and Guengerich, 1993).

An extension of this issue is that all of these enzymes (or actually the genes) show genetic polymorphism, which can influence the levels of expression, enzyme stability, and catalytic properties. In one sense, each allelic variant is a distinct enzyme. The majority of polymorphisms in the coding sequences do not affect the catalytic properties of enzymes, and one should remember that in a heterozygote the second copy should be rather normal. However, in cases in which the therapeutic index is narrow, a significant alteration of the catalytic activity can have a significant role, as exemplified by the R144C and I359L polymorphisms with P450 2C9 and warfarin (Daly et al., 2002).

Many of the enzymes discussed here are inducible, and another chapter will deal with the issues. Induction can be the result of the drug under consideration for metabolism, another drug, or a separate exposure in the diet or smoking. Most of the enzymes discussed here can also be inhibited. One of the considerations with new drugs is that they may inhibit the metabolism of other drugs and lead to undesirable (and unexpected) drug interactions. A special problem, which is not unusual, is that some inhibitors are mechanism based and irreversible, particularly with the P450s.

In modern pharmaceutical development, early screening is done with new candidate drugs for the nature of metabolites, identification of enzymes involved, induction, and inhibition. A general goal is to have only limited induction and inhibition and to have several enzymes involved in the initial attack on the drug (but slowly enough to maintain acceptable bioavailability and drug half-life). A major contribution of a highly polymorphic enzyme is considered undesirable.

A final point is that most of the reactions have the effect of deactivating drugs. In a few cases a product will have more of the desired pharmacological activity (i.e., "prodrugs"). A concern is that some reactions yield electrophilic "reactive" metabolites that can produce toxicity.

2.4 FRACTIONAL CONTRIBUTIONS OF DIFFERENT ENZYMES

The number of enzymes involved in drug metabolism is large and the number of possibilities with a new drug can be bewildering. A perspective on the roles of enzymes is shown in Fig. 2.1a (Williams et al., 2004). Of the enzymes participating in the metabolism of drugs, the dominant players (~75%) are the P450 enzymes, followed by UDP-glucuronosyltransferases (covered in the next chapter) and esterases. Together, these reactions account for ~95% of drug metabolism. A further breakdown of the P450 enzymes is presented in Fig. 2.1b, with five of the P450s accounting for ~90% of the reactions. A single P450, P450 3A4, is involved in about half of these reactions (Guengerich, 1999; Rendic, 2002). Although the contributions of other enzymes are less, they cannot be ignored and may be critical in many cases. The overall patterns shown in Fig. 2.1 have not changed considerably in recent years (Wrighton and Stevens, 1992), although a diminution of the roles of the polymorphic P450s

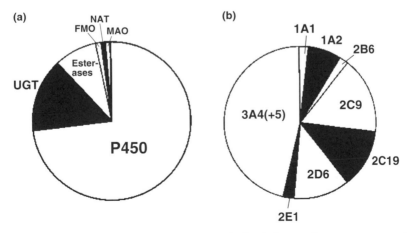

FIGURE 2.1 (a) Fractions of drugs metabolized by various enzyme systems. (b) Fractions of drugs that are P450 substrates metabolized by individual P450s. (Adapted with permission from Williams et al., 2004).

2C19 and 2D6 might be expected in the future as development strategies select against these.

In the following sections, the abbreviation used for nomenclature and searching is given at the start of each section. Some of these are not unique (e.g., CYP, COX) and are shared with other genes/enzymes, and alternative searches are recommended.

2.5 OXIDATION ENZYMES

2.5.1 Cytochrome P450 (P450, CYP)

The general stoichiometry of P450 reactions is mixed-function oxidation (R = substrate).

$$NADPH + O_2 + RH \rightarrow ROH + NADP^+ + H_2O$$

In some cases the stoichiometry is less obvious because of rearrangements of the product (carbinolamines in N-dealkylation) or variations on the general mechanism (desaturations).

The human genome contains 57 P450 genes. For updates on the genes (http://drnelson.utmem.edu/CytochromeP450.html) and polymorphisms (http://www.imm.ki.se/CYPallelesi) see the indicated Web sites. One approach to the functions of the human P450s is presented in Table 2.1. Many of the P450s have specific roles in the metabolism of steroids, eicosanoids, and fat-soluble vitamins. About one quarter have some roles in the metabolism of

TABLE 2.1   Classification of human P450s based on major substrate class (Guengerich, 2005; Guengerich et al., 2006).

Sterols	Xenobiotics	Fatty acids	Eicosanoids	Vitamins	Unknown
1B1	1A1	2J2	4F2	2R1	2A7
7A1	1A2	4A11	4F3	24A1	2S1
7B1	2A6	4B1	4F8	26A1	2U1
8B1	2A13	4F12	5A1	26B1	2W1
11A1	2B6		8A1	26C1	3A43
11B1	2C8			27B1	4A22
11B2	2C9				4F11
17A1	2C18				4F22
19A1	2C19				4V2
21A2	2D6				4X1
27A1	2E1				4Z1
39A1	2F1				20A1
46A1	3A4				27C1
51A1	3A5				
	3A7				

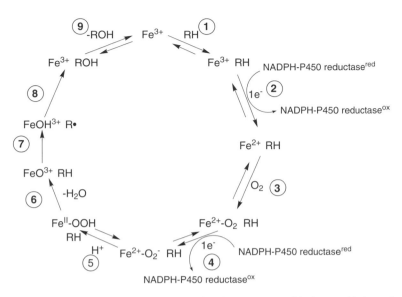

FIGURE 2.2 Generalized catalytic cycle for P450 oxidations. (Adapted with permission from Guengerich, 2001).

xenobiotics, with about five dominant in drug metabolism (Fig. 2.1b). The functions of $\sim 1/4$ of the P450s are still largely uncharacterized, although these "orphans" are not expected to have major roles in drug metabolism (Guengerich, 2005; Guengerich et al., 2006).

The generalized catalytic cycle for P450s is shown in Fig. 2.2. Steps 1–3, and 9 can be studied but the details of steps 4–8 are more difficult to observe directly (Schlichting et al., 2000), or are inferred from physical organic considerations. Electrons are transferred from an auxiliary enzyme, NADPH-P450 reductase, or possibly cytochrome b_5 in some cases. The reaction chemistry can generally be rationalized in the context of processes involving FeO^{3+} (formally $Fe^V = O$ or $Fe^{IV} = O$ plus a porphyrin radical) (Ortiz de Montellano and De Voss, 2005). Alternate dioxygen complexes (FeO_2^+, FeO_2H^{2+}) (Chandrasena et al., 2004) and multiple forms of FeO^{3+} chemistry (Shaik et al., 2005) have been proposed to contribute to P450 reactions. However, the dominant factors in determining catalytic selectivity are generally agreed to be the interactions of the substrate with the amino acids of the protein and, within the context of a set of similar compounds (Burka et al., 1985; Macdonald et al., 1989), the ease of abstraction of hydrogen atoms or nonbinding electrons. Structures of some P450–substrate complexes have now become available and are useful, although even with these there is an issue of whether the juxtaposition is that relevant to the step in which oxidation occurs (Fig. 2.2).

Some of the major types of P450 reactions are shown in Fig. 2.3. These have common chemistry and can be rationalized in terms of odd-electron oxidations. The basic mechanisms can be extended to a variety of other reactions, which are treated in more detail elsewhere (Guengerich, 2001; Ortiz

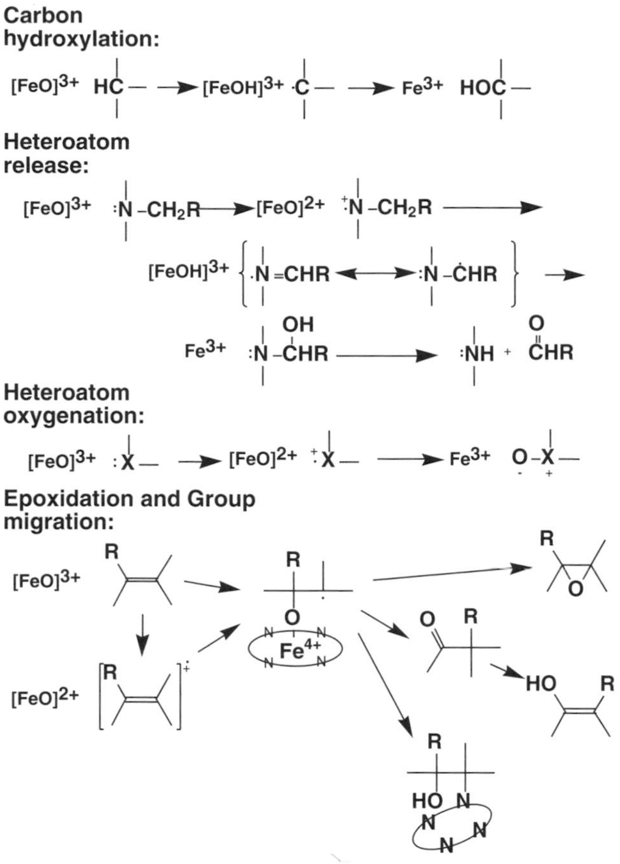

Carbon hydroxylation:

$[FeO]^{3+}$ HC— ⟶ $[FeOH]^{3+}$ ·C— ⟶ Fe^{3+} HOC—

Heteroatom release:

$[FeO]^{3+}$:N–CH$_2$R ⟶ $[FeO]^{2+}$ ⁺N–CH$_2$R ⟶

$[FeOH]^{3+}$ { ·N=CHR ⟷ :N–ĊHR } ⟶

Fe^{3+} :N–CHR(OH) ⟶ :NH + O=CHR

Heteroatom oxygenation:

$[FeO]^{3+}$:X— ⟶ $[FeO]^{2+}$ ⁺:X— ⟶ Fe^{3+} O–X—

Epoxidation and Group migration:

$[FeO]^{3+}$, $[FeO]^{2+}$ ... Fe^{4+} ...

FIGURE 2.3 Some major categories of the P450 reactions (adapted with permission from Guengerich and Macdonald, 1984).

de Montellano and De Voss, 2005). Some of these transformations are shown in Fig. 2.4, but the list is certainly not comprehensive. Finally, many of the most interesting new reactions have come out of practical issues in understanding the metabolism of new drug candidates in the pharmaceutical industry (Evans et al., 2004; Kalgutkar et al., 2005).

2.5.2 Flavin-Containing Monooxygenase (FMO)

The cellular location (endoplasmic reticulum) reaction stoichiometry is the same, mixed-function oxidation, as for P450,

$$NADPH + O_2 + RX \rightarrow RXO + NADP^+ + H_2O$$
$$(X = N, X, P)$$

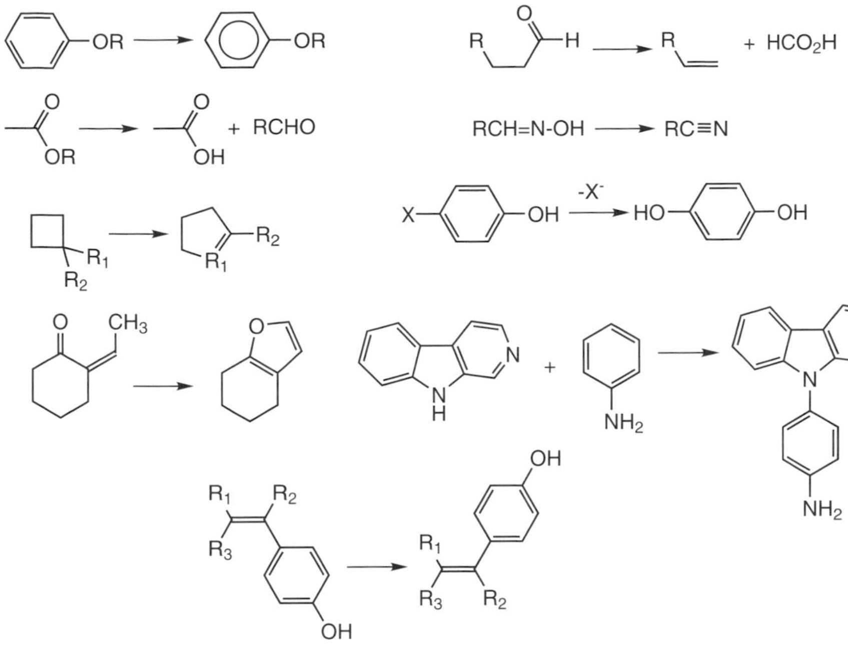

FIGURE 2.4 Some uncommon P450 oxidations (Guengerich, 2001).

although the range of substrates is much more restricted. The substrates are generally soft nucleophiles, almost always N, S, and P atoms (phosphines in the case of P). Of the reactions ascribed to P450 (Figs. 2.3 and 2.4), FMO is generally restricted to heteroatom oxygenation (Fig. 2.3).

Five forms of FMO are found in humans (and several experimental animals). Important physiological substrates have not been established, although some possibilities have been proposed (Ziegler, 1993). An interesting polymorphism involves defects in FMO 3, giving rise to deficient metabolism of trimethylamine and a resulting "fish-odor syndrome" (Al-Waiz et al., 1987). In contrast to P450, FMOs appear not to be inducible or readily inhibited.

FMO is a single-protein mixed-function oxidase, in contrast to P450. The flavin (FMN) acts as both the entry point for electrons and the terminal oxidant, using a C-4a-hydroperoxide. The catalytic mechanism (Fig. 2.5) provides some insight into the properties of this enzyme. The general "resting state" has the 4a-hydroperoxide ready to react with substrates, and the rate-limiting step is generally the breakdown of the flavin 4a-alcohol (step involving H_2O release in Fig. 2.5). The manifestation of this mechanism is that k_{cat} (V_{max}) does not vary considerably but K_m does, due to rates of reactivity more than substrate affinity.

A practical consideration in working with tissues is that FMOs are not very heat stable, particularly in the absence of pyridine nucleotides (Ziegler and

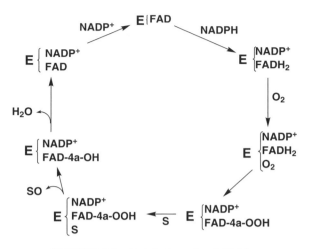

FIGURE 2.5   Catalytic cycle of FMOs.

Mitchell, 1972), and care should be taken so as not to overlook possible contributions of this enzyme. The C-4a-hydroperoxide is not a strong electrophile, and the range of reaction is limited. In other flavoproteins, this hydroperoxide can produce epoxides or Baeyer–Villiger products (Massey, 2000) but none have been reported for FMOs. FMOs can hydroxylate activated phenyl groups, for example, phenols and anilines (Fig. 2.6) (Frederick et al., 1982).

2.5.3   Monoamine Oxidase (MAO)

MAO oxidizes amines, particularly biogenic amines involved as neurotransmitters (Cashman, 1997). Like FMO, this is a flavoprotein oxidase, but the mechanism is different:

$$RCH_2NH_2 + O_2 \rightarrow RCHO + H_2O_2 + NH_3$$

Two forms of MAO are known, A and B, which are slightly different. The enzymes are found in the outer mitochondrial membrane of liver and some neurogenic tissues.

The endogenous substrates of this enzyme are biogenic amines, for example, dopamine and tryptamine, and MAO has been a target for inactivating drugs (Van Houten et al., 1998). Although the enzyme had long been considered only to use primary amines as substrates, an interesting discovery was the oxidation of N-methyl, 4-phenyl-1,2,5,6-tetrahydropyridine (MPTP) to poison mitochondria in the substantia nigra of the brain, yielding a Parkinson's disease-type syndrome (Chiba et al., 1985) (Fig. 2.7).

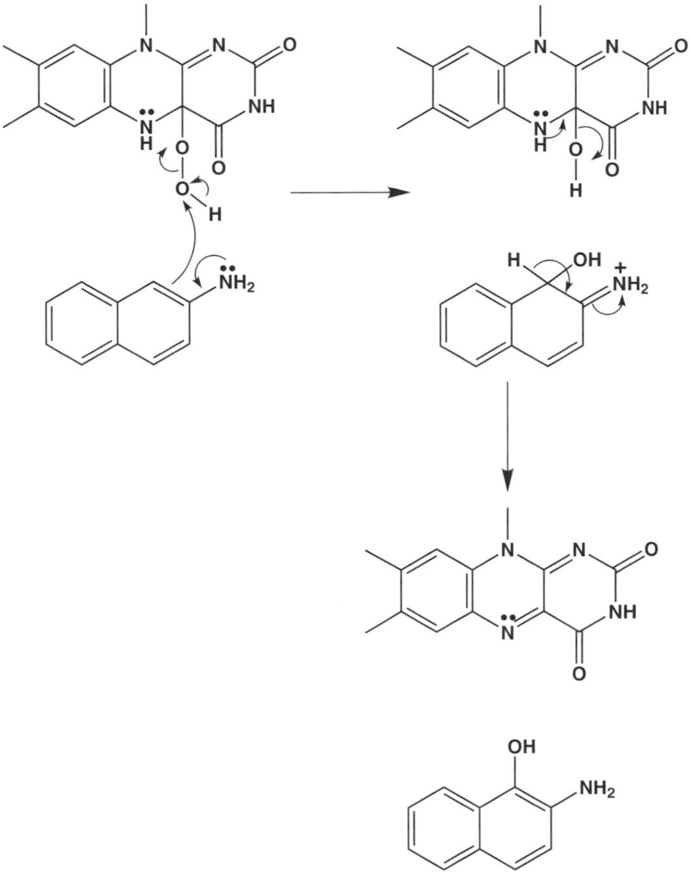

FIGURE 2.6 Proposed mechanism of hydroxylation of activated aromatic ring by FMO (Frederick et al., 1982).

2.5.4 Aldehyde Oxidase and Xanthine Dehydrogenase

These two enzymes are found in the cytosol of the liver (and other tissues) and also oxidize substrates (Panoutsopoulos et al., 2004; Rajagopalan, 1997), with a very different stoichiometry and mechanism:

$$RH + H_2O \rightarrow ROH + 2e^-$$

The two electrons may be transferred to an oxidized pyridine nucleotide or to O_2 (yielding H_2O_2).

These enzymes contain flavin, a molybdenum center, and Fe–S clusters in the ratio 1 : 1 : 4. The pathway of electron flow is from Mo to Fe–S to FAD.

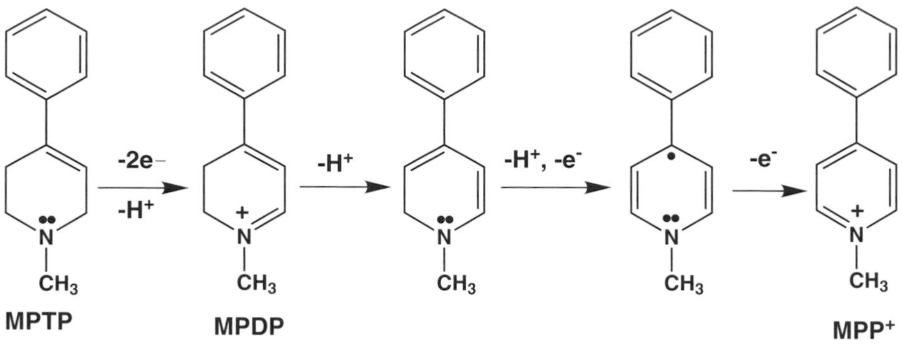

FIGURE 2.7 Oxidation of *N*-methyl, 4-phenyl-1,2,5,6-tetrahydropyridine by MAO (Chiba et al., 1985).

Thus, the Mo center is involved in the oxidation of the substrate. In the mechanism, the source of the oxygen atom inserted into the substrate is H_2O instead of O_2 (as with P450 and FMO). The list of substrates includes some endogenous and xenobiotic aldehydes, and various heterocycles, including purines, pyridines, pyrimidines, pteridines, and others. Xanthine is a substrate for xanthine dehydrogenase but not aldehyde oxidase; purines are substrates for both.

As with FMO, this enzyme has some inherent instability and care should be exercised in processing tissue samples of the presence of this enzyme is an issue. Aldehyde oxidase is a true "oxidase," in that it transfers electrons to O_2 (to form H_2O_2). The literature contains numerous reports about (milk) xanthine oxidase; subsequent work revealed that this is really xanthine dehydrogenase (Rajagopalan, 1997), which is readily converted to an oxidase by proteolysis or modification of sulfides.

Rodents contain additional genes in this family but humans have only the two genes mentioned here (Kurosaki et al., 2004).

2.5.5 Peroxidases

A variety of peroxidases are found in various mammalian cells, including prostaglandin synthases, myeloperoxidase, lipoxygenases, and eosinophile peroxidases (Marnett et al., 1997). The reactions have some similarity to events in P450 chemistry in the context that reactions involve high rodent FeO chemistry. With regard to drug metabolism, the basic reaction of interest is

$$RX + RX^\bullet \rightarrow \ldots$$

and the reactions of the radical $RX^\bullet$ include radical propagation, dimer formation, dealkylation, and so on (Marnett et al., 1997). Other possible reactions include oxygenation ($R \rightarrow RO$) and the formation of reactive halides, for example, $X^-(+H^+) \rightarrow HOX$, which can do various types of chemistry and other peroxidases acting on acetaminophen (to generate the quinoneimine), isoxicam, cyclophophamide, procainamide, and arylamines (Marnett et al., 1997). These reactions are generally extrahepatic. Some of the products may be associated with lupus and autoimmune diseases (Miyamoto et al., 1997).

These reactions begin with the formation of a high valent iron complex, FeO^{3+} (as with P450). Much of the understanding of these systems has been developed with plant and microbial models, for example, horseradish peroxidase and (fungal) chloroperoxidase. An initial reaction with an alkyl hydroperoxide (ROOH) or H_2O_2 generates "Compound I" (formally FeO^{3+}), which then abstracts an electron (or possibly a hydrogen atom) from a substrate to yield "Compound II," which has a similar oxidation–reduction potential (Hayashi and Yamazaki, 1979) and may also abstract electrons.

2.5.6 Alcohol Dehydrogenases (ADH)

The general reaction is reversible:

$$RCH_2OH + NAD^+ \leftrightarrow RCH{=}O + NADH + H^+$$

ADHs are concentrated in the liver, where these can collectively account for ~3% of the total protein (Edenberg and Bosron, 1997). The enzymes have rather broad specificity for primary and some secondary alcohols. ADHs play the major role in the metabolism of ethanol in humans.

ADHs are dimers, either heterodimers or homodimers composed of 40 kDa subunits. Zn^{2+} is a requirement. At least seven ADH genes are known in humans (Duester et al., 1999), with other reports of dehydrogenases appearing (Deng et al., 2002). However, these may be considered a subset of an even larger superfamily of dehydrogenases (Gonzalez-Duarte and Albalat, 2005). The different ADH subunits differ in their specificity. Further, a number of polymorphisms are known that give rise to attenuated function. Some of these show strong racial linkages. ADHs are inhibited by pyrazole and some similar compounds.

2.5.7 Aldehyde Dehydrogenases (ALDH)

ALDHs catalyze the oxidation of aldehydes to carboxylic acids, which is usually not reversible:

$$RCHO + NAD^+ \rightarrow RCO_2H + NADH + H^+$$

The reaction can be compared with that of the aldehyde oxidase (*vide supra*)

$$RCHO + H_2O + O_2 \rightarrow RCO_2H + H_2O_2$$

The ALDHs are concentrated in the liver, mainly in the mitochondria (Petersen and Lindahl, 1997). The human genome contains 19 putatively functional ALDH genes (Vasiliou and Nebert, 2005). In animal models, some of the ALDHs are inducible by barbiturates or by ligands of the Ah receptor, for example, ALDH3 (Takimoto et al., 1992). A number of polymorphisms have been identified that result in attenuation of catalytic activity; as with ADH, many of these show racial linkages.

The reaction mechanism is ordered bi-bi, with $NAD(P)^+$ binding first, and then, NAD(P)H being the last product to leave the enzyme (Petersen and Lindahl, 1997). Inhibitors include disulfiram and some others that overlap with ADH and P450 2E1, two other enzymes that process ethanol and acetaldehyde.

One substrate is *trans, trans*-muconaldehyde, an oxidation product of benzene. Examples of drugs in which ALDH has a major contribution are cyclophosphamide and procarbazine. Other substrates include 4-hydroxynonenal and other products of lipid peroxidation (Petersen and Lindahl, 1997).

2.6 REDUCTION

2.6.1 P450, ADH

The general tendency is for these enzymes to oxidize substrates, but they also catalyze reductions.

The reaction with ADH is relatively straightforward and simply a $2e^-$ reduction of an aldehyde or imine.

$$RCHO + NAD(P)H + H^+ \rightarrow RCH_2OH + NADP^+$$

The P450-catalyzed reductions are certainly not as common as the oxidations, and the question can be raised as to why these occur at all, given the proclivity of ferrous P450 to react with O_2. However, some parts of the body have low oxygen tension, such as the venous sections of the liver, and the reductions can be documented. Some of the reductions that have been documented to involve P450s include (Baker and Van Dyke, 1984; Wislocki et al., 1980):

$$RNO_2 \rightarrow R-N{=}O$$

$$R-N{=}O \rightarrow RNHOH$$

$$RNHOH \rightarrow RNH_2$$

$$R_3N^+O^- \rightarrow R_3N$$

$$R-N{=}N-R' \rightarrow R-NH_2 + H_2N-R'$$

$$CCl_4 \rightarrow CCl_3 \cdot + Cl^-$$

Benzo[a]pyrene 4,5-oxide → benzo[a]pyrene
Halothane → 2-Cl-l,l,l-F_3 ethane + 2-C1-1,l-F_2 ethylene

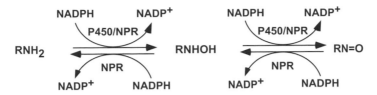

FIGURE 2.8 Abortive redox cycling and oxidation of NADPH in a system involving a heterocyclic arylamine (Kim et al., 2004). NPR = NADPH-P450 reductase.

2.6.2 NADPH-P450 Reductase

The normal functions of this flavoprotein are to transfer electrons from NADPH to P450 (and to heme oxygenase, the enzyme that degrades heme, after breakdown of proteins to release free heme, to biliverdin and CO). However, the enzyme can react with oxidized molecules, in what is probably generally a second-order reaction, to reduce them. Some of the substrates for reduction overlap with the nitrogen-containing molecules listed above under P450. In the scheme shown in Fig. 2.8, P450 1A2 catalyzes the oxidation steps and NADPH-P450 reductase catalyzes the reduction.

Because of the nature of the electron transfer processes in the flavins of this protein, it can catalyze either $1e^-$ or $2e^-$ reductions. As pointed out later under the heading of NAD(P)H–quinone reductase, this is an important issue in that $1e^-$ reduction processes can generate radicals. For instance, NADPH-P450 reductase reduces the herbicide paraquat (methyl viologen) to a radical, which rapidly reacts with O_2 to form superoxide anion (Fig. 2.9).

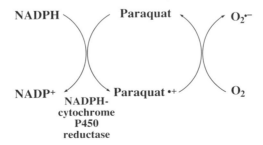

FIGURE 2.9 Abortive redox cycling, oxidation of NADPH, and superoxide production with paraquat.

2.6.3   Aldo-Keto Reductases (AKR)

The AKRs constitute a large family of enzymes involved in reduction of various aldehydes and ketones:

$$H^+ + NAD(P)H + \ \diagup\!\!\!=\!O \longrightarrow \diagdown\!\!\!\diagup\!\!\stackrel{OH}{\underset{H}{}} + NAD(P)^+$$

The reaction is usually in favor of reduction, and these enzymes are distinct from the ADHs. The enzymes in the group, which is now quite large (Jez et al., 1997; Penning, 2004), share ~40% sequence identity. Substrates include endobiotics and xenobiotics. The proteins are generally 30–40 kDa in size. The mechanism is ordered bi-bi, where the cofactor (NAD(P)H) binds first and leaves last, as with ADH and ALDH. Several crystal structures are now available. A conserved triad appears to be involved in at least many of these enzymes, involving Tyr55, Asp50, Lys84, and His1 (17 in the case of AKR1C9) (Penning, 2004). It is of interest to note that even the "housekeeping" enzymes, lactic dehydrogenase and D-glyceraldehyde 3-phosphate dehydrogenase are now classified as members of the superfamily.

Substrates include sugar aldehydes, and some of the enzymes (e.g., aldose reductase) are targets in the treatment of diabetes and glaucoma. Other substrates include steroids, prostaglandins, lipid aldehydes (e.g., 4-hydroxynonenal and its glutathione conjugate), aflatoxin B_1 dialdehyde (Ellis et al., 1993; Guengerich et al., 2001), the tobacco-specific nitrosamine 4- (methylnitrosamino)-1-(3-pyridyl)-1-butanone (NNK), and polycyclic aromatic hydrocarbon (PAH) dihydrodiols (Flynn and Kubiseski, 1997; Penning, 2004). In most of these reactions the reaction results in detoxication, but in the latter case activation can also occur with the formation of $O_2^{-\bullet}$ and reactive catechol quinones (Penning, 2004).

Some of the AKR enzymes are inducible. Systems involved include an osmotic stress system, AP1:fos/jun (where several forms of each of the two components may be involved), and ARE:Keap1/Nrf2/small Maf. The latter of these has been clearly shown to respond to oxidative and electrophilic stresses, and the responses can be considered part of an adaptive protective mechanism.

2.6.4   Quinone Reductase (NQO)

These enzymes catalyze the reduction of quinones to hydroquinones (both *para* and *ortho*)

Similarly, iminoquinones can be reduced, and the enzymes can also reduce nitro compounds and azo dyes (Ross, 1997).

NQO1 was first characterized as "DT diaphorase," with diaphorase being an older term for an enzyme that catalyzes electron transfer from pyridine nucleotides. "DT" indicated that this enzyme could accept electrons from NADH (formerly termed DPNH) or NADPH (formerly TPNH). Some of the other early work was unclear but the location of the enzyme is cytosolic (Huang et al., 1979). This is a flavoprotein that reduces substrates. In contrast to NADPH-P450 reductase, most NQO reactions are 2-electron reductions, avoiding the generation of radicals. However, the product hydroquinones may react with O_2 to generate $O_2^{-\bullet}$.

NQO1 is inducible via both the Ah receptor and the ARE pathways (Ross, 1997). Dicoumarol is a long-established potent inhibitor of NQO. A second gene, NQO2; is found in humans (Jaiswal, 1994). A related p53-regulated gene with considerable sequence identity has been discovered (PIG3) (Nicholls et al., 2004) but no redox function has yet been identified.

2.6.5   Glutathione Peroxidase (GPX)

These enzyme systems reduce potentially toxic organic hydroperoxides, including H_2O_2:

$$2GSH + ROOH \rightarrow GSSG + ROH$$

The reaction is coupled to the regeneration of GSH by GSH reductase

$$NADPH + H^+ + GSSG \rightarrow NADP^+ + 2GSH$$

The net process can be written as

$$NADPH + H^+ + ROOH \rightarrow NADP^+ + ROH$$

Six human GSH peroxidases have been identified. Most of the enzymes contain selenocysteine in the active site, with the exceptions of GPX 5 and 6 (Burk, 1997). Some GSH transferase enzymes also display low GSH peroxidase activity (Ketterer and Meyer, 1989). Thioredoxin, a low M_r iron–sulfur protein, can also transfer electrons (instead of GSH in some cases). Some GSH peroxidases are extracellular and function in plasma.

2.7   HYDROLYSIS

2.7.1   Epoxide Hydrolase

Epoxides are reactive electrophiles by nature of their ring strain and change localization; they vary considerably in stability and reactivity (Guengerich,

2003). P450s convert many olefins and aromatic compounds to epoxides that, unless hydrolyzed or conjugated, can react with tissue nucleophiles to initiate injury.

Epoxide hydrolases catalyze the simple addition of H_2O to epoxides:

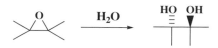

In this way, the chemistry is more related to the enzymes that catalyze conjugations than to those that oxidize and reduce. The major epoxide hydrolase is a microsomal enzyme, abundant at a relatively high concentration in the liver and many other tissues. This is often referred to simply as the microsomal epoxide hydrolase. At least three other epoxide hydrolases are also found in the endoplasmic reticulum, but these all have specialized functions and do not act on xenobiotic chemicals (Hammock et al., 1997).

A soluble epoxide hydrolase also exists, which is a distinct protein. This enzyme hydrolyzes some model and xenobiotic epoxides although its repertoire is limited. The enzyme does hydrolyze squalene epoxide, lanosterol oxide, and some epoxide derivatives of polyunsaturated fatty acids. Interestingly, transgenic mice devoid of either soluble or microsomal epoxide hydrolase do not show an unusual phenotype, although more subtle changes cannot be ruled out and transgenic animals missing soluble epoxide hydrolase had lower blood pressure (Sinal et al., 2000). Transgenic studies also show the requirement for the microsomal epoxide hydrolase in the activation of some polycyclic hydrocarbons (Miyate et al., 1999).

The mechanism of these epoxide hydrolases, first established with the microsomal enzyme, was surprising in that it involves an acyl intermediate, more akin to an esterase than a system with activation of a water or hydroxide (Lacourciere and Armstrong, 1994). The mechanism was establish using an elegant single turnover ^{18}O experiment; the lack of H_2O incorporation in the first product is only consistent with an acyl intermediate (Lacourciere and Armstrong, 1993). Subsequent studies have shown that the rate of decay of the acyl intermediate is rate-limiting in the mechanism (Tzeng et al., 1996). This is not a particularly fast process ($\sim 1\,s^{-1}$) and places a limit on k_{cat} (epoxide hydrolase is clearly an enzyme in which the mechanism yields a case with $K_m \ll K_d$). This knowledge is useful in understanding why the hydrolysis of very labile epoxides such as PAH diol-epoxides (Gozukara et al., 1981) and aflatoxin B_1 8,9-*exo*-epoxide shows little enhancement by epoxide hydrolase (Johnson et al., 1997).

2.7.2   Esterases and Amidases

This is a rather large and heterogeneous family of proteins, from multiple gene families. Collectively they constitute the third largest group of reactions on drugs, behind P450s and UGTs (Fig. 2.1a). These enzymes are collectively found in many places, including the liver and plasma.

One group of esterases has an α,β-fold and is prominent in the liver cytosol (Quinn, 1997). Acetylcholinesterase, butyl cholinesterase, and lipases have been used as models for these esterases. Generally esterases also have amidase activity (and *vice versa*, due to the basic mechanisms). All esterases appear to use a catalytic triad to activate a nucleophile, which is used to form an enzyme-acyl intermediate. The triad consists of a nucleophile, a general base catalyst, and an acidic residue.

Some esterases are loosely bound to the endoplasmic reticulum and constitute a separate family, using ester and amide substrates. At least six of these have been reported in humans (Sone and Wang, 1997). Some evidence exists for induction and other regulation of this group. In general, the soluble esterases do not seem to be inducible.

Not every ester cleavage is due to esterases. For instance a P450 can catalyze an oxidative ester cleavage via carbon hydroxylation (Guengerich, 1987). Esterases reactions are prominent in the metabolism of drugs and pesticides. The inhibitors of esterases are generally strong electrophiles, some of which are intermediates generated during the metabolism of pesticides (e.g., the inhibition of acetylcholinesterase is a classic case in this area (Quinn, 1997)).

2.8 SUMMARY

The purpose of this section has been an overview of the major enzymes systems. Although this group of reactions has often historically been treated together as Phase I metabolism, the group is very heterogeneous in many respects and individual reactions have been treated separately. Knowledge of these systems has been fundamental in influencing how drug development is conducted. However, it should still be emphasized that predictions are difficult in this area, based only on drug structures, and that experiments remain a necessary part of the process, in terms of establishing pathways and enzymes involved.

REFERENCES

Al-Waiz M, Ayesh R, Mitchell SC, Idle JR, Smith RL. A genetic polymorphism of the N-oxidation of trimethylamine in humans. Clin Pharmacol Therap 1987;42:588–594.

Baker MT, Van Dyke RA. Reductive halothane metabolite formation and halothane binding in rat hepatic microsomes. Chem-Biol Interact 1984;49:121–132.

Burk RF. Selenium-dependent glutathione peroxidases. In: Guengerich FP, editor. Biotransformation, Vol. 3, Comprehensive Toxicology, first ed. Elsevier Science, Oxford; 1997. p 229–242.

Burka LT, Guengerich FP, Willard RJ, Macdonald TL. Mechanism of cytochrome P-450 catalysis. Mechanism of N-dealkylation and amine oxide deoxygenation. J Am Chem Soc 1985;107:2549–2551.

Cashman JR. Monoamine oxidase and flavin-containing monooxygenases. In: Guengerich FP, editor. Biotransformation, Vol. 3, Comprehensive Toxicology, first ed. Elsevier Science, Oxford; 1997. p 69–96.

Chandrasena RE, Vatsis KP, Coon MJ, Hollenberg PF, Newcomb M. Hydroxylation by the hydroperoxy-iron species in cytochrome P450 enzymes. J Am Chem Soc 2004;126:115–126.

Chiba K, Peterson LA, Castagnoli KP, Trevor AJ, Castagnoli N Jr. Studies on the molecular mechanism of bioactivation of the selective nigrostriatal toxin 1-methyl-4-phenyl-1,2,3,6-tetrahydropyridine. Drug Metab Dispos 1985;13:342–347.

Daly AK, Day CP, Aithal GP. CYP2C9 polymorphism and warfarin dose requirements. Br J Clin Pharmacol 2002;53:408–409.

Deng Y, Wang Z, Gu S, Ji C, Ying K, Xie Y, Mao Y. Cloning and characterization of a novel human alcohol dehydrogenase gene (ADHFel). DNA Seq 2002;13:301–306.

Distlerath LM, Reilly PEB, Martin MV, Davis GG, Wilkinson GR, Guengerich FP. Purification and characterization of the human liver cytochromes P450 involved in debrisoquine 4-hydroxylation and phenacetin O-deethylation, two prototypes for genetic polymorphism in oxidative drug metabolism. J Biol Chem 1985;260:9057–9067.

Duester G, Jarrks J, Felder MR, Holmes RS, Hoijg J-O, Paris X, Plapp BV, Yin S-J, Jornvall H. Recommended nomenclature for the vetebrate alchol dehydrogenase gene family. Biochem Pharmacol 1999;58:389–395.

Edenberg HJ, Bosron WF. Alcohol dehydrogenases. In: Guengerich FP, editor. Biotransformation, Vol. 3, Comprehesive Toxicology, first ed. Elsevier Science, Oxford; 1997. p 119–131.

Ellis EM, Judah DJ, Neal GE, Hayes JD. An ethoxyquin-inducible aldehyde reductase from rat liver that metabolizes aflatoxin B_1 defines a subfamily of aldo–keto reductases. Proc Natl Acad Sci USA 1993;90:10350–10354.

Flynn TG, Kubiseski TJ. Aldo–keto reductases: structure, mechanism, and function. In: Guengerich FP, editor. Biotransformation, Vol. 3, Comprehensive Toxicology. first ed. Elsevier Science, Oxford; 1997. p133–147.

Frederick CB, Mays JB, Ziegler DM, Guengerich FP, Kadlubar FF. Cytochrome P450 and flavin-containing monooxygenase-catalyzed formation of the carcinogen N-hydroxy-2-aminofluorene, and its covalent binding to nuclear DNA. Cancer Res 1982;42:2671–2677.

Gonzalez-Duarte R, Albalat R. Merging protein, gene and genomic data: the evolution of the MDR-ADH family. Heredity 2005;95:184–197.

Gozukara EM, Belvedere G, Robinson RC, Deutsch J, Coon MJ, Guengerich FP, Gelboin HV. The effect of epoxide hydratase on benzo[a]pyrene diol epoxide hydrolysis and binding to DNA and mixed-function oxidase proteins. Mol Pharmacol 1981;19:153–161.

Guengerich FP. Oxidative cleavage of carboxylic esters by cytochrome P-450. J Biol Chem 1987;262:8459–8462.

Guengerich FP. Biotransformation, Vol. 3, Compr Toxicology, first ed. Elsevier Science, Oxford; 1997.

Guengerich FP. Human cytochrome P-450 3A4: regulation and role in drug metabolism. Annu Rev Pharmacol Toxicol 1999;39:1–17.

Guengerich FP. Common and uncommon cytochrome P450 reactions related to metabolism and chemical toxicity. Chem Res Toxicol 2001;14:611–650.

Guengerich FP. Cytochrome P450 oxidations in the generation of reactive electrophiles: epoxidations and related reactions. Arch Biochem Biophys 2003;409:59–71.

Guengerich FP. Human cytochrome P450 enzymes, In: Ortiz de Montellano PR, editor. Cytochrome P450: Structure, Mechanism, and Biochemistry, third ed. Kluwer Academic/Plenum Publishers; New York: 2005. p 377–531.

Guengerich FP, Macdonald TL. Chemical mechanisms of catalysis by cytochromes P-450: a unified view. Acc Chem Res 1984;17:9–16.

Guengerich FP, Cai H, McMahon M, Hayes JD, Sutter TR, Groopman JD, Deng Z, Harris TM. Reduction of aflatoxin B, dialdehyde by rat and human aldo–keto reductases. Chem Res Toxicol 2001;14:727–737.

Guengerich FP, Wu Z-L, Bartleson CJ. Function of human cytochrome P450s: characterization of the remaining orphans. Biochem Biophys Res Commun 2005;338:465–469.

Hammock BD, Grant DF, Storms DH. Epoxide hydrolases. In: Guengerich FP, editor. Biotransforrnation, Vol. 3, Comprehensive Toxicology, first ed. Elsevier Science, Oxford; 1997. p 283–305.

Hayashi Y, Yamazaki I. The oxidation-reduction potentials of compound I/compound II and compound II/ferric couples of horseradish peroxidases A2 and C. J Biol Chem 1979;254:9101–9106.

Huang M-T, Miwa GT, Cronheim N, Lu AYH. Rat liver cytosolic azoreductase. Electron transport properties and the mechanism of dicoumarol inhibition of the purified enzyme. J Biol Chem 1979;254:11223–11227.

Jaiswal AK. Human NAD(P)H-quinone oxioredutase. Gene structure, activity, and tissue-specific expression. J Biol Chem 1994;269:14502–14508.

Jez JM, Flynn TG, Penning TM. A new nomenclature for the aldo–keto reductase superfamily. Biochem Pharmacol 1997;54:639–647.

Johnson WW, Ueng Y-F, Yamazaki H, Shimada T, Guengerich FP. Role of microsomal epoxide hydrolase in the hydrolysis of aflatoxin B_1 8,9-epoxide. Chem Res Toxicol 1997;10:672–676.

Josephy PD, Guengerich FP, Miners JO. Phase 1 and phase 2 drug metabolism: terminology that we should phase out. Drug Metab Dispos 2005;37:579–584.

Kalgutkar AS, Gardner I, Obach RS, Shjaffer CL, Callegari E, Henne KR, Mutlib AE, Dalvie DK, Jee JS, Nakai Y, O'Donnell JP, Boer, Harriman SP. A comprehensive listing of bioactivation pathways of organic functional groups. Curr Drug Metab 2005;6:161–225.

Ketterer B, Meyer DJ. Glutathione transferases: a possible role in the detoxication and repair of DNA and lipid hydroperoxides. Mut Res 1989;214:33–40.

Kim D, Kadlubar FF, Teitel CH, Guengerich FP. Formation and reduction of aryl and heterocyclic nitroso compounds and significance in the flux of hydroxylamines. Chem Res Toxicol 2004;17:529–536.

Kurosaki M, Terao M, Barzago MM, Bastone A, Bemardinello D, Salmona M, Garattini E. The aldehyde oxidase gene cluster in mice and rats. Aldehyde oxidase homologue 3, a novel member of the molybdo-flavoenzyme family with selective expression in the olfactory mucosa. J Biol Chem 2004;279:50482–50498.

Lacourciere GM, Armstrong RN. The catalytic mechanism of microsomal epoxide hydrolase involves an ester intermediate. J Am Chem Soc 1993;115:10466–10467.

Lacourciere GM, Armstrong RN. Microsomal and soluble epoxide hydrolases are members of the same family of C–X bond hydrolase enzymes. Chem Res Toxicol 1994;7:121–124.

Macdonald TL, Gutheim WG, Martin RB, Guengerich FP. Oxidation of substituted N,N-dimethylanilines by cytochrome P450: estimation of the effective oxidation–reduction potential of cytochrome P450. Biochemistry 1989;28:2071–2077.

Marnett LJ, Landino LM, Reddy GR. Peroxidases. In: Guengerich FP, editor. Biotransformation, Vol. 3, Comprehensive Toxicology, first ed. Elsevier Science, Oxford; 1997. p 149–163.

Massey V. The chemical and biological versatility of riboflavin. Biochem Soc Trans 2000;28:283–296.

Miyamoto G, Zahid N, Uetrecht JP. Oxidation of diclofenac to reactive intermediates by neutrophils, myeloperoxidase, and hypochlorous acid. Chem Res Toxicol 1997;10:414–419.

Miyate M, Kudo G, Lee Y-H, Y.mg TJ, Gelboin HV, Fernandez-Salguero, Kimura S, Gonzalez FJ. Targeted disruption of the microsomal epoxide hydrolase gene. Microsomal epoxide hydrolase is required for the carcinogenic acitvity of 7,12-dimethylbenz[a]anthracene. J Biol Chem 1999;274:23963–23968.

Nicholls CD, Shields MA, Lee PWK, Robbins SM, Beattie TL. UV-dependent alternative splicing uncouples p53 activity and PIG3 gene function through rapid proteolytic degradation. J Biol Chem 2004;279:24171–24178.

Ortiz de Montellano PR, De Voss JJ. Substrate oxidation by cytochrome P450 enzymes. In: Ortiz de Montellano PR, editor. Cytochrome P450: Structure, Mechanism, and Biochemistry, third ed. Kluwer Academic/Plenum Publishers; New York: 2005. p 183–245.

Panoutsopoulos GI, Kouretas D, Beedham C. Contribution of aldehyde oxidase, xanthine oxidase, and aldehyde dehydrogenase on the oxidation of aromatic aldehydes. Chem Res Toxicol 2004;17:1368–1376.

Penning TM. Introduction and overview of the aldo–keto reductase superfamily. In: Penning TM, Petrash JM, editors. Aldo-Keto Reductases and Toxicant Metabolism. first ed. Washington, DC: American Chemical Society; 2004. p 3–20.

Petersen D, Lindahl R. Aldehyde dehydrogenases. In: Guengerich FP, (editor). Biotransformation, Vol. 3, Comprehensive Toxicology, first ed. Elsevier Science, Oxford; 1997. p 97–118.

Quinn DM. Esterases of the α,β hydrolase fold family. In: Guengerich FP, editor. Biotransformation, Vol. 3, Comprehensive Toxicology. first ed. Elsevier Science, Oxford; 1997. p 243–264.

Rajagopalan KV. Xanthine dehydrogenase and aldehyde oxidase. In: Guengerich FP, (editor). Biotransformation, Vol. 3, Comprehensive Toxicology. first ed. Elsevier Science, Oxford; 1997. p 165–178.

Rendic S. Summary of information on human CYP enzymes: human P450 metabolism data. Drug Metab Rev 2002;34:83–448.

Ross D. Quinone reductases. In: Guengerich FP, (editor). Biotransformation, Vol. 3, Comprehensive Toxicology, first ed. Elsevier Science, Oxford; 1997. p 179–197.

Schlichting I, Berendzen J, Chu K, Stock AM, Maves SA, Benson DE, Sweet BM, Ringe D, Petsko GA, Sligar SG. The catalytic pathway of cytochrome P450$_{cam}$, at atomic resolution. Science 2000;287:1615–1622.

Seto Y, Guengerich FP. Partitioning between N-dealkylation and N-oxygenation in the oxidation of N,N-dialkylarylamines catalyzed by cytochrome P450 2B1. J Biol Chem 1993;268:9986–9997.

Shaik S, Kumar D, de Visser SP, Altun A, Thiel W. Theoretical perspective on the structure and mechanism of cytochrome P450 enzymes. Chem Rev 2005;105:2279–2328.

Sinal CJ, Miyate M, Tohkin M, Nagata K, Bend JR, Gonzalez FJ. Targeted disruption of soluble epoxide hydrolase reveals a role in blood pressure regulation. J Biol Chem 2000;275:40504–40510.

Sone T, Wang CY. Microsomal amidases and caraboxylesterases. In: Guengerich FP, (editor). Biotransformation, Vol. 3, Comprehensive Toxicology. first ed. Elsevier Science, Oxford; 1997. p 265–281.

Takimoto K, Lindahl R, Pitot HC. Regulation of 2,3,7,8-tetrachlorodibenzo-p-dioxin-inducible expression of aldehyde dehydrogenase in hepatoma cells. Arch Biochem Biophys 1992;298:492–497.

Tzeng H-F, Laughlin LT, Lin S, Armstrong RN. The catalytic mechanism of microsomal epoxide hydrolase involves reversible formation and rate-limiting hydrolysis of the alkyl-enzyme intermediate. J Am Chem Soc 1996;118:9436–9437.

Van Houten KA, Kim JM, Bogdan MA, Feni DC, Mariano PS. A new strategy for the design of monoamine oxidase inactivators. Exploratory studies with tertiary allylic and propargylic amino alcohols. J Am Chem Soc 1998;120:5864–5872.

Vasiliou V, Nebert DW. Analysis and update of the human aldehyde dehydrogenase (ALDH) gene family. Hum Genomic 2005;2:138–143.

Williams JA, Hyland R, Jones BC, Smith DA, Hust S, Goosen TC, Peterkin V, Koup JR, Ball SE. Drug–drug interactions for UDP-glucuronosyltransferase substrates: a pharmacokinetic explanation for typically observed low exposure (AUC$_i$/AUC) ratios. Drug Metab Dispos 2004;32:1201–1208.

Wislocki PG, Miwa GT, Lu AYH. Reactions catalyzed by the cytochrome P-450 system. In: Jakoby WB, (editor). Enzymatic Basis of Detoxication, Vol. I. first ed. New York: Academic Press; 1980. p 135–182.

Wrighton SA, Stevens JC. The human hepatic cytochromes P450 involved in drug metabolism. CRC Crit Rev Toxicol 1992;22:1–21.

Ziegler DM. Recent studies on the structure and function of multisubstrate flavin-containing monooxygenases. Annu Rev Pharmacol Toxicol 1993;33:179–199.

Ziegler DM, Mitchell CH. Microsomal oxidase IV: properties of a mixed-function amine oxidase isolated from pig liver microsomes. Arch Biochem Biophys 1972; 150:116–125.

3

CONJUGATIVE METABOLISM OF DRUGS

Rory Remmel, Swati Nagar and Upendra Argikar

3.1 UDP-GLUCURONOSYLTRANSFERASES

Glucuronidation is the addition (conjugation) of glucuronic acid to various functional groups. Compounds may be glucuronidated directly or after oxidative metabolism (Phase II metabolism). Reactions can occur on alcohols (ROH), phenols (Ar–OH), amines (RNH_2), tertiary and heterocylic amines ($RNR'R''$), amides ($R–CO–NH_2$) thiols (RSH), and acidic carbon atoms (Fig. 3.1). Addition of glucuronic acid results in conjugates that are

1. more polar
2. are ionized at physiologic pH ($pK_a \sim 4$)
3. have an increase molecular weight ($+176$).

These features facilitate excretion of glucuronides via the kidney either by glomerular filtration or by active secretion or both. In addition, glucuronides are commonly excreted by the liver via the bile into the small intestine. Glucuronides are too polar to diffuse through cell membranes and therefore specific transporters are necessary for their movement across membranes.

Monoglucuronides are generally considered to be final metabolites, although there is recent evidence with some of the nonsteroidal anti-inflammatory agents, that glucuronides may be substrates for CYP2C9 oxidation. Diglucuronides of bilirubin, some steroids, and di-OH chrysene (a polycyclic aromatic hydrocarbon) have also been observed.

Drug Metabolism in Drug Design and Development, Edited by Donglu Zhang, Mingshe Zhu and W. Griffith Humphreys
Copyright © 2008 John Wiley & Sons, Inc.

Overall reaction:

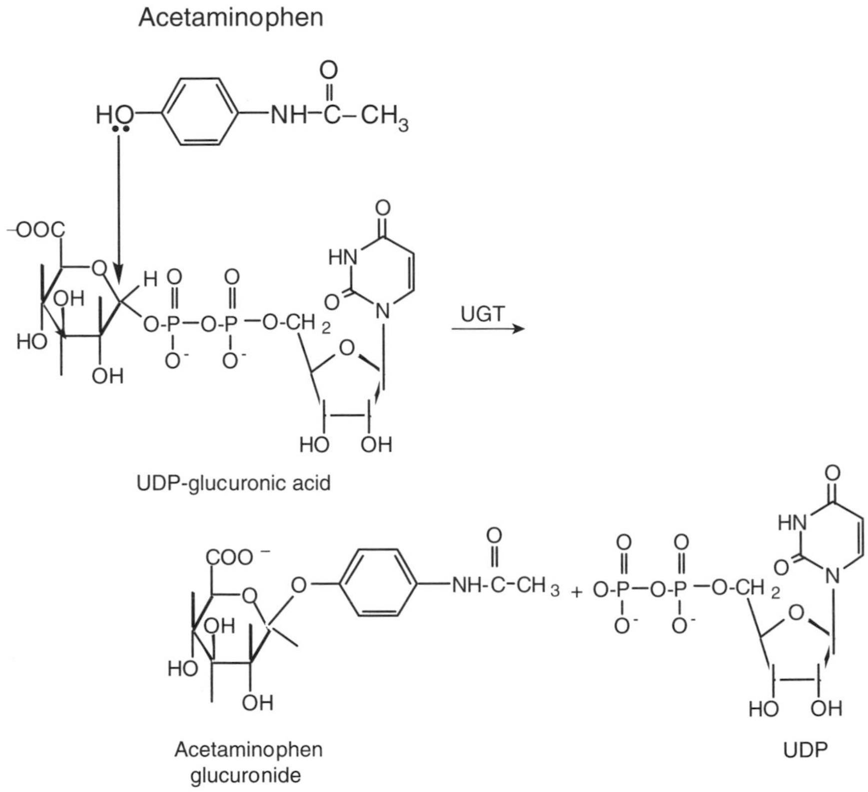

Acetaminophen

UGT

UDP-glucuronic acid

Acetaminophen   UDP
glucuronide

FIGURE 3.1 The glucuronidation reactions. Enzyme: UDP glucuronosyltransferase (UGT or UDPGT); Cosubstrate: uridene diphosphoglucuronic acid (UDPGA)-activated cosubstrate.

3.1.1 Location Within the Cell

The active site of the UGTs face the lumen of the endoplasmic reticulum (Fig. 3.2), whereas the active site of P450, also microsomal enzymes, faces the cytosolic side. Nonpolar substrates can diffuse through the ER membrane and be conjugated in the ER lumen. However, UDPGA must be transported into the ER and the resulting glucuronide products generally need to be transported out of the ER into the cytosol (Bossuyt, 1994a and 1994b). There is *trans*-stimulation of UDPGA influx by UDP-*N*-acetylglucosamine (UDPGlcNAC), UDP-xylose, and UDP-glucose (Bossuyt, 1996). Transport proteins for glucuronides have been characterized on the hepatocyte sinusoidal membrane, bile canalicular membrane, and in kidney tubules. In the hepatocyte, glucuronides are transported through the sinusoidal membrane by MRP3. On the bile canalicular membrane, the major transporter is MRP2. Based on

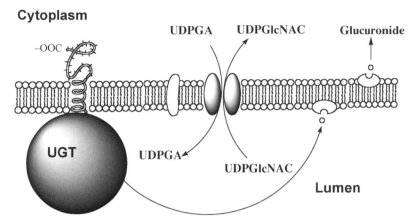

FIGURE 3.2 Localization of UGT enzymes in the endoplasmic reticulum. The active site is located on the inside of the ER with a single transmembrane domain and a 25 a.a. COO⁻ cytosolic tail. UDPGA is transported into the ER and is *trans*-stimulated by UDPGlcNAC. Once the glucuronides are formed, they must be transported out of the ER by a separate transport protein (depicted by flipping through the membrane). Source: Adapted from Clarke DJ and Burchell B, The uridine diphosphate glucuronosyltransferase multigene family: function and regulation. In: Conjugation-Deconjugation Reactions in Drug Metabolism and Toxicity, F.C. Kaufman, editor, New York, Springer-Verlag, 3.8: 1994.

substrate kinetics, it appears that there may be multiple glucuronide transporters on the ER membrane to move the glucuronide products from the lumen to the cytoplasm by facilitated diffusion (ATP-independent) (Csala, 2004).

3.1.2 Endogenous Substrates

There are a large number of endogenous substrates for the UGTs. These include bilirubin, the primary degradation product of heme (Fig. 3.3). In the native state, bilirubin exists as a highly nonpolar conformation with internal hydrogen bonds sequestering the two carboxyl groups. Once inside the

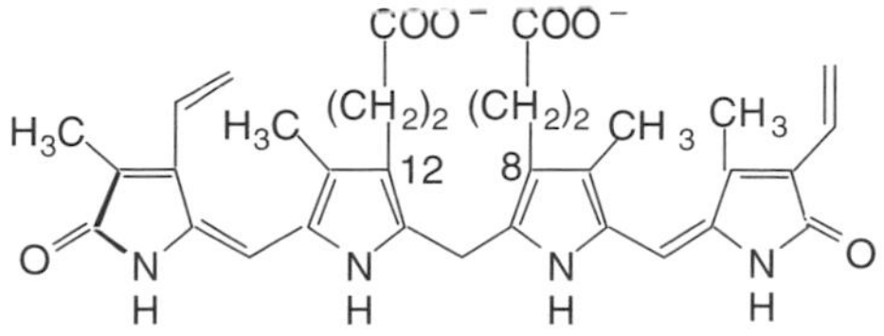

Bilirubin

FIGURE 3.3 The structure of the UGT1A1 endogenous substrate, bilirubin.

UGT1A1 binding pocket, the carboxyl groups appear to be accessible to glucuronidation. Monoglucuronides at the 8 or 12 position and a diglucuronide are the major products. The enzyme is also capable of forming xylose and glucose conjugates with the corresponding UDP-sugar.

3.1.2.1 Steroids   Estrogens, androgens, progestogens are all substrates for the enzyme. Estrogens are conjugated at both the 3-OH and 17-OH positions by either glucuronic acid or sulfate. These conjugates (particularly sulfates) are the major circulating forms and may serve as a reservoir for the body. Glucuronides are excreted into urine and bile. Estradiol and ethinyl estradiol are conjugated by a UGT1A1 at the 3- position (Ebner, 1993). An important detoxification function of UGTs is to glucuronidate the catechol estrogens, for example 2-OH and 4-OH estradiol. Catechol estrogens (potential mutagens and carcinogens) are glucuronidated by UGT1A1, UGT1A3, UGT2B4, and UGT2B7. UGT1A1 and UGT1A3 have much better activity for 2-OH estradiol whereas UGT2B4 and UGT2B7 prefer 4-OH-estradiol (Cheng, 1998). These UGT2B enzymes are present in breast tissue and in breast tumor lines such as MCF-7 cells (Turgeon, 2001) and may help prevent mutations leading to breast cancer. Aldosterone (a glucocorticoid) and its metabolites are glucuronidated by UGT2B7. Glucuronidation of androgens, for example, androsterone, androstane-3α, 17β-diol are catalyzed by UGT2B15 and UGT2B17, two forms that are present in the prostate and in liver. UGT2B17 has a greater glucuronidation efficiency for dihydrotestosterone than UGT2B15.

Other endogenous substrates include lipids, especially pharmacologically active arachidonic acid metabolites such as 12 and 15-HETE (hydroxyeicosatetraenoic acid), 13-HODE (hydroxyoctadecaenoic acid), and leukotriene B4 that are substrates for UGT2B7, UGT2B10, UGT2B11, as well as other enzymes. Bile acids such as lithocholic acid and hyodeoxycholic acid are catalyzed by specific UGT2B forms (UGT2B4 and UGT2B7). UGT1A3 also has activity for some bile acids (via conjugation at the COOH group). Other endogenous substrates include vitamin D and its metabolites and vitamin A analogs (retinoic acid). Finally, thyroid hormones such as thyroxine are also glucuronidated to inactive forms. Positions of glucuronidation of bile acids and steroids are shown in Fig. 3.4.

3.1.2.2 Glycolipids   Sulfated, glucuronic acid-containing glycolipids (ceramides) are present in the peripheral nerves and cauda equina (Chou et al., 1991).

3.1.3 Enzyme Multiplicity

Like the P450 multigene family, there are several different isozymes present in two gene families. There are three major gene families:

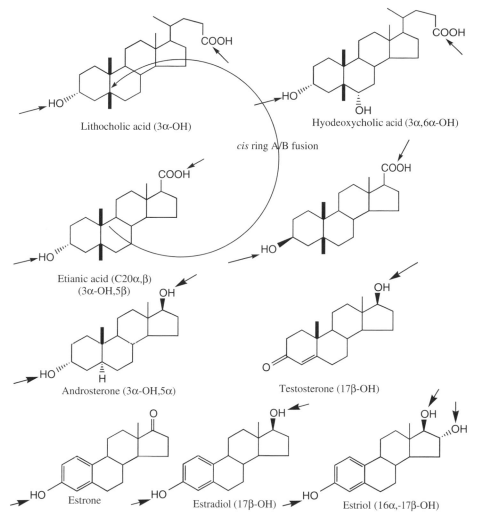

FIGURE 3.4   Structures of endogenous compounds involved in glucuronidation: bile acids, lithocholic acid (LA) and hyodeoxycholic acid (HDCA); short chain bile acids, etianic acid and isoetianic acid, steroid hormones, androsterone, testosterone, estrone, estradiol, estriol.

- UGT1—various forms catalyze conjugation of planar phenols, bulky phenols, amines, tertiary amines, and bilirubin. (Nine active human forms now cloned are expressed, i.e., 1A1, 1A3–1A10).
- UGT2A—olfactory (nasal) UGTs.
- UGT2B—xenobiotics, steroids and bile acids (≥4 human active enzymes, i.e., 2B4, 2B7, 2B10, 2B15, 2B17).

A highly conserved N-terminal region containing the UDPGA binding site is found in both gene families. For the UGT1 family, the N-terminal region is identical due to alternative splicing of the constant region gene locus with different variable regions. The nomenclature for the UGT1A enzymes is the same for all species (Mackenzie, 2005). Individual enzymes are named based on the proximity of their variable region exon to the constant region exons 2–5. Because the UGT2B enzymes are encoded by separate genes on chromosome 4, they have been named in the order of their submission to the nomenclature committee. Additional information can be obtained at the UGT Web site maintained at Flinders University in Australia—http://som.flinders.edu.au/FUSA/ClinPharm/UGT/.

Tables 3.1 and 3.2 define the nomenclature for these two major families and contain information on endogenous substrates and tissue distribution. Tables 3.3 and 3.4 contain information on drugs and other xenobiotic substrates, inhibitors, and inducers.

3.1.4 Inducibility

Like the P450 gene family, the UGT genes are independently regulated. It appears that the regulatory elements responsible for cytochrome P450 induction are also present for the UGT genes. Induction is responsible for a number of drug–drug interactions.

3.1.4.1 Aryl Hydrocarbon (Ah) Receptor   3-Methylcholanthrene, TCDD, and β-naphthoflavone (aromatic hydrocarbon inducers) induce metabolism of planar phenols (specially of polycyclic aromatic hydrocarbons) by UGT1A6 and UGT1A7. Xenobiotic response elements (XRE) have been identified in the rat UGT16 (Aeyeung, 2003) and human UGT1A6 (Munzel, 1998). UGT1A1 is also induced by Ah receptor agonists and bilirubin itself appears to activate the Ah receptor.

3.1.4.2 Constitutive Androstane Receptor (CAR) Activation   Phenobarbital and phenytoin are known to induce metabolism of bilirubin, some bulky phenols, borneol, morphine, and steroidal UGTs catalyzed by UGT1A1, UGT1A9, and UGT2B7. Phenobarbital response elements (PBREMs) have been identified in these genes.

3.1.4.3 Pregnane-X- Receptor (PXR) Activation   PXR acivators such as rifampin and cholestatic bile acids are known to increase the glucuronidation of a number of drugs including glucuronidation of AZT and morphine (UGT2B7), acetaminophen (UGT1A6), and lamotrigine (UGT1A4) in humans. Several UGTs contain PXR response elements or respond to rifampin induction including UGT1A1, UGT1A3, UGT1A4, and UGT2B7 (Gardner-Stephen, 2004).

TABLE 3.1   UGT1A nomenclature: *see* **http://som.flinders.edu.au/FUSA/ ClinPharm/UGT/.**

UGT1 gene	Trivial names	Endogenous substrates	Tissue expression	Comments
UGT1A1	Rat B1 HUG$_{Br-1}$ UgtBr1 (mouse)	Bilirubin Estradiol (3-OH) Morphine (3-OH) Catechol estrogens (2OH > 4OH) Retinoic acid leukotriene B4 12 and 15-HETE 13-HODE	Liver, intestine, mammary gland	Polymorphic regulation Defective genes—Crigler–Najjar syndrome Lowered expression—Gilbert's syndrome *Trans*-retinoic acid is a substrate for rat UGT1A1 but not human?
UGT1A2	Rat B2	Bilirubin		Pseudogene in humans Rapidly declines in tissue culture of rat hepatocytes
UGT1A3		Bile acid Catechol estrogens (2OH > 4OH) leukotriene B4 12 and 15-HETE 13-HODE	Liver, intestine, testes, prostate	
UGT1A4	HP3 HUG$_{Br-2}$	12 and 15-HETE 13-HODE	Liver, intestine	Pseudogene in rats and mice
UGT1A6	HP1 HlugP1 Ugt1-6 Rat4NP		Liver, kidney, intestine, brain, ovary, testes Spleen, skin	Pseudogene in cats Polymorphic in humans
UGT1A7	A2 1g		Gastric epithelium esophagus	
UGT1A8		Leukotriene B4	Intestine, esophagus	
UGT1A9	HP4	Catechol estrogens (4OH > 2OH) estradiol (low) 12 and 15-HETE 13-HODE	Liver, kidney ovary, testes Spleen, skin, esophagus	
UGT1A10 *UGT1A11- 1A13*			Intestine, lung	Pseudogenes

TABLE 3.2　UGT2B nomenclature: *see* **http://som.flinders.edu.au/FUSA/ ClinPharm/UGT/.**

UGT1 gene	Species	Endogenous substrates	Tissue expression	Comments
UGT2B1	Rat	Estradiol (17-OH); morphine (3OH)	Liver, low in kidney, intestine, testes	97% similar to UGT2B2
UGT2B2	Rat		Liver	Gene deletion in adrosterone-deficient Wistar rats
UGT2B3	Rat	Bile acid		
UGT2B4	Human	Hyodeoxycholic acid; 3α-OH-pregnanes and 3α, 16α, 17β-androgens 12-HETE, 15-HETE and 13-HODE	Liver	Activity of a variant (described as UGT2B11) toward steroids described by Jin, Mackenzie, and Miners, 1997
UGT2B5	Rabbit			
UGT2B6	Rat			
UGT2B7	Human	Estradiol (17β-OH), estriol, 3α-OH-pregnanes 3α, 16α, 17β -androgens *trans*-retinoic acid, leukotriene B4, morphine 3/6-OH 12-HETE, 15-HETE, and 13-HODE	Liver, kidney, espophagus, intestine, brain (cerebellum)	Polymorphism at 268 (Tyr or His)
UGT2B8	Rat		Liver	
UGT2B9	Cyno-mologus monkey	C_{18}, C_{19}, C_{20} steroids, fatty acids (C_6–$C_{12)}$ morphine 3/6-OH	Liver	89% identical to UGT2B7
UGT2B10	Human	12-HETE, 15-HETE, and 13-HODE (prostaglandins)	Liver, adrenals, prostate	
UGT2B11	Human	12-HETE, 15-HETE, and 13-HODE	RNA present in liver, kidney, mammary gland, prostate, skin, adipose, adrenal, and lung	91% identical to UGT2B10 76% identical to UGT2B15 and UGT2B17

TABLE 3.2 (*Continued*)

UGT1 gene	Species	Endogenous substrates	Tissue expression	Comments
UGT2B12	Rat		Liver, kidney, intestine	
UGT2B13	Rabbit	17β-Estradiol (low)	Adult liver	4-Hydroxybiphenyl is selective substrate. 66% identical to rat UGT2B1
UGT2B14	Rabbit	Estriol	Adult liver	73% identical to UGT2B13
UGT2B15	Human	Testosterone, dihydro-testosterone	Liver, prostate, testes, esophagus	
UGT2B16	Rabbit			
UGT2B17	Human	Testosterone, androsterone	Liver, kidney, prostate, testes, uterus, placenta, mammary glands, adrenals, skin	
UGT2B18	Monkey			
UGT2B19	Monkey			
UGT2B20	Monkey			
UGT2B21	Guinea pig			

3.1.4.4 Glucocorticoid Activation A glucocorticoid response element was identified near the CAR/PXR binding site in human UGT1A1 (Sugatani, 2005). Low concentrations of dexamethasone enhanced PXR activation of this gene when the glucorticoid receptor and GRIP1 were coexpressed into HepG2 cells.

3.1.4.5 Peroxisome Proliferator–Activated Receptor (PPAR) Activation Clofibrate, ciprofibrate, WY-14643, perfluorodecanoate:
    Fibrates are relatively specific inducers of bilirubin glucuronidation (UGT1A1), UGT1A3, and UGT1A9 (Barbier, 2003).

3.1.4.6 Antioxidant Response Element (ARE) UGTs are upregulated by treatment with a number of chemopreventative agents such as sulforphane. This occurs through the formation of reactive electrophilic metabolites that bind to selected cysteine residues in the cytosolic sensor protein Keap1. Upon adduction of these cysteines, Kea1p is signaled for ubiquination resulting in the release of the orphan transcription factor Nrf2. Nrf2 is transported to the nucleus where it binds to antioxidant response elements in a wide number of genes including heme oxygenase, NQO1 (quinone reductase), glutathione-S-traneferases (GSTs) that play a protective role in oxidative damage via elimination of these metabolites. Ritter and coworkers found no

TABLE 3.3 Selective substrates for individual UGT1A isozymes.

Isoenzyme	Endogenous substrates	Reported K_m	Drug or xenobiotic substrates	Inducers	Inhibitors
UGT1A1	*Bilirubin, estradiol* (3-hydroxy), 2-hydroxyestrone, 2-hydroxyestradiol trans-retinoic acid, catechol estrogens (2- and 4-hydroxy)		*Ethinyl estradiol, morphine* (3-hydroxy), buprenorphine ferulic acid, genistein naltrexone (low), naloxone (low), SN-38 (active metabolite of irinotecan) alizarin, quinalizarin	Bilirubin, chlorophen-oxypropionic acid, chrysin, clofibrate 3-methylcholanthrene, phenylpropionic acid, phenobarbital, etc. pregnenolone-16α-nitrile and dexamethasone, clotrimazole, rifampin, and St. John's wort	Atazanavir
UGT1A2	—				—
UGT1A3	Bile acids (carboxyl functional group), catechol estrogens (2-OH > 4-OH), 2-OH-estrone, 2-hydroxyestradiol, decanoic acid, dodecanoic acid, bilirubin (low)	—	*Cyproheptadine, alizarin, buprenorphine,* norbuprenorphine, bropirimine, diphenylamine, diprenorphine, emodin, esculetin, eugenol, ezetimibe, fisetin, genestein, 3-hydroxydesloratadine, 7-hydroxyflavone, hydromorphone, 4-methylumbelliferone, morphine, nalorphine, naloxone, naltrexone, *naringenin,* quercehtin, scopoletin, thymol, umbelliferone.	β-Naphthoflavone, rifampin(?)	—

| UGT1A4 | *Estrogens:* 2-hydroxy-estrone and 2-hydroxy estradiol, 4-hydroxy catechol estrogens (low), estriol, *Progestins:* 5α-pregnan-3α, 20α -diol, 16α -hydroxy pregennolone, 19- hydroxy and 21- hydroxy pregnenolone, pregneneolone, androsterone, epiandrosterone, etiocholanone, *Androgens:* dehydroepiandrosterone, dihydrotestosterone, epitestosterone, testostereore, 5×-androstan-3α, 17β-diol, 5β-androstan-3α, 11α, 17β-triol; bilirubin (very low), F_6-1α,23S,25(OH)$_3$D$_3$—a hexafluorinated Vit D$_3$ analog | *Carboxyl group:* clofibrate, ciprofibrate, etodolac, fenoprofen, ibuprofen, ketoprofen, naproxen (racemic > S), valproic acid and formation of simvastatin and atorvastatin lactones via an intermediate acyl glucuronide *Tertiary amines:* amitriptyline, chlorpheniramine, chlorpromazine, clozapine, cyproheptadine, diphenylamine, doxepin, imipramine, ketotifen, loxapine, promethazine, tripellennamine, trifluoperazine *Aromatic heterocyclic amines:* croconazole, lamotrigine, nicotine (30X velocity than UGT1A3), 1-phenylimidazole, posaconazole, retigabine *Primary and secondary amines:* 2- and 4-aminobiphenyl, diphenylamine, desmethylclozapine *Alcoholic and phenolic substrates:* borneol, carveol, carvacrol diosgenin, *hecogenin*, isomenthol, menthol, neomenthol, 1- and 2-naphthol (low), *p*-nitrophenol (low), nopol, tigogenin | Trifluoperazine: 6 μM (Ebner, 1993) Hecogenin: 10 μM (Uchaichipat, 2006; Green, 1998) | phenobarbital, phenytoin, and carbamazepine |

TABLE 3.3 *(Continued)*

Isoenzyme	Endogenous substrates	Reported K_m	Drug or xenobiotic substrates	Inducers	Inhibitors
UGT1A6	*Serotonin*, 3-hydroxy methyl DOPA	Serotonin: 6 μM (Krishnaswamy, 2003)	*Phenols*: acetaminophen, 2-amino-5-nitro-4-trifluoromethylphenol (flutamide metabolite), BHA, BHT, 7-hydroxy coumarin, 4-hydroxy-coumarin (low), dobutamine, 4-ethylphenol, 3-ethylphenol, 4-fluorocatechol, 2-hydroxybiphenyl, 4-iodophenol, 4-isopropylphenol (low), 4-methylcatechol, 4-methylphenol, methylsalicylate, 4-methylumbelliferone, 4-nitrophenol, 4-nitrocatechol, octylgallate, phenol, 4-propylphenol (low), *cis*-resveratrol, salicylate, 4-*tert*-butylphenol (low), tetrachlorocatechol, vanillin *Amines*: 4-aminobiphenyl, 1-naphthylamine > 2-naphthylamine, N-OH-2-naphthylamine.	TCDD, β-naphtoflavone, 3-methyl chloranthrene	α-Napthol, 4-t-butyl phenol, 4-methylum-belliferone, 7-hydroxy coumarin

(continued)

UGT1A7	Estriol, 2-OH-estradiol, 4-OH-estrone	*Drugs:* acetaminophen, beta blocking adrenergic agents (low activity) such as atenolol, labetolol, metoprolol, pindolol, propranolol, naproxen ($R \gg S$ for rat 1A6), salicylate, valproic acid. *Flavonoids:* chrysin, 7-hydroxy flavone, naringenin	
		Benzo(*a*)pyrene phenols (7-OH $\gg$ 9-OH $>$ -OH), Benzo(a)pyrene-t-7, 8-dihydrodiol (*7R*-glucuronide, low affinity), 2-OH-biphenyl, 4-methylumbelliferone, 1- and 2-naphthol, 4-nitrophenol, octylgallate, vanillin	TCDD
UGT1A8/9*	2-Hydroxy estrone, 4-Hydroxy estrone, 2-Hydroxy estradiol, 4-Hydroxy estradiol, estrone, dihydrotestosterone, *trans*-retinoic acid, 4-Hydroxy retinoic acid, hyocholic acid, hyodeoxycholic, testosterone, leukotriene B4,	Propofol: 300 μM	
		Alizarin, anthraflavic acid, apigenin, Benzo(*a*)pyrene-*t*-7, 8-dihydrodiol (*7R*- and 8*S*-glucuronides), emodin, fisetin, flavoperidol, genistein, naringenin, quercetin, quinalizarin, 4-methylumbelliferone, scopoletin, carvacrol, eugenol, 1-naphthol, *p*-nitrophenol, 4-aminobiphenyl,	3-Methyl cloranthrene

49

TABLE 3.3 *(Continued)*

Isoenzyme	Endogenous substrates	Reported K_m	Drug or xenobiotic substrates	Inducers	Inhibitors
			2-hydroxy, 3- hydroxy, and 4- hydroxy biphenyl, buprenorphine (low), morphine (low), naloxone, naltrexone, ciprofibrate, diflunisal, diphenylamine, furosemide, mycophenolic acid (high), phenolphthalein, propofol, valproic acid, nandrolone, 1-methyl-5α-androst-1-en-17β-ol-3-one (metabolite of metenolone), 5α-androstane-3α,17β-diol (metabolite of testosterone), (−)-epigallocatechin gallate (tea phenol), SN-38 (low)[metabolite of irinotecan], troglitazone (moderate), raloxifene (both 6β- and 4′-β-glucuronides), quercetin, luteolin		
		Entecapone: 10 μM (Lautala, 2000)			
UGT1A9/8*	Retinoic acid, thyroxine (T4), tri-iodothyronine (T3; minor), 4-hydroxyestrone, 4-hydroxyestradiol (major)		*Planar phenols:* Phenol, acetaminophen, 2-hydroxybiphenyl, 4-iodophenol, 4-propylphenol, 4-isopropylphenol (low), 4-ethylphenol,	TCDD, tetrabutyl hydroquinone, clofibric acid	High concentrations of propofol

3-ethylphenol,
4-methylphenol,
4-nitrophenol,
4-*tert*-butylphenol
(low), methylsalicylate,
salicylate, mono(ethylhexyl)
phthalate, **BHA**, **BHT**, vanillin,
7-hydroxycoumarin,
4-hydroxycoumarin
(low), 4-methylumbelliferone
Bulky phenols: phenol red,
phenolphthalein,
fluorescein
Simple catechols: *octyl*
gallate, propyl gallate
Primary amines:
4-aminobiphenyl
Xenobiotics: acetaminophen,
p-**HPPH** (phenytoin metabolite),
retigabine fenofibric acid,
gemfibrozil, ciprofibric acid,
clofibric acid, troglitazone
(10-fold lower activity than
fibrates), SN-38
(active metabolite of irinotecan),
mycophenolic acid, *propfol*,
atenolol, labetolol,
metoprolol, pindolol, propranolol,
diflunisal, fenoprofen,

(*continued*)

TABLE 3.3 (*Continued*)

Isoenzyme	Endogenous substrates	Reported K_m	Drug or xenobiotic substrates	Inducers	Inhibitors
			ibuprofen, ketoprofen, mefenamic acid, naproxen (low activity against all NSAIDs), bumetanide, furosemide, (dapsone) ethinyl estradiol -minor), dobutamine, dopamine, levodopa, carbidopa, entacapone, R –oxazepam, emodin, chrysin, 7-hydroxyflavone, galangin, naringenin, quercetin carveol, nopol, citronellol, 6-hydroxychrysene		
UGT1A10	2-OH-estrone (low), 4-OH estrone (low), dihydrotestosterone, testosterone		Alizarin, anthraflavic acid, apigenin, Benzo(a)pyrene-t-7, 8-dihydrodiol (7R- and 8S-glucuronides, high affinity), emodin, fisetin, genistein, naringenin, quercetin, quinalizarin, 4-methylumbelliferone, scopoletin, carvacrol, eugenol, mycophenolic acid, 17β-methyl-5β–androst-4-ene-3α,17α-diol (metabolite of metadienone), nandrolone, 1-methyl-5α-androst-1-en-17β–ol-3-one (metabolite of metenolone), 5α-androstane-3α, 17β-diol (metabolite of testosterone), SN-38 (minor), raloxifene (4′-β-glucuronide only)		

UGT1A11 — — —
UGT1A12 — — —

UGT1A8 and UGT1A9 were previously mislabeled. The old UGT1A8 is now UGT1A9 and vice versa.
AhR activators in humans (aromatic hydrocarbon receptor)–tetrachlorodibenzodioxin (TCDD), β-naphthoflavone, 3-methylcloranthrene.
PXR (pregnenolone-16α-nitrile-X-receptor) activators in rodents–pregnenolone-16α-nitrile (PCN), dexamethasone.
PXR (pregnenolone-16α-nitrile-X-receptor) activators in humans–clotrimazole, rifampin, and St. John's wort.
CAR (constitutive androstane receptor) activators in humans–3-methylcholanthrene, phenylpropionic acid, phenobarbital, phenytoin, carbamazepine.
PPARα (peroxisome proliferated-activated receptor-α) activator in humans–clofibric acid, fenobibric acid, pirinixic acid.
PPARγ (peroxisome proliferated-activated receptor-γ) activator in humans–rosiglitazone.
FXR (farnesoid-X-receptor) activators in humans – chenodeoxycholic acid.
LXR (liver-X-receptor).
RXR (retinod-X-receptor).
Underlinied substrates denote the most commonly used probes for enzymatic activity.

TABLE 3.4 Selective substrates for individual UGT2B isozymes.

Isoenzyme	Endogenous substrates	Reported K_m	Drug or xenobiotic substrates	Inducers	Inhibitors
UGT2B1	*Estradiol (17-hydroxy)*		Morphine		
UGT2B2					
UGT2B3					
UGT2B4	*Bile acids* 6α hydroxy bile acids, 3α-hydroxy pregnanes, 3α-, 16α-, 17β-androgens, metabolites of poly unsaturated fatty acids (PUFA), arachidonic and linoleic acids, estriol, 2-hydroxy estriol, 4-hydroxy estrone		*Phenols:* Eugenol, 4-nitrophenol, 2-aminophenol, 4-methyl umbelliferone, morphine	Fenofibric acid, chenodeoxycholic acid-activated FXR	
UGT2B5					
UGT2B6					
UGT2B7	*Arachidonic acid metabolites:* Leukotriene B4 (LTB4), 5-hydroxyeicosatetraenoic acid (HETE), 12-HETE, 15-HETE, and 13-hydroxyoctadecadienoic acid (HODE)		*R-Oxazepam*, naproxen, menthol, *AZT (zidovudine)*, abacavir, acetaminophen, almokalant, carvedilol, chloramphenicol, epirubicin, 1′-hydroxy estragole, 5-hydroxy rofecoxib, lorazepam, menthol, 4-methylumbelliferone, 1-naphthol (low), 4-nitrophenol, octylgallate, propranolol, temazepam, maxipost	Rifampin, Phenobarbital, HNFHNF1α	*R*-Oxazepam and zidovudine (competitive), Flunitrazepam relatively potent ($K_i \sim 50$–$90\ \mu M$), but also inhibits UGT1A3 ($K_i = 20$–$30\ \mu M$ for 2-hydroxy estrogens) and UGT1A1 ($K_i > 200\ \mu M$). diclofenac, etonitazenyl

Enzyme	Substrates			
UGT2B7 (cont.)	*Bile acids:* hyodexycholic acid *Estrogens:* Estriol, estradiol (17β-hydroxy), 4-hydroxy estrone (high) 2- hydroxy estrone, 2- hydroxy estriol *Pregnanes:* 3α-hydroxy pregnanes, *Androgens:* 3α-. 16α-, 17β-androgens, *Others:* 5α- and 5β-dihydroaldosterone, trans-retinoic acid, *Carboxylic acid-containing drugs:* benoxaprofen, ciprofibrate, clofibric acid, diflunisal, dimethylxanthenone-4-acetic acid (DMXAA), fenoprofen, ibuprofen, indomethacin, ketoprofen, naproxen, pitavastatin, simvastatin acid, tiaprofenic acid, valproic acid, zaltoprofen, zomepirac. *Opioids:* morphine 3OH > 6OH, buprenorphine, nalorphine, naltrexone, codeine (low) and naloxone.	*Morphine*	*4-Hydroxy biphenyl*	4-Hydroxy biphenyl (competitive)
UGT2B8				
UGT2B9	C18, C19, C20 steroids, fatty acids (C6–C12)			
UGT2B10	No known substrates			
UGT2B11	No known substrates			
UGT2B12				
UGT2B13	*17β-Estradiol* (low)			
UGT2B14	*Estriol*			
UGT2B15	*Testosterone,* dihydrotestosterone	S-oxazepam: 30 μM (Court, 2002)		
UGT2B16				
UGT2B17	*Testosterone,* androsterone			
UGT2B18				
UGT2B19				
UGT2B20				
UGT2B21				

evidence of an ARE in rat UGT1A6 but were able to show that oltipraz, which normally activates AREs was acting through an XRE (Auyeung, 2003) indicating that oltipraz is a mixed ARE/XRE agonist like beta-naphthoflavone (Kohle and Bock, 2006). Munzel et al. reported that there were three ARE-like motifs in human 1A6 including one very close to the XRE (Munzel, 2003).

3.1.5 Pharmacogenetics

Genetic polymorphisms have been identified in all of the UGT enzymes that have been extensively evaluated. The most well-known polymorphism occurs in UGT1A1. In Caucasians, a TA insertion into a TATA box (*UGT1A1*28*) results in lowered expression of the enzyme and mild hyperbilirubinemia (Gilbert's syndrome) in most subjects who carry two variant alleles. The frequency of the variant in Caucasians is approximately 15%. Neonates who are homozygous *UGT1A1*28/*28* have a high rate of neonatal jaundice requiring light therapy. Gilbert's patients also experience a significantly higher rate of neutropenia from irinotecan, and the FDA has approved a genotyping test for patients taking this anticancer medication. Incidence of Grade 2 or greater hyberbilirubinemia is significantly higher in Gilbert's patients taking atazanavir or indinavir, two HIV protease inhibitors known to inhibit UGT1A1 *in vitro*. Polymorphisms in the UGTs and their relationship to cancer incidence or treatment has been recently reviewed by the authors (Nagar and Remmel, 2006). Guillemette published an extensive review on the pharmacogenetics of the UGTs in 2003, but this is a highly active area of research that is rapidly maturing. Information on polymorphisms for the UGT1A and UGT2B families can be found on the UGT Web site with appropriate references at http://galien.pha.ulaval.ca/labocg/alleles/alleles.html.

3.1.6 Experimental Considerations

3.1.6.1 Microsomal Incubation Conditions   Incubations in animal or human liver microsomes are the most common way to determine activity in the presence of added substrate, UDPGA, Mg^{2+}, and a buffer. As there is no method available to directly determine enzyme concentration, the incubations are standardized by addition of the same amount of protein (typically 0.25–1.0 mg protein/1 mL) after determination of linearity of product formation with respect to protein concentration and time. In general, the enzyme is stable up to 45 min to 1 h. Because of the location of the enzyme, a portion of the microsomal vesicle will be obtained in the normal configuration with the enzyme active site entrapped within the vesicle. Since UDPGA must have access to the active site, and the UDPGA influx transporter is not operative without ATP, it may be necessary to "activate" or "remove latency" of the enzyme. In the past this has been achieved by a variety of methods, but most commonly by addition of detergents such as Brij 58, Lubrol, or Triton X

100. Unfortunately, detergents may differentially affect the individual enzymes and there is often a sharp concentration optimum for each preparation and substrate. More recently, the pore-forming antibiotic alamethacin has been widely used, which has a broader optimum concentration (typically 25–100 μg/mg protein) (Fulceri, 1994). Unfortunately, alamethacin is not very soluble and may require small amounts of methanol or ethanol for solubilization. Organic solvent concentrations should be kept at less than 0.5% in the final incubation. Alamethacin is incubated with the microsomes and buffer on ice for 30–35 min prior to addition of substrate and UDPGA. Addition of UDP-N-acetylglucosamine (UDP-GlucNAC) and ATP also stimulates activity (Fulceri, 1994), and it has been suggested that this is due to transport of UDP-GlucNAC out of the lumen in exchange for transport of UDPGA into the ER lumen by the transporter (trans-stimulation) (Bossuyt, 1996). Other additives to incubations may include a beta-glucuronidase inhibitor such as saccharo-1,4-lactone and an esterase inhibitor. Esterase inhibitors may be important for substrates containing carboxylic acids (acyl glucuronide formation). Addition of these inhibitors can increase product formation especially if long incubation periods are needed to increase assay sensitivity.

The choice of pH and buffers are also important considerations. Tris-containing buffers are commonly used for glucuronidation reactions. However, Tris has a relatively high pK_a 8.06 and a relatively narrow pH range for adequate buffering capacity. Tris-maleate buffers have a broader pH buffering range as maleate has a pK of 6.2, providing good buffering capacity from 5.7 to 8.5. Tris buffers are temperature sensitive, and thus a pH 7.7 buffer prepared at 25°C has a pH of 7.4 at 37°C. Phosphate (pK_a 7.2) is a good buffer, but binds to Mg^{2+} ions that are present in the incubation and interact with the phosphate backbone of UDPGA in the active site. Final buffer pH values vary widely in the literature. Owens and coworkers have consistently used a pH of 6.5 for bilirubin glucuronidation generating higher turnover rates whereas higher activities may be demonstrated for weakly basic or phenolic substrates at pH values of 8.0 or greater. The transferase reaction requires nucleophilic attack of the high energy phosphate bond, so one would expect that amine substrates (typical pK_a s of 8–10) would be better nucleophiles at higher pH.

Prior to conducting enzyme kinetic studies, it is important to determine linearity of glucuronide production with respect to time and protein concentration. This should be done at saturating substrate concentrations ($5 \times K_m$) or under conditions where less than 10% of the substrate is lost. The K_m values for many substrates are in the low millimolar range, thus solubility and solvent concentrations are important considerations. In general, a solvent concentration of <1% is desirable, preferably ≤0.1%. Unfortunately, many drug substrates are hydrophobic and poorly water soluble. Uchaipichat et al. recently studied the effects of different solvents on 4-methylumbelliferone and 1-naphthol glucuronidation by individual UGT enzymes. UGT2B15 (inhibition by ≥0.5% acetonitrile) and UGT2B17 (inhibition by ≥0.5% DMSO and ethanol) appeared to be most affected by organic solvents.

3.1.6.2 Incubations with Cloned, Expressed Enzymes Individual UGT enzymes have been expressed in a wide variety of systems including insect cells (Supersomes or Baculosomes), *Escherichia coli*, yeast, and mammalian cells. Zakim and Dannenberg have demonstrated that the lipid membrane composition can influence activity (Zakim, 1992). There tends to be excellent protein expression insect cells transfected with baculovirus, but when activity is measured compared to mammalian cells systems, there appears to be significant amounts of inactive protein due to either poor membrane insertion or improper folding (lack of chaperones?). Bacteria do not have an ER, but alteration of the signal sequence results in active membrane bound preparations. Yeast and mammalian cells such as HEK293 or V79 cells have a more typical membrane environment and may be preferable for expression of ER proteins.

In insect cells, yeast or mammalian cells, microsomal preparations can be prepared, however, the yield of microsomal protein from cultured cells is often low. Sonication of a frozen whole cell lysate appears to provide a fully activated preparation and is easier than preparing microsomes. If intact microsomes are needed from transfected mammalian cells, a recent procedure delineated by Sukodhub and Burchell is recommended for their preparation (Sukodhub, 2005).

When comparing the activities of individual UGT enzymes for a single substrate (isoenzyme screening), it is important to normalize for expression. This can be done by correction for protein expression by Western blotting in the UGT1A family with antibodies directed against the UGT1A constant region (encoded by shared exons 2–5). However, it is not possible to correct for protein expression when comparing between the UGT1A family and UGT2B enzymes because the purified proteins are not available. Thus, velocity measurements are usually based on a per milligram of protein basis, which must be interpreted with caution.

3.1.6.3 Analytical Methods There are three basic methods used to measure glucuronidation rates in microsomes:

(1) radiometric methods with radiolabeled substrate of ^{14}C-UDPGA; (2) fluorescence disappearance with fluorescent substrates, such as 4-methylumbelliferone; (3) chromatographic methods, most commonly HPLC-UV or liquid chromatography–mass spectrometry LC–MS.

Radiometric methods have been widely employed for substrate screening assays. The most common general method is to use ^{14}C-UDPGA as the cosubstrate, resulting in labeled glucuronide product(s). The glucuronides are easily separable from the ^{14}C-UDPGA by thin-layer chromatography (TLC) or by high pressure liquid chromatography (HPLC). Radioactivity on TLC plates can be counted on a plate scanner or by densitometric quantitation on film or a phosphoimager. This method can have significantly more variability than stand-alone HPLC or LC/MS methods. Alternatively, radiolabeled substrate can be employed, requiring separation of the more nonpolar aglycone from the polar glucuronide. In some cases, this can be accomplished by simple solvent extraction, for example, with ^{3}H-steroids or ^{14}C-naphthol. Alternatively, a radiometric HPLC detector or fraction collection and liquid scintillation counting can be employed. The latter method is more sensitive especially for tritiated compounds.

Direct HPLC assays for glucuronide production can be easily done on reversed-phase columns if an authentic standard of the glucuronide is available, however, for many drug candidates this may not be possible. Since most glucuronides have identical absorbance spectra as their aglycones (the sugar does not contribute to absorption at wavelengths above 210 nm), one can construct a standard curve with the aglycone for quantitation of the glucuronide, assuming there are no matrix interferences. To insure that the product is a glucuronide, it is usually wise to collect the peak and treat with beta-glucuronidase to ensure that the aglycone peak is observed after hydrolysis. Alternatively, glucuronidase treatment of an incubation (lacking saccharo-1,4-lactone) with subsequent loss of the glucuronide peak is sufficient evidence. However, some N-glucuronides are often resistant to beta-glucuronidase hydrolysis, for example, lamotrigine (Sinz, 1991) and olanzapine (Kassahun, 1997). Primary and secondary amino glucuronides are generally easily cleaved with dilute acid, whereas acyl (ester) glucuronides are easily cleaved with dilute base (0.1 N NaOH and mild heating).

LC–MS methods on either single quadrupole or time-of-flight instruments or tandem mass spectrometers are sensitive, and efficient procedures to quantitate glucuronide production if an authentic standard is available. Glucuronides readily fragment with a neutral loss of 176 amu to produce the aglycone fragmentation product. Glucuronides have a pK_a of around 4, and thus are readily detected under negative ion conditions at pH 4.5 or higher by electrospray or assisted electrospray, so ammonium acetate buffers are commonly employed. Trifluoroacetic acid can cause ion suppression especially in negative ion mode, but may improve glucuronide retention on C8 or C18 columns by protonating the carboxylic acid group. Ion pairing of charged carboxyl groups with triethylamine or morpholine can be used to improve retention. If the drug substrate is a weak base, positive ion mode may also be used.

3.1.7 Enzyme Selective Substrates and Inhibitors

Enzyme selective substrates and inhibitors have been widely used to distinguish individual cytochrome P450 activities, but this has been problematic for glucuronidation because of the overlapping substrate specificity for some of these enzymes. This is especially true for the closely homologous enzymes UGT1A3 and UGT1A4, and UGT1A7–UGT1A10. However, selective substrates for some of the enzymes have been well characterized.

UGT1A3 and UGT1A4 are >95% homologous and share many common substrates especially tertiary amine substrates. There are some substrate differences. The general UGT substrate 4-methylumbelliferone is a substrate for 1A3 but not 1A4. Bile acid conjugation at the 24-COOH group and fulvestrant are much more efficiently catalyzed by 1A3 versus 1A4, whereas trifluoperazine, hecogenin, and tamoxifen N-glucuronidation appear to be selective for 1A4. Trifluoperazine may be somewhat problematic due to nonspecific binding to microsomal proteins and surfaces. UGT1A7–1A10 are highly homologous (>90%) and have overlapping substrate selectivity for a number of phenolic substrates, but differ by their tissue distribution. UGT1A9 is expressed highly in liver and kidney, but not in intestine. UGT1A7 is expressed in the esophagus and gastric epithelium but not in liver. Both UGT1A8 and UGT1A10 are expressed in intestine, but not in liver. UGT1A8 is also expressed in lung. In liver tissue, propofol and entacapone are selective substrates for 1A9, but 1A8 and 1A10 can glucuronidate these bulky phenolic compounds. Entecapone is more selective due to a lower K_m and may be preferred, but is not widely available. Zidovudine (AZT, azidothymidine) appears to be fairly selective for UGT2B7, but is also turned over by UGT2B4 with similar K_m values. Maxipost was the first substrate discovered for the UGT2B7-catalyzed amide N-glucuronidation with a K_m of 13 μM (Zhang et al., 2004). Carbamazepine N-glucuronidation appears to be a UGT2B7 selective substrate, but has a high K_m.

Several of these compounds can be employed as selective inhibitors for screening purposes, such as bilirubin for 1A1, hecogenin for UGT1A4, serotonin for 1A6, but they should be employed at the proper concentrations ($2-4\times$ than K_m) as they may affect other enzymes at higher concentration. Unfortunately, selective inhibitory antibodies have not been developed. Atazanavir appears to be a potent UGT1A1 inhibitor (Zhang et al., 2005). Fluconazole, a nonsubstrate, appears to be a selective UGT2B7 inhibitor (Uchaipichat, 2006). Valproic acid and probenecid inhibit multiple UGTs and may be useful as general inhibitors, but are not selective.

3.1.8 Drug–Drug Interactions and Glucuronidation

A common conception of interactions involving glucuronidation are that they are not important because of either the availability of multiple enzymes catalyzing the same reaction or the relatively high K_m values for many substrates compared to the P450 enzymes. However, there are several important clinical interactions that have been delineated. Interactions are likely to occur via induction by AhR, PXR, CAR, and PPAR activators with individual enzymes resulting in increased clearance and lowered drug exposure (AUC values). Inhibitory interactions can occur when glucuronidation is a predominant metabolic elimination pathway, when the glucuronidation is catalyzed by a single enzyme and when the therapeutic concentrations of the inhibitor are close to the K_i of the target UGT. These principles have been discussed recently in a review, but are the same for any inhibitory drug interaction. Drug–drug interactions involving glucuronidation have been exhaustively reviewed recently by Kiang et al. (Kiang, 2005) and in a monograph by one of the co-authors (Remmel, 2000). It is beyond the scope to list all of the drug interactions in this short review, but some key examples will be discussed.

3.1.8.1   Common   Drug   Metabolizing   Enzyme   Lamotrigine   (LTG)   is   a
diaminotriazine antiepileptic drug that is widely used and promoted due to
its lack of interaction with other AEDs that are oxidatively metabolized by
P450s. Greater than 90% of LTG is excreted as *N*-glucuronide metabolites.
However, LTG is subject to inductive interactions with carbamazepine,
phenytoin, and phenobarbital that reduce the half-life from 24 h to
approximately 12 h necessitating dose adjustment. In contrast, patients who
are taking valproate have a half-life of 60–80 h due to inhibition of LTG
metabolism. LTG is a UGT1A4 substrate and recently Miners and coworkers
suggested that *N*-glucuronidation may also be catalyzed UGT2B7 based on
inhibition with valproic acid. Valproic acid is given at high doses and is a
substrate for multiple UGT enzymes including UGT1A3–1A10, and 2B7
(Argikar, 2005; Staines, 2004). In a transgenic mouse expressing the human
UGT1A complex, we have found induction of lamotrigine glucuronidation
with pregnenolone-16a-carbonitrile (a mouse PXR activator) and TCDD (an
Ah receptor ligand) (Chen, 2005). *UGT1A4* is a pseudogene in mice producing
an inactive protein. In humans, rifampin increases LTG clearance (Ebert,
2000), suggesting that PXR activators will increase UGT1A4 expression.

Zidovudine (AZT) is an HIV reverse transcriptase inhibitor and chain
terminator that is extensively glucuronidated (70% of the dose) primarily by
UGT2B7. Metabolism of AZT is induced by rifampin (PXR), ritonavir,
tipranavir, and efavirenz. Zidovudine clearance is inhibited by methadone
(McCance-Katz, 1998) (opiates like codeine and morphine are UGT2B7
substrates), fluconazole Trapnell, 1998, atovaquone (Lee, 1996), and valproate
(Lertora, 1994). Rifampin increased the formation clearance to AZT-
glucuronide by twofold (Gallicano, 1999).

3.1.8.2 Metabolic Switching   Inhibition of glucuronidation may result in
drug toxicity by directly increasing parent compound concentrations or can
lead to metabolic switching to oxidative pathways that produce reactive
metabolites. In acetaminophen acute hepatotoxicity, glucuronidation, and
sulfation pathways are saturated resulting in more of the drug metabolized to
NAPQI, the reactive quinone imine metabolite. Once glutathione is depleted,
NAPQI reacts with cellular macromolecules leading to cell death. In our
laboratory, we have studied the bioactivation of naltrexone and interactions
with NSAIDs. In the course of an open label trial of patients taking naltrexone
for pathologic gambling and who were also exposed to NSAIDs, a high
percentage of patients had elevated liver enzymes. Subsequent studies have
demonstrated that some NSAIDs, specially fenamate derivatives, are potent
UGT2B7 inhibitors, the enzyme responsible for naltrexone glucuronidation.
Naltrexone was metabolized by CYP3A4 in the presence of glutathione to two
glutathione conjugates presumably through a catechol/quinone intermediate
(Kalyanaraman, MS thesis, 2005). Thus, inhibition of glucuronidation leading
to greater production of the catechol metabolite may be responsible for this
interaction observed in the clinical studies.

3.1.8.3   Inhibition of Glucuronide Renal Clearance   Glucuronides are cleared by either biliary or renal clearance. Biliary secretion of glucuronides is largely mediated by MRP2 (ABCC2) whereas renal clearance is due to a combination of glomerular filtration and active secretion. Probenecid, an anion transport inhibitor, is known to inhibit the secretion of glucuronides and can result in high concentrations of circulating glucuronides. In some cases, especially acyl glucuronides, plasma esterases or tissue beta-glucuronidase can cleave the glucuronide back to parent resulting in reduced apparent clearance and elevated parent drug concentrations. This reversible metabolism is also observed in renal impairment and is sometimes termed as a "futile cycle" of metabolism. Similar drug interactions would be expected with inhibitors of MRP2 in terms of reduced biliary secretion of glucuronides in the liver.

3.1.8.4   Inhibition   of   P450-Mediated   Metabolism   by   Glucuronide Metabolites   Recently, Ogilvie et al (2006) reported that the glucuronide of gemfibrozil was a potent CYP2C8 inhibitor. The interaction of gemfibrozil with cerivastatin resulted in multiple cases of severe rhabdomyolysis and several deaths resulting in the withdrawal of cerivastatin from the US market. Gemfibrozil itself is a weak CYP2C8 inhibitor and also inhibits the glucuronidation of statins, a necessary step in the formation of the lactone metabolites. Cerivistatin is glucuronidated and oxidatively metabolized by CYP2C8 and CYP3A4. The finding that the gemfibrozil glucuronide was an even more potent inhibitor of CYP2C8 than gemfibrozil, coupled with the high local hepatic concentrations in the liver, suggest that glucuronidation had a major role in the observed interaction.

3.1.9   Summary

Glucuronidation is a critical Phase II conjugation reaction for many endogenous and exogenous compounds. The UGT enzyme superfamily catalyzes this reaction. UGT substrates, inducers and inhibitors, and analytical techniques to measure glucuronidation are discussed. Drug glucuronidation is an important contributor to drug–drug interactions via various mechanisms. The pharmaco-genomic knowledge concerning these enzymes is rapidly maturing and clearance of some important anticancer and HIV drugs is affected by polymorphisms. In an effort to reduce the contribution of P450 enzymes in the elimination of drug candidates to avoid interactions, one must be aware of potential problems that can arise by interactions in this important conjugation pathway.

3.2   CYTOSOLIC SULFOTRANSFERASES

Sulfonation is a common "Phase II" conjugation reaction that occurs across species. Numerous endobiotics as well as xenobiotics are sulfonated in the human body. This process of sulfonation was discovered by Baumann with the isolation of the phenol sulfate conjugate from human urine. Several years later, the cosubstrate 3′-phosphoadenosine 5′- phosphosulfate (PAPS) was discovered (Robbins and Lipmann, 1957), and the superfamily of enzymes namely the sulfotransferases (SULTs) has now been characterized to a great extent (Coughtrie, 2002). The sulfonation reaction is described in detail here, along with a summary of the SULT superfamily nomenclature and family members. Analysis of sulfonate conjugates with examples is also discussed.

Sulfonation of chemicals involves the conjugation of the substrate with a sulfonyl (SO_3^-) group (Fig. 3.5). The cosubstrate PAPS acts as the sulfonyl donor and the reaction is catalyzed by a SULT enzyme. Conjugation can occur at –C–OH, –N–OH, and –NH side chains to yield O-sulfates and N-sulfates. PAPS is synthesized from inorganic sulfate and ATP by the enzymes sulfurylase and adenosine 5′-phosphosulfate kinase in prokaryotes, and a bifunctional enzyme PAPS synthetase (PAPSS) in higher organisms including humans. Figure 3.6 exhibits sulfonation reactions with some common substrates. The sulfonation reaction follows a random or ordered bi–bi mechanism depending on the substrate and specific SULT isozyme studied, via SN2 displacement (Gamage et al., 2005). Sulfonate transfer occurs without the formation of intermediates. Substrate inhibition is commonly observed in this reaction at high substrate concentrations, probably due to the formation of dead-end complexes.

Overall reaction:

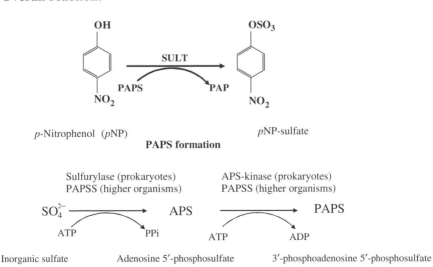

FIGURE 3.5 The overall sulfonation reaction, and formation of the cosubstrate PAPS. PAPSS: PAPS synthetase—"fused" bifunctional enzyme, varients PAPSS1 and PAPSS2 found in humans.

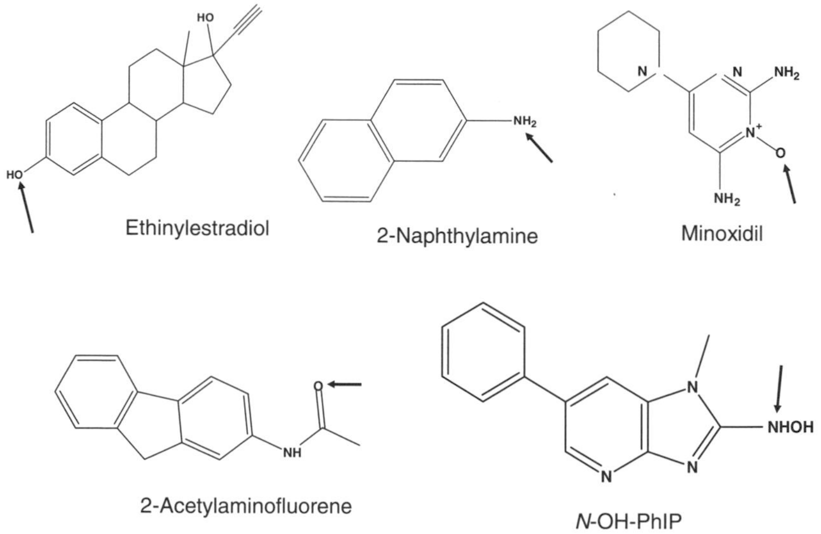

FIGURE 3.6 Common SULT substrates. Arrows indicate site of sulfonation for each substrate.

Sulfonation is generally a detoxification pathway whereby the conjugated product has greater water solubility and is therefore excreted more readily from the body. However, several chemicals have been shown to form mutagenic and carcinogenic reactive electrophiles upon sulfonation (Glatt, 2000). Additionally, it is the sulfonated forms of minoxidil and cholecystokinin that elicit biologic activity (Weinshilboum et al., 1997).

3.2.1 Cellular Location and Tissue Expression

Cytosolic SULTs exist as homodimers or heterodimers in solution. Cytosolic SULTs are responsible for the conjugation of endogenous substrates such as steroids, bile acids, and neurotransmitters, as well as several xenobiotics. Membrane—bound SULTs are present in the Golgi apparatus of cells, and catalyze sulfonation and posttranslational modification of peptides, proteins, lipids, and glycosaminoglycans. Sulfonation activity is found to be the highest in the liver and small intestine, although other organs also express the SULTs. Endogenous steroids are sulfonated in most mammalian tissues, with high activity in the adrenal tissue, kidney, and brain.

3.2.2 The SULT Superfamily of Cytosolic Enzymes

The cytosolic SULTs form a large superfamily of genes. Members of this superfamily are assigned families and subfamilies based on amino acid homology (Blanchard et al., 2004). Thus, members of a family share at least

45% amino acid sequence identity, while subfamily members share at least 60% homology. Each family is designated by an Arabic numeral following the "SULT" abbreviation, followed by an alphabetical subfamily, and the unique isoform identified by an Arabic numeral. Thus, SULT4A1 belongs to family 4 subfamily A. To date, 11 SULT isoforms have been discovered in humans, and these are listed in Table 3.5. Each isoform catalyzes the sulfonation of specific substrates, although overlapping substrate specificity has been noted. The SULT1A1 isoform catalyzes sulfonation of several small planar phenols, while SULT1A3 is responsible for the conjugation of catecholamines. The main substrates of the SULT1B subfamily are thyroid hormones. Human SULT1C catalyzes the conjugation of some procarcinogens, while SULT1E1 is the main isoform responsible for estrogen sulfation. Although SULT4A1 expression has been reported in the brain, its function is as yet undetermined. Recent research indicated a possible association of the SULT4A1 gene with schizophrenia susceptibility (Brennan and Condra, 2005).

The crystal structures of some rodent and human SULTs have been characterized (Gamage et al., 2005). Generally, SULT proteins have a highly conserved PAPS—binding region and a variable hydrophobic substrate—binding site. The substrate—binding pocket is variable and can therefore accommodate different types of substrates; this flexibility is in the substrate—binding region also explains overlapping substrate—specificity among SULT family members.

3.2.3 Inducibility

Very little is understood about SULT gene regulation and tissue-specific expression. The cloned human *SULT1A* genes were shown to lack TATA or CCAAT boxes near their putative transcriptional start sites (Aksoy and Weinshilboum, 1995; Her et al., 1996; Raftogianis et al., 1996). Studies have indicated that there is a marked difference in SULT gene regulation among species. Thus, several glucocorticoids and glucocorticoid-like chemicals induce bovine and rat SULT1A1 mRNA levels as well as protein levels, but no such effect is observed in human cells (Beckmann et al., 1994; Duanmu et al., 2000; Liu and Klaassen, 1996; Runge-Morris et al., 1996). Inducibility in the rat is probably via glucocorticoid receptor or PXR (Duanmu et al., 2002; Runge-Morris et al., 1999; Sonoda et al., 2002). Recent studies with the novel human CAR activator Citco have yielded contradictory results in human hepatocytes, with SULT1A1 mRNA induction reported in one sample, but no induction in an independent study (Maglich et al., 2003; Pacifici and Coughtrie, 2005).

3.2.4 SULT Pharmacogenetics

Much research has been conducted on characterization of genetic polymorphisms in *SULT* genes (Coughtrie, 2002; Blanchard et al., 2004; Raftogoanis et al.,

TABLE 3.5 Human SULT isoforms, substrates, and genetic polymorphisms.

Human SULT cDNA	Substrates	Common genetic polymorphisms
SULT1A1	Simple phenols 17β-estradiol Iodothyronines Acetaminophen Minoxidil 17α-ethinylestradiol Isoflavones Hydroxy-tamoxifen	SULT1A1*2 (Arg213His) SULT1A1*3 (Met223Val)
SULT1A2	Catecholestrogens Simple phenols	SULT1A2*2 (Ile7Thr) SULT1A2*3 (Pro19Leu)
SULT1A3	Dopamine (catecholamines) Tyramine Serotonin Salbutamol Isoprenaline Dobutamine Hydroxylated tibolone 4-Hydroxypropranalol	Lys234Asn
SULT1B1	Simple phenols Catechols Iodothyronines 0-desmethylnaproxen	
SULT1C2	N-hydroxy-2-acetylaminfluorene	
SULT1C4	N-hydroxy-2-acetylaminfluorene	
SULT1E1	Estrone 17β-Estradiol 17α-Ethinylestradiol Equilenin Diethylstilbestrol Thyroxine 0-desmethylnaproxen 3-OH-benozo[a]pyrene Phytoestrogens	Asp22Tyr Ala32Val Pro253His
SULT2A1	DHEA Pregnolone Cholesterol Cortisol Testosterone Bile salts PAHs (benzylic alcohols) Hydroxy-tamoxifen	Ala63Pro Lys227Glu Ala261Thr
SULT2B1_v1	DHEA Pregnenolone 3β-Hydroxy steroids	
SULT2B1_v2	DHEA Pregnenolone 3β-Hydroxy steroids	
SULT4A1	?	

1999; Nagar et al., 2006). Genetic polymorphisms have been reported in *SULT1A1, SULT1A2, SULT2A1, SULT1C2,* and *SULT1E1* genes (Table 3.5). The functional significance of this genetic variation has been studied, and there are several reports of association, for example, between *SULT1A1* genotype and breast or colorectal cancer (Bamber et al., 2001; Zheng et al., 2001).

3.2.5 Analytical Detection of Sulfonated Metabolites

There are typically three methods to detect sulfonated metabolites: (1) radiometric assays employing S^{35}-labeled PAPS; (2) HPLC-UV detection; (3) LC/MS detection. The protein source can be tissue cytosol or purified recombinant SULT protein.

3.2.5.1 Radiometric Assays Traditionally, sulfonated metabolites for many small chemicals have been detected with a radiometric assay that utilizes S^{35}-labeled PAPS (Anderson and Weinshilboum, 1980; Foldes and Meek, 1973). The reaction involves incubation of the substrate, cosubstrate, and enzyme in an appropriate buffer. The incubation is terminated by the addition of barium hydroxide, barium acetate, and zinc sulfate, which cause the unreacted PAPS to precipitate out. Thus, the unprecipitated radioactivity is associated with the sulfonated product and can be quantitated with liquid scintillation counting. A variation of this assay has been developed for larger molecules such as flavonoids, where the incubation is terminated by the addition of ethyl acetate under acidic pH conditions and in the presence of an ion-pairing agent, whereby the sulfonated product can then be detected in the organic phase upon liquid–liquid phase separation (Varin et al., 1987).

3.2.5.2 High Pressure Liquid Chromatography (HPLC) Sulfonated drug conjugates have also been detected with HPLC and LC/MS techniques. Since authentic standards are not available for most sulfonated products, reversed-phase HPLC assays have typically relied on deconjugation of the product with the enzyme sulfatase, detection of the parent compound, and subsequent calculation of the product formed. An example of sulfonate detection with HPLC is an assay developed for curcuminoid sulfate quantitation (Asai and Miyazawa, 2000). Here, the parent compounds (curcumin, demethoxycurcumin, and bidemethoxycurcumin) were separated on a TSKgel-ODS 80Ts 250 mm × 4.6 mm column at 40°C. The mobile phase consisted of acetonitrile/water (48 : 52 v/v), containing 50 mM phosphoric acid at 1 mL/min. The eluent was monitored with a UV–VIS detector at 425 nm. For detection of sulfated curcuminoids, the sample was incubated with 100 U sulfatase type VIII at 37°C for 4 h. Quercetin sulfates have similarly been quantitated with HPLC with detection at 370 nm (van der Woude, et al., 2004). Deconjugation of sulfates was achieved enzymatically by the use of sulfatase. Separation of chemicals was achieved on a C18 5U 150 mm × 4.6 mm column, with a gradient run of the mobile phase at 20–80% acetonitrile and water containing 0.1% trifluoroacetic acid over 2 min, at a flow rate of 1 mL/min. Detection was performed at 370 nm.

Deconjugation is achieved by the use of the enzyme sulfatase, which cleaves the sufonate group from the conjugate to release the parent compound. Several types of sulfatases are commercially available, and the choice is usually dictated by the multiple conjugation pathways involved. Thus, sulfatases with or without glucuronidase activity can be utilized. A major limitation of deconjugation is that the assay relies on quantitation of the parent chemical—the difference between the parent peak area at the start and end of the sulfatase reaction gives an estimate of the product formed. This is not as reliable as direct quantitation of the product itself against an authentic standard. Another potential drawback is incomplete desulfation of the product, leading to inaccurate estimation of sulfonate formation.

3.2.5.3   Liquid Chromatography–Mass Spectrometry (LC/MS)

3.2.5.3   Liquid Chromatography–Mass Spectrometry (LC/MS)   Sulfonate conjugates are easily detected by mass spectrometry with constant neutral loss scans that are characteristic for the presence of sulfate (80 Da) conjugated metabolites. Thus, several sulfonated conjugates of steroids and drugs have been characterized with LC/MS techniques. Figure 3.7 depicts some examples of sulfonated products detected by MS. Additional examples of LC/MS studies to detect and quantify sulfonate conjugates are listed in Table 3.6.

3.2.6   SULT Inhibitors (Pacifici and Coughtrie, 2005)

One of the first SULT1A1 inhibitors identified in the rat liver was 2, 6-dichloro-4-nitrophenol (DCNP) (Mulder and Scholtens, 1977). DCNP is a dead-end inhibitor, and exhibits low IC_{50} values toward SULT1A1 and SULT1A3 (Seah and Wong, 1994). Hydroxylated polychlorinated biphenyls (HPCBs) are potent inhibitors of recombinant human SULT1E1. HPCBs exhibit low micromolar IC_{50} values toward thyroid hormones (Schuur et al., 1998). Several dietary chemicals such as quercetin, curcumin, and flavones are known to inhibit SULTs. Some commonly used drugs that inhibit SULT1A1 and SULT1A3 activity include NSAIDs such as mefenamic acid, naproxen, and salicylic acid.

3.2.7   Drug–Drug Interactions and Sulfonation

Drug–drug and food–drug interactions have been reported due to inhibition of SULT activity in the human liver as well as the duodenum. Salbutamol sulfonation was inhibited by salicylic acid more potently in the liver than the

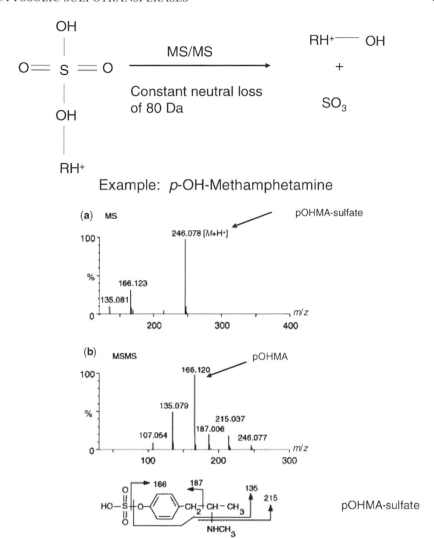

FIGURE 3.7 Detection of sulfonated conjugates with mass spectrometry. Adopted from Shima et al., J Chromatog B Analyt Technol Biomed life Sci 2005 Oct 28, epub.

duodenum (Vietri et al., 2000). The dietary flavonoid quercetin inhibits metabolism of dopamine, minoxidil, salbutamol, acetaminophen, and apomorphine (Marchetti et al., 2001; Vietri et al., 2002). Apomorphine and salbutamol sulfonation is additionally inhibited by mefenamic acid. Salicylic acid and apomorphine are known to interact in a similar fashion (Vietri et al., 2000; Vietri et al., 2002).

TABLE 3.6 Examples of LC/MS assays employed to study sulfonate conjugates.

Substrate	LC/MS assay details	Ion	References
Curcuminoids	System: LC–MS on a PE Mariner MS system LC: TSKgel-ODS 80Ts column, 40 °C; 5% folic acid, 5 mM ammonium acetate in acetonitrile (ACN): water 50 : 50 v/v at 1 mL/min MS detection: ES, positive mode	$[M + H]^+$: Curcumin $m/z = 369.13$ curcumin sulfate $m/z = 449.09$	Asai and Miyazawa (2000)
Quercetin	System: LC/MS on a Finnigan MAT 95 system LC: C18 5 U column; 0.1% acetic acid in 20–80% ACN over 30 min at 0.2 mL/min MS detection: ES, negative mode, source 4.5 kV, capillary 180 °C	Quercetin m/z 301 Monosulfate m/z 381	van der Woude et al. (2004)
Genistein	System: LC/MS/MS on a PE series 200 HPLC system with PE Sciex API 3000 triple quadrupole MS, ESI interface LC: C18 column; 5 mM ammonium acetate pH 7.0 and ACN 5–95% over 11 min at 0.2 mL/min MS detection: MRM, negative mode IS: chrysin	Genistein m/z 269.3–133.1 Genistein sulfate m/z 349.2–269.3	Soucy et al. (2005)
p-OH-methamphetamine (pOHMA)	(Metabolite identification) System: LC/MS/MS on a Shimadzu LCMS-IT-TOF system, ESI interface LC: L-column ODs semi-micro; 10 mM ammonium formate pH 3.5: methanol (95 : 5 v/v) MS detection: ESI, positive mode, source 4.5 kV, capillary 200 °C	pOHMA m/z 166 pOHMA-sulfate m/z 246	Shima et al. (2005)

Compound	Method	Detection	Reference
Androsterone	(Metabolite quantitation) System: LC–MS on a Shimadzu LCMS QP2010A LC: L-column ODs semi-micro; 10 mM ammonium formate pH 3.5 with 5–40% methanol MS detection: SIM mode System: LC/MS/MS on Micromass Quatro micro triple stage quadrupole, z spray ESI interface LC: C18 column; 2% aqueous formic acid, methanol and water in a nonlinear gradient MS detection: MRM, negative mode; source 3.5 kV, capillary 200 °C	$[M − 1]^+$ Androsterone sulfate 369 IS: d_4 – androsterone sulfate 373	Cawley et al. (2005)
Anisodamine	System: LC-MS and LC/MS/MS on an LCQ Duo quadrupole ion trap MS, ESI interface LC: C18 column; 0.01% triethylamine pH 3.5: methanol 40 : 60 v/v at 0.2 mL/min MS detection: ESI, positive mode, source 5 kV, capillary 200 °C	$[M + H]^+$ Anisodamine m/z 306 Anisodamine sulfate m/z 386	Chen et al. (2005)
2-amino-1-methyl-6-phenylimidazo[4,5-b]pyridine (PhIP)	System: LC/MS/MS on a Finnigan LCQ, ESI interface LC: YMC basic column; gradient of water:methanol:acetic acid at 0.2 mL/min MS detection: ESI positive mode, source 4.5 kV, capillary 240 °C	$[M + H]^+$ PhIP m/z 241 PhIP-sulfate m/z 321	Kulp et al. (2000)
Dehydroepiandrosterone (DHEA)	System: LC–MS on a Hitachi M 1000 LC/APCI MS LC: C18 column at 40 °C; 5 mM ammonium acetate and ACN: water 17 : 3 to 13 : 7 v/v over 16 min at 1 mL/min MS detection: SIM, negative mode, source 2 kV, capillary 399 °C IS: $^2H_4$-DHEA-sulfate m/z 371	DHEA m/z 287 DHEA-sulfate m/z 367	Nakajima et al. (1998)

3.2.8 Summary

Sulfotransferases are important Phase II enzymes that catalyze the conjugation of numerous endogenous chemicals as well as several xenobiotics. SULTs exhibit tissue-specific expression, but their gene regulation is poorly understood. SULT pharmacogenetics have been evaluated and may account for some interindividual variability to xenobiotic exposure. Several chemicals are known to inhibit SULTs, and this leads to numerous drug–drug and food–drug interactions.

3.3 GLUTATHIONE-S-TRANSFERASES

3.3.1 General Overview

GSTs catalyze the attack of the endogenous nucleophile, glutathione, to electrophilic molecules. Glutathione is a tripeptide of glycine–cysteine–gamma-glutamic acid. Glutathione is the major protective nucleophile in the body. GSTs also catalyze the SN2 displacement of halogen atoms on halogenated alkanes and aromatic halogens. The enzyme activates the cysteine thiol into a more nucleophilic thiolate anion that then attacks an electrophilic center held in close proximity to the thiolate. Glutathione transferases are active in their dimeric forms. They can form either homodimers or heterodimers with two catalytic centers. Heterodimers typically have intermediate activity between the two homodimers. There are several major classes of glutathione-S-transferase. In general, glutathione serves to protect against a number of carcinogens, but in some cases can actually bioactivate compounds, for example, 1,2-dibromoethane and sevoflurane. Kidney toxicity of sevoflurane (via Compound A) occurs via the GST-mercapturate-beta lyase pathway (Goldberg et al., 1999). A number of anticancer agents (nitrogen mustards, busulfan) and inhalation anesthetics (halothane, enflurane, sevoflurane) are metabolized by GST enzymes. Glutathione conjugates are excreted via transporters into the bile or are converted to their mercapturic acid derivatives (N-acetyl cysteine conjugates) and excreted into the urine. The GSTs have broad substrate specificity. They also play an important role in arachidonic acid metabolism resulting in the formation of leukotrienes. Prostaglandin A2 (PGA2), PGJ2, and D12PGJ2 (containing alpha-beta unsaturated ketones) are substrates for GSTs (Boogaard et al., 1989). The attenuation of PGJ_2 signaling results in upregulation of Nrf2-dependent and PPARγ-dependent gene expression and down regulation of NF-κB-dependent gene expression (Park et al., 2004). There are excellent recent reviews on GSTs in *Methods of Enzymology* (V. 401, 2005) and by Hayes et al. 2005.

3.3.2 Classification of the GST Enzymes

There are six major classes of the human cytosolic GST enzymes, namely, alpha, mu, omega, pi, theta, and the newly discovered zeta class (Mannervik et al., 2005). The sigma class is involved in prostaglandin metabolism (isomerization of PGH_2 to PGD_2) (Jowsey et al., 2001). Kappa class enzymes are expressed in mitochondria (Pemble et al., 1996). GST classes are abbreviated in Roman capitals and individual members are designated with a number. There may be more than one enzyme within a class (Table 3.7). GSTs are soluble cytosolic proteins existing as homo- or heterodimers and each monomer weighs approximately 25,000 Da for example, GSTA1 has 221 amino acids, MW = 25,500 Da. Heterodimers are named according to their subunit composition, for example, GSTA1-2. The amino acid sequence identity within each class is greater than 50%.

3.3.3   Localization and Expression

Soluble glutathione-S-transferases are found in most cells in the body. They are cytosolic enzymes, and all of the GSTs have been crystallized. In the liver, they represent a large percentage of the cytosolic protein (up to 5%) and may play a role in shuttling highly lipophilic molecules such as bilirubin within the cell. A trivial name for this function is ligandin. The alpha class is expressed highly in liver, intestine, testes, kidney, and adrenals. GSTA1 release into the plasma is a highly sensitive marker of liver toxicity. GSTM2-2 is found primarily in skeletal muscle. GSTM3 is in brain, lung, testes, GSTM4 in lymphoblasts, and GSTM5 in brain. GSTP1-1 is widely expressed in all tissues except the liver. It is the major form in fetal liver. GSTT enzymes are also widely expressed. Human sigma class enzymes are expressed most highly in macrophages, placenta, and adipose tissue. There are also membrane bound GSTs—a microsomal GST (MGST1) conjugates chlorodinitrobenzene (CDNB) like most other GSTs (Morgenstem, 2005).

3.3.4   Reactions Catalyzed by GSTs

Glutathione-S-transferases catalyze the reaction of glutathione with electrophiles, for example, epoxides, quinones, quinonemethides, to form glutathione conjugates by activating the thiol of glutathione into the more reactive thiolate anion within the active site. This greatly enhances the nucleophilicity of the sulfur atom and significantly enhances reaction rates. Reactive electrophilic metabolites of drugs, often formed by the cytochrome P450 system, are common substrates for the GSTs. GSTs also catalyze SN2 displacement reactions of halogen atoms with glutathione, for example, conjugation of chlorodinitrobenzene (CDNB) or dichloronitrobenzene (DCNB), two commonly used general substrates. In addition, several anticancer alkylating agents such as nitrogen mustards, busulfan, melphalan, and cyclophosphamide are also substrates for these enzymes. The newly identified zeta class enzyme (GSTZ1) functions as a maleylacetoacetate isomerase, a biochemical step in the catabolism of tyrosine, by activating haloacetic acids (Board and Ander, 2005; Tong et al., 1998). Two typical reactions are shown in Fig. 3.9.

3.3.5   Regulation of GSTs

GST alpha, mu, and pi expression is increased when cells are exposed to electrophiles (e.g., diethymaleate, paraquat) that generate reactive oxygen

TABLE 3.7 GST nomenclature and tissue expression (Mannervik et al, 2005).

Enzyme subunit	Class	Trivial names	Tissue expression	Comments
GSTA1	Alpha	B1B1, GST-type 1, ε	Liver, intestine testis ≫ kidney, adrenal > pancreas ≫ lung, brain > heart	Inducible by oltipraz, sulforophane, and phenobarbital. GSTalpha comprises about 5% of liver cytosolic proteins. Δ^5 steroid isomerase
GSTA2	Alpha	B2B2, GST2-type2, γ	Lung, pancreas, testis > kidney > adrenal > brain, lung, heart	Only 11 a.a. changes compared to GSTA1
GSTA3	Alpha		Lung, stomach, trachea, placenta	Δ^5 steroid isomerase
GSTA4	Alpha		All tissues, small intestine > spleen	
GSTM1a	Mu	GST1-type 2, μ	Liver ≫ testis > brain, adrenal, kidney, pancreas > lung, heart	Polymorphic—about 40–60% of Caucasians have a gene deletion (M1*0/*0).
GSTM1b	Mu	GST1-type 1,	Liver ≫ testis > brain, adrenal, kidney, pancreas > lung, heart	Allelic variation at codon 173 between (M1*A, M1*B)
GSTM2	Mu	GST4	Brain > testis > heart > pancreas > kidney > adrenal > lung, liver	84% identity with GSTM1-1 No polymorphisms identified
GSTM3	Mu	GST5	Testis ≫ brain, lung Spleen ≫ others	Polymorphism resulting in 3 bp deletion in intron 6 that generate a YY1 response element
GSTM4	Mu		Liver, skeletal muscle > heart,brain ≫ pancreas	
GSTM5	Mu		Brain, testis, lung	
GSTO1	Omega		Liver, heart, skeletal muscle, kidney	Monoethylarsonic acid, dehydroascorbic acid; modulates Ca^{2+} release by ryanodine receptors in ER. Polymorphisms ↑ the risk of stroke and vascular dementia. 20% homology to other GSTs
GSTO2	Omega		Testis	64% identity with GSTO1 Dehydroascorbic acid reductase
GSTP1	Pi	GST3, π	Brain > lung, heart Testis > adrenal, kidney, pancreas > liver macrophages	Large variations in expression. Highly expressed in tumors Four variants
GSTS1	Sigma	S1-1	Testis	Converts PGH_2 to PGD_2
GSTT1	Theta	φ	Kidney, liver > small intestine > brain, spleen, testis, pancreas, prostate	Null allele is 30–40% in Germans Partial gene deletion in 15–25% other Caucasians. 60% in Asians
GSTT2	Theta		Liver?	Truncation in C-terminal region: possibly a pseudogene
GSTZ1	Zeta	+ four allelic variants	liver	Activates dichloroacetic acid. Functions as maleylacetoacetate isomerase. (tyrosine degradation)

species (hydrogen peroxide, OH radicals). There is an antioxidant response element (ARE) in the *GSTA1*, *GSTA3*, *GSTM1*, and *GSTP1* promoter regions. The transcription factor Nrf2 binds to the ARE. Normally, Nrf2 is confined to the cytoplasm and is bound to the actin-binding protein Keap1. Electrophiles cause Nrf2 to dissociate from Keap1 (by reacting with thiols or activating kinases) allowing Nrf2 to go to the nucleus and form heterodimers with small Maf protein. The heterodimeric complex then binds to AREs (Hototashi and Yamamoho, 2004). Other ARE containing proteins include UGT1A1, NQO1 (quinone reductase), γ-glutamylcysteine synthase (necessary for GSH production), cysteine-glutamate exchange transporter (responsible for cysteine uptake), heme oxygenase-1, thioredoxin, and MRP1. Nrf2-knockout mice are more sensitive to acetaminophen and butylated hydroxytoluene (BHT) toxicity and have increased cancer rates when exposed to diesel exhaust, aflatoxin, or benzo(*a*)pyrene (Itoh et al., 2004). Several chemopreventative agents also upregulate GSTs via Nrf2 release such as sulforophane (in broccoli), oltipraz, and isothiocyanates such as 3-methylsulfinylpropyl isothiocyanate (in broccoli) or glucosinolates present in the seeds of cruciferous vegetables (McWalter et al., 2004).

3.3.6 GST Alpha Class

3.3.6.1 GSTA Substrates GSTA1-1 – Δ^5-androstene-3,17-dione (isomerase), Δ^5-pregnane 3,20-dione, PGA$_2$, PGJ$_2$ (R-isomer conjugate favored) *Busulfan*, Bay region diol epoxides of PAHs (*syn* or $R > S$), for example B(*a*)P-diol epoxide, moderate chlorodinitrobenzene (CDNB), low 1,2-dichloro-4-nitrobenzene (DCNB) activity, various nitrogen mustards, such as chlorambucil and melphalan, busulfan, doxorubicin, vincristine, etoposide, mitoxantrone, *N*-acetoxy-PhiP, fatty acid hydroperoxides.

> GSTA2-2—*Cumene hydroperoxide*, fatty acid hydroperoxides, dibenzo(a)-pyrene diol epoxide, CDNB-moderate, DCNB-moderate, 7-chloro-4-nitro-2-oxa-1,3-diazole, ethacrynic acid (low–moderate), PEITC, sulforophane.
>
> GSTA3-3—Δ^5-androstene-3,17-dione, Δ^5-pregnene-3,20-dione, DBPDE.
>
> GSTA4-4—*4-hydroxynonenal* (HNE), 4-hydroxydecenal, crotonyloxy-methyl-2-cylohexenone (COMC-6), CDNB—low, ethacrynic acid—low.

3.3.6.2 Polymorphisms in the Alpha Class GSTs Board and coworkers identified several SNPs by analysis of an expressed sequence tag (EST) database containing sequence from 300 human cDNA libraries (Tetlow et al., 2001) Ten polymorphisms were identified in the coding region of the GSTA1 and GSTA2 genes, and six were verified by sequence analysis (Table 3.8). PCR/RFLP analysis identified three variants in GSTA2 that were present in Caucasians (Australian), Bantu and Creole Africans, and Chinese populations (Table 3.9). One was a silent mutation (365G > A) and two encoded for amino

TABLE 3.8   GSTA1 and GSTA2 polymorphisms identified by BLAST and EST analysis.

GST	Sequence change	Amino acid change	Effect
GSTA1	115G > T	Nonsense	
GSTA1	159G > T	Q35H	
GSTA1	365G > A	Silent	None
GSTA1	598G > C	Q199H	
GSTA1	669A > C	Stop codon to Tyr	
GSTA2	322C > T	L108F	BLAST and EST only
GSTA2	335C > G	T112S	No effect on kinetics
GSTA2	389A > T	N130I	BLAST and EST only
GSTA2	517A > G	S173G	BLAST and EST only
GSTA2	588G > T	K196N	
GSTA2	629C > A	E210A	No effect on kinetics

acid changes T122S (conservative) and E210A (at start of external helix). Neither residue was in the active site. When expressed in *E. coli*, the kinetic properties of the GSTA2-2 enzyme were not affected.

Coles et al. found variants in the promoter region of the human GSTA1 gene with four SNPs: The combination of −631T or G, −567T, −69C, −52G was designated as hGSTA1*A and −631G, −567G, −69T, −52A was designated as hGSTA1*B (Coles et al., 2001). Genotyping has been done for the C–69T mutation and this relatively common in Caucasian, Asian, and African populations (Table 3.10). Decreased expression was observed in a bank of 55 liver samples that were heterozygous or homozygous for the *B variant. However, Bredschneider could find no relationship between haplotypes in a separate liver bank (Bredschreider et al., 2002). They reported several other polymorphisms in the GSTA1 gene. Busulfan, a chemotherapeutic agent, is an index substrate for the enzyme. Conjugation of busulfan varied seven–eight fold in 48 human liver samples. Expression of GSTA1-1 as determined by Western blotting also was variable (25–205 pmole/mg cytosolic protein). Eight SNPs were identified: a silent mutation (A365G) in exon 5 (Tetlow et al., 2001)

TABLE 3.9   Frequencies of GSTA1 polymorphisms in Caucasians, Africans, and Chinese populations.

Polymorphism	Australian Caucasians (*n* = 200)	African (Bantu and Creole) (*n* = 99)	Chinese (*n* = 96)
GSTA1 (K135K)	365G = 0.42	365G = 0.23	365G = 0.12
	365A = 0.58	365A = 0.77	365A = 0.88
GSTA2 (T122S)	S = 0.995	S = 0.995	S = 1
	T = 0.005	T = 0.005	T = 0
GSTA2 (E210A)	629A = 0.84	629A = 0.606	629A = 0.812
	629C = 0.16	629C = 0.394	629C = 0.188

TABLE 3.10 Distribution of the C–69T hGSTA1 genotypes in control populations.

Population	N	A/A	A/B	B/B
African-American	70	0.61	0.26	0.13
African-American women	52	0.63	0.26	0.12
Caucasian	278	0.38	0.48	0.14
Caucasian men	162	0.35	0.52	0.13
Caucasian women	116	0.41	0.44	0.16
Hispanic women	53	0.36	0.51	0.13

and seven sequence changes in the promoter region. No relationship between any of the identified SNPs was observed in the liver bank with either GSTA1 function or expression. In 48 Caucasian human liver samples, this variant was present in 47.9% of the samples. Matsuno et al. genotyped 147 healthy Japanese subjects and found 70.7% were *A/*A; 26.5% *A/*B, and 2.7% *B/*B (allele frequency 0.16) (Table 3.11) (Matsuno et al., 2004).

3.3.7 GST Mu Class

3.3.7.1 Overview The GSTM1 through GSTM5 gene locus is on chromosome 1p spanning 100 kB. Deletion of the GSTM1 gene (null phenotype) is relatively common ($\sim$70% allele frequency in Caucasians and Asians, 50% in Africans). Given the protective role of GST against environmentally bioactivated carcinogens, many genetic polymorphism studies have been done with regard to cancer incidence. Polymorphisms in other GSTM enzymes have also been identified, especially with GSTM3. One mutation of GSTM3 is an insertion of three base pair sequence that generates a transcription factor binding site for ying yang 1 (GSTM3 AA). It is believed that the presence of the YY1 transcription factor acts as an inducer. GSTM3 is expressed in higher amounts in lung than GSTM1. It is also expressed in significant amounts in astrocytes in the blood-brain barrier and in brain tissue.

TABLE 3.11 Frequency distributions of GST1A1 and GSTT1 genotypes 147 Japanese subjects.

	GSTA1			GSTT1		
	*A/*A	*A/*B	*B/*B	*A/*A	*A/*0	*A/*0
N	104	39	4	9	67	71
Observed frequency (%)	70.7	26.5	2.7	6.1	45.6	48.3
95% CI	62.9–77.5	20.1–34.2	1.1–6.8	3.3–11.2	37.7–53.6	40.4–56.3
Expected frequency (%)	71.2	26.3	2.4	8.4	41.1	50.6

(Adapted from Coles et al., 2001).

3.3.7.2 Substrates GSTM1-1—*Trans-stilbene oxide,* DCNB-high, CDNB-moderate, Aflatoxin B1-*exo* 8,9-epoxide, androstene 3,17-dione, B(*a*)P-diol epoxide, B(*a*)P-4,5-oxide, chrysene diol epoxide, cumene hydroperoxide, ethacrynic acid, *p*-nitrophenyl acetate, PGA2, PGJ2, styrene 7,8-oxide, *trans*-4-phenyl-3-buten-2-one.

GSTM2-2—*Dopa o-quinones,* PGH_2 to PGE_2, DCNB-high, CDNB-high, aminochrome, 2-cyano-1,3-dimethyl-1nitroso-guanidine.

GSTM3-3—BCNU, PGH_2 to PGE_2, CDNB-low, DCNB-low, cumene hydroperoxide-low, ethacrynic acid-low.

GSTM4-4—CDNB-low, DCNB-low, cumene hydroperoxide-low, ethacrynic acid-low.

GSTM5-5—CDNB-moderate to low.

3.3.7.3 Polymorphisms Approximately 50% of Caucasians and E. Asians do not have a gene for GSTM1. This is termed the null allele (*GSTM1∗0*). In some Polynesian and Micronesian populations, the allele frequency is very high ($\geq$90%), whereas in Africans, the null allele occurs much less frequently ($\sim$20–25% homozygous for the null genotype) (Table 3.12).

3.3.7.4 Polymorphisms and Effects on Drug Metabolism or Drug Toxicity Cisplatin is an important chemotherapeutic agent with dose-limiting side effects of ototoxicity, nephrotoxicity, and peripheral neuropathies. Peters et al. examined GST polymorphisms and cisplatin-induced ototoxicity (Peters et al., 2000). A significant protective effect was observed for the GSTM3*B allele (GSTM3 AA). The frequency of this allele was 0.18 in a group with normal hearing ($n = 19$) treated with cisplatin versus 0.025 in the group with early hearing impairment ($n = 20$).

3.3.7.5 Role in Cancer Since glutathione conjugation is a primary means of detoxifying electrophiles, its role in cancer has been extensively investigated. The *GSTM1* gene deletion deficiency (present in 50% of Caucasians) has been associated with a moderate increased risk for lung and bladder cancer (RR = 1.5–2.0). Many studies on other types of cancers show a minimal effect. The relationship between *GSTM1* variants, alone or in combination with CYP1A1 (a bioactivating enzyme, expressed in epithelial tissues) and polycyclic aromatic hydrocarbon DNA adducts is still controversial. More recent studies have shown a clear dependence upon CYP1A1 genotype especially in GSTM1-deficient smokers, despite their relatively low expression in the lung. Benzo(*a*) pyrene 7,8-diol-9,10-epoxide (BPDE) adducts were 100× higher in *GSTM1 0/0* smokers with a CYP1A1 high inducibility genotype compared to subjects with active GSTM1 (Bartsch, 1996). Several other studies also indicate a strong relationship with GSTM1 and DNA adducts (reviewed by Bartsch, 2000).

Linkage studies have failed to identify a role for GSTM3 in lung cancer, adenocarcinomas, small cell lung carcinoma (Risch, 2001), or in astrocytomas. A positive risk was identified in bladder cancer (OR = 2.31, CI 1.7–2.82) for GSTM3 AA genotype and for the GSTM1 null genotype (OR = 3.54,

TABLE 3.12 Frequency of the GSTM1*0/*0 (null allele) in various ethnic groups

Ethnic group	#Studied	Frequency
United States (Caucasian)	1751	0.737
Swedish	544	0.747
Finnish	482	0.684
German	734	0.718
French	1184	0.731
Scottish	42	0.673
English	1122	0.760
Russian	1000	0.697
Spanish	312	0.705
Saudis	895	0.750
Chinese	96	0.765
Japanese	639	0.689
Korean	165	0.721
Singapore	244	0.749
Filipino	80	0.814
Micronesian	37	1.00
Melanesian	49	0.795
Polynesian	49	0.904
Indian	43	0.560
African men	292	0.483
African women	187	0.566

Data from compilation of Gerte (2001).

CI = 2.99–4.41) (Schnakenberg E, 2000). Approximately 5–10% of chemical workers exposed to diisocyanates develop asthma and the combination of the GSTM3AA genotype and the GSTM1 null genotype resulted in a strong (OR = 11.0, CI 2.2–55.3) correlative effect (Piirila et al., 2001).

3.3.8 GST Pi Class

3.3.8.1 Substrates GSTP1-1—*PAH fjord region and bay region diol epoxides*, acrolein, base propenals, chlorambucil, ethacrynic acid, CNDB-moderate, DCNB-low, ethacrynic acid-high, 7-chloro-4-nitrobenzo-2-oxa-1,3-diazole, thiotepa, COMC-6, 4-oxo-nonenal, PGA2 PGJ$_2$ (*S*-isomer conjugates favored).

3.3.8.2 Polymorphisms GSTP1 is the most abundant GST in human lung, with strong expression in the bronchial epithelium. A major role for this enzyme may be to protect the body from inhaled toxicants such as PAHs. GSTP1-1 can be translocated into the nucleus and may serve a protective role against electrophiles and cancer drugs resulting in drug resistance (Goto et al., 2002). Exposure to cancer drugs such as doxorubicin and etoposide have been shown to increase nuclear transport and transport can be inhibited by a mushroom (*Agaricus bisporus*) lectin (Goto et al., 2002). Four

TABLE 3.13   Frequency of GSTP1 Polymorphisms in German control and cancer patients (Risch et al., 2001).

Gene	Percentage in German Caucasians (controls) (%)	Lung cancer (%)	Adenocarcinoma (%)
GSTP1*1/*1 (AA313)	47.3 (n = 167)	45.4 (n = 176)	46.4 (n = 70)
GSTP1*1/*2 (GA313)	42.8 (n = 151)	44.9 (n = 174)	45.7 (n = 69)
GSTP1*2/*2 (GG313)	9.9 (n = 35)	9.8 (n = 38)	7.9 (n = 12)

GSTP1 variants have been identified: P1*A (104I, 113A), P1*B (104V, 113A), P1*C (104V, 113V), and P1*D (104I, 113V) (Table 3.13). The GSTP1*2 and GSTP1*3 alleles produce variant enzymes that have a catalytic efficiency (k_{cat}/K_m) that is three to four times lower than the most common GSTP1*A allele. The A313G polymorphism in GSTP1 results in Ile104Val conversion and is relatively common (allele frequency = 0.26) (Table 3.14). The GSTP1*3 allele is rare. Ile104 is in the hydrophobic substrate binding pocket of GSTP1-1.

3.3.8.3   GSTP1 Polymorphism and Cancer   Cancer treatment regimens often include alkylating agents such as busulfan, thiotepa, chlorambucil, and melphalan. These drugs undergo glutathione conjugation that inactivates these alkylating drugs. The major enzyme involved in this inactivation is the GSTalpha and GSTpi class enzymes. The glutathione conjugates are removed from tumor cells by transport proteins. Some tumors have increased expression of these GSTs, a possible resistance mechanism. The glutathione conjugates can also result in product inhibition of the GSTs and failure to remove the GSTs by down regulation of the transporter expression may also play a role in drug resistance in tumor cells. Several epidemiological studies have reported associations of higher rates of testicular, oral pharyngeal, and bladder cancer with these variant alleles, but not with breast, colon, or lung cancers. A recent meta-analysis for prostate cancer has been completed examining the effects of GST mu, pi, and theta polymorphisms (Ntais et al., 2005). Overall, no significant random effects odd ratio (1.05 overall) were observed in these studies, although individual studies have reported higher risks. The authors concluded that the GST polymorphism were unlikely to be major risk factors for prostate cancer (Table 3.15).

TABLE 3.14   Frequency in GSTP1 allele frequencies in control groups of Caucasians (Goto et al., 2002).

Gene	Heterozygous	Homozygous	Allele frequency
GSTP1*1	0.493	0.438	0.685
GSTP1*2	0.442	0.413	0.262
GSTP1*3	0.126	0.0057	0.0687

1138 Controls tested for *1 and *2. 878 Controls tested for *3.

TABLE 3.15   Frequency of GSTT1 null genotypes in German control and cancer patients.

Gene	Percentage in German Caucasians (controls)	Lung cancer (%)	Adenocarcinoma (%)
GSTT1 null	18.8 (n = 65)	12.8 (n = 49)	14.7 (n = 22)
GSTT1 non-null	81.2 (n = 281)	87.2 (n = 334)	85.3 (n = 128)

3.3.9   GST Theta Class

3.3.9.1   Substrates   GSTT1-1—*Dichloromethane*, ethylene dibromide, butadiene diepoxide, cumene hydroperoxide (moderate-high), 1,2-epoxy-3 (4-nitrophenoxy-propane, ethylene oxide, methyl bromide, methylene chloride, 4-nitrobenzyl chloride, 4-nitrophenethylbromide, phenethylisothiocyanate.

GSTT2-2—*Cumene hydroperoxide, menaphthyl sulfate.*

3.3.9.2   Polymorphisms   A null allele for the GSTT1 gene (*GSTT1*0/*0*) is also fairly common. In Caucasians, approximately 19.7% (range 13–26%) are homozygous for the null allele. The null allele is much more common in Asians (mean 47%, range 35%–52%). In Nigerians, 38% were homozygote null compared to 24% in African-Americans (Garte et al., 2001).

Several epidemiological studies comparing persons with the GSTT1 null phenotype (*GSTT1*0*) have been completed. Of the completed studies with at least 100 cases and 100 controls, three found an association with the null phenotype and increased risk for cancer, two found a higher rate of the null phenotype in controls, and found no significant change (Risch et al., 2001). One study suggested an increased risk for myelodysplastic diseases but another study failed to find an association. There may be an increased risk for some types of brain tumors. In a recent Danish study, there was a stronger association with early-onset lung cancer (OR = 9.6) and null genotype with an OR = 2.4 overall (Sorensen et al., 2004).

3.3.10   GST Zeta Class

The GST Zeta class was initially identified by a BLAST homology search (Tong et al., 1998). It appears to play a role in acetoacetate metabolic pathways. Its role in the metabolism of drugs is not known (Board and Anders 2005).

3.3.10.1   Substrates   GSTZ1-1—*Dichloroacetic acid* (conversion to glyoxylate) and other haloacetic acids, maleylacetoacetate *cis–trans* isomerase (penultimate step in tyrosine degradation), ethacrynic acid (low).

3.3.11   Incubation Conditions and Analytical Methods

For most general purposes, one incubates substrates in the presence of cytosol and added glutathione. A more purified preparation of the major cytosolic

GSTs in liver (GSTA, GSTM, GSTP) can be prepared by affinity chromatography on S-linked glutathione agarose (Sigma G4510) or immobilized *S*-hexylglutathione. After loading of the cytosol and washing with 0.15 M NaCl in a 0.04 M phosphate buffer, pH 7.0, the GSTs can be eluted with 50 mM GSH in a 0.1 M Tris buffer, pH 9.6 (Coles and Kadlubar, 2005). Hepatocytes may also be used, but depending on the cell–cell contact and transporter function, excretion of the conjugates out of the cells may not occur, thus requiring lysis of the cells for measurement. In cases where the cells are well polarized (e.g., hepatocyte spheroids), significant canalicular transport can occur resulting in the appearance of glutathionyl and cysteinyl–glycine conjugates. It has been reported that cryopreserved hepatocytes have low glutathione concentrations upon thawing and may lead to underreporting of GSH conjugate production compared to freshly isolated cells (Sohlenius-Sternbeck and Schmidt, 2005). Cloned, expressed GSTs (GSTA1-1, GSTM1-1, and GSTP1-1 homodimers) are also commercially available (Panvera/InVitrogen or Sigma) and can be used to determine the activity of individual enzymes or for detailed kinetic studies with individual substrates. With purified enzymes or cloned, expressed preparations glutathione is added. Typical GSH concentrations range from 0.3 to 5 mM. The K_m for GSTA1-1 has been reported as 120 μM for CDNB and with rat GSTA1-1 (Schramm et al., 1984) the K_m was 0.16 mM for CDNB and 0.81 mM for *trans*-4-phenyl-3-buten-2-one (Warholm et al., 1983). Intracellular GSH concentrations vary depending upon the tissue but in hepatocytes and leukocytes the concentration is ≥ 5 mM. However, at high GSH and substrate concentrations, nonenzymatic formation of GS-conjugates may occur. A variety of buffers have been used: Tris buffers at pH 7.5 are commonly employed but pH optima may vary from 6.5 to 8. Tris-maleate buffers can extend the buffering capacity to cover lower pHs if desired.

Glutathione conjugates are highly polar, ionized conjugates that are generally analyzed by either HPLC or LC/MS methods. In organic/water mobile phases, the conjugates will elute in the solvent front. In acidic mobile phases containing acetic acid or trifluoroacetic acid (TFA), the amino acid carboxyl groups are unionized, and the protonated amino groups can form ion pairs with the organic acids, resulting in significant retention. In the case of acetaminophen, we have found that both the glutathione conjugate and acetaminophen elute after acetaminophen in the presence of 0.05% TFA in the mobile phase. TFA is commonly used in peptide analysis, but can cause some ion suppression in electrospray mass spectrometry, particularly in the negative ion mode. In the aforementioned article by Stern et al., acetaminophen and its glutathione-derived conjugates were analyzed by atmospheric pressure positive ion MS (Stern et al., 2005). While fragmentation patterns can be quite complex, typical fragmentation of peptide bonds is generally observed and easy to follow especially with an ion trap mass spectrometer, where the bonds that are easiest to break can be followed down in an MS^n analysis (e.g., neutral loss of the gamma-glutamyl (M-130) and glycine moieties from glutathione).

3.3.12   Glutathione Conjugate Metabolism (Mercapturic Acid Pathway)

GSH conjugates are highly polar, charged conjugates that cannot passively diffuse through membranes. They can be further metabolized by dipeptidase enzymes that metabolize GSH. The breakdown products are glutamate, glycine, and the cysteine conjugate. In the liver, glutathione conjugates and cysteinyl–glycine conjugates can be directly transported into the bile, Cysteine conjugates can be transported into either the bile or the blood by ABC transporters. Cysteine conjugates can be further acetylated by a specialized N-acetyltransferase to form the mercapturic acid conjugate (N-acetylcysteine conjugate) (Fig. 3.8), a reaction that occurs primarily in the kidney. Mercapturates are often the major metabolites of glutathione-related metabolism excreted in the urine of humans. Another pathway of metabolism of the cysteine conjugate is cleavage of the conjugate by beta-lyase to release a free thiol. Cysteine β-lyase is present in high concentrations in the kidney.

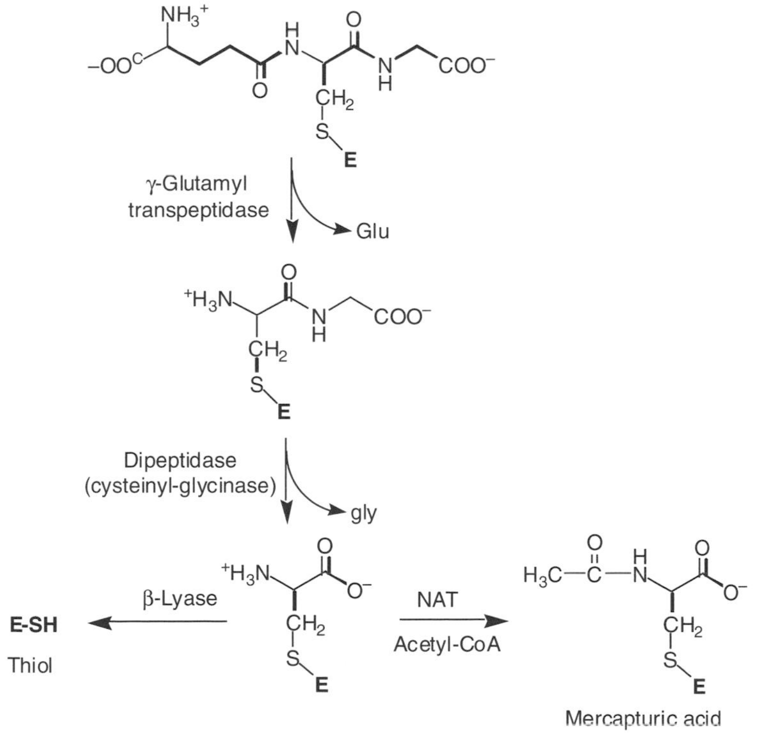

FIGURE 3.8   GSH conjugate metabolism: formation of mercapturic acid, N-acetyl cysteine and thiol conjugates.

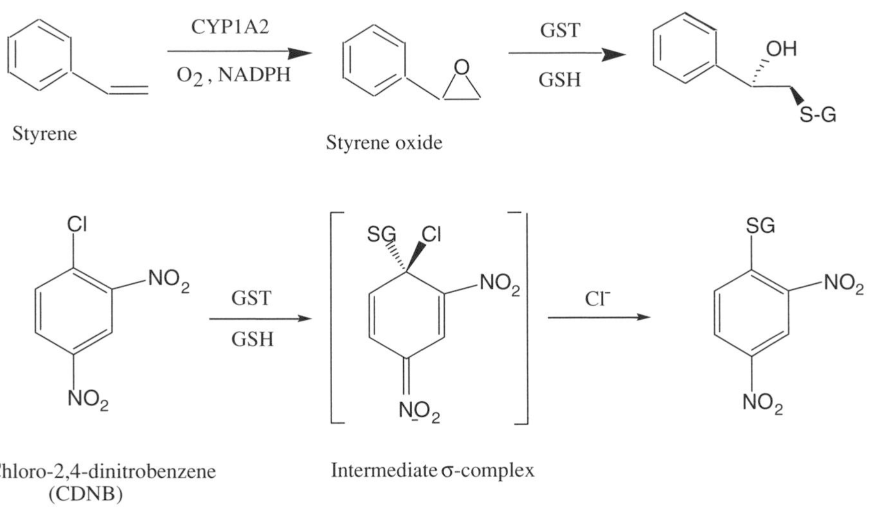

FIGURE 3.9   Two typical reactions catalyzed by GSTs.

REFERENCES

Aksoy IA, Weinshilboum RM. Human thermolabile phenol sulfotransferase gene (STM): molecular cloning and structural characterization. Biochem Biophys Res Commun 1995;208(2):786–795.

Anderson R, Weinshilboum R. Phenolsulfotransferase in human tissue: radiochemical enzymatic assay and biochemical properties. Clin Chem Acta 1980;103:79–90.

Asai A, Miyazawa T. Occurrence of orally administered curcuminoid as glucuronide and glucuronide/sulfate conjugates in rat plasma. Life Sci 2000;67(23):2785–2793.

Bamber D, Fryer A. et al. Phenol sulphotransferase SULT1A1*1 genotype is associated with reduced risk of colon cancer. Pharmacogenetics 2001;11:679–685.

Bartsch H. DNA adducts in human carcinogenesis: etiological relevance and structure-activity relationship. Mutat Res 1996;340(2–3):67–79.

Beckmann JD, Illig M. et al. Regulation of phenol sulfotransferase expression in cultured bovine bronchial epithelial cells by hydrocortisone. J Cell Physiol 1994;160(3):603–610.

Blanchard et al., 2004 Blanchard RL, Freimuth RR. et al. A proposed nomenclature system for the cytosolic sulfotransferase (SULT) superfamily. Pharmacogenetics 2004;14(3):199–211.

Board PG, Anders MW. Human glutathione transferase zeta. Methods Enzymol 2005;401:61–77.

Boogaard PJ, Mulder GJ, Nagelkerke JF. Isolated proximal tubular cells from rat kidney as an *in vitro* model for studies on nephrotoxicity. II. Alpha-methylglucose uptake as a sensitive parameter for mechanistic studies of acute toxicity by xenobiotics. Toxicol Appl Pharmacol 1989;101(1):144–157.

Bossuyt X, Blanckaert N. Carrier-mediated transport of intact UDP-glucuronic acid into the lumen of endoplasmic-reticulum-derived vesicles from rat liver. Biochem J 1994a;302(1):261–269.

Bossuyt X, Blanckaert N. Functional characterization of carrier-mediated transport of uridine diphosphate N-acetylgylucosamine across the endoplasmic reticulum membrane. Eur J Biochem 1994b;223(3):981–988.

Bossuyt X, Blanckaert N. Uridine diphosphoxylose enhances hepatic microsomal UDP-glucuronosyltransferase activity by stimulating transport of UDP-glucuronic acid across the endoplasmic reticulum membrane. Biochem J 1996;315(1):189-193.

Bredschneider M, Klein K, Murdter TE. et al. Genetic polymorphisms of glutathione S-transferase A1, the major glutathione S-transferase in human liver: consequences for enzyme expression and busulfan conjugation. Clin Pharmacol Ther 2002;71(6):479–487.

Brennan MD, and Condra J. Transmission disequilibrium suggests a role for the sulfotransferase-4A1 gene in schizophrenia. Am J Med Genet B Neuropsychiatr Genet 2005;139(1):69–72.

Cawley AT, Kazlauskas R. et al. Determination of urinary steroid sulfate metabolites using ion paired extraction. J Chromatogr B Analyt Technol Biomed Life Sci 2005;825(1):1–10.

Chen H, Wang H. et al. Liquid chromatography-tandem mass spectrometry analysis of anisodamine and its phase I and II metabolites in rat urine. J Chromatogr B Analyt Technol Biomed Life Sci 2005;824(1–2):21–29.

Chou DKH, Flores S, Jungalwala FB. J Biol Chem1991;266:17941–17949.

Coles BF, Kadlubar FF. Human alpha class glutathione S-transferases: genetic polymorphism, expression, and susceptibility to disease. Methods Enzymol 2005;401:9–42.

Coles BF, Morel F, Rauch C. et al. Effect of polymorphism in the human glutathione S-transferase A1 promoter on hepatic GSTA1 and GSTA2 expression. Pharmacogenetics 2001;11(8):663–669.

Coughtrie M. Sulfation through the looking glass – recent advances in sulfotransferase research for the curious. Pharmacogenomics J 2002;2:297–308.

Duanmu Z, Dunbar J. et al. Induction of rat hepatic aryl sulfotransferase (SULT1A1) gene expression by triamcinolone acetonide: impact on minoxidil-mediated hypotension. Toxicol Appl Pharmacol 2000;164(3):312–320.

Duanmu Z, Locke D. et al. Effects of dexamethasone on aryl (SULT1A1)- and hydroxysteroid (SULT2A1)-sulfotransferase gene expression in primary cultured human hepatocytes. Drug Metab Dispos 2002;30(9):997–1004.

Foldes A, Meek JL. Rat brain phenolsulfotransferase: partial purification and some properties. Biochim Biophys Acta 1973;327(2):365–374.

Gamage N, Barnett A. et al. Human sulfotransferases and their role in chemical metabolism. Toxicol Sci 2005.

Garte S, Gaspari L, Alexandrie AK. et al. Metabolic gene polymorphism frequencies in control populations. Cancer Epidemiol Biomarkers Prev 2001;10(12): 1239–1248.

Glatt H. Sulfotransferases in the bioactivation of xenobiotics. Chem Biol Interact 2000;129(1–2):141–170.

Goldberg ME, Cantillo J, Gratz I. et al. Dose of compound A, not sevoflurane, determines changes in the biochemical markers of renal injury in healthy volunteers. Anesth Analg 1999;88(2):437–445.

Goto S, Kamada K, Soh Y, Ihara Y, Kondo T. Significance of nuclear glutathione S-transferase pi in resistance to anti-cancer drugs. Jpn J Cancer Res 2002;93(9):1047–1056.

Hayes JD, Flanagan JU, Jowsey IR. Glutathione transferases. Annu Rev Pharmacol Toxicol 2005;45:51–88.

Her C, Raftogianis R. et al. Human phenol sulfotransferase STP2 gene: molecular cloning, structural characterization, and chromosomal localization. Genomics 1996;33(3):409–420.

Itoh K, Tong KI, Yamamoto M. Molecular mechanism activating Nrf2-Keap1 pathway in regulation of adaptive response to electrophiles. Free Radic Biol Med 152004;36(10):1208–1213.

Jowsey IR, Thomson AM, Flanagan JU. et al. Mammalian class Sigma glutathione S-transferases: catalytic properties and tissue-specific expression of human and rat GSH-dependent prostaglandin D2 synthases. Biochem J 2001;359(Pt 3):507–516.

Kulp KS, Knize MG. et al. Identification of urine metabolites of 2-amino-1-methyl-6-phenylimidazo[4,5-b]pyridine following consumption of a single cooked chicken meal in humans. Carcinogenesis 2000;21(11):2065–2072.

Liu L, Klaassen CD. Regulation of hepatic sulfotransferases by steroidal chemicals in rats. Drug Metab Dispos 1996;24(8):854–858.

Maglich JM, Parks DJ. et al. Identification of a novel human constitutive androstane receptor (CAR) agonist and its use in the identification of CAR target genes. J Biol Chem 2003;278(19):17277–17283.

Mannervik B, Board PG, Hayes JD, Listowsky I, Pearson WR. Nomenclature for mammalian soluble glutathione transferases. Methods Enzymol 2005;401:1–8.

Marchetti F, De Santi C. et al. Differential inhibition of human liver and duodenum sulphotransferase activities by quercetin, a flavonoid present in vegetables, fruit and wine. Xenobiotica 2001;31(12):841–847.

Matsuno K, Kubota T, Matsukura Y, Ishikawa H, Iga T. Genetic analysis of glutathione S-transferase A1 and T1 polymorphisms in a Japanese population. Clin Chem Lab Med 2004;42(5):560–562.

McWalter GK, Higgins LG, McLellan LI. et al. Transcription factor Nrf2 is essential for induction of NAD(P)H:quinone oxidoreductase 1, glutathione S-transferases, and glutamate cysteine ligase by broccoli seeds and isothiocyanates. J Nutr 2004;134(12 Suppl):3499S–3506S.

Morgenstern R. Microsomal glutathione transferase 1. Methods Enzymol 2005;401: 136–146.

Motohashi H, Yamamoto M. Nrf2-Keap1 defines a physiologically important stress response mechanism. Trends Mol Med 2004;10(11):549–557.

Mulder GJ, Scholtens E. Phenol sulphotransferase and uridine diphosphate glucuronyltransferase from rat liver *in vivo* and *vitro*. 2,6-Dichloro-4-nitrophenol as selective inhibitor of sulphation. Biochem J 1977;165(3):553–559.

Nagar S, Walther S. et al. Sulfotransferase (SULT) 1A1 polymorphic variants *1, *2, and *3 are associated with altered enzymatic activity, cellular phenotype, and protein degradation. Mol Pharmacol 2006;69(6):2084–2092.

Nakajima M, Yamato S. et al. Determination of dehydroepiandrosterone sulphate in biological samples by liquid chromatography/atmospheric pressure chemical ionization-mass spectrometry using [7,7,16,16-2H4]-dehydroepiandrosterone sulphate as an internal standard. Biomed Chromatogr 1998;12(4):211–216.

Ntais C, Polycarpou A, Ioannidis JP. Association of GSTM1, GSTT1, and GSTP1 gene polymorphisms with the risk of prostate cancer: a meta-analysis. Cancer Epidemiol Biomarkers Prev 2005;14(1):176–181.

Ogilvie BW, Zhang D. et al. Glucuronidation converts gemfibrozil to a potent, metabolism-dependent inhibitor of CYP2C8: implications for drug-drug interactions. Drug Metab Dispos 2006;34(1):191–197.

Pacifici GM, Coughtrie MWH. Human Cytosolic Sulfotransferases. Boca Raton FL: CRC Press; 2005.

Park EY, Cho IJ, Kim SG. Transactivation of the PPAR-responsive enhancer module in chemopreventive glutathione S-transferase gene by the peroxisome proliferator-activated receptor-gamma and retinoid X receptor heterodimer. Cancer Res 152004;64(10):3701–3713.

Pemble SE, Wardle AF, Taylor JB. Glutathione S-transferase class Kappa: characterization by the cloning of rat mitochondrial GST and identification of a human homologue. Biochem J Nov 11996;319(Pt 3):749–754.

Peters U, Preisler-Adams S, Hebeisen A. et al. Glutathione S-transferase genetic polymorphisms and individual sensitivity to the ototoxic effect of cisplatin. Anticancer Drugs 2000;11(8):639–643.

Piirila P, Wikman H, Luukkonen R. et al. Glutathione S-transferase genotypes and allergic responses to diisocyanate exposure. Pharmacogenetics 2001;11(5):437–445.

Raftogianis RB, Her C. et al. Human phenol sulfotransferase pharmacogenetics: STP1 gene cloning and structural characterization. Pharmacogenetics 1996;6(6):473–487.

Raftogianis RB, Wood TC. et al. Human phenol sulfotransferases SULT1A2 and SULT1A1:genetic polymorphisms, allozyme properties, and human liver genotype-phenotype correlations. Biochem Pharmacol 1999;58(4):605–616.

Risch A, Wikman H, Thiel S. et al. Glutathione-S-transferase M1, M3, T1 and P1 polymorphisms and susceptibility to non-small-cell lung cancer subtypes and hamartomas. Pharmacogenetics 2001;11(9):757–764.

Robbins PW, Lipmann F. Isolation and identification of active sulfate. J Biol Chem 1957;229(2):837–851.

Runge-Morris M, Rose K. et al. Regulation of rat hepatic sulfotransferase gene expression by glucocorticoid hormones. Drug Metab Dispos 1996;24(10):1095–1101.

Runge-Morris M, Wu W. et al. Regulation of rat hepatic hydroxysteroid sulfo-transferase (SULT2-40/41) gene expression by glucocorticoids: evidence for a dual mechanism of transcriptional control. Mol Pharmacol 1999;56(6):1198–1206.

Schuur AG, Legger FF. et al. In vitro inhibition of thyroid hormone sulfation by hydroxylated metabolites of halogenated aromatic hydrocarbons. Chem Res Toxicol 1998;11(9):1075–1081.

Schramm VL, McCluskey R, Emig FA, Litwack G. Kinetic studies and active site-binding properties of glutathione S-transferase using spin-labeled glutathione, a product analogue. J Biol Chem 251984;259(2):714–722.

Seah VM, Wong KP. 2,6-Dichloro-4-nitrophenol (DCNP), an alternate-substrate inhibitor of phenolsulfotransferase. Biochem Pharmacol 1994;47(10):1743–1749.

Shima N, Tsutsumi H. et al. Direct determination of glucuronide and sulfate of p-hydroxymethamphetamine in methamphetamine users' urine. J Chromatogr B Analyt Technol Biomed Life Sci 2005.

Sohlenius-Sternbeck AK, Schmidt S. Impaired glutathione-conjugating capacity by cryopreserved human and rat hepatocytes. Xenobiotica 2005;35(7):727–736.

Sonoda J, Xie W. et al. Regulation of a xenobiotic sulfonation cascade by nuclear pregnane X receptor (PXR). Proc Natl Acad Sci U S A 2002;99(21):13801–13806.

Sorensen M, Autrup H, Tjonneland A, Overvad K, Raaschou-Nielsen O. Glutathione S-transferase T1 null-genotype is associated with an increased risk of lung cancer. Int J Cancer 102004;110(2):219–224.

Soucy NV, Parkinson HD. et al. Kinetics of genistein and its conjugated metabolites in pregnant sprague–dawley rats following single and repeated genistein administration. Toxicol Sci 2005.

Stern ST, Bruno MK, Hennig GE, Horton RA, Roberts JC, Cohen SD. Contribution of acetaminophen-cysteine to acetaminophen nephrotoxicity in CD-1 mice: I. Enhancement of acetaminophen nephrotoxicity by acetaminophen-cysteine. Toxicol Appl Pharmacol 152005;202(2):151–159.

Tetlow N, Liu D, Board P. Polymorphism of human Alpha class glutathione transferases. Pharmacogenetics 2001;11(7):609–617.

Tong Z, Board PG, Anders MW. Glutathione transferase zeta-catalyzed biotransformation of dichloroacetic acid and other alpha-haloacids. Chem Res Toxicol 1998;11(11):1332–1338.

Trapnell CB, Klecker RW. et al. Glucuronidation of 3'-azido-3'-deoxythymidine (zidovudine) by human liver microsomes: relevance to clinical pharmacokinetic interactions with atovaquone, fluconazole, methadone, and valproic acid. Antimicrob Agents Chemother 1998;42(7):1592–1596.

van der Woude H, Boersma MG. et al. Identification of 14 quercetin phase II mono- and mixed conjugates and their formation by rat and human phase II *in vitro* model systems. Chem Res Toxicol 2004;17(11):1520–1530.

Varin L, Barron D. et al. Enzymatic assay for flavonoid sulfotransferase. Anal Biochem 1987;161(1):176–180.

Vietri M, Pietrabissa A. et al. Differential inhibition of hepatic and duodenal sulfation of (−)-salbutamol and minoxidil by mefenamic acid. Eur J Clin Pharmacol 2000;56(6–7):477–479.

Vietri M, Vaglini F. et al. Sulfation of R(−)-apomorphine in the human liver and duodenum, and its inhibition by mefenamic acid, salicylic acid and quercetin. Xenobiotica 2002;32(7):587–594.

Warholm M, Guthenberg C, Mannervik B. Molecular and catalytic properties of glutathione transferase mu from human liver: an enzyme efficiently conjugating epoxides. Biochemistry 191983;22(15):3610–3617.

Weinshilboum RM, Otterness DM. et al. Sulfation and sulfotransferases 1: sulfotransferase molecular biology: cDNAs and genes. FASEB J 1997;11(1):3–14.

Zhang D, Zhao W, Roongta V, Mitroka J, Klunk L, Zhu M. Amide N-glucuronidation of maxipost catalyzed by UGP-glucuronosyltransferase 2B7 in humans. Drug Metab Dispos 2004;32:545-551.

Zhang D, Chando TJ, Everett DW, Patten CJ, Dehal SS, Humphreys WG. In vitro inhibition of UDP glucuronosyltransferases by atazanavir and other protease inhibitors and the relationship of this properties to in vivo bilirubin glucuronidation. Drug Metab Dispos 2005;33:1729-1739.

Zheng W, Xie D. et al. Sulfotransferase 1A1 polymorphism, endogenous estrogen exposure, well-done meat intake, and breast cancer risk. Cancer Epidemiol Biomarkers Prev 2001;10:89–94.

4

ENZYME KINETICS

Timothy S. Tracy

4.1 INTRODUCTION

Determination of enzyme kinetic parameters for the metabolism of a new chemical entity has become an important part of the drug discovery and development process. Estimation of kinetic parameters such as K_m and V_{max} is critical in determining the degree of involvement of a particular enzyme in the metabolism of a compound under development since these parameters allow for determination of the intrinsic clearance of an enzyme in the drug's metabolism and thus, comparison among enzymes involved. Furthermore, these parameters determined *in vitro* may be used to predict the human dose to be given as well as the predicted *in vivo* pharmacokinetics. Thus, it is crucial that enzyme kinetic parameters, such as K_m and V_{max}, be determined as accurately as possible. This chapter will discuss the various types of kinetic profiles (both typical and atypical) that are commonly observed with drug metabolizing enzymes and the equations that allow estimation of the kinetic parameters. In addition, enzyme inhibition kinetics will be discussed as it is important in assessment of the drug interaction potential of a new chemical entity. It should be noted that the outstanding works of Segel (1975) and Cornish-Bowden (2004) are highly recommended for anyone conducting enzyme kinetics experiments and served as the primary reference sources in the preparation of this chapter.

Drug Metabolism in Drug Design and Development, Edited by Donglu Zhang, Mingshe Zhu and W. Griffith Humphreys
Copyright © 2008 John Wiley & Sons, Inc.

4.2   ENZYME CATALYSIS

The catalysis by enzymes involved in oxidative drug metabolism reactions (e.g., the cytochromes P450) typically follow the kinetic scheme as outlined in Scheme 4.1. In this case, in theory all reactions are reversible and an enzyme–substrate complex [E-S] must be formed before product formation and subsequent release can occur. In this case, k_{cat}, which is the capacity of the enzyme–substrate complex to generate product, is equal to k_2. For clarification, it should be noted that the Michaelis constant (K_m, see below) is derived from the microscopic rate constants in Scheme 4.1 ($K_m = (k_{-1} + k_2)/k_1$) using the steady state assumption.

Though not depicted kinetically here, the reader should be aware that conjugation enzyme reactions (such as the glucuronosyl transferases) are terreactant systems that involve an enzyme and two cosubstrates and are generally more complicated kinetically but can readily be described in the same fashion.

4.3   MICHAELIS–MENTEN KINETICS

The fundamental cornerstone of the kinetic characterization of enzymatic reactions has been and remains the Michaelis–Menten equation (Eq. 4.1).

$$v = \frac{V_{max}[S]}{K_m + [S]} \tag{4.1}$$

In this case, v is the velocity of the reaction, [S] is the substrate concentration, V_{max} (also known as V or V_m) is the maximum velocity of the reaction, and K_m is the Michaelis constant. From this equation quantitative descriptions of enzyme-catalyzed reactions, in terms of rate and concentration, can be made. As can be surmised by the form of the equation, data that is described by the Michaelis–Menten equation takes the shape of a hyperbola when plotted in two-dimensional fashion with velocity as the y-axis and substrate concentration as the x-axis (Fig. 4.1). Use of the Michaelis–Menten equation is based on the assumption that the enzyme reaction is operating under both steady state and rapid equilibrium conditions (i.e., that the concentration of all of the enzyme–substrate intermediates (see Scheme 4.1) become constant soon after initiation of the reaction). The assumption is also made that the active site of the enzyme contains only one binding site at which catalysis occurs and that only one substrate molecule at a time is interacting with the binding site. As will be discussed below, this latter assumption is not always valid when considering the kinetics of drug metabolizing enzymes.

$$E + S \underset{k_{-1}}{\overset{k_1}{\rightleftharpoons}} E\text{-}S \xrightarrow{k_2} E + P$$

SCHEME 4.1   Kinetics of catalysis.

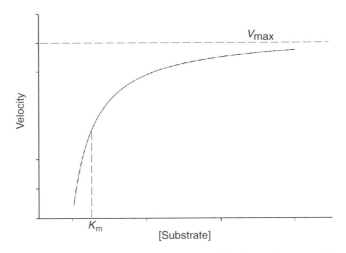

FIGURE 4.1 Representative hyperbolic plot indicative of a reaction following Michaelis–Menten kinetics. K_m = substrate concentration necessary to achieve one-half maximum velocity. V_{max} = maximum velocity.

4.3.1 Meanings of K_m, V_{max} and Their Clinical Relevance

The kinetic parameters K_m and V_{max} are estimated from the Michaelis–Menten equation and provide quantitative information regarding enzyme function. K_m or the Michaelis constant is operationally defined as the concentration of substrate at which half-maximal velocity of the reaction is achieved (Fig. 4.1). With respect to the single substrate reaction scheme (Scheme 4.1), it should be realized that K_m is equal to $(k_1 + k_2)/k_1$ and thus is the amalgamation of several rate constants. With respect to affinity, unfortunately, K_m is frequently (and incorrectly) used interchangeably with K_S, which is the substrate dissociation constant. Though K_m may sometimes approximate K_S, the two do not have to be equal and numerous examples exist where these parameter values vary dramatically.

V_{max} is an estimation of the maximum velocity of the reaction (Fig. 4.1) and is the product of k_{cat} and e_0, where k_{cat} is the capacity of the enzyme–substrate complex to form product and e_0 is the enzyme concentration. The k_{cat} parameter is also known as the *catalytic constant* or the *turnover number* and refers to the number of catalytic cycles or the number of molecules of substrate that one molecule of enzyme can convert to product per unit time. As stated above, V_{max} is only an "estimation" of the maximum velocity of the reaction, since the true maximum velocity is never reached at a finite substrate concentration.

Apart from simply obtaining kinetic parameters and understanding reaction rates, the estimation of K_m and V_{max} has several important applications in the drug discovery and development processes. Typically, it is assumed that *in vivo*, enzymatic reactions take place when substrate concentrations are much lower than the K_m. In this case, the [S] term drops out of the denominator of Eq. 4.1 and the equation reduces to

$$v \approx \frac{V_m[S]}{K_m} \qquad (4.2)$$

such that velocity becomes proportional to [S]. It is also well established that the velocity (or rate of metabolism) of the reaction is equal to the product of intrinsic clearance and substrate concentration (Eq. 4.3).

$$v = CL_{int}[S] \qquad (4.3)$$

This intrinsic clearance (CL_{int}) is analogous to the *in vivo* intrinsic clearance in that it is the ability of the enzyme to clear (metabolize) drug in the absence of blood flow or protein binding restrictions. Equation 4.2 is analogous to Equation 4.3 and thus equality is frequently assumed such that

$$CL_{int} = \frac{V_{max}}{K_m} = \frac{v}{[S]} \qquad (4.4)$$

For a complete discussion of these assumptions, see the excellent report of Houston (1994).

The ability to estimate CL_{int} from *in vitro* data thus becomes important in making *in vitro–in vivo* correlations to estimate the "first dose in man" and *in vivo* drug disposition. The more accurate the prediction from *in vitro* data, the less likely that either therapeutic failure or toxicity will occur when the new chemical entity is administered to humans or animals.

4.4 GRAPHICAL KINETIC PLOTS

Though less frequently used given the advent of personal computer programs capable of performing nonlinear regression analysis of data quickly and efficiently, the use of graphical plots in analyzing kinetic data can give quick estimates of relevant kinetic parameters. In addition, graphical replots of kinetic data can have diagnostic value in helping to deduce kinetic mechanisms with multiple substrates or with inhibition (see below). Nonlinear regression analysis of the data with the appropriate model (e.g., Michaelis–Menten model) will most accurately generate estimates of enzyme kinetic parameters.

Probably the most commonly used graphical plot in analyzing enzyme kinetic data is the double-reciprocal plot (Lineweaver–Burk plot) (Fig. 4.2). In this analysis the Michaelis–Menten equation is rewritten so that the results can be plotted as a straight line where $1/v$ is plotted along the y-axis and $1/[S]$ along the x-axis. In this plot, the slope of the line best fitting the data points is equal to K_m/V_{max}, the y-intercept $= 1/V_{max}$ and the x-intercept $= -1/K_m$. It should be noted that this type of plotting suffers from potentially large errors in $1/v$. At

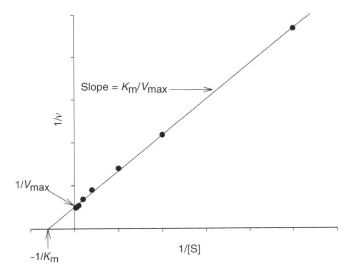

FIGURE 4.2 Lineweaver–Burk (double-reciprocal) plot of a reaction that follows Michaelis–Menten kinetics. Kinetic parameters are as defined previously.

very small values of v, even small errors lead to very large errors in $1/v$; this is not noticeable at large values of v (Dowd and Riggs, 1965). Weighting of the data can help somewhat to overcome these shortcomings.

Another type of graphical plot used to analyze kinetic data is the Hanes–Woolf plot (Fig. 4.3). In this case, $[S]/v$ is plotted along the y-axis and $[S]$ is plotted along the x-axis. For this plot, the x-intercept $= -K_m$, the y-intercept $= K_m/V_{max}$ and the slope of the best-fit line estimates $1/V_{max}$. The Hanes–Woolf

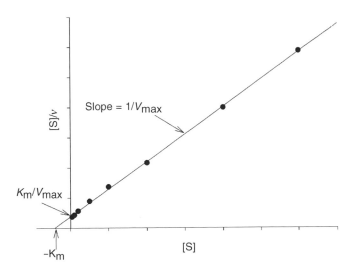

FIGURE 4.3 Hanes–Woolf plot of a reaction that follows Michaelis–Menten kinetics. Kinetic parameters are as defined previously.

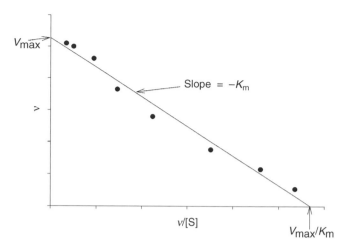

FIGURE 4.4   Eadie–Hofstee plot of a reaction that follows Michaelis–Menten kinetics. Kinetic parameters are as defined previously.

plot suffers less from the error issues discussed above regarding the double-reciprocal plot and thus, may be the most preferred graphical plot for estimating kinetic values.

Finally, investigators also have used the Eadie–Hofstee plot to estimate enzyme kinetic parameters (Fig. 4.4). In this last case, v is plotted along the y-axis and $v/[S]$ along the x-axis. The slope of the best-fit line is equal to $-K_m$, the y-intercept $= V_{max}$ and the x-intercept $= V_{max}/K_m$. As opposed to the double-reciprocal plot, the Eadie–Hofstee plot can make good data look worse (Dowd and Riggs, 1965). Interestingly, another use of the Eadie–Hofstee plot is to diagnose atypical kinetic profiles based on the shape of the data-fit obtained. This will be discussed in detail later in the chapter.

4.5   ATYPICAL KINETICS—ALLOSTERIC EFFECTS

4.5.1   Overview of Atypical Kinetic Phenomena

For the purposes of this discussion, the term "atypical kinetics" is used to describe an *in vitro* drug metabolism kinetic profile that does not fit the standard hyperbolic function when velocity is plotted versus substrate concentration. Though examples of atypical kinetic phenomena have existed for many years, it has only been in the past 10 years that these types of kinetic profiles in drug metabolism enzyme kinetics have been extensively documented and studied in a systematic manner. Though the primary focus of these studies has been on the cytochrome P450 enzymes, increasing numbers of examples of atypical phenomena in glucuronosyl transferase enzymes, transporters, and other drug metabolizing and disposition enzymes are becoming evident. The term allosterism is also used frequently in the P450 literature to refer to the ability of a ligand (a.k.a. effector molecule other than the substrate molecule being metabolized) that binds in a distinct, noncatalytic site on the enzyme and affects the rate of substrate metabolism. This effector can be a second (or potentially third or fourth, etc.) substrate molecule binding to the same enzyme molecule or it can be a structurally distinct molecule. Several studies suggest that these "allosteric" sites may be distinct binding regions within the same enzyme pocket that the substrate occupies, leading to the conclusion that multiple molecules may bind within the enzyme active site pocket simultaneously (Dabrowski et al., 2002; Hummel et al., 2004; Hutzler et al., 2003; Korzekwa et al., 1998; Rock et al., 2003; Shou et al., 1994), though the possibility of effectors binding outside the active site pocket cannot be excluded. Though this does not provide a mechanistic explanation for atypical kinetics, this simultaneous binding of substrate and effector provides a reasonable foundation for explaining the phenomena of atypical kinetic profiles and cooperativity. In the following sections, various types of atypical kinetic profiles will be discussed including homotropic cooperativity (sigmoidal, biphasic, and substrate inhibition kinetics) wherein two molecules of the same substrate are responsible for the kinetics observed and heterotropic cooperativity (activation and substrate dependent inhibition) where two structurally distinct molecules are involved.

4.5.2 Homotropic Cooperativity

4.5.2.1 Sigmoidal Kinetics (with a single substrate)   Cooperativity resulting in sigmoidicity of substrate–velocity plots (Fig. 4.5) has been most commonly associated with multimeric enzymes, such as hemoglobin, where the substrate

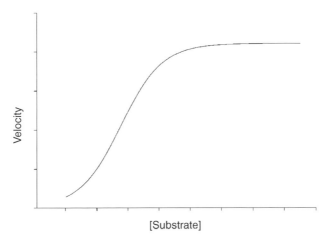

FIGURE 4.5 Representative plot depicting sigmoidal kinetics (homotropic cooperativity).

binds to equivalent sites on each protein subunit. However, the advent of more sensitive analytical techniques allowing both accurate and precise metabolite quantification at very low amounts has resulted in the identification and reporting of numerous examples of sigmoidicity in drug metabolizing enzymes.

In this case, the homotropic cooperativity occurs with a monomeric enzyme through substrate binding to both a catalytic and an effector site, and these sites may not be equivalent (Shou et al., 1999). This results in two K_m parameter estimates and two V_{max} parameter estimates, as discussed below. One of the most commonly used equations describing sigmoidal kinetic profiles is the Hill equation (Eq. 4.5):

$$v = \frac{V_m[S]^n}{K' + [S]^n} \tag{4.5}$$

This equation allows a description of sigmoidal kinetics, while requiring the estimation of only a few parameters. In this case, V_m has the same meaning as in the Michaelis–Menten equation and though K' is analogous, it is not equivalent to K_m (unless $n = 1$) since K' is comprised of both K_m and the interaction factors (Shou et al., 1999). The coefficient n is called the Hill coefficient since it measures relative cooperativity and reflects the degree of sigmoidicity observed (i.e., the greater the n value, the more cooperativity and sigmoidicity are observed). It should be noted that the n value does not necessarily equal the number of binding sites for substrate since it frequently is not an integer. However, the Hill equation and its resulting parameters are useful in comparing substrates that induce sigmoidal kinetics since the efficiency of the reaction (V_{max}/K') and cooperativity (n) can be compared for different substrates of the same enzyme.

To gain additional insight into the substrate–enzyme interactions occurring during sigmoidal kinetics, including the events occurring at each of the substrate binding sites, other equations have been developed for describing sigmoidal kinetics of drug metabolizing enzymes that also allow estimation of two K_m values (K_{m1} and K_{m2}) and two V_{max} values (V_{m1} and V_{m2}) (Korzekwa et al., 1998). This equation (Eq. 4.6) allows estimation of the events occurring at each of the distinct subsites within the active site binding pocket.

$$v = \frac{\frac{V_{m1}[S]}{K_{m1}} + \frac{V_{m2}[S]^2}{K_{m1}K_{m2}}}{1 + \frac{[S]}{K_{m1}} + \frac{[S]^2}{K_{m1}K_{m2}}} \tag{4.6}$$

As implied above, this equation allows for estimation of the K_m and V_{max} parameters associated with each binding site, and as such, is more informative. It should be realized, however, that in order to accurately estimate these four parameters, substantially more data points may be needed than are needed to estimate parameters using Equation 4.5.

4.5.2.2 Biphasic Kinetics (Nonasymptotic)   For the purposes of this discussion, a biphasic kinetic profile is defined as one in which the kinetic profile does not follow saturation kinetics and has two distinct phases (Fig. 4.6). Note that sigmoidal kinetics may also be biphasic but exhibits saturation.

At lower substrate concentrations the kinetic profile exhibits curvature similar to a hyperbolic profile, however, as the substrate reaches higher concentrations the profile increases linearly (instead of becoming asymptotic) with no evidence of saturation. In multienzyme systems, this type of profile may be observed when one of the operable enzymes exhibits a hyperbolic profile while another enzyme's actions result in a linear profile. However, when this type of profile occurs using single enzyme systems (e.g., expressed enzyme preps) the substrate is either binding in multiple productive orientations within the enzyme active site, or there are two substrate molecules bound within the enzyme active site. In this case, one of the binding orientations results in a low K_m, low V_{max} component (responsible for the semihyperbolic nature of the profile) while the other binding orientation produces the high K_m–high V_{max} component that results in the linear portion of the profile. As with modeling of sigmoidal kinetics, one must use an equation that describes the kinetics of multiple molecules of the same substrate interacting with multiple binding regions within the enzyme active site, such as Equation 4.7.

$$v = \frac{(V_{m1}[S]) + (CL_{int})[S]^2}{(K_{m1} + [S])} \tag{4.7}$$

In this case the parameters K_{m1} and V_{m1} represent the initial curved portion of the plot that occurs at lower substrate concentrations (i.e., the low K_m and low V_m portions of the profile). The CL_{int} term is used to describe the linear

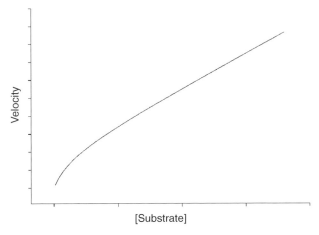

FIGURE 4.6   Representative plot depicting biphasic kinetics (two phases with one not achieving saturation).

portion of the profile that occurs at higher substrate concentrations. In this case, CL_{int} is the ratio of V_{max2}/K_{m2} but since the profile never saturates (becomes asymptotic) one cannot estimate the individual parameters and thus, represents the ratio of these two parameters. Thus, CL_{int} is the slope for this linear portion of the kinetic profile.

4.5.2.3 Substrate Inhibition
Substrate inhibition represents another example of cooperativity in enzyme kinetic reactions, but of a different profile than described to this point. With substrate inhibition kinetics, the velocity of a reaction increases (as expected for hyperbolic profiles) to an apex, however, beyond this point the velocity of the reaction decreases with increasing substrate concentrations (Fig. 4.7).

Again, it is believed that this scenario occurs due to simultaneous binding of a second substrate molecule within the active site resulting in a type of uncompetitive inhibition. It remains unclear whether binding of the second substrate molecule results in a partial inhibition of binding of the first substrate molecule or whether the binding of this second molecule causes a conformational change in the enzyme active site that inhibits substrate turnover.

Through application of the appropriate equation (Eq. 4.8) that is derived from the uncompetitive inhibition model using an extra substrate molecule instead of inhibitor, one can approximately estimate the parameters K_m and V_{max}, as well as K_{si} (the inhibition constant for the second substrate molecule).

$$\frac{V_m[S]}{K_m + [S] + \frac{[S]^2}{K_{si}}} \tag{4.8}$$

It should be realized that these are only approximations since V_{max} is never truly reached and thus, K_m is also not exactly discernible. In the past,

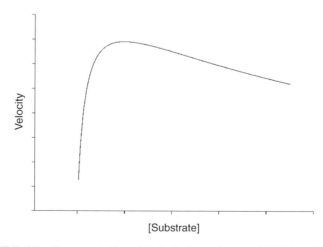

FIGURE 4.7   Representative plot depicting substrate inhibition kinetics.

researchers have incorrectly fit these types of data by either ignoring (and thus, omitting) the data points after the apex in which the velocity is declining or they have simply applied the Michaelis–Menten equation to the entire data set. Lin et al. (2001) in some comprehensive work have demonstrated the necessity to fit the entire data set to the proper equation and have also presented the hazards of failing to fit the data properly. This equation is extremely useful and can be used with datasets containing a limited (<15) number of data points. If the researcher wishes to obtain additional information about the processes occurring during substrate inhibition kinetics, Lin et al. (2001) have provided a more complex equation (Eq. 4.9) that requires substantially more data points but that provides much greater detailed kinetic parameter estimates regarding this type of kinetics.

$$v = \frac{V_m\left(\frac{1}{K_S} + \frac{\beta[S]}{\alpha K_i K_S}\right)}{\frac{1}{S} + \frac{1}{K_S} + \frac{1}{K_i} + \frac{[S]}{\alpha K_S K_i}} \tag{4.9}$$

It should be noted that in this case, K_s is approximately equal to K_m and K_i is the dissociation constant of the substrate binding to the "inhibitory" region within the enzyme active site. Also, α represents the factor by which the dissociation (K_s and K_i) of substrate at both sites changes when a second substrate molecule is bound, whereas β is the factor by which V_{max} changes upon binding of the second substrate molecule. Thus, one gains additional information as to the degree to which the binding of the second substrate molecule alters K_m and V_{max}.

One must also realize that ternary complex situations exist, such as with the glucuronosyl transferases that involve the enzyme–aglycone–glucuronic acid complex and that substrate inhibition may also occur in these systems. In this case, a different equation is required to adequately describe substrate inhibition (Eq. 4.10).

$$\frac{V_m[A][B]}{K_{iA}K_{mB} + K_{mB}[A] + K_{mA}[B] + [A][B]\left(1 + \frac{[B]}{K_{sib}}\right)} \tag{4.10}$$

However, in this case, K_{sib} is a constant that defines the strength of inhibition and is not a dissociation constant.

4.5.3   Heterotropic Cooperativity

4.5.3.1   Heterotropic Activation   In the case of heterotropic activation, the addition of another compound to the incubation mixture results in an increased velocity for substrate turnover without increasing the amount of enzyme present. This result is in contrast to the inhibition interaction that is typically expected when two substrates are coincubated. Again, this phenomenon is thought to occur due to simultaneous binding of both substrate and effector molecule within the enzyme active site simultaneously. One of the first conclusive demonstrations of this phenomenon was the observation that the CYP3A4 substrate 7,8-benzoflavone increased the V_{max} of phenanthrene metabolism without changing the K_m and conversely, phenanthrene decreased the V_{max} of 7,8-benzoflavone metabolism, again with no effect on K_m (Shou et al., 1994). Later, it was shown that dapsone increased the V_{max} and decreased the K_m of CYP2C9-mediated flurbiprofen metabolism, while only minimal changes in dapsone metabolism were noted (Hutzler et al., 2001). Numerous examples of heterotropic activation of cytochrome P450 and glucuronosyl transferase mediated metabolism, as well as P-glycoprotein transport have been noted (Egnell et al., 2003; Hutzler and Tracy, 2002; Kenworthy et al., 2001; Pfeiffer et al., 2005; Taub et al., 2005; Witherow and Houston, 1999).

In the simplest case, this activation is of a nonessential nature and maintains the hyperbolic shape of the kinetic profile (Fig. 4.8). Though K_m and V_{max} may change, the velocity versus substrate curve remains hyperbolic.

Certainly, one can fit each of the individual velocity versus substrate curves at each effector concentration and determine multiple K_ms and V_{max}s for the process of activation. However, it may be more informative to fit all the data together to ascertain additional information regarding the phenomenon. In this case, equation (Eq. 4.11) may be used to estimate K_m and V_{max} for the

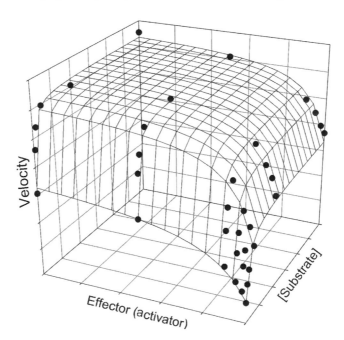

FIGURE 4.8 Representative plot depicting heterotropic cooperativity (activation kinetics) where velocity of the reaction increases with increasing effector (activator) concentration.

heteroactivation, as well as additional parameters that shed light into the changes that have occurred in the enzyme kinetics.

$$v = \frac{V_{\max}[S]}{K_m \dfrac{\left(1 + \frac{[B]}{K_B}\right)}{\left(1 + \frac{\beta[B]}{\alpha K_B}\right)} + [S] \dfrac{\left(1 + \frac{[B]}{\alpha K_B}\right)}{\left(1 + \frac{\beta[B]}{\alpha K_B}\right)}} \qquad (4.11)$$

In this case, K_m and $V_{\max}$ represent the values obtained in the absence of effector and K_B is the binding constant of the effector. Additionally, interaction factors including α, which is the effect on K_m due to the presence of effector ($\alpha < 1$ is indicative of activation and $\alpha > 1$ is indicative of inhibition) and β, which is the effect on $V_{\max}$ due to the presence of effector ($\beta < 1$ is indicative of inhibition and $\beta > 1$ is indicative of activation) can be estimated.

4.6   GRAPHICAL ANALYSIS OF ATYPICAL KINETIC DATA

As mentioned previously, use of the Eadie–Hofstee plot can be beneficial in discerning whether atypical kinetic profiles are operational. When plotted in this fashion, data that is of a hyperbolic nature (i.e., follows Michaelis–Menten kinetics) will give a straight line (Fig. 4.9a). However, in the case of sigmoidal kinetics, an Eadie–Hofstee style plot will yield a characteristic hook (Fig. 4.9b) whereas a plot of this type to data that is exhibiting biphasic kinetics will give a concave shape (Fig. 4.9c). Though not necessarily obvious from the standard v versus [S] plot, these Eadie–Hofstee graphical representations of the data can help to discern whether the data is nonhyperbolic. Finally, in the case of substrate inhibition, the Eadie–Hofstee plot of these types of data will be of a convex nature, analogous to the v versus [S] plot (Fig. 4.9d).

4.7   ENZYME INHIBITION KINETICS

4.7.1   Overview

When a compound is added to an incubation reaction mixture and its presence decreases the rate of substrate turnover, the process is termed inhibition. In drug discovery and development, the study of inhibition potential *in vitro* is of great importance as the results may help to screen out compounds with significant *in vivo* inhibition potential and correspondingly, the potential to elicit drug–drug interactions. This chapter will deal primarily with reversible inhibition as the concepts of irreversible inhibition are more extensively covered in a later chapter. If the enzyme has no residual catalytic activity in the presence of saturating concentrations of the inhibitor, this is termed complete inhibition. However, more often than not, some residual enzyme catalytic

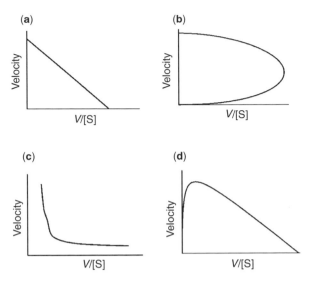

FIGURE 4.9 Eadie–Hofstee plots useful to diagnose the type of kinetics occurring in a reaction for (**a**) hyperbolic (Michaelis–Menen) kinetics, (**b**) Sigmoidal kinetics, (**c**) Biphasic kinetics with no saturation of second phase, and (**d**) Substrate inhibition kinetics.

activity remains, even at saturating concentrations of the inhibitor, and thus partial inhibition is observed. This partial inhibition is, in fact, analogous to the activation phenomenon described earlier in that residual enzyme activity exists whether effector is present at saturating levels, or not present at all. Graphical representations discussed later, allow a simple way to visualize whether partial inhibition is occurring. Though several types of reversible inhibition are discussed below, regardless, proper characterization and estimation of kinetic parameters such as the inhibition constant K_i are essential to arriving at appropriate conclusions concerning the inhibition potential of a new chemical entity.

4.7.2 Competitive Inhibition

The most commonly observed type of enzyme inhibition is that of competitive inhibition and it is also the simplest. As represented in Scheme 4.2, in this case

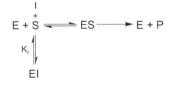

SCHEME 4.2 Competitive inhibition.

both substrate and inhibitor are competing for the enzyme active site and their binding is mutually exclusive. When the inhibitor is bound to the enzyme, the complex is incapable of reacting (turning over substrate).

In this case, formation of EI is a dead-end complex and the only way to generate catalytically active enzyme is for the reformation of E + I. Because competitive inhibition has no effect on V_{max}, a change (increase) in K_m must occur and thus this type of inhibition is characterized as "an increase in K_m." However, to assist in characterization of inhibitors it may be more straightforward and simpler to describe competitive inhibition as a decrease in V_{max}/K_m with no change in the apparent V_{max}. The equation for defining competitive inhibition is listed below as Equation 4.12.

$$v = \frac{V_{max}[S]}{K_m(1 + \frac{[I]}{K_i}) + [S]} \qquad (4.12)$$

In this equation, V_{max} and K_m have their usual meanings and [I] is the free inhibitor concentration. K_i is the inhibition constant and is equal to $K_i = [E][I]/[EI]$. Note that the equation is of the same form as the Michaelis–Menten equation and can be rewritten as the following (Eq. 4.13):

$$v = \frac{V_{max}^{app}[S]}{K_m^{app} + [S]} \qquad (4.13)$$

where V_{max}^{app} and K_m^{app} are the apparent values of V_{max} and K_m, respectively, and are the values as they appear in the presence of the inhibitor. Thus, $V_{max}^{app} = V_{max}$ and $K_m^{app} = K_m(1 + [I]/K_i)$, then (Eq. 4.14) follows:

$$\frac{V_{max\ app}^{app}}{K_m^{app}} = \frac{\frac{V_{max}}{K_m}}{1 + \frac{[I]}{K_i}} \qquad (4.14)$$

From Equation 4.14, it thus becomes apparent that in the case of competitive inhibition, the apparent value of V_{max}/K_m is decreased by the factor $(1 + [I]/K_i)$ and V_{max} remains unchanged. It should be noted, that this is frequently also stated as "competitive inhibition results in an increase in K_m with no change in V_{max}."

4.7.3 Mixed Inhibition

In the case of mixed inhibition, both specific and catalytic effects are present and thus, both V_{max}^{app}/K_m^{app} and V_{max}^{app} are altered. Thus, both the maximal velocity and the concentration at which half-maximal velocity is achieved are changed as compared to the absence of inhibitor. As can be seen in Scheme 4.3, the mechanism that produces mixed inhibition is much more complex than that of competitive inhibition since the formation of an enzyme–inhibitor complex is not necessarily a dead-end reaction. In this case, substrate can bind to the

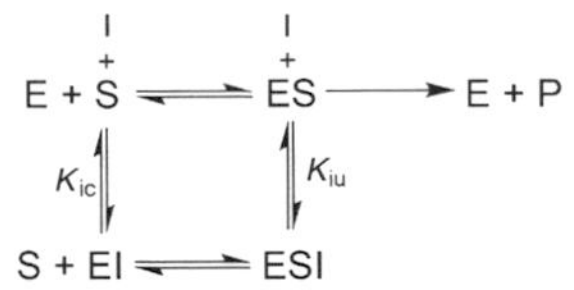

$$E + \overset{|}{\overset{+}{S}} \;\rightleftharpoons\; \overset{|}{\overset{+}{ES}} \longrightarrow E + P$$

$$K_{ic} \Big\updownarrow \qquad\qquad \Big\updownarrow K_{iu}$$

$$S + EI \rightleftharpoons ESI$$

SCHEME 4.3   Mixed inhibition ($K_{ic} = K_i$ competitive and $K_{iu} = K_i$ uncompetitive).

enzyme–inhibitor complex that can subsequently dissociate to form an enzyme–substrate complex that can undergo catalysis to produce product.

It is apparent from Scheme 4.3 that the inhibitor cannot only "compete" with substrate for binding to the enzyme but bind to an enzyme molecule that subsequently binds a substrate molecule also or to an enzyme-substrate complex to affect catalytic turnover. These multiple binding mechanisms help explain the effects of mixed inhibition on both V_{max} and K_m. The effects of mixed inhibition on the velocity of the reaction can be described by the following mixed inhibition equation (Eq. 4.15):

$$v = \frac{V_{max}[S]}{K_m(1 + \frac{[I]}{K_i}) + [S](1 + \frac{[I]}{K_i})} \tag{4.15}$$

where V_{max}, K_m and K_i have their usual meanings. Following the format used for competitive inhibition, one can also develop equations for describing the apparent K_m and V_{max} resulting from mixed inhibition of the reaction. In this case, the V_{max}^{app} resulting from mixed inhibition can be described with (Eq. 4.16):

$$V_{max}^{app} = \frac{V_{max}}{1 + \frac{[I]}{K_i}} \tag{4.16}$$

Likewise, the resulting K_m^{app} can be described with (Eq. 4.17):

$$K_m^{app} = \frac{K_m(1 + \frac{[I]}{K_{ic}})}{1 + \frac{[I]}{K_{iu}}} \tag{4.17}$$

where K_{ic} is the competitive portion of the inhibition and K_{iu} is the uncompetitive portion of the inhibition.

And finally, the ratio of V_{max}^{app} to K_m^{app} can be described with (Eq. 4.18):

$$\frac{V_{max}^{app}}{K_m^{app}} = \frac{\frac{V_{max}}{K_m}}{1 + \frac{[I]}{K_{ic}}} \tag{4.18}$$

4.7.4 Noncompetitive Inhibition

Pure noncompetitive inhibition (decrease in V_{max} with no change in K_m) is seldom observed in enzyme kinetics studies, except in the case of very small inhibitors, such as protons, metal ions, and small anions. For noncompetitive inhibition to occur, one would have to imagine a situation in which the inhibitor affected the catalytic properties of the enzyme but did not affect substrate binding. This implies that the free enzyme and the enzyme–substrate complex would have to exhibit equal binding constants for the inhibitor, which is unlikely. Upon careful examination, most situations originally thought to involve noncompetitive inhibition are in fact, mixed inhibition. Thus, noncompetitive inhibition will not be discussed here.

4.7.5 Uncompetitive Inhibition

The final type of inhibition that will be discussed is that of uncompetitive inhibition. In this case, the inhibitor decreases the apparent value of V_{max} but has no effect on V_{max}/K_m; in other words, both V_{max} and K_m decrease proportionately. Uncompetitive inhibition differs from competitive inhibition in that the inhibitor only binds to the enzyme–substrate complex (Scheme 4.4) and not to the free enzyme.

As with the other types of inhibition, equations have been derived that permit estimation of the inhibition constant as well as K_m and V_{max} for uncompetitive inhibition (Eq. 4.19).

$$ v = \frac{\left(\dfrac{V_{max}}{1+\frac{[I]}{K_i}}\right)[S]}{\left(\dfrac{K_m}{1+\frac{[I]}{K_i}}\right)+[S]} \tag{4.19}$$

That both catalytic (V_{max}) and specific (K_m) effects are noted in uncompetitive inhibition is noted in the following equations. Both V_{max}^{app} and K_m^{app} are reduced by the same factor (Eqs. 4.20 and 4.21):

$$ V_{max}^{app} = \frac{V_{max}}{1+\frac{[I]}{K_i}} \tag{4.20}$$

$$ K_m^{app} = \frac{K_m}{1+\frac{[I]}{K_i}} \tag{4.21}$$

such that the following equality is derived from the V/K ratios (Eq. 4.22).

$$ \frac{V_{max}^{app}}{K_m^{app}} = \frac{V_{max}}{K_m} \tag{4.22}$$

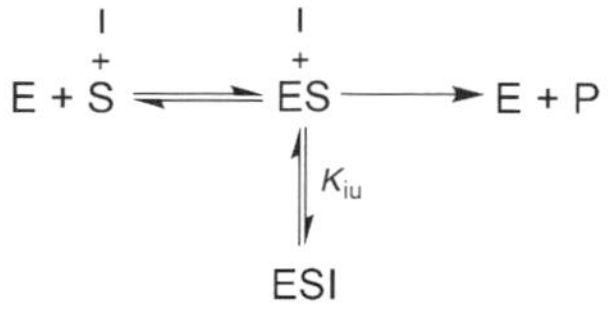

$$ E + S \rightleftharpoons ES \longrightarrow E + P $$
$$ \big\updownarrow K_{iu} $$
$$ ESI $$

SCHEME 4.4 Uncompetitive inhibition.

**TABLE 4.1   Alterations in kinetic parameters caused
by various types of inhibition.**

Type of inhibition	$V_{\mathrm{max}}^{\mathrm{app}}$	$V_{\mathrm{max}}^{\mathrm{app}}/K_{\mathrm{m}}^{\mathrm{app}}$	$V_{\mathrm{m}}^{\mathrm{app}}$
Competitive	$\leftrightarrow$	$\uparrow$	$\uparrow$
Mixed	$\downarrow$	$\updownarrow$	$\uparrow$
Pure noncompetitive	$\downarrow$	$\downarrow$	$\leftrightarrow$
Uncompetitive	$\downarrow$	$\leftrightarrow$	$\downarrow$

4.7.6   Summary of Effects of Various Inhibition Types of Kinetic Parameters

A summary of the effects on K_{m} and V_{max} resulting from the various types of inhibition is presented in Table 4.1, for reference.

4.7.7   Meanings of IC$_{50}$ and K_{i} Parameters

The parameters of IC$_{50}$ and K_{i} are frequently used to describe the inhibitory potential of a compound. However, it should be realized that the two parameters are not interchangeable and differ in the information provided by their estimation. Estimation of the IC$_{50}$ provides the concentration of drug that results in 50 % inhibition of substrate metabolism. Yet, the IC$_{50}$ parameter is dependent on the concentration of both substrate and enzyme studied. An estimation of IC$_{50}$ for a given inhibitor is only comparable to the IC$_{50}$ estimate of another inhibitor when studied at the same substrate and enzyme concentrations. As long as the incubation conditions are consistent, IC$_{50}$ values can be compared for inhibition potency differences. The IC$_{50}$ does have the advantage of being relatively easy to obtain as it requires the use of a single substrate concentration and fewer data points than determinations of K_{i} (see below). In contrast, K_{i} values represent the dissociation constant of the enzyme–inhibitor complex (not the concentration producing 50 % inhibition as is sometimes mistakenly believed). To properly estimate the K_{i} for a given inhibitor one typically uses three to four substrate concentrations and four to five inhibitor concentrations (including zero) in the incubations. However, the K_{i} has the advantage of being both enzyme and substrate concentration independent. Thus, K_{i} values can be compared across inhibitors, regardless of the conditions studied (substrate concentrations and enzyme concentrations).

4.8   INHIBITION KINETICS GRAPHICAL PLOTS

As with estimation of enzyme kinetic parameters, the most accurate method for determining inhibition kinetic parameters is through nonlinear regression fitting of the data set. However, it may be beneficial to plot the data graphically for a quick estimation of the inhibition parameters but also to help determine the type of inhibition that is occurring. Sometimes, differences in measures of model differences from nonlinear regression fitting (e.g., F-tests, Akaike Information Criterion) may be so close as to make determinations between two models difficult. In this case, graphical plots may serve to help distinguish between models of inhibition.

Two of the most common plots for determining inhibition constants and the type of inhibition are those as described by Dixon (1953) and Cornish-Bowden (1974). The two methods are actually very complementary and can serve as a double check on determinations of the type of inhibition. In addition, they each make up for the inability of the other method to estimate the inhibition constant in all types of inhibition.

In the method of Dixon, plots are constructed with $1/v$ plotted along the y-axis and [I] plotted along the x-axis (Fig. 4.10a–c). Experiments are typically conducted over a range of substrate concentrations (3–4 concentrations of substrate at values bracketing the K_m) and at a range of inhibitor concentrations (3–4 inhibitor concentrations bracketing the expected K_i) as well as a control set of incubations with only substrate and no inhibitor present. In the case of competitive inhibition (Fig. 4.10a), the plotted lines intersect at a point in the upper left quadrant of the plot. A perpendicular from the point of intersection to the y-axis gives an estimation of $1/v$, while a perpendicular drawn from the intersection point to the x-axis gives an estimate of the negative of K_i $(-K_i)$. Mixed inhibition gives a slightly different plot in that the point of intersection is nearer to the x-axis (Fig. 4.10b) and a perpendicular drawn from the intersection point to the x-axis gives $-K_i$ and a perpendicular drawn to the y-axis gives $1/(1 - K_i)V_{max}$. Finally, a Dixon plot of uncompetitive inhibition results in parallel lines with no intersection, and thus the inability to graphically estimate any kinetic parameters (Fig. 4.10c).

As mentioned above, the graphical methods proposed by Cornish-Bowden are complementary as will be demonstrated below. In these types of plots (Fig. 4.10d–f), $[S]/v$ is plotted along the y-axis and the inhibitor concentration [I] is plotted along the x-axis. In the case of competitive inhibition (Fig. 4.10d) a series of parallel lines is obtained that can be used to confirm the type of inhibition, but no kinetic parameter estimates can be made. For mixed inhibition (Fig. 4.10e) the series of lines intersect in the lower left quadrant of the plot and can be used to distinguish mixed inhibition from competitive (note the ambiguity of mixed inhibition plots versus competitive inhibition plots with the Dixon method). For mixed inhibition, a perpendicular drawn from the intersection point to the x-axis allows an estimation of $-K_i$ and a perpendicular drawn to the x-axis is equal to $K_m(1 - K_i)/V_{max}$. Finally, in contrast to the Dixon plot, application of the method of Cornish-Bowden to a case of uncompetitive inhibition (Fig. 4.10f) results in a series of lines intersecting in the upper left quadrant of the graph with a perpendicular from the intersection point to the x-axis giving an estimate of $-K_i$ and one drawn from this same intersection to the y-axis giving an estimate of K_m/V_{max}.

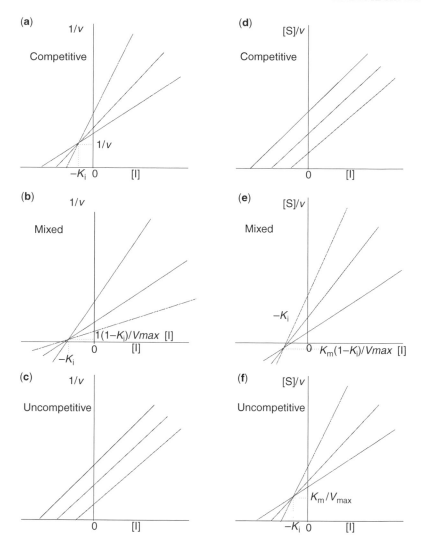

FIGURE 4.10 Diagnostic plots to determine the type of inhibition occurring in a kinetic reaction. Left side, panels (**a**), (**b**), and (**c**) are Dixon plots representative of competitive, mixed, and uncompetitive inhibition, respectively. Right side, panels (**d**), (**e**), and (**f**) are Cornish-Bowden plots representative of competitive, mixed, and uncompetitive inhibition, respectively.

4.9   MECHANISM-BASED ENZYME INACTIVATION KINETICS

Mechanism-based enzyme inactivation kinetics refers to the irreversible inhibition of an enzyme via a catalytically formed reactive intermediate that binds covalently (typically) to the enzyme active site prior to release and causes permanent inactivation of the enzyme. This type of inhibition is also known as

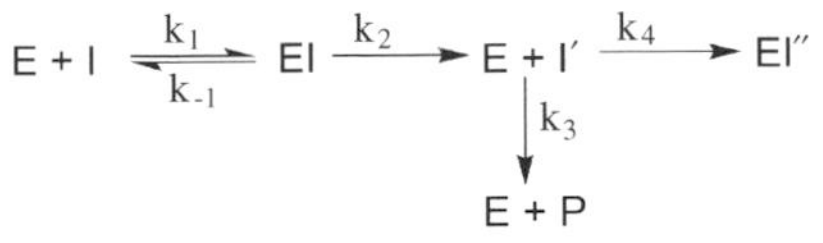

$$E + I \xrightleftharpoons[k_{-1}]{k_1} EI \xrightarrow{k_2} E + I' \xrightarrow{k_4} EI''$$

$$\downarrow k_3$$

$$E + P$$

SCHEME 4.5 Mechanism-based enzyme inactivation.

suicide inactivation, enzyme-activated irreversible inhibition and other terms. An excellent primer on mechanism-based enzyme inactivation is provided by Silverman (1998). The enzymatic scheme for this kind of inactivation is shown in Scheme 4.5.

Due to the nature of this process (necessary reactive intermediate formation [I'] and subsequent permanent inactivation of the enzyme [EI'']), the effects of this inhibition not only tend to reach a maximum effect more slowly but also last for a longer period of time (i.e., until the inactivated enzyme is replaced by synthesis of a new, functional enzyme molecule. Furthermore, since not every reactive intermediate [I'] goes on to form an inactivator complex, some molecules (generally the majority) go on to form product (Scheme 4.5), again adding to the length of time for full inactivation. Unlike competitive inhibition, withdrawal of the drug causing the inactivation does not immediately result in restoration of enzyme function. In fact, it can take several days for complete turnover of an enzyme pool and restoration of full metabolic function.

Silverman (1998) has defined seven criteria that must be satisfied in order for a process to be characterized as involving mechanism-based inactivation. These are time dependence, saturation, substrate protection, irreversibility, stoichiometry of inactivation, involvement of a catalytic step, and inactivation occurs prior to release of the activated species. This review will not expound on each

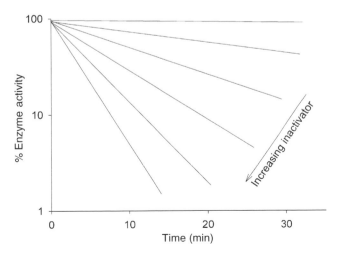

FIGURE 4.11 Plots representative of the decreasing enzyme activity with respect to the time and concentration dependence of mechanism-based inactivation kinetics.

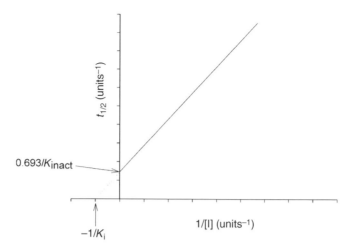

FIGURE 4.12   Kitz–Wilson plot of the half-lives (rate constants) for mechanism-based inactivation at each inactivator concentration.

of the criteria here as they are covered extensively in the above reference. In addition, the reader is encouraged to consult the reference Silverman (1998) for the proper procedures to follow in carrying out studies of mechanism-based enzyme inactivation.

As a first step in determining the kinetics of mechanism-based enzyme inactivation, the time-dependence of the reaction is typically studied. In these experiments, the effects of preincubation (primary incubation) with increasing concentrations of the inactivator over various time intervals is studied for their effects on catalytic activity of the enzyme toward the reporter substrate. This typically results in a plot similar to that represented in Figure 4.11.

From this plot the half-lives (rate constants) for the inactivation by each concentration of inactivator can be calculated from the slopes of the individual lines. These half-life values are then plotted along the y-axis versus $1/[I]$ plotted along the x-axis. This plot is also known as a Kitz–Wilson plot (Fig. 4.12). In the case of a saturation reaction, (i.e., at infinite inactivator concentration there is a finite half-life) the point where the plotted line intersects the y-axis is equal to $0.693/k_{inact}$, where k_{inact} is the rate of inactivation and represents a complex mixture of k_2, k_3 and k_4 (see Scheme 4.5). The dissociation constant for the enzyme-mechanism-based inactivator complex (K_I) can also be estimated from this plot as the x-intercept of the line represents $-1/K_I$ (Fig. 4.12).

ACKNOWLEDGMENT

Portions of this work were funded by a grant from the National Institutes of Health (#GM 063215).

REFERENCES

Cornish-Bowden A. A simple graphical method for determining the inhibition constants of mixed, uncompetitive and non-competitive inhibitors. Biochemistry J 1974; 137:143–144.

Cornish-Bowden A. Fundamentals of Enzyme Kinetics. London: Portland Press; 2004.

Dabrowski MJ, Schrag ML, Wienkers LC, Atkins WM. Pyrene.pyrene complexes at the active site of cytochrome P450 3A4: evidence for a multiple substrate binding site. J Am Chem Soc 2002;124:11866–11867.

Dixon M. The determination of enzyme inhibitor constants. Biochemistry 1953;55: 170–171.

Dowd JE, Riggs DS. A comparison of estimates of Michaelis–Menten kinetic constants from various linear transformations. J Biol Chem 1965;240:863–869.

Egnell AC, Houston JB, Boyer CS. *In vivo* CYP3A4 heteroactivation is a possible mechanism for the drug interaction between felbamate and carbamazepine. J Pharmacol Exp Ther 2003;305:1251–1262.

Houston JB. Utility of *in vitro* drug metabolism data in predicting *in vivo* metabolic clearance. Biochem Pharmacol 1994;47:1469–1479.

Hummel MA, Gannett PM, Aguilar JS, Tracy TS. Effector-mediated alteration of substrate orientation in cytochrome P450 2C9. Biochemistry 2004;43:7207–7214.

Hutzler JM, Hauer MJ, Tracy TS. Dapsone activation of CYP2C9-mediated metabolism: evidence for activation of multiple substrates and a two-site model. Drug Metab Dispos 2001;29:1029–1034.

Hutzler JM, Tracy TS. Atypical kinetic profiles in drug metabolism reactions. Drug Metab Dispos 2002;30:355–362.

Hutzler JM, Wienkers LC, Wahlstrom JL, Carlson TJ, Tracy TS. Activation of cytochrome P450 2C9-mediated metabolism: mechanistic evidence in support of kinetic observations. Arch Biochem Biophys 2003;410:16–24.

Kenworthy KE, Clarke SE, Andrews J, Houston JB. Multisite kinetic models for CYP3 A4:simultaneous activation and inhibition of diazepam and testosterone metabolism. Drug Metab Dispos 2001;29:1644–1651.

Korzekwa KR, Krishnamachary N, Shou M, Ogai A, Parise RA, Rettie AE, Gonzalez FJ, Tracy TS. Evaluation of atypical cytochrome P450 kinetics with two-substrate models: evidence that multiple substrates can simultaneously bind to cytochrome P450 active sites. Biochemistry 1998;37:4137–4147.

Lin Y, Lu P, Tang C, Mei Q, Sandig G, Rodrigues AD, Rushmore TH, Shou M. Substrate inhibition kinetics for cytochrome P450-catalyzed reactions. Drug Metab Dispos 2001;29:368–374.

Pfeiffer E, Treiling CR, Hoehle SI, Metzler M. Isoflavones modulate the glucuronidation of estradiol in human liver microsomes. Carcinogenesis 2005;26:2172–2178.

Rock DA, Perkins BN, Wahlstrom J, Jones JP. A method for determining two substrates binding in the same active site of cytochrome P450B M3: an explanation of high energy omega product formation. Arch Biochem Biophys 2003;416:9–16.

Segel IH. Rapid Equilibrium Bireactant and Terreactant Systems. New York: John Wiley & Sons, Inc.; 1975.

Shou M, Grogan J, Mancewicz JA, Krausz KW, Gonzalez FJ, Gelboin HV, Korzekwa KR. Activation of CYP3A4: Evidence for the simultaneous binding of two substrates in a cytochrome P450 active site. Biochemistry 1994;33:6450–6455.

Shou M, Mei Q, Ettore MW, Jr, Dai R, Baillie TA, Rushmore TH. Sigmoidal kinetic model for two co-operative substrate-binding sites in a cytochrome P450 3A4 active site: an example of the metabolism of diazepam and its derivatives. Biochem J 1999; 340 (Pt 3):845–853.

Silverman RB. Mechanism-Based Enzyme Inactivation: Chemistry and Enzymology. Boca Raton: CRC Press; 1998.

Taub ME, Podila L, Ely D, Almeida I. Functional assessment of multiple P-glycoprotein (P-gp) probe substrates: influence of cell line and modulator concentration on P-gp activity. Drug Metab Dispos 2005;33:1679–1687.

Witherow LE, Houston JB. Sigmoidal kinetics of CYP3A substrates: an approach for scaling dextromethorphan metabolism in hepatic microsomes and isolated hepatocytes to predict *in vivo* clearance in rat. J Pharmacol Exp Ther 1999;290:58–65.

5

METABOLISM-MEDIATED DRUG–DRUG INTERACTIONS

HONGJIAN ZHANG, MICHAEL W. SINZ, AND A. DAVID RODRIGUES

5.1 INTRODUCTION

Serious drug–drug interactions (DDIs) are major liabilities for any new drug entering the pharmaceutical marketplace. In recognition of the importance of such drug interactions, pharmaceutical companies employ rapid *in vitro* screening methods in drug discovery to eliminate problematic compounds, in addition to conducting more detailed studies in drug development to fully characterize and assess the potential liability. Eventually, the drug interaction (or lack thereof) is fully evaluated in patients or volunteers during early clinical trails. Although drug interactions can lead to changes in the pharmacologic (pharmacodynamic) profile of drugs, the present discussion will focus on metabolic drug interactions leading to altered pharmacokinetics. Classical metabolic drug interactions involve inhibition or induction of drug-metabolizing enzyme activities. This is especially critical when a given drug is predominately cleared by a polymorphically expressed drug-metabolizing enzyme, or by a single enzyme susceptible to inhibition or induction.

In the proceeding paragraphs, illustrations of metabolic drug interactions will be described. Throughout, drugs will be described as either "victims" or "perpetrators" of drug interactions. A perpetrator is a drug causing a change in enzyme activity (increase or decrease) and a victim is a drug affected by altered metabolism/pharmacokinetics (either through increased or through decreased metabolism). Although most of the examples described herein focus

Drug Metabolism in Drug Design and Development, Edited by Donglu Zhang, Mingshe Zhu and W. Griffith Humphreys
Copyright © 2008 John Wiley & Sons, Inc.

on the cytochrome P450 (CYP) family of enzymes, many of the same concepts more or less apply equally to other enzymes (e.g., nonmicrosomal and Phase II drug-metabolizing enzymes) and drug transporters.

5.2 ENZYME INHIBITION

5.2.1 Types of Inhibition

The simplest way to describe CYP enzyme catalysis is Michaelis–Menten (M–M) kinetics, where the relationship between the rate of catalysis and substrate concentration is best characterized by a single-binding site and a hyperbolic curve. Most often, characterization of CYP inhibition is conducted when CYP catalysis follows M–M kinetics to avoid complications of data interpretation. Recently, atypical kinetics have been demonstrated for CYP catalysis, where two or more enzyme–substrate binding sites may be involved (Houston and Kenworthy, 2000; Shou, 2002; Shou et al., 2001). Modeling of CYP inhibition using atypical kinetics has also been described (Houston and Galetin, 2005; Zhang and Wong, 2005).

Mechanisms of CYP inhibition can be broadly divided into two categories; reversible inhibition and mechanism-based inactivation. Depending on the mode of interaction between CYP enzymes and inhibitors, reversible CYP inhibition is further characterized as competitive, noncompetitive, uncompetitive, and mixed (Ito et al., 1998b). Evaluation of reversible inhibition of CYP reactions is often conducted under conditions where M–M kinetics is obeyed. Based on the scheme illustrated in Fig. 5.1, various types of reversible inhibition are summarized in Table 5.1. Figure 5.1 depicts a simple substrate–enzyme complex during catalysis. In the presence of a reversible inhibitor, such a complex can be disrupted leading to enzyme inhibition.

Mechanism-based CYP inhibition or irreversible inhibition, involves permanent inactivation of CYP enzymes during catalysis, where reactive intermediate(s) are formed, leading to apoprotein or heme-ion center modification. Typical characteristics of mechanism-based enzyme inhibition include time-dependent loss of enzyme activity, a rate of inactivation generally following saturation kinetics, enzyme activity that cannot be recovered after

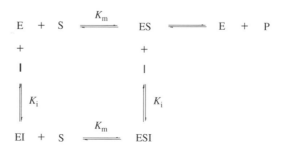

FIGURE 5.1 Reversible inhibition scheme.

TABLE 5.1 Types of inhibition (when catalysis conforms to Michaelis–Menten kinetics)[a].

	Mode of inhibitor binding	Kinetic equation	IC_{50} versus K_i
Reversible			
Competitive	Unbound enzyme	$V = V_{max}/(1 + K_m/S/(1 + I/K_i))$	$IC_{50} = K_i \times (1 + S/K_m)$
Uncompetitive	Substrate-bound enzyme	$V = V_{max}/(1 + I/K_i + K_m/S)$	$IC_{50} = K_i \times (1 + S/K_m)$
Noncompetitive	Unbound and bound enzyme at different site(s) of substrate–enzyme binding site	$V = V_{max}/(1 + K_s/S)/(1 + I/K_i)$	$IC_{50} = K_i$
Mixed	Unbound and bound enzyme at the same and different site(s) of substrate–enzyme binding site	$V = V_{max}/((1 + I/K'_i) + (1 + K_s/S) \times (1 + I/K_i))$	$IC_{50} = K_i \times (1 + S/K_m)/ (1 + K_i/K'_i) \times (S/K_m))$
Metabolism-based			
Irreversible		$k_{abs} = k_{inact} \times I/(K_i + I)$	
MI-complex			
Time-dependent formation of potent reversible inhibitor			

Abbreviations: V, velocity at a given substrate concentration; V_{max}, maximum velocity; K_m, the binding affinity between substrate and enzyme; K_s, dissociation constant of substrate–enzyme complex; K_i, dissociation constant of inhibitor–enzyme complex; K_{obs}, rate of inactivation at a given inhibitor concentration; k_{inact}, maximal rate of inactivation; K_I, half maximal rate of inactivation (exact physical meaning is not defined); MI, metabolite–intermediate; K'_i, dissociation constant of inhibitor–enzyme complex in the presence of substrate; S, substrate concentration; IC_{50}, concentration of inhibitor that gives rise to a 50% decrease in activity.
[a]Segel (1993); Silverman (1988); Zhang and Wong, 2005.

dialysis of the incubate, a suitable (1 : 1) stoichiometry of radioactively labeled inactivator to active site, and a demonstrable catalytic step for the conversion of the inactivator to the reactive intermediate (Kent et al., 2001; Silverman, 1988). However, recent reports have suggested that mechanism-based CYP inactivation does not necessarily conform to 1 : 1 stoichiometry (Koenigs et al., 1999; Koenigs and Trager, 1998).

Because of the experimental manifestation of time-dependency and require-ment for CYP catalysis, mechanism-based CYP inactivation is often referred to as "time-dependent," "metabolism-dependent," or "preincubation-dependent" inhibition. A detailed description of the kinetic characteristics of this type of inhibition has been published (Silverman, 1988) and a simplified kinetic equation is presented in Table 5.1. In cases where CYP activity can be recovered by dialysis, the term "quasi-irreversible" inhibition has been proposed (Ma et al., 2000). In addition, a time-dependency of CYP inhibition can result from the formation of potent yet reversible metabolites (Ma et al., 2000; Zhao et al., 2002; Zhout et al., 2005). Formation of a metabolite–intermediate (MI) complex has been described as another cause for time-dependent CYP inhibition by many quasi-irreversible inhibitors; in this situation the metabolite or intermediate coordinates with the heme-ion thus decreasing the rate of catalysis.

5.2.2 *In vitro* Evaluation of Inhibition

Potential CYP inhibition by new chemical entities (NCEs) can be evaluated using cDNA-expressed human CYP enzymes (Crespi and Penman, 1997), human liver microsomes (HLM) (Bjornsson et al., 2003), and cell-based systems (Yueh et al., 2005a, 2005b;). Within a screening environment, cDNA-expressed CYP systems with probe substrates that form fluorescent metabolites have provided the throughput needed to rank order potency and characterize the structure–activity relationship (SAR), whereas HLM-based assays are often used for detailed characterization of CYP inhibition and the data are often included in regulatory submissions. Cell-based assays are used less frequently due to the lack of availability and reproducibility of cells as well as factors (cellular uptake or secondary metabolic processes) that may confound the interpretation of CYP enzyme kinetics. On the contrary, human liver microsomes and cDNA-expressed CYP enzymes do not represent a true physiological condition, and therefore, experimental artifacts may arise also. It is known that discrepancies exist in the generation of inhibition parameters (e.g., IC_{50} or K_i) when using one system over another, and some of the underlying causes (e.g., nonspecific binding and protein concentration) have been well described (Margolis and Obach, 2003; Tran et al., 2002; Walsky and Obach, 2004; Wienkers, 2001; Wienkers and Heath, 2005).

5.2.3 Prediction of CYP Inhibition Using *In vitro* Data

Quantitative predictions of *in vivo* CYP inhibition-mediated drug–drug interactions have been attempted using mathematical models with various

TABLE 5.2   Pharmacokinetic model describing an inhibitory drug–drug interaction.[a]

General equation	$$\frac{AUC_{(i)}}{AUC} = \frac{1}{\left[\dfrac{f_m \times f_{m,\,P450}}{CL_{int}/CL_{int,i}}\right] + 1 - (f_m \times f_{m,\,P450})}$$
	$$\frac{CL_{int}}{CL_{int,i}} =$$
Reversible competitive and non-competitive inhibition	$$1 + \frac{[I]}{K_i}$$
Reversible uncompetitive inhibition	$$1 + \left[\frac{[I]}{K_i}\right]\left[\frac{[S]}{[S] + K_m}\right]$$
Mechanism-based inhibition	$$\frac{K_{deg} + \dfrac{[I] \times k_{inact}}{[I] + K_I}}{K_{deg}}$$

Abbreviations: AUC_i, area under the curve in the presence of an inhibitor; AUC, area under the curve in the absence of an inhibitor; $CL_{int,i}$, intrinsic clearance in the presence of an inhibitor; CL_{int}, intrinsic clearance in the absence of an inhibitor; f_m, fraction of dose metabolized via CYPs; $f_{m,P450}$, fraction of total CYP-mediated metabolism catalyzed by inhibited CYP form; [I], inhibitor concentration; [S], substrate concentration; K_i, dissociation constant of inhibitor–enzyme complex; K_m, dissociation constant of substrate enzyme complex; k_{inact}, rate constant of CYP inactivation; K_{deg}, rate constant of CYP degradation or turnover; K_I, half maximal rate of inactivation (exact physical meaning is not defined).
[a]Ito et al. (1998b).

degrees of success (Table 5.2). General assumptions underpinning *in vivo* predictions using *in vitro* data include (1) liver is the organ for drug clearance; (2) a well-stirred model of hepatic clearance; and (3) *in vivo* conditions are similar to those *in vitro* (Ito et al., 1998a). A simple and well-accepted rule is the $[I]/K_i$ ratio (Bjornsson et al., 2003), where [I] is the inhibitor concentration at the CYP enzyme and K_i is the inhibition constant that can be experimentally determined *in vitro*. A ratio of $[I]/K_i > 1$ suggests an interaction is highly likely, whereas ratios of $0.1 < [I]/K_i < 1$ or $[I]/K_i < 0.1$ indicate that an interaction is likely or not likely, respectively. Although the quality of *in vitro* CYP inhibition data is a determining factor for accurate predictions (Wienkers and Heath, 2005), inhibitor concentrations at the enzyme site and the interplay of different clearance pathways *in vivo* are two of the most important unknowns that often derail successful predictions. Because the actual [I] cannot be measured, various correction factors and assumptions have been considered and specific examples are summarized in Table 5.3. Moreover, dynamic [I] as opposed to static [I] has also been examined using physiologically-based pharmacokinetic models (Chien et al., 2003; Kanamitsu et al., 2000). In addition, other factors, such as time-dependent inhibition, involvement of intestinal CYP enzymes, and other clearance pathways should also be considered. For example, Fig. 5.2 describes the change in area under the plasma concentration curve (AUC)

TABLE 5.3   Considerations and assumptions for predicting *in vivo* drug–drug interactions via inhibition.

Choice of [I]	Key considerations	References
Static [I]	Different [I]: steady-state plasma C_{max}; $C_{max,fu}$; portal vein I_{max}; or $I_{max,fu}$	Chien (2003), Ito et al. (2004), Ito et al. (2002), Ito (1998b), Obach et al. (2006)
	Parallel pathways of drug elimination	Ito et al. (2005), Obach (2006)
	Time-dependent CYP inhibition	Galetin et al. (2006), Obach (2006)
	Inhibitor absorption rate	Brown et al. (2005), Obach (2006)
	Intestinal inhibition	Galetin (2006), Obach (2006)
Dynamic [I]	PBPK, simulated [I]	Chien (2003), Kanamitsu (2000)

Abbreviations: [I], inhibitor concentration; C_{max}, maximum plasma concentration; $C_{max,fu}$, maximum free plasma concentration; I_{max}, maximum portal vein concentration; $I_{max,fu}$, maximum free portal vein concentration; PBPK, physiologically-based pharmacokinetics.

versus fraction metabolized in relation to the ratio of $[I]/K_i$. The lines labeled 1, 2, and 3 represent increasing ratios (i.e., increasing drug interaction potential). At a given $[I]/K_i$ ratio, the change in AUC is significant (>2-fold) only when the inhibited CYP contributes to greater than 60% of total clearance ($f_m \times f_{m,P450} > 0.6$). If additional pathways are involved in clearance

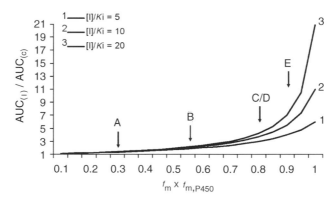

FIGURE 5.2   Representative plot of how AUC changes in regard to $f_m \times f_{m,P450}$ and $[I]/K_i$. K_i, dissociation constant of inhibitor-enzyme complex; [I], inhibitor concentration at the enzyme; f_m, fraction of dose metabolized via all CYPs; $f_{m,P450}$, fraction of total CYP-mediated metabolism catalyzed by inhibited CYP form; the product of $f_m \times f_{m,P450}$ represents the contribution of a specific CYP enzyme to overall clearance.

TABLE 5.4 Examples of CYP substrates and inhibitors used in clinical DDI studies.[a]

CYP isoform	Substrates	Inhibitors	Inducers
1A2	Caffeine, theophyline	Fluvoxamine	Smoking
2A6	Not available	Not available	Not available
2B6	Efavirenz	Not specified	Rifampin
2C8	Repaglinide, rosiglitazone	Gemfibrozil	Rifampin
2C9	Warfarin, tolbutamide	Fluconazole (use of PM subjects)	Rifampin
2C19	Omeprazole	Omeprazole, fluvoxamine (use of PM subjects)	Rifampin
2D6	Desipramine, dextromethorphen	Quinidine, paroxetine (use of PM subjects)	Not identified
2E1	Chlorzoxazone	Disulfirum	Ethanol
3A4/3A5	Midazolam, buspirone, felodipine, simvastatin	Ketoconazole, itraconazole	Rifampin, carbamazepine

[a]Adapted from (Bjornsson, 2003; Tucker, 2001), November 1999 FDA Guidance for Industry, and 2004 Preliminary Concept paper (www.fda.gov).
PM: Poor metabolizers.

(e.g., Phase II or renal excretion), then, the likelihood of a drug interaction is reduced and even large $[I]/K_i$ ratios become irrelevant.

Use of *in vitro* data to prospectively predict DDIs is difficult and some have proposed a relatively simple "rank-order" approach (Obach et al., 2005). In this instance, a NCE is evaluated as an inhibitor of different CYP forms *in vitro*. Standardized incubation conditions are employed, so that the IC_{50}s or K_is for each CYP form are ranked in order of increasing potency. The CYP form with the lowest IC_{50} or K_i is evaluated first in the clinic. For example, the inhibition is greatest with CYP3A4, and then a midazolam clinical DDI study is initiated first. Other CYPs are followed up with suitable probe drugs as needed (Table 5.4).

5.2.4 Clinical Evaluation of Inhibition

Treatment modalities employing combinations or cocktails of several drugs have become more commonplace. Therefore, clinically relevant drug–drug interactions are sometimes inevitable, and have become an important safety and regulatory concern. The pharmaceutical industry has, therefore, responded and it is now routine to conduct DDI studies during clinical development in order to identify and characterize potential CYP liabilities, and to provide such information to health care providers should a NCE become a marketed drug.

Clinical evaluation of CYP inhibition is most valuable if studies are conducted within the context of clinical relevance. For drugs with a wide therapeutic window, pharmacokinetic DDIs, to some degree, can be tolerated. On the contrary, a small increase in systemic exposure may lead to detrimental effects for a drug with a narrow therapeutic window. With the expectation to harmonize and/or standardize clinical DDI studies, a minimal best practice for *in vivo* CYP inhibition-mediated DDIs has been proposed, and a list of common CYP substrates and inhibitors for clinical DDI studies is shown in Table 5.4 (Bjornsson, 2003; Tucker et al., 2001). *In vitro* CYP inhibition data are important for selecting appropriate CYP probe substrates and inhibitors as well as guiding DDI study designs. Often, clinical DDI studies are conducted in healthy volunteers. However, depending on the therapeutic indication or mechanism of clearance, clinical DDI studies can also be conducted with specific subpopulations of subjects, such as patients with cancer, HIV infection, hepatic or renal impairment. Studies with genotyped subjects (e.g., CYP2D6, CYP2B6, CYP2C9, and CYP2C19) are also possible.

5.3 ENZYME INDUCTION

5.3.1 Enzyme and Pharmacokinetic Changes

Enzyme induction or the process of creating excess enzyme in a biological system can give rise to pharmacokinetic situations whereby drug interactions occur. To fully understand the fundamental mechanism by which greater concentrations of enzyme can cause pharmacokinetic disequilibrium, we must describe the concept of intrinsic clearance. The maximum rate of metabolic clearance is the ratio of V_{max}/K_m referred to as intrinsic metabolic clearance (derived from M–M kinetics and assuming [substrate] $< K_m$) (Rowland, 1988). K_m is a constant for any enzyme–substrate pair, however, V_{max} is a composite term with enzyme concentration being a variable: $V_{max} = k_{cat} \times [E]_T$, where k_{cat} is the catalytic constant and $[E]_T$ is the total concentration of enzyme (Segel, 1993). In most cases, induction does not cause a change in K_m or k_{cat}. Under normal situations, the pool of total enzyme exists in equilibrium between the constitutive production and degradation of enzyme. When greater amounts of enzyme are produced and the degradation rate of enzyme does not change, the total pool of enzyme concentration increases (i.e., $[E]_T$ increases) and the intrinsic metabolic clearance increases. Whereas the overall hepatic metabolic clearance of a drug is a function of blood flow, unbound drug concentration, and intrinsic clearance; increasing the intrinsic clearance elevates the overall hepatic elimination of a drug (Rowland, 1988).

Pharmacokinetically, the victims of enzyme induction generally demonstrate reduced AUC, C_{max}, trough concentrations, and half-life as a reflection of increased clearance. Table 5.5 illustrates several perpetrators and victims of CYP3A4 enzyme induction and the effects on pharmacokinetic parameters. In each of the cases illustrated in Table 5.5, there are significant biological consequences to a reduction in exposure. For example, reduced ethinyl estradiol levels can lead to unexpected pregnancies and reduced cyclosporine levels can lead to organ transplant rejection.

TABLE 5.5 **Percent reduction in oral pharmacokinetic parameters due to enzyme induction by rifampicin or St. John's wort.**

Victims	Rifampicin			St. John's wort			References
	AUC	C_{max}	$t_{1/2}$	AUC	C_{max}	$t_{1/2}$	
Ethinyl estradiol	64	42	42	24	min[a]	48	Hall et al. (2003), LeBel et al. (1998)
Midazolam	96	94	58	41	21	26	Backman et al. (1996), LeBel (1998)
Cyclosporine	73	48	30	15	9	NR	Bauer et al. (2003), Hebert et al. (1992)
Statins	A[c]:80	A:40	A:74	P[d]:71	P:74	P:min	Backman et al. (2005), Sugimoto et al. (2001)
Protease inhibitors	S[e]:70	S:65	S:17	I[f]:57	I:28/81[g]	I:NR	Grub et al. (2001), Piscitelli (2000)

[a]Minimal change.
[b]Not reported.
[c]Atorvastatin.
[d]Pravastatin.
[e]Saquinavir.
[f]Indinavir.
[g]Percent reduction in C_{max} and C_{8h}.

St. John's wort is a common herbal remedy for the treatment of depression and contains a complex mixture of components that can be perpetrators of drug interactions. Most common of these interactions is the enzyme induction effects of the St. John's wort component, hyperforin (Zhou et al., 2004). Administration of hyperforin leads to an increase in the transcription, translation, and enzyme activity of CYP3A4. Interactions with St. John's wort include: cyclosporine, oral contraceptives, midazolam, anticonvulsants, and HMG-CoA reductase inhibitors (Izzo, 2004; Zhou, 2004). When coadministered with these CYP3A4 substrates (victims), the plasma levels of concomitantly administered drugs fall below their efficacious levels and become ineffective. For example, the plasma exposure (AUC) of indinavir when coadministered with St. John's wort decreased by 57%. In addition, the 8 h trough concentration of indinavir fell by 81%, well below the efficacious concentration sebsequently leading to increases in HIV RNA viral load (Piscitelli et al., 2000). Xenobiotics, whether herbal or prescribed medicines, such as rifampicin another potent CYP3A4 enzyme inducer, must all be carefully assessed for their ability to alter the activity of drug-metabolizing enzymes.

In addition to the victim-perpetrator scenario where two drugs are involved, there exists a situation where a single drug acts as both victim and perpetrator called autoinduction. Carbamazepine (CBZ) is an anticonvulsant that causes CYP3A4 enzyme induction in patients and this induction increases the clearance of CBZ itself because CBZ is also a CYP3A4 substrate. In the case of CBZ, the clearance increases from Day 1 to Day 17 (0.028 to 0.056 L/h/kg) and the steady state plasma concentrations of CBZ decline over this period, reaching a lower plateau after several weeks (Bertilsson et al., 1980; Eichelbaum et al., 1975). After the autoinduction phase has reached its lower plateau, doses of CBZ are then readjusted (increased) to achieve new efficacious concentrations of drug. Autoinduction is a phenomenon that can occur anytime a drug induces an enzyme that is also predominately involved in its own metabolic clearance.

5.3.2 Mechanisms of Enzyme Induction

There are two general mechanisms by which enzyme induction occurs; stabilization of mRNA (or enzyme) or increased gene transcription (Okey, 1992). Induction of CYP2E1 is an example where enzyme stabilization is known to occur and results in increased activity due to a decrease in enzyme degradation (Chien et al., 1997). However, the most common mechanism of CYP enzyme induction is transcriptional gene activation. Transcriptional activation is mediated via nuclear receptors (NRs) that function as transcription factors, such as aromatic hydrocarbon receptor (AhR) (Mandal, 2005), constitutive androstane receptor (CAR) (Qatanani and Moore, 2005), farnesoid X receptor (FXR) (Westin et al., 2005), and pregnane X receptor or SXR-steroid X receptor (PXR) (Honkakoski et al., 2003; Mangelsdorf et al., 1995; Moore et al., 2002; Tirona and Kim, 2005; Wang and LeCluyse, 2003). These NRs are activated either through ligand (drug) binding or other mechanisms of activation, such as regulation of coactivators or corepressors. Table 5.6 illustrates several NRs, their target genes, and common human activators. A more comprehensive list of target genes can be found in recent reviews (Handschin and Meyer, 2003; Puga et al., 2000; Wang and LeCluyse, 2003).

The general concept for NR signaling is that in the absence of a ligand, the NR is associated with NR corepressor complexes, conferring a basal level of transcription. Ligand binding to the ligand binding domain (LBD) of the NR induces conformational changes that lead to the release of corepressors and recruitment of coactivators (Glass et al., 1997; Glass and Rosenfeld, 2000; Mangelsdorf and Evans, 1995). The LBDs of many NRs are different among various animal species, especially between laboratory animals and humans (Blumberg et al., 1998; LeCluyse, 2001b; Moore et al., 2002). Therefore, *in vitro* animal or *in vivo* animal models of enzyme induction can be misleading and are generally not employed to study or predict the potential induction effect in humans. With PXR as an example, recruitment of coactivators and the dimerization partner retinoid X receptor (RXR) contributes to chromatin remodeling and subsequent transcriptional activation via specific response elements (RE) that consist of a core DNA consensus sequence that can be configured into a variety of structured motifs (Blumberg, 1998; Tirona and Kim, 2005). Regulation is achieved by the binding of the NR complex through the DNA binding domain (DBD) to their respective RE present in the promoter region of target genes (drug-metabolizing enzymes or drug transporters). This process is illustrated in Fig. 5.3, using PXR as an example. PXR, which resides in

TABLE 5.6 Nuclear receptors, their activated genes, and common activators for each receptor.

Nuclear receptor	Target Genes	Activators	References
AhR	CYP1A1/2, CYP1B1, CYP2S1, UGT1A1, UGT1A6, 3-hydroxybutyrate dehydrogenase, SULT2B	TCDD,[a] BNF,[b] 3MC,[c] Cigarette smoking	Mandal (2005), Runge-Morris and Kocarek (2005), Zhou et al. (2005)
CAR	CYP2B6, CYP3A4, UGT1A1, UGT2B1, SULT2A1, MRP2, MRP3, MRP4	Phenytoin, CITCO,[d] 5β-pregnane-3,20-dione, phenobarbital	Honkakoski (2003), Klaassen and Slitt (2005), Moore (2002), Runge-Morris and Kocarek (2005), Zhou (2005)
PXR	CYP3A4,CYP2B6, CYP2C9, GST, UGT1A1, UGT1A6, UGT1A9, SULT2A1, MRP2, MRP3, MRP5, MDR1	Rifampicin, SR12813, hyperforin, clotrimazole, forskolin	Carnahan and Redinbo (2005), Ding and Staudinger (2005), Honkakoski (2003), Klaassen and Slitt (2005), Moore (2002), Runge-Morris and Kocarek (2005), Zhou (2005)
FXR	CYP7A1, UGT2B4, MDR3, BSEP	GW4064, fexaramine, TTNPB[e]	Dixit et al. (2005), Klaassen and Slitt (2005), Zhou (2005)

[a]2,3,7,8-Tetrachlorodibenzo-*p*-dioxin.
[b]β-Naphthoflavone.
[c]3-Methylcholanthrene.
[d]6-(4-Chlorophenyl)imidazo[2,1-*b*][1,3]thiazole-5-carbaldehyd *O*-(3,4-dichlorobenzyl)oxime.
[e](*E*)-4-[2-(5,6,7,8-Tetrahydro-5,5,8,8-tetramethyl-2-naphthylenyl)-1-propenyl] benzoic acid.

the cytosol and nucleus, is activated once a ligand binds, then corepressors (R) are removed and coactivators (A) recruited. At this point, the activated receptor dimerizes with RXR (translocates to the nucleus) and binds to the response elements (PXR RE) found within the CYP3A4 promoter region. Recruitment of additional transcription factors, such as, RNA polymerase, initiates transcription of the CYP3A4 mRNA, followed by subsequent translation into active enzyme.

The mechanism of CAR-mediated gene transcription differs somewhat from that of PXR. CAR is always activated and is therefore sequestered in the

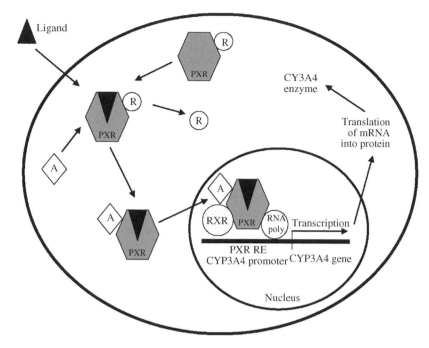

FIGURE 5.3   Mechanism of PXR ligand activated induction of CYP3A4.

cytoplasm by chaperone proteins, such as HSP90, in order to keep its transcriptional activity at constitutive levels. Translocation of CAR to the nucleus is thought to be initiated by direct agonist binding to the receptor, or through a partially elucidated ligand-independent mechanism involving kinases that dephosphorylate CAR. Phenobarbital is an example of a drug that does not bind to CAR, yet causes nuclear translocation and transcriptional activation of a target gene, CYP2B6 (Qatanani and Moore, 2005). From this point, CAR dimerizes with RXR and other transcription factors, binds to its respective RE in the promoter region of target genes and initiates transcription.

Activation of AhR target genes can also occur through two distinct mechanisms. The inactive AhR resides in the cytoplasm in a complex with HSP90 and other proteins. The receptor can be activated upon ligand binding (ligands such as TCDD, 3MC, BNF) and translocates to the nucleus where it dimerizes with the AhR nuclear translocator. This dimer complex recruits additional transcription factors and binds to the promoter region of AhR target gene response elements and initiates transcription. Another mechanism by which AhR target genes can be activated is through a ligand-independent mechanism involving protein tyrosine kinases. Omeprazole-mediated induction of CYP1A is thought to proceed through this particular signal transduction pathway (Backlund and Ingelman-Sundberg, 2005).

5.3.3 Induction Models

The number of models to evaluate induction of drug-metabolizing enzymes has increased tremendously over the past decade. As mentioned previously, animal models are not amenable to assessing the induction potential of humans due to differences in the LBD of most nuclear receptors. As a result, nearly all of the current induction models are based on human reagents. One of several high throughput models (cell free) involves a measurement of ligand binding to the expressed NR in scintillation proximity type assays (Zhu et al., 2004). A simple cell-based model involves the coexpression of a nuclear receptor along with an appropriate gene promoter region coupled to a reporter gene. An example of this type of model utilizes HepG2 cells where human PXR is expressed along with a second vector expressing a partial CYP3A4 promoter fused to the luciferase gene. Activation of PXR proceeds as previously described except that the luciferase gene is induced and a direct luminescent readout can be measured (Luo et al., 2002, 2004). In the case of CAR activation, measurement of CAR translocation into the nucleus of primary hepatocytes has been used successfully to identify activators of this receptor (Wang et al., 2004). Another cell-based system that has the ability to measure multiple drug-metabolizing enzymes and transporters within a single cell system is the immortalized hepatocyte cell line, Fa2N-4 (Mills et al., 2004). Ultimately, the present "gold standard" for assessing human induction of drug-metabolizing enzymes are primary cultures of fresh or cryopreserved human hepatocytes (LeCluyse, 2001a). In this situation, all of the NR, corepressors/activators, dimerization partners, response elements, and target genes are present in their natural environment. Although there are issues with the timely acquisition of human hepatocytes from multiple donors, this model remains the most representative of the *in vivo* situation in humans.

Transgenic animal models of human induction have received greater attention as a means of moving away from the static *in vitro* models to a more dynamic *in vivo* situation where concentration–time effects, and how other absorption–distribution–metabolism–excretion (ADME) properties affect the overall level of enzyme induction (Xie and Evans, 2002). As an example, there are mouse strains where the mouse PXR or CAR has been knocked out and the human NR inserted. It has been shown in the case of the humanized PXR mouse, that rifampicin (a human PXR ligand) activates the hPXR *in vivo* and initiates transcription of mouse *cyp3a11* (Ma et al., 2007; Xie et al., 2000). At similar doses of rifampicin in wild type mice, no induction of *cyp3a11* is found because rifampicin is not an effective ligand for mouse PXR. One caveat exists for these models, only one gene has been humanized (i.e., the NR) and all other aspects of the mouse remain normal. Drugs evaluated in these *in vivo* systems will be absorbed, distributed, metabolized, and eliminated according to mouse physiology and not human physiology, albeit they are sometimes similar. Additional studies and validation will be necessary to fully support the use of these humanized animal models to predict human induction.

5.4 REACTION PHENOTYPING

For any established drug or a NCE, it is important to understand the routes of elimination, the contribution of each route to overall clearance and the enzyme(s) involved in metabolism. A human radiolabeled study provides critical information related to the fraction of the dose cleared via CYP-dependent metabolism (f_m). *In vitro* reaction phenotype data are also important (Rodrigues, 1999). In this instance, one uses different approaches to determine the contribution or each CYP form toward total CYP-mediated metabolism ($f_{m,P450}$). The two data sets are then integrated because the product of f_m and $f_{m,P450}$ (i.e., $f_m \times f_{m,P450}$) is one of the key factors that determines the magnitude of a DDI (Table 5.2). As shown in Table 5.7, drugs largely metabolized by a single CYP enzyme are far more susceptible to drug–drug interactions. Itraconazole is a CYP3A4 inhibitor and gemfibrozil-*O*-glucuronide (major circulating metabolite of gemfibrozil) is a mechanism-based inhibitor of CYP2C8 (Niemi et al., 2003, 2005; Ogilvie et al., 2006). This particular combination would significantly impact metabolic clearance mediated by both CYP3A4 and CYP2C8. As a result, a significant drug–drug interaction is observed for repaglinide, a CYP3A4 and CYP2C8 substrate ($f_m \times f_{m,P450}$ approaching 1.0). In comparison, the two enzymes play minor roles in the overall clearance of netaglinide and the magnitude of the DDI is reduced.

Likewise, marked variability in pharmacokinetics can be anticipated for drugs that are primarily metabolized by polymorphic CYP enzymes such as CYP2D6, CYP2C9 and CYP2C19 (Ma et al., 2004; Nagata and Yamazoe, 2002). For example, the incidence of CYP2C19 poor metabolizers (PMs) is as high as 38% in Asians (e.g., Japanese and Chinese) and less than 15% in Caucasians (Rodrigues and Rushmore, 2002). Aside from pharmacokinetic variability associated with CYP2C19 PM subjects, inhibition of alternative clearance

TABLE 5.7 Effect of gemfibrozil and itraconazole on the pharmacokinetics of two insulin secretogogues.[a]

	Netaglinide		Repaglinide	
f_m	0.80		0.97	
	$f_{m,P450}$	$f_m \times f_{m,P450}$	$f_{m,P450}$	$f_m \times f_{m,P450}$
$f_{m,CYP2C9}$	0.70	0.56		
$f_{m,CYP3A4}$	0.30	0.24	0.53–1.0	0.51–0.97
$f_{m,CYP2C8}$			≤ 0.47	≤ 0.45
Effect of gemfibrozil + itraconazole on AUC (fold-increase)	1.5		~20	

Abbreviations: f_m, fraction of the dose metabolized by all CYPs; $f_{m,P450}$, fraction of total CYP-dependent metabolism catalyzed by each CYP; $f_m \times f_{m,P450}$, contribution of a specific CYP enzyme to overall clearance.
[a]Bidstrup et al. (2003), Niemi et al. (2003), Niemi et al. (2005).

pathways (such as CYP3A4 by ketoconazole) would render this subpopulation at a much greater risk for adverse events when a drug is cleared largely via two CYPs only (e.g., CYP2C19 and CYP3A4). Therefore, CYP reaction phenotyping data are extremely useful when predicting pharmacokinetic-based DDIs, designing clinical studies, and understanding pharmacokinetic variability in human subjects.

Ideally, definitive CYP reaction phenotyping should be available before the initiation of clinical development. Unfortunately, more accurate data can only be obtained once clearance pathways are identified in human subjects, and human radiolabeled studies are generally not conducted as the first set of clinical studies for NCEs. In this context, CYP reaction phenotyping, performed using various human *in vitro* systems is expected to be as complete as possible (Bjornsson, 2003).

Figure 5.2 describes *in vitro* reaction phenotype data for five CYP3A4 substrates that have gone into development. In each case, the impact of ketoconazole on their pharmacokinetics has been evaluated clinically. As expected, CYP3A4 played a major role in the overall clearance of compounds "C," "D," and "E" and ketoconazole elicited a marked impact on their pharmacokinetics (> 4-fold increase in AUC). In comparison, the enzyme played a minor role in the clearance of compounds "A" and "B" and the AUC increase due to ketoconazole was less marked (< 2-fold).

5.4.1 Experimental Considerations

During drug discovery and early development, initial CYP-reaction phenotyping studies are generally conducted using cDNA-expressed human CYP systems by monitoring substrate depletion, due to the lack of synthetic metabolites or radiolabeled material. Results from this type of study often yield useful and qualitative information related to the CYP(s) responsible for metabolic clearance. Once synthetic standards of metabolites become available, cDNA-expressed human CYP systems can then be optimized and definitive kinetic parameters such as K_m and V_{max} (or k_{cat}) can be determined by monitoring metabolite formation. The choice of substrate concentration range is an important consideration in early CYP reaction phenotyping studies because one has to ensure that the experimental conditions enable the identification of low K_m ($\leq 20\ \mu M$) CYP forms. In addition to cDNA-expressed human CYP enzymes, metabolism of NCEs can be investigated with human liver microsomes in the presence of specific chemical inhibitors or inhibitory antibodies to further define the role of specific CYP(s) (Rodrigues, 1999). Here, enzyme kinetic parameters can also be obtained with pooled microsomal preparations (multiple organ donors) or microsomes from individual (genotyped) human liver preparations. Activity correlation studies are sometimes conducted to further elucidate the role of key CYP enzymes using a panel of human liver microsomes from at least 10 different organ donors. In this instance, metabolite formation is correlated with CYP form-specific activities and/or immunoquantitated levels of individual CYP forms. More integrated systems, such as primary hepatocytes or precision-cut liver slices, are also used to identify additional (e.g., Phase II) pathways and enable initial estimates of f_m. As much as possible, a complete CYP reaction phenotyping package should contain data obtained from cDNA-expressed human CYP systems, human liver microsomes, and primary human hepatocytes (Rodrigues, 1999).

5.4.2 Data Interpretation and Integration

As discussed above, various *in vitro* systems can yield useful information and increase one's understanding of the individual roles each CYP enzyme plays in drug clearance. Each system has unique properties that must be considered when analyzing and interpreting data. For example, bridging metabolism data generated from cDNA-expressed CYP systems to human liver microsomes requires use of certain scaling factors and at least two approaches (e.g., relative activity factor, RAF and abundance normalized rates, NR) have been proposed (Crespi, 1995; Remmel and Burchell, 1993; Rodrigues, 1999; Venkatakrishnan et al., 2000).

Although cDNA-expressed human CYP enzymes are "cleaner" *in vitro* systems for studying CYP-mediated reactions, human liver microsomes are preferred due to the existence of multiple CYP enzymes and other native components that allow for consideration of drug–protein interactions and parallel or sequential metabolism. As such, data obtained from human liver microsomes are the "net" outcome of CYP-mediated metabolism (Houston and Kenworthy, 2000). Although complications may arise due to differing effects on enzyme kinetics at varying substrate concentrations, the role of CYPs in metabolism of a particular drug can be reasonably assessed with specific chemical inhibitors, specific inhibitory antibodies, and sensitive analytical methods (Gelboin et al., 1999; Lu et al., 2003; Walsky and Obach, 2004). As discussed above, activity correlation studies using a panel of human liver microsomes with specific probe substrates are additional means to delineate CYP-mediated metabolism (Lu et al., 2003).

Cell- or tissue-based *in vitro* systems possess the advantage of simultaneous assessment of both Phase I and Phase II drug-metabolizing enzymes. Although freshly isolated human hepatocytes are the preferred cell-based system to evaluate drug metabolism, the availability of such cells has limited its broad utility in the industrial setting. As a result, cryopreserved human hepatocytes have become a valuable alternative (Gebhardt et al., 2003; Gomez-Lechon et al., 2004). In addition, precision-cut liver slices have been used to investigate drug metabolism and liver toxicity (de Kanter et al., 2002). Identification of non-CYP-mediated metabolism using these integrated *in vitro* systems has been critical to understand the overall metabolic clearance of NCEs and a certain degree of success has been achieved in predicting *in vivo* metabolic clearance using *in vitro* hepatocyte data (Hallifax et al., 2005; Ito et al., 1998a; Niro et al., 2003).

5.4.3 Clinical Evaluation

The cost of clinical investigations, safety and regulatory concerns would likely prohibit the development of any NCEs at high risk for significant drug–drug interactions. In this regard, the determination of each CYP's contribution to overall clearance is highly desirable or required in some therapeutic areas. Therefore, the most important utility of *in vitro* CYP reaction phenotype data is the ability to prioritize drug–drug interaction studies. This is illustrated in Fig. 5.4 by analogy with the rank order approach described for inhibitors of CYP (Obach et al., 2005). For instance, a ketoconazole or rifampicin interaction study would be required if a NCE is primarily metabolized ($f_{m,P450} > 0.5$) by CYP3A4. Likewise, the pharmacokinetics of a NCE would have to be evaluated in specific populations if the compound were largely metabolized by polymorphic CYPs, such as CYP2D6, CYP2C9, or CYP2C19.

Prospectively, there are always instances where *in vivo* human outcomes could not be fully appreciated from *in vitro* data. Retrospectively, however, *in vitro* systems are valuable tools to understand clinical drug–drug interactions. Cerivastatin, a potent HMG–CoA reductase inhibitor, was thought to have a low propensity for CYP-mediated drug–drug interactions because the compound was metabolized by both CYP3A4 and CYP2C8; and the lack of clinically relevant interactions with commonly used drugs appeared to support this hypothesis (Muck, 2000). However, a significant interaction with gemfibrozil, and concerns over severe rhabdomyolysis associated with this type of interaction, led to its eventual withdrawal from the market (Backman et al., 2002; Roca et al., 2002). Subsequent investigative studies indicated that gemfibrozil-*O*-glucuronide, a major metabolite of gemfibrozil, is an inhibitor of cerivastatin hepatic uptake transport via the organic anion transporter protein, as well as a

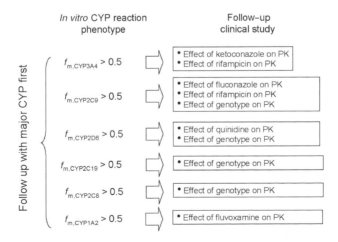

FIGURE 5.4 Proposed clinical drug interaction studies based on *in vitro* CYP reaction phenotyping data. It is assumed that the drug in question is cleared via CYP-dependent metabolism ($f_m \sim 1.0$).

mechanism-based inactivator of CYP2C8 (Ogilvie et al., 2006; Shitara et al., 2004). Therefore, the impact of gemfibrozil on cerivastatin pharmacokinetics is likely caused by a combination of hepatic uptake and CYP2C8 inhibition. This example clearly demonstrates the need to thoroughly examine the roles of CYP-mediated metabolism and drug transporters in drug disposition.

REFERENCES

Backlund M, Ingelman-Sundberg M. Regulation of aryl hydrocarbon receptor signal transduction by protein tyrosine kinases. Cell Signal 2005;17:39–48.

Backman JT, Kyrklund C, Neuvonen M, Neuvonen PJ. Gemfibrozil greatly increases plasma concentrations of cerivastatin. Clin Pharmacol Ther 2002;72:685–691.

Backman JT, Luurila H, Neuvonen M, Neuvonen PJ. Rifampin markedly decreases and gemfibrozil increases the plasma concentrations of atorvastatin and its metabolites. Clin Pharmacol Ther 2005;78:154–167.

Backman JT, Olkkola KT, Neuvonen PJ. Rifampin drastically reduces plasma concentrations and effects of oral midazolam. Clin Pharmacol Ther 1996;59:7–13.

Bauer S, Stormer E, Johne A, Kruger H, Budde K, Neumayer HH, Roots I, Mai I. Alterations in cyclosporin a pharmacokinetics and metabolism during treatment with St John's wort in renal transplant patients. Br J Clin Pharmacol 2003;55: 203–211.

Bertilsson L, Hojer B, Tybring G, Osterloh J, Rane, A. Autoinduction of carbamazepine metabolism in children examined by a stable isotope technique. Clin Pharmacol Ther 1980;27:83–88.

Bidstrup TB, Bjornsdottir I, Sidelmann UG, Thomsen MS, Hansen KT. CYP2C8 and CYP3A4 are the principal enzymes involved in the human in vitro biotransformation of the insulin secretagogue repaglinide. Br J Clin Pharmacol 2003;56: 305–314.

Bjornsson TD, Callaghan JT, Einolf HJ, Fischer V, Gan L, Grimm S, Kao J, King SP, Miwa G, Ni L, Kumar G, McLeod J, Obach RS, Roberts S, Roe A, Shah A, Snikeris F, Sullivan JT, Tweedie D, Vega JM, Walsh J, Wrighton SA. The conduct of in vitro and in vivo drug–drug interaction studies: a pharmaceutical research and manufacturers of America (PHARMA) perspective. Drug Metab Dispos 2003;31: 815–832.

Blumberg B, Sabbagh W Jr, Juguilon H, Bolado J Jr, vanMeter CM, Ong ES, Evans, RM. SXR, a novel steroid and xenobiotic-sensing nuclear receptor. Genes Dev 1998;12:3195–3205.

Brown HS, Ito K, Galetin A, Houston JB. Prediction of in vivo drug–drug interactions from in vitro data: impact of incorporating parallel pathways of drug elimination and inhibitor absorption rate constant. Br J Clin Pharmacol 2005;60:508–518.

Carnahan VE, Redinbo MR. Structure and function of the human nuclear xenobiotic receptor PXR. Curr Drug Metab 2005;6:357–367.

Chien JY, Mohutsky MA, Wrighton SA. Physiological approaches to the prediction of drug–drug interactions in study populations. Curr Drug Metab 2003;4: 347–356.

Chien JY, Thummel KE, Slattery JT. Pharmacokinetic consequences of induction of CYP2E1 by ligand stabilization. Drug Metab Dispos 1997;25:1165–1175.

Crespi CL. Xenobiotic-metabolizing human cells as tools for pharmacological and toxicological research. Advance in Drug Research. New York: Academic Press; 1995. p 179–235.

Crespi CL, Penman BW. Use of cDNA-expressed human cytochrome P450 enzymes to study potential drug–drug interactions. Adv Pharmacol 1997;43:171–188.

de Kanter R, Monshouwer M, Meijer DK, Groothuis GM. Precision-cut organ slices as a tool to study toxicity and metabolism of xenobiotics with special reference to nonhepatic tissues. Curr Drug Metab 2002;3:39–59.

Ding X, Staudinger JL. Induction of drug metabolism by forskolin: the role of the pregnane x receptor and the protein kinase a signal transduction pathway. J Pharmacol Exp Ther 2005;312:849–856.

Dixit SG, Tirona RG, Kim RB. Beyond CAR and PXR. Curr Drug Metab 2005;6:385–397.

Eichelbaum M, Ekbom K, Bertilsson L, Ringberger VA, Rane A. Plasma kinetics of carbamazepine and its epoxide metabolite in man after single and multiple doses. Eur J Clin Pharmacol 1975;8:337–341.

Galetin A, Burt H, Gibbons L, Houston JB. Prediction of time-dependent CYP3A4 drug–drug interactions: impact of enzyme degradation, parallel elimination pathways, and intestinal inhibition. Drug Metab Dispos 2006;34:166–175.

Gebhardt R, Hengstler JG, Muller D, Glockner R, Buenning P, Laube B, Schmelzer F, Ullrich M, Utesch D, Hewitt N, Ringel M, Hilz BR, Bader A, Langsch A, Koose T, Burger HJ, Maas J, Oesch F. New hepatocyte *in vitro* systems for drug metabolism: metabolic capacity and recommendations for application in basic research and drug development, standard operation procedures. Drug Metab Rev 2003;35:145–213.

Gelboin HV, Krausz KW, Gonzalez FJ, Yang TJ. Inhibitory monoclonal antibodies to human cytochrome P450 enzymes: a new avenue for drug discovery. Trends Pharmacol Sci 1999;20:432–438.

Glass CK, Rose DW, Rosenfeld MG. Nuclear receptor coactivators. Curr Opin Cell Biol 1997;9:222–232.

Glass CK, Rosenfeld MG. The coregulator exchange in transcriptional functions of nuclear receptors. Genes Dev 2000;14:121–141.

Gomez-Lechon MJ, Donato MT, Castell JV, Jover R. Human hepatocytes in primary culture: the choice to investigate drug metabolism in man. Curr Drug Metab 2004;5:443–462.

Grub S, Bryson H, Goggin T, Ludin E, Jorga K. The interaction of saquinavir (soft gelatin capsule) with ketoconazole, erythromycin and rifampicin: comparison of the effect in healthy volunteers and in HIV-infected patients. Eur J Clin Pharmacol 2001;57:115–121.

Hall SD, Wang Z, Huang SM, Hamman MA, Vasavada N, Adigun AQ, Hilligoss JK, Miller M, Gorski JC. The interaction between St John's wort and an oral contraceptive. Clin Pharmacol Ther 2003;74:525–535.

Hallifax D, Rawden HC, Hakooz N, Houston JB. Prediction of metabolic clearance using cryopreserved human hepatocytes: kinetic characteristics for five benzodiazepines. Drug Metab Dispos 2005;33:1852–1858.

Handschin C, Meyer UA. Induction of drug metabolism: the role of nuclear receptors. Pharmacol Rev 2003;55:649–673.

Hebert MF, Roberts JP, Prueksaritanont T, Benet LZ. Bioavailability of cyclosporine with concomitant rifampin administration is markedly less than predicted by hepatic enzyme induction. Clin Pharmacol Ther 1992;52:453–457.

Honkakoski P, Sueyoshi T, Negishi M. Drug-activated nuclear receptors CAR and PXR. Ann Med 2003;35:172–182.

Houston JB, Galetin A. Modelling atypical CYP3A4 kinetics: principles and pragmatism. Arch Biochem Biophys 2005;433:351–360.

Houston JB, Kenworthy KE. *In vitro–in vivo* scaling of CYP kinetic data not consistent with the classical Michaelis–Menten model. Drug Metab Dispos 2000;28:246–254.

Ito K, Brown HS, Houston JB. Database analyses for the prediction of *in vivo* drug–drug interactions from *in vitro* data. Br J Clin Pharmacol 2004;57:473–486.

Ito K, Chiba K, Horikawa M, Ishigami M, Mizuno N, Aoki J, Gotoh Y, Iwatsubo T, Kanamitsu S, Kato M, Kawahara I, Niinuma K, Nishino A, Sato N, Tsukamoto Y, Ueda K, Itoh T, Sugiyama Y. Which concentration of the inhibitor should be used to predict *in vivo* drug interactions from *in vitro* data? AAPS Pharm Sci 2002;4:E25.

Ito K, Hallifax D, Obach RS, Houston JB. Impact of parallel pathways of drug elimination and multiple cytochrome P450 involvement on drug–drug interactions: CYP2D6 paradigm. Drug Metab Dispos 2005;33:837–844.

Ito K, Iwatsubo T, Kanamitsu S, Nakajima Y, Sugiyama Y. Quantitative prediction of *in vivo* drug clearance and drug interactions from *in vitro* data on metabolism, together with binding and transport. Annu Rev Pharmacol Toxicol 1998a;38:461–499.

Ito K, Iwatsubo T, Kanamitsu S, Ueda K, Suzuki H, Sugiyama Y. Prediction of pharmacokinetic alterations caused by drug–drug interactions: metabolic interaction in the liver. Pharmacol Rev 1998b;50:387–412.

Izzo AA. Drug interactions with St. John's wort (*Hypericum perforatum*): a review of the clinical evidence. Int J Clin Pharmacol Ther 2004;42:139–148.

Kanamitsu S, Ito K, Sugiyama Y. Quantitative prediction of *in vivo* drug–drug interactions from *in vitro* data based on physiological pharmacokinetics: use of maximum unbound concentration of inhibitor at the inlet to the liver. Pharm Res 2000;17:336–343.

Kent UM, Juschyshyn MI, Hollenberg PF. Mechanism-based inactivators as probes of cytochrome P450 structure and function. Curr Drug Metab 2001;2:215–243.

Klaassen CD, Slitt AL. Regulation of hepatic transporters by xenobiotic receptors. Curr Drug Metab 2005;6:309–328.

Koenigs LL, Peter RM, Hunter AP, Haining RL, Rettie AE, Friedberg T, Pritchard MP, Shou M, Rushmore TH, Trager WF. Electrospray ionization mass spectrometric analysis of intact cytochrome P450: identification of tienilic acid adducts to P450 2C9. Biochemistry 1999;38:2312–2319.

Koenigs LL, Trager WF. Mechanism-based inactivation of cytochrome P450 2B1 by 8-methoxypsoralen and several other furanocoumarins. Biochemistry 1998;37:13184–13193.

LeBel M, Masson E, Guilbert E, Colborn D, Paquet F, Allard S, Vallee F, Narang PK. Effects of rifabutin and rifampicin on the pharmacokinetics of ethinylestradiol and norethindrone. J Clin Pharmacol 1998;38:1042–1050.

LeCluyse EL. Human hepatocyte culture systems for the *in vitro* evaluation of cytochrome P450 expression and regulation. Eur J Pharm Sci 2001a;13:343–368.

LeCluyse EL. Pregnane X Receptor: molecular basis for species differences in cyp3a induction by xenobiotics. Chem Biol Interact 2001b;134:283–289.

Lu AY, Wang RW, Lin JH. Cytochrome P450 *in vitro* reaction phenotyping: a re evaluation of approaches used for P450 isoform identification. Drug Metab Dispos 2003;31:345–350.

Luo G, Cunningham M, Kim S, Burn T, Lin J, Sinz M, Hamilton G, Rizzo C, Jolley S, Gilbert D, Downey A, Mudra D, Graham R, Carroll K, Xie J, Madan A, Parkinson A, Christ D, Selling B, LeCluyse E, Gan LS. CYP3A4 induction by drugs: correlation between a pregnane x receptor reporter gene assay and CYP3A4 expression in human hepatocytes. Drug Metab Dispos 2002;30:795–804.

Luo G, Guenthner T, Gan LS, Humphreys WG. CYP3A4 induction by xenobiotics: biochemistry, experimental methods and impact on drug discovery and development. Curr Drug Metab 2004;5:483–505.

Ma B, Prueksaritanont T, Lin JH. Drug interactions with calcium channel blockers: possible involvement of metabolite–intermediate complexation with CYP3A. Drug Metab Dispos 2000;28:125–130.

Ma JD, Nafziger AN, Bertino JS Jr. Genetic polymorphisms of cytochrome P450 enzymes and the effect on interindividual, pharmacokinetic variability in extensive metabolizers. J Clin Pharmacol 2004;44:447–456.

Ma X, Shah Y, Cheung C, Guo GL, Feigenbaum L, Krausz KW, Idle JR, Gonzalez FJ. The pregnane X receptor gene-humanized mouse: a model for investigating drug–drug interactions mediated by cytochromes P450 3A. Drug Metab Dispos 2007;35:194–200.

Mandal PK. Dioxin: a review of its environmental effects and its aryl hydrocarbon receptor biology. J Comp Physiol [B] 2005;175:221–230.

Mangelsdorf DJ, Evans RM. The RXR heterodimers and orphan receptors. Cell 1995; 83:841–850.

Mangelsdorf DJ, Thummel C, Beato M, Herrlich P, Schutz G, Umesono K, Blumberg B, Kastner P, Mark M, Chambon P, Evans RM. The nuclear receptor superfamily: the second decade. Cell 1995;83:835–839.

Margolis JM, Obach RS. Impact of nonspecific binding to microsomes and phospholipid on the inhibition of cytochrome P4502D6: implications for relating *in vitro* inhibition data to *in vivo* drug interactions. Drug Metab Dispos 2003;31:606–611.

Mills JB, Rose KA, Sadagopan N, Sahi J, deMorais SM. Induction of drug metabolism enzymes and mdr1 using a novel human hepatocyte cell line. J Pharmacol Exp Ther 2004;309:303–309.

Moore LB, Maglich JM, McKee DD, Wisely B, Willson TM, Kliewer SA, Lambert MH, Moore JT. Pregnane x receptor (PXR), constitutive androstane receptor (CAR), and benzoate x receptor (BXR) define three pharmacologically distinct classes of nuclear receptors. Mol Endocrinol 2002;16:977–986.

Muck W. Clinical pharmacokinetics of cerivastatin. Clin Pharmacokinet 2000;39:99–116.

Nagata K, Yamazoe Y. Genetic polymorphism of human cytochrome P450 involved in drug metabolism. Drug Metab Pharmacokinet 2002;17:167–189.

Niemi M, Backman JT, Juntti-Patinen L, Neuvonen M, Neuvonen PJ. Coadministration of gemfibrozil and itraconazole has only a minor effect on the pharmacokinetics of the CYP2C9 and CYP3A4 substrate nateglinide. Br J Clin Pharmacol 2005;60:208–117.

Niemi M, Backman JT, Neuvonen M, Neuvonen PJ. Effects of gemfibrozil, itraconazole, and their combination on the pharmacokinetics and pharmacodynamics of repaglinide: potentially hazardous interaction between gemfibrozil and repaglinide. Diabetologia 2003;46:347–351.

Niro R, Byers JP, Fournier RL, Bachmann K. Application of a convective-dispersion model to predict *in vivo* hepatic clearance from *in vitro* measurements utilizing cryopreserved human hepatocytes. Curr Drug Metab 2003;4:357–369.

Obach RS, Walsky RL, Venkatakrishnan K, Gaman EA, Houston JB, Tremaine LM. The utility of *in vitro* cytochrome P450 inhibition data in the prediction of drug–drug interactions. J Pharmacol Exp Ther 2006;316:336–348.

Obach RS, Walsky RL, Venkatakrishnan K, Houston JB, Tremaine LM. *In vitro* cytochrome P450 inhibition data and the prediction of drug–drug interactions: qualitative relationships, quantitative predictions, and the rank-order approach. Clin Pharmacol Ther 2005;78:582–592.

Ogilvie BW, Zhang D, Li W, Rodrigues AD, Gipson AE, Holsapple J, Toren P, Parkinson A. Glucuronidation converts gemfibrozil to a potent, metabolism-dependent inhibitor of CYP2 C8: implications for drug–drug interactions. Drug Metab Dispos 2006;34:191–197.

Okey A. Enzyme induction in the cytochrome P450 system. In: Kalow W Editor. Pharmacogenetics of Drug Metabolism. New York: Pergamon Press; 1992. p 549–608.

Piscitelli SC, Burstein AH, Chaitt D, Alfaro RM, Falloon J. Indinavir concentrations and St John's wort. Lancet 2000;355:547–548.

Puga A, Maier A, Medvedovic M. The transcriptional signature of dioxin in human hepatoma HEPG2 cells. Biochem Pharmacol 2000;60:1129–1142.

Qatanani M, Moore DD. CAR, the continuously advancing receptor, in drug metabolism and disease. Curr Drug Metab 2005;6:329–339.

Remmel RP, Burchell B. Validation and use of cloned, expressed human drug-metabolizing enzymes in heterologous cells for analysis of drug metabolism and drug–drug interactions. Biochem Pharmacol 1993;46:559–566.

Roca B, Calvo B, Monferrer R. Severe rhabdomyolysis and cerivastatin-gemfibrozil combination therapy. Ann Pharmacother 2002;36:730–731.

Rodrigues AD. Integrated cytochrome P450 reaction phenotyping: attempting to bridge the gap between cDNA-expressed cytochromes P450 and native human liver microsomes. Biochem Pharmacol 1999;57:465–480.

Rodrigues AD, Rushmore TH. Cytochrome P450 pharmacogenetics in drug development: *in vitro* studies and clinical consequences. Curr Drug Metab 2002;3:289–309.

Rowland MT, Tozer TN. Clinical Pharmacokinetics: Concepts and Applications. Philadelphia: Lea & Febiger; 1988.

Runge-Morris M, Kocarek TA. Regulation of sulfotransferases by xenobiotic receptors. Curr Drug Metab 2005;6:299–307.

Segel I. Enzyme Kinetics: Behaviour and Analysis of Rapid Equilibrium and Steady State Enzyme Systems. New York: Wiley Interscience; 1993.

Shitara Y, Hirano M, Sato H, Sugiyama Y. Gemfibrozil and its glucuronide inhibit the organic anion transporting polypeptide 2 (oatp2/oatp1b1:Slc21a6)-mediated hepatic uptake and CYP2C8-mediated metabolism of cerivastatin: analysis of the mechanism of the clinically relevant drug–drug interaction between cerivastatin and gemfibrozil. J Pharmacol Exp Ther 2004;311:228–236.

Shou M. Kinetic analysis for multiple substrate interaction at the active site of cytochrome P450. Methods Enzymol 2002;357:261–276.

Shou M, Lin Y, Lu P, Tang C, Mei Q, Cui D, Tang W, Ngui JS, Lin CC, Singh R, Wong BK, Yergey JA, Lin JH, Pearson PG, Baillie TA, Rodrigues AD, Rushmore TH. Enzyme kinetics of cytochrome P450-mediated reactions. Curr Drug Metab 2001;2:17–36.

Silverman R. Mechanism-Based Enzyme Inactivation: Chemistry and Enzymology. Boca Raton, FL: CRC Press; 1988.

Sugimoto K, Ohmori M, Tsuruoka S, Nishiki K, Kawaguchi A, Harada K, Arakawa M, Sakamoto K, Masada M, Miyamori I, Fujimura A. Different effects of St John's wort on the pharmacokinetics of simvastatin and pravastatin. Clin Pharmacol Ther 2001;70:518–524.

Tirona RG, Kim RB. Nuclear receptors and drug disposition gene regulation. J Pharm Sci 2005;94:1169–1186.

Tran TH, Von Moltke LL, Venkatakrishnan K, Granda BW, Gibbs MA, Obach RS, Harmatz JS, Greenblatt DJ. Microsomal protein concentration modifies the apparent inhibitory potency of CYP3A inhibitors. Drug Metab Dispos 2002;30: 1441–1445.

Tucker GT, Houston JB, Huang SM. Optimizing drug development: strategies to assess drug metabolism/transporter interaction potential-toward a consensus. Clin Pharmacol Ther 2001;70:103–114.

Venkatakrishnan K, von Moltke LL, Court MH, Harmatz JS, Crespi CL, Greenblatt DJ. Comparison between cytochrome P450 (CYP) content and relative activity approaches to scaling from cDNA-expressed CYPs to human liver microsomes: ratios of accessory proteins as sources of discrepancies between the approaches. Drug Metab Dispos 2000;28:1493–1504.

Walsky RL, Obach RS. Validated assays for human cytochrome P450 activities. Drug Metab Dispos 2004;32:647–660.

Wang H, Faucette S, Moore R, Sueyoshi T, Negishi M, LeCluyse E. Human constitutive androstane receptor mediates induction of CYP2B6 gene expression by phenytoin. J Biol Chem 2004;279:29295–29301.

Wang H, LeCluyse EL. Role of orphan nuclear receptors in the regulation of drug-metabolising enzymes. Clin Pharmacokinet 2003;42:1331–1357.

Westin S, Heyman RA, Martin R. FXR, a therapeutic target for bile acid and lipid disorders. Mini Rev Med Chem 2005;5:719–727.

Wienkers LC. Problems associated with in vitro assessment of drug inhibition of CYP3A4 and other P-450 enzymes and its impact on drug discovery. J Pharmacol Toxicol Methods 2001;45:79–84.

Wienkers LC, Heath TG. Predicting in vivo drug interactions from in vitro drug discovery data. Nat Rev Drug Discov 2005;4:825–833.

Xie W, Barwick JL, Downes M, Blumberg B, Simon CM, Nelson MC, Neuschwander-Tetri BA, Brunt EM, Guzelian PS, Evans RM. Humanized xenobiotic response in mice expressing nuclear receptor SXR. Nature 2000;406:435–439.

Xie W, Evans RM. Pharmaceutical use of mouse models humanized for the xenobiotic receptor. Drug Discov Today 2002;7:509–515.

Yueh MF, Kawahara M, Raucy J. Cell-based high-throughput bioassays to assess induction and inhibition of CYP1A enzymes. Toxicol In Vitro 2005a;19:275–287.

Yueh MF, Kawahara M, Raucy J. High volume bioassays to assess CYP3A4-mediated drug interactions: induction and inhibition in a single cell line. Drug Metab Dispos 2005b;33:38–48.

Zhang ZY, Wong YN. Enzyme kinetics for clinically relevant CYP inhibition. Curr Drug Metab 2005;6:241–257.

Zhao XJ, Jones DR, Wang YH, Grimm SW, Hall SD. Reversible and irreversible inhibition of cyp3a enzymes by tamoxifen and metabolites. Xenobiotica 2002;32:863–878.

Zhou J, Zhang J, Xie W. Xenobiotic nuclear receptor-mediated regulation of udp-glucuronosyltransferases. Curr Drug Metab 2005;6:289–298.

Zhou S, Chan E, Pan SQ, Huang M, Lee EJ. Pharmacokinetic interactions of drugs with St john's wort. J Psychopharmacol 2004;18:262–276.

Zhou S, Yung Chan S, Cher Goh B, Chan E, Duan W, Huang M, McLeod HL. Mechanism-based inhibition of cytochrome p450 3a4 by therapeutic drugs. Clin Pharmacokinet 2005;44:279–304.

Zhu Z, Kim S, Chen T, Lin JH, Bell A, Bryson J, Dubaquie Y, Yan N, Yanchunas J, Xie D, Stoffel R, Sinz M, Dickinson K. Correlation of high-throughput pregnane x receptor (pxr) transactivation and binding assays. J Biomol Screen 2004;9:533–540.

6

DRUG TRANSPORTERS IN DRUG DISPOSITION, DRUG INTERACTIONS, AND DRUG RESISTANCE

CINDY Q. XIA, JOHNNY J. YANG, AND SURESH K. BALANI

6.1 INTRODUCTION

Evolution has created ways for living beings to adapt to changes in nature, and to defend themselves against threats to their existence. Only in the past century have we started to grasp the complex nature of the biology and biochemical processes occurring in human beings. Transporter proteins are involved in some of these processes. Essentially, the body cannot function well without the transporters that allow uptake of, for example, nutrients and efflux of harmful substances, like cytotoxins, from cells. The role of transporters was realized well over three decades ago (Juliano and Ling, 1976), much after the emergence of knowledge on CYPs. We are currently following in the footsteps of CYPs knowledge-based development, and steadily increasing our understanding of transporter functions, types, locations, and their roles in a qualitative fashion. Only in the past decade have we found that some CYPs work in concert with transporters. It is just a matter of time till we have means of quantitative assessment of the role of transporters in drug disposition and interactions in whole bodies. It is a welcome change to see, for the first time, considerations for transporters-based drug–drug interactions (DDI) appearing in the FDA's "draft guidance on DDI studies" in September 2006 (www.fda.gov/cder/guidelines.htm). This should stimulate further standardization of processes and eventually harmonization of ways and when to conduct DDI studies.

Drug Metabolism in Drug Design and Development, Edited by Donglu Zhang, Mingshe Zhu and W. Griffith Humphreys
Copyright © 2008 John Wiley & Sons, Inc.

This chapter provides our current understanding of transporter types, methods involved in the assessment of their function, and examples of transporter substrates and inhibitors causing changes in pharmacokinetics and/or pharmacodynamics of various drugs in animal models and/or humans. This should not be considered as an exclusive review but rather an updated understanding. References are provided for some of the past reviews in this area for extended reading.

Classically, transporters are proteins that translocate endogenous compounds (such as bile acids, lipids, sugars, amino acids, steroids, hormones, and electrolytes) and xenobiotics (such as drugs and toxins) across biological membrane to maintain cellular and physiologic solute concentrations and fluid balance as well as to provide a mechanism of detoxification for any potentially harmful foreign substances in cells. Transporter proteins are divided into the adenosine triphosphate (ATP)-binding cassette (ABC) transporter superfamily and the solute carrier (SLC) family of proteins. SLC transporters act by facilitating the uptake of their substrates into the cells. This family of transporters contains 46 subfamilies and 360 transporters including sodium-bile acid cotransporters (NTCP, SLC10 family), proton oligopeptide cotransporters (PEPT, SLC15 family), organic anion transporting polypeptides (OATP, SLC21 family), organic cation/anion/zwitterion transporters (OCT/OAT, SLC22 family), and nucleoside transporters (NT, SLC29 family). SLC transporters are divided into facilitative transporter and active transporter classes. Facilitative transporters are not coupled to any energy source and passively facilitate the diffusion of molecules across the membrane down their concentration gradients allowing a rapid equilibrium across the membrane. The active SLC transporters use an energy source that is provided by an ion-exchanger that causes pH alteration in the microenvironment of the cell surface, or is indirectly coupled to Na^+/K^+ ATPase that can create an intracellular negative membrane potential due to the imbalance in charge movement. ABC efflux membrane transporters consist of transmembrane domains (TMDs) and nucleotide binding domains (NBDs). They are directly coupled to ATPase activity and hydrolyze ATP to derive energy for pumping substrates across the cell membrane. The full efflux transporters, such as P-glycoprotein (P-gp) and multidrug resistance protein (MRP), possess two NBDs in one polypeptide chain. The half transporters, such as breast cancer resistance protein (BCRP), only contain one NBD (Borst and Elferink, 2002). The half transporters function as a dimer or tetramer bridged by specific linkages. Among 49 human genes in seven subfamilies of ABC transporters, P-gp (also known as multidrug resistance 1 (MDR1) protein) in ABCB family, MRP1 and MRP2 in ABCC family and BCRP (also known as mitoxantrone resistance protein (MXR), ABCG2, ABCP) in ABCG family are the major ABC transporters to confer resistance in the tumor cells and to efflux xenobiotics (such as drugs or toxins) out of normal tissues. Uptake (SLC family) and efflux (ABC family) transporters interact dynamically to mediate the accumulation and translocation of drugs or endogenous substrates into a cell. The gene nomenclature, protein name, tissue distribution, driven force, substrate properties and the substrates, inhibitors and inducers of the common drug-related transporters are listed in Table 6.1 (Haimeur et al., 2004; Kong et al., 2004; Xia et al., 2005b).

6.2 ROLES OF TRANSPORTERS IN DRUG DISPOSITION AND TOXICITY

Transporter proteins affect drug absorption in small intestine and drug elimination in liver and/or kidney by governing drug substance in and out of the intestinal enterocytes, hepatocytes, or renal tubular cells. Transporters can also limit or facilitate penetration of drugs into brain, placenta, tumor, T-cells, and others. The inhibition or lack of transporter functions can alter the exposure of drugs to tissues and potentially result in either lack of efficacy or increased toxicity, a classic example of this is provided by studies on antiparasitic agent avermectin that caused neurotoxicity in CF-1 mice deficient in P-gp (Lankas et al., 1997). The roles of transporters in drug disposition have been evaluated by using transporter knockout or deficient animals or by using transporter inhibitors in both animals and humans. The localizations of major transporters in human intestine, liver, kidney, and brain are illustrated in Fig. 6.1.

6.2.1 Transporters in Drug Absorption

Drug efflux transporters of the ABC family can restrict drug absorption by pumping drugs out of intestinal epithelial cells. Of the known ABC drug efflux transporters, P-glycoprotein, localized on the mucosal membrane of intestines, is well documented for its involvement in oral drug absorptions. Immunohistological studies showed high P-gp protein levels on the apical surface of columnar epithelial cells but not in crypt cells in human jejunum and colon (Thiebaut et al., 1987). The mRNA expression of P-gp increased longitudinally along the gastrointestinal (GI) tract in humans (stomach< duodenum< jejunum/ileum <colon) (Fojo et al., 1987; Thoern et al., 2005). Similar to P-gp, strong staining of BCRP was observed on the luminal surface of the intestine (Jonker et al., 2000). However, BCRP mRNA expression was maximal in human duodenum and decreased continuously down to the rectum (terminal ileum 93.7%, ascending colon 75.8%, transverse colon 66.6%, descending colon 62.8%, and sigmoid colon 50.1%, as compared to the level in duodenum) (Gutmann et al., 2005). The role of P-gp or BCRP in limiting xenobiotics absorption has been directly evident in Mdr1a/1b (−/−) or Bcrp1(-/-) mice (Chen et al., 2003a; Xia et al., 2005a). By comparing the oral drug exposure in wild type (wt) mice and knockout mice, P-gp and Bcrp1 have been shown to play major roles in the reduction of absorption such as of several HIV protease inhibitors, topotecan, etoposide, tacrolimus, paclitaxel,

TABLE 6.1 Tissue distribution and substrate properties of major ABC and SLC transporters.

Gene	Protein name	Tissue distribution	Substrate properties	Selected inhibitor and (inducer)	Driving force
ABC transporters					
ABCB1	MDR1, P-glycoprotein	Intestine, liver, kidney, brain, placenta, adrenal, testes, cancer cells	Lipophilic, amphiphilic with weak organic cation, containing hydrogen bond donor and acceptor, such as digoxin, talinolol, vinblastine, paclitaxel, fexofenadine, quinidine, loperamide, topotecan, gleevec, colchicines, daunorubicin, Calceine-AM, Rhodamine123	ritonavir, ketoconazole, cyclosporine, verapamil, erythromycin, quinidine, PSC833, GF918120, LY335979 (rifampin, St John's wort)	Intrinsic ATPase activity and ATP hydrolysis
ABCB11	BSEP, SPGP	Liver	Bile salts and paclitaxel	All major bile salts, CsA, bosentan	Intrinsic ATPase activity and ATP hydrolysis
ABCC1	MRP1	Ubiquitous (mainly in lung, kidney, brain, colon, testis, peripheral blood mononuclear cells), cancer cells	Glutathione, glucuronide and sulfate conjugates. Hydrophilic with organic anion. Substrates overlap between MRP1, MRP2, and MRP3, such as calcein, LTC$_4$, methotrexate, and vinblastine.	Probenecid, indomethacin, MK571, cyclosporine A (chlorambucil, epirubicin)	Intrinsic ATPase activity and ATP hydrolysis

ABCC2	MRP2, CMOAT	Intestine, liver, kidney, brain, placenta, cancer cells	Cisplatin is a substrate for MRP2 but not for MRP1. Folic acid and monovalent bile salts such as cholate, taurocholate, and glycocholate are substrates for MRP3 but not for MRP1 and MRP2.	Probenecid, indomethacin, MK571, cyclosporine (dexamethasone, St. John's wort)	Intrinsic ATPase activity and ATP hydrolysis
ABCC3	MRP3	Intestine, pancreas, placenta, adrenal cortex, liver, kidney, prostate, cancer cells		Benzbromarone (phenobarbital)	Intrinsic ATPase activity and ATP hydrolysis
ABCC4	MRP4	Prostate, lung, adrenals, ovary, testis, pancreas, small intestine, cancer cells	Nucleoside analogues and cyclic nucleotides (cGMP and cAMP) Unlike MRP4, MRP5 transport GS conjugates but not E17βG. MRP4: adefovir, zidovudine, monophospate MRP5: adefovir, mercaptopurine	Probenecid, slidenafil	Intrinsic ATPase activity and ATP hydrolysis

(continued)

TABLE 6.1 (*Continued*)

Gene	Protein name	Tissue distribution	Substrate properties	Selected inhibitor and (inducer)	Driving force
ABCC5	MRP5	Ubiquitous (mainly in skeletal muscle, heart and brain), cancer cells		Probenecid, slidenafil	Intrinsic ATPase activity and ATP hydrolysis
ABCG2	BCRP, MXR, ABCP	Placenta, intestine, liver, breast, brain, cancer cells	Broad substrate specificity, partly overlapping between P-gp and MRP substrates. Substrates of BCRP can be either hydrophobic or hydrophilic, negatively or positively charged, xenobiotics or endobiotics, and unconjugated or conjugated, such as estrone sulfate, lysotracker, methotrexate, sulfasalazine, topotecan and imitinib	Ko143, fumitremorgin, HIV inhibitors, novobiocin, imitinib (Gleevac), gefitinib (Iressa)	Intrinsic ATPase activity and ATP hydrolysis
Solute Carrier Transporters					
SLC10A1	NTCP	Liver	Bile salts, such as taurocholate	All major bile salts, BQ-123, CsA	Na$^+$ dependent

Gene	Protein	Tissue	Substrates	Inhibitors/Notes
SLC10A2	ASBT	Intestine, kidney	Bile salts, such as taurocholate	All major bile salts, BQ-123, CsA
SLC15A1	PEPT1	Intestine, Kidney	Dipeptides, tripeptides, and peptidomimetic drugs such as glyclysarcrosine, valacyclovir, and β-lactam antibiotics (cephalexin, ceftibuten)	Glycylsarcrosine, protonophore, such as carbonyl cyanide 4-trifluoromethoxy-phenylhydrazone (FCCP) and inhibitors of the Na^+/H^+ exchanger, such as amiloride	H^+ dependent
SLC15A2	PEPT2	Kidney		
SLC21A8/ SLCO1B3	OATP-8, OATP1B3	Liver	Digoxin, bile acids, BQ123, E17βG, DHEAS, estrone sulfate	Digoxin
SLC21A3/ SLCO1A2	OATP-A, OATP, OATP1A2	Brain, liver, intestine, testis, prostate	Relative bulky and hydrophobic organic anions (including bile acid, bilirubin, prostaglandin E_2, tetraiodothyronine, and triiodothyronine), neutral compound ouabain, and organic cations such as *N*-methyl quinidine and rocuronium.	CsA, verapamil, rifampin, ketoconazole, HIV inhibitors, Quinidine, indocyanine green	ND
SLC21A6/ SLCO1B1	OATP-C, LST-1, OATP2, OATP1B1	Liver	Estrone Sulfate, E17βG, DHEAS, bromsulfophtalein (BSP), fexofenadine, DNP-s-glutathione, LTC_4, pravastatin, rifampicin	

(*continued*)

TABLE 6.1 (Continued)

Gene	Protein name	Tissue distribution	Substrate properties	Selected inhibitor and (inducer)	Driving force
SLC21A9/ SLCO2B1	OATP-B, OATP-RP2, OATP2B1	Liver, intestine, pancreas, lung, ovary, testes, spleen			
SLC21A11/ SLCO3A1	OATP-D, OATP3A1	Ubiquitous (strong expression in leukocytes spleen), cancer cells			
SLC21A12/ SLCO4A1	OATP-E, OATP-RP1, OATP4A1	Ubiquitous (mainly in skeletal muscles), cancer cells			
SLC22A1	OCT1	Liver	Small organic cations Tetraethylmethylammonium (TEA), MPP$^+$, metformin, azidothymine (AZT), choline	Tetraethylmethylammonium, cimetidine, and HIV inhibitors	
SLC22A2	OCT2	Kidney, brain	Small organic cations		OCT2 is driven by membrane potential
SLC22A3	OCT3, EMT	Skeletal muscle, liver, placenta, kidney, heart	TEA, dopamine, guanidine		
SLC22A4	OCTN1	Kidney, skeletal muscle, prostate, placenta, heart	Small organic cations	TEA, verapamil, imipramine, nicotine, procainamide	OCTN1 is driven by H$^+$

144

Gene	Transporter	Tissue distribution	Substrates	Inhibitors/Remarks
SLC22A5	OCTN2, CT1	Kidney, skeletal muscle, prostate, lung, heart, pancreas, small intestine, liver	TEA, carnitine, quinidine, verapamil; Small organic cations; Na$^+$ dependent: carnitine, phaloridine; Na$^+$ independent: verapamil, quinidine, TEA	TEA, cimetidine, acetylcholine, choline, serotonine, quinidine, verapamil
SLC22A6	OAT1	Kidney, brain	Organic anions. Although cimetidine is an organic cation, it is a substrate that is recognized by both organic cation and anion transporters. PAH, methotrexate, estrone sulfate, prostaglandin, DHEAS	
SLC22A7	OAT2	Liver, kidney		
SLC22A8	OAT3	Kidney, brain		
SLC29A1	ENT1	Ubiquitous, cancer cells	Nucleosides and nucleoside analogues	Facilitated transport; Nitrobenzylthioinosine and dipyridamole for ENT1
SLC29A2	ENT2	Ubiquitous, cancer cells	Adenosine, uridine, inosine	
SLC28A1	CNT1	Liver, kidney, intestine, brain	Purine nucleosides, uridine	Na$^+$ free buffer; Na$^+$ dependent

(continued)

TABLE 6.1 (*Continued*)

Gene	Protein name	Tissue distribution	Substrate properties	Selected inhibitor and (inducer)	Driving force
SLC28A2	CNT2	Kidney, heart, liver, skeletal muscle, pancreas, placenta, brain, cervix, prostate, small intestine, rectum, colon, lung	Pyrimidine nucleosides, adenosine	Na$^+$ free buffer	
SLC28A3	CNT3	Mammary gland, pancreas, bone marrow, trachea, intestine, liver, lung, placenta, prostate, testis, brain, heart	Nucleosides and nucleoside analogues	Na$^+$ free buffer	

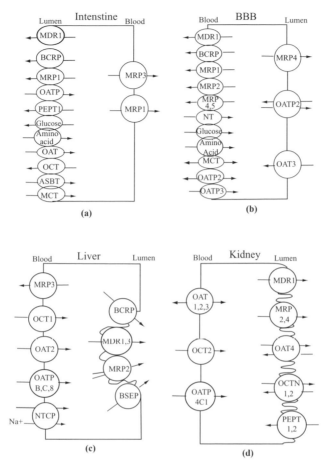

FIGURE 6.1 Localization of transporters in intestinal enterocytes, brain endothelial cells, hepatocytes and kidney tubular epithelial cells.

ivermectin, loperamide and UK-224671 (P-gp substrates (Chen et al., 2003a)), as well as topotecan, nitrofurantoin, ME-3229, GV-196771, and sulfasalazine (BCRP substrates) (Xia et al., 2005b; Zaher et al., 2006). Sparreboom et al. have used Mdr1a(-/-) mice to demonstrate the effect of gut P-gp on the pharmacokinetics of paclitaxel (Sparreboom et al., 1997). The area under the plasma concentration time curves (AUC) was two- and six-fold higher in Mdr1a(-/-) mice than in wt mice after IV and oral drug administration, respectively. Consequently, the oral bioavailability in mice receiving 10 mg paclitaxel per kg body weight increased from only 11% in wt mice to 35% in Mdr1a(-/-) mice. The cumulative fecal excretion (0–96 h) was markedly reduced from 40% (after IV administration) to 87% (after oral administration) of the administered dose in wt mice to below 3% in Mdr1a(-/-) mice. Biliary excretion was not markedly different in wt and Mdr1a(-/-) mice. After IV administration of paclitaxel (10 mg/kg) to mice with a cannulated gall bladder, 11% of the dose was recovered within 90 min in the intestinal contents of wt mice while less than 3% was recovered in Mdr1a(-/-) mice. All these data clearly suggest that P-gp limits the oral uptake of paclitaxel and mediates direct excretion of the drug from the systemic circulation into the intestinal lumen.

In general, the influence of efflux transporters on the intestinal drug absorption is significant for the substrates with either low solubility (Wu and Benet, 2005) or low permeability with high affinity to efflux transporters (Ogihara et al., 2006). For a substrate with low permeability and high affinity, efflux transporter may contribute more to the membrane clearance (V_{max}/K_m) than the passive diffusion, and thus the changes in the efflux activity may significantly alter their intestinal absorption rate. After oral administration of [^{3}H] vinblastine, a P-gp substrate with low permeability and high affinity, to mice, the maximum concentration (C_{max}) and the $AUC_{0-24\,h}$ in Mdr1a/1b (-/-) mice were 1.5 times greater than those in wild type mice, whereas these parameters were not significantly different between the two strains in the case of [^{3}H] verapamil, a P-gp substrate with high permeability and low affinity (Ogihara et al., 2006). A low solubility drug, whether it is high or low permeable, tends to have low concentrations coming into enterocytes and less chance to saturate the efflux transporters. Therefore, the efflux transporters are likely to affect the absorption rate and oral bioavailability (Weller et al., 1993).

Besides P-gp, other efflux pumps such as BCRP can affect drug absorption. Pretreatment of wt mice with gefitinib (Iressa), an oral epidermal growth factor receptor tyrosine kinase inhibitor, increased oral absorption and decreased systemic clearance of topotecan (a substrate for both P-gp and BCRP). Gefitinib inhibited the efflux of BCRP and MDR1 substrates and restored vincristine sensitivity in MDR1-expressing cells. Although gefitinib inhibited BCRP more potently than MDR1 (10-fold), the inhibition of both transporters occurred at clinically relevant concentrations (e.g., 1–5 μM) (Leggas et al., 2006).

Uptake transporter prodrug substrates have been used to improve drug absorption through GI tract. The most successful example is an antiviral prodrug valacyclovir, which shows oral bioavailability three to five times greater than its parent drug acyclovir (Weller et al., 1993). The increased oral bioavailability is attributed to PEPT1-mediated absorption, which was demonstrated by *in situ* rat perfusion model, Caco-2 cells, and PEPT1-transfected CHO cells (Balimane et al., 1998).

6.2.2 Transporters in Drug Distribution

Transporters also contribute to the drug distribution in certain tissues. Most statins are taken up into the hepatocytes by OATP, excreted into the bile by efflux transporters, and reabsorbed in the intestine, thereby effectively undergoing enterohepatic circulation and maintaining high concentrations in the liver (Shitara and Sugiyama, 2003). Metformin, a biguanide antidiabetic drug, is distributed into the liver via Oct1 and into the kidney via Oct2 (Shitara and Sugiyama, 2006).

One of the important tissue barriers for xenobiotics is blood–brain barriers (BBB). Currently, several ABC transporters, P-gp, BCRP, MRP1, and SLC transporters, amino acid, glucose, organic cation, and organic anion transporters have been identified in the BBB (Fig. 6.1b) (Loescher and Potschka, 2005). The role of the transporters in the BBB has been extensively investigated in transporter knockout mice. P-gp is highly expressed in the luminal membrane of brain endothelia cells and plays a critical role in restricting the passage of lipophilic compounds into the brain (Loescher and Potschka, 2005). Antihistamines have found their greatest therapeutic potential in the treatment and management of various allergic disorders, including seasonal and perennial rhinitis, urticaria, and dermatological conditions (Hindmarch, 2002). However, the most problematic aspect of their use is sedation, which can severely compromise the safe performance of cognitive and psychomotor tasks of everyday living. The third generation antihistamine drug fexofenadine, which is a P-gp substrate (Tahara et al., 2005) and does not cross the blood–brain barrier, has no objective evidence of sedation even at doses two to three times those normally used for seasonal allergic rhinitis (Hindmarch, 2002). Wolff et al. reported that imatinib (Gleevec, STI-571), a potent and selective tyrosine kinase inhibitor, effectively controlled the systemic proliferation of transduced bone marrow cells in mice, but many of the mice unexpectedly developed progressive neurological deficit due to leukemia infiltration of the brain and leptomeninges after 2–4 months although imatinib has shown an effective inhibition on glioblastoma cell growth in preclinical *in vitro* and *in vivo* studies (Kilic et al., 2000), which suggested that there was inadequate imatinib penetration of the drug into the central nervous system (CNS) (Wolff et al., 2003). Imatinib concentrations in mouse cerebrospinal fluid (CSF) were less than 1% than that in plasma (Wolff et al., 2003). A limited penetration of imatinib into the brain was also observed both preclinically and clinically (Pfeifer et al. 2003; Neville et al. 2004; Leis et al., 2004). Both Mdr1(-/-) and Bcrp1(-/-) knockout mice demonstrated that P-gp and Bcrp could limit imatinib brain penetration (Breedveld et al., 2005). Thus, BCRP and P-gp inhibitors may improve delivery of imatinib to malignant gliomas.

P-gp, BCRP, and MRPs are present in placenta and contribute to the maternal–fetal barrier. Their role in restricting placenta penetration of drugs can be demonstrated by using efflux transporter inhibitors and pregnant transporter genetic knockout mice. With the pretreatment of GF120918 (an inhibitor of both P-gp and BCRP) 2 h before intravenous administration of topotecan to pregnant Mdr1a/1b(-/-) mice, the plasma levels of topotecan were 3.2- and 1.6-fold higher in fetuses and dams, respectively. These data suggest that Bcrp1 plays an important role in protecting fetuses from potentially toxic xenobiotics (Jonker et al., 2000).

6.2.3 Transporters in Drug Metabolism

It has been recognized that there is a huge overlap of the substrate specificity and tissue distributions between P-gp and cytochrome P450 (CYP) 3A. Most of the CYP3A4 inhibitors also inhibit P-gp (Benet et al., 2003). Inducers of CYP3A4, such as St. John's Worts (SJW), reserpine, rifampicin, phenobarbital, and triacetyloleandomycin, can also induce P-gp expression in cells and in man (Benet et al., 2003). The spatial relationship of P-gp traversing the plasma membrane and CYP locating on the endoplasmic reticulum inside the cells suggests that P-gp may cooperatively influence metabolism by controlling the exposure of a substrate to CYP enzymes. Drug is absorbed into enterocytes, hepatocytes, and renal tubular cells where it can be metabolized by CYP3A and subjected to be pumped out of the cells by P-gp as well. This process allows CYP3A to have repeated access to the drug molecule, resulting in low bioavailability of CYP3A substrates, less accumulation of parent compound inside the cells, and more metabolite formation (Benet et al., 2003). By using CYP3A4-expressing Caco-2 cell monolayers, the P-gp inhibitor LY335979 (zosuquidar trihydrochloride) (0.5 mM, apical), compared with vehicle, increased saquinavir cell content and its metabolite M7 formation rate, but decreased the intestinal first-pass extraction (E_i) by approximately 50% (Mouly et al., 2004).

Efflux pumps also help to eliminate the metabolites of drugs from systemic circulation. For example, most drug glucuronide-conjugates are MRP2 and/or BCRP substrates (Adachi et al., 2005) while most sulfate conjugates are BCRP substrates (Adachi et al., 2005; Zamek-Gliszczynski et al., 2005).

6.2.4 Transporters in Drug Excretion

Transporters are involved in biliary and renal excretions that are the two common routes of drug elimination. In the liver, a drug is first taken up into hepatocytes, then either secreted back to the systemic circulation or excreted to the bile in an intact form or as metabolites via Phase I and/or Phase II enzymes. Given the involvement of transporters in both uptake at the sinusoidal membrane and efflux at the sinusoidal and canulicular membranes (Fig. 6.1c), the hepatic clearance can be expressed based on well-stirred model as the following equation (Shitara et al., 2005; Yamazaki et al., 1996):

$$CL_{H,int} = \frac{PS_{influx}(CL_{H,met} + CL_{Biliary})}{PS_{efflux} + (CL_{H,met} + CL_{Biliary})} \tag{6.1}$$

where $CL_{H, int}$, $CL_{H, met}$, and $CL_{Biliary}$ represent the overall hepatic intrinsic clearance, hepatic metabolism clearance, and biliary excretion, respectively. PS_{influx} and PS_{efflux} are the membrane transport clearance across the sinusoidal membrane. When PS_{efflux} is much smaller than $CL_{H, met}$ and $CL_{Biliary}$ ($PS_{efflux} \ll CL_{H, met} + CL_{Biliary}$), the overall hepatic clearance will be equal to PS_{influx}. At steady state the hepatic intrinsic clearance of pravastatin, a substrate for OATP2 and MRP2 (Tokui et al., 1999; Yamazaki et al., 1997), was regulated by the uptake process, followed by rapid metabolism and/or biliary excretion with minimal efflux to the systemic circulation in rats after infusion. The total hepatic elimination rate at steady state exhibited Michaelis–Menton saturation with the drug concentration and the K_m and V_{max} obtained in rats with different mathematical models (i.e., well stirred, parallel tube, and dispersion models) were comparable with the initial uptake velocity measured from *in vitro* hepatocytes (Tokui et al., 1999).

Uptake transporters, such as OCT1, OAT2, OATPB, OATP2, OATP8, and NTCP, in the sinusoidal membrane of hepatocytes affect PS_{influx}. Efflux transporters affect PS_{efflux} in the sinusoidal membrane (e.g., MRP3) and $CL_{biliary}$ in the canulicular membranes (e.g., P-gp, BCRP, and MRP2). After IV administration of nitrofurantoin (a BCRP substrate), AUC in Bcrp1(-/-) mice was almost two-fold higher than in wild type mice (139.2 min* µg/mL versus 73.9 min* µg/mL). Hepatobiliary excretion of nitrofurantoin was almost abolished in Bcrp1(-/-) mice (9.6% versus 0.2% in wild type and Bcrp1 knockout mice, respectively) (Merino et al., 2005). The impact of P-gp on biliary excretion of its substrate was first indicated by canalicular membrane vesicle studies and isolated perfused rat liver. Mdr1a(-/-) or Mdr1a/1b(-/-) mice provide more direct evidence to demonstrate the involvement of P-gp in biliary elimination of its substrates such as digoxin, doxorubicin, and vinblastine. Mrp2 appears to have a less profound impact on the intestinal absorption of its substrates than on their biliary excretion. In order to assess Mrp2-mediated biliary excretion and oral absorption respectively, wild type Sprague–Dawley (SD) rats and Eisai hyperbilirubinemic Sprague–Dawley rats (EHBR) received an IV infusion or an oral dose of furosemide, probenecid, or methotrexate (MTX) (Chen et al., 2003b). The biliary clearance of probenecid and MTX was reduced approximately 40-fold in EHBR rats as compared to control rats. Biliary clearance of furosemide was similar in EHBR and control rats. In all cases, no significant difference in absorption was observed between EHBR and control rats. This study demonstrated that Mrp2 mediates the biliary excretion of probenecid and MTX but not furosemide.

Similar to hepatocytes, a drug needs to be taken up through the basolateral membrane of renal epithial cells before the drug is excreted into urine. Metabolism may also occur in the kidney. Efflux transporters on the luminal brush-border membrane can pump an intact drug or its metabolites into the urine (Fig. 6.1d). Renal excretion of drugs can be described by three processes: glomerular filtration, renal tubular secretion (basolateral uptake transporters and apical efflux transporters are involved in this process), and reabsorption from the renal tubular lumen (apical uptake transporters are involved in this process). Generally, renal clearance can be expressed by the following equation:

$$CL_R = (1 - FR)(f_u GFR + CL_{sec})$$ (6.2)

where FR, f_u, GFR, and CL_{sec} are the reabsorbed fraction, protein unbound fraction in the blood, glomerular filtration rate and secretion rate, respectively. GFR is a passive process by which only unbound drugs can be filtered, whereas reabsorption and secretion often involve active transporters. Technically, it is difficult to quantify each process of renal excretion. However, the excretion ratio (ER, which is the $CL_R/(f_uGFR)$ ratio) reflects the relative contribution of each process to the overall renal excretion. If the ER of the drug is greater than unity, the tubular secretion is more dominant. In contrast, when the ER is less than unity, tubular reabsorption is more significant.

Uptake transporter knockout mice, such as Oat(-/-), Oct(-/-), and Pept2(-/-) mice, have been available to evaluate the role of these transporters in renal clearance of selected substrates. In the kidney, basolaterally localized OAT1-3 and OCT1-2 (Fig. 6.1d) are important for renal tubular secretion. OAT4 and PEPT1-2 localized in the brush-border membrane are mainly responsible for renal reabsorption. After IV administration of [^{14}C]glycylsarcosine (GlySar) (0.05 μmol/g of body weight) to wild type and Pept2(-/-) mice, both total and renal clearance of GlySar increased two fold in Pept2(-/-) mice, resulting in concomitantly lower systemic concentrations compared to wild type mice (Ocheltree et al., 2005). In addition, the ER of GlySar was 0.54 in wild type versus 0.94 in Pept2(-/-) mice, suggesting that in Pept2(-/-) mice the renal reabsorption of GlySar was almost abolished and GlySar was mainly eliminated by glomerular filtration. By combinatorial usage of wild type and Pept2 (-/-) mice the relative contribution of Pept1 and Pept2 on the kidney reabsorption of GlySar was assessed using Equation 6.2. Of the 46% of GlySar that was reabsorbed in wild type mice, Pept2 accounted for 86% and Pept1 accounted for 14% of the reabsorption.

P-gp is localized at the apical brush-border membrane of the proximal renal tubule in the kidney that implies its function in renal secretion. Hori et al. (1993) demonstrated that digoxin was actively secreted in the isolated perfused rat kidney with the ER of 2.5. P-gp inhibitors, quinidine, and verapamil decreased the ER of digoxin to unity, suggesting that digoxin is actively secreted into urine by P-gp. The function of Bcrp1 expressed on the brush-border membrane of proximal tubule cells of the kidneys in the renal secretion was first demonstrated by Mizuno et al. The renal clearance of E3040S (6-hydroxy-5,7-dimethyl-2-methylamino-4-(3-pyridylmethyl)benzothiazole sulfate) was 2.4-fold lower in Bcrp1(-/-) mice compared to that in wild type mice (Mizuno et al., 2004). However, BCRP expression is negligible in the human kidney (Hori et al., 1993). Given that species differences in renal localization of BCRP could lead to different routes of elimination of BCRP substrates, one must take the interspecies differences into consideration when predicting human PK and route of elimination using preclinical PK data.

BCRP plays an important role in transporting drug into human milk. BCRP expression is strongly induced in lactating, but not in virgin or nonlactating, mammary gland epithelia of mice, cows, and humans (Jonker et al., 2005). Nitrofurantoin, a commonly used urinary tract antibiotic prescribed to lactating woman was reported to be secreted into human milk (Merino et al., 2005). The underlying mechanism was demonstrated in mice where milk-to-plasma ratio of nitrofurantoin in wild type was 80 times higher compared to that in Bcrp1 knockout mice. Elimination of nitrofurantoin via milk, however, was somewhat higher than its hepatobiliary elimination (7.5–15% versus 9% of the dose 1 h after IV injection) in lactating mice. Most of other BCRP substrates such as PhIP, topotecan, acyclovir, cimetidine, doxorubicin, and doxorubicinol but not folic acid, DHEAS, or porphyrin vitamin B12 concentrated in the milk and have high milk/plasma ratios (Jonker et al., 2005). The identification of the BCRP function in mammary gland has provided insight into the proper usage of BCRP substrate drugs in lactating woman.

6.2.5 Transporters in Toxicity

Generation of bile flow is regulated by ATP-dependent process and depends on the coordinated action of a number of transporter proteins in the sinusoidal and canalicular domains of the hepatocyte. Dysfunction or inhibition of any of these proteins can lead to retention of substrates or their metabolic products, with hyperbilirubinemia or cholestasis as a result. Bilirubin, the end product of heme catabolism, is taken up from blood into hepatocytes by passive diffusion and the sinusoidal membrane transporter, OATP2, conjugated by UGT in the hepatocyte and then excreted into the bile through the bile canallicular membrane transporter, MRP2, mainly as bilirubin glucuronides (Cui et al., 2001; Muller and Jansen, 1997, Kamisako et al., 2000). Mrp2 deficient rats, GY/TR⁻ (transporter deficient Wistar rats) and EHBR, suffer hyperbilirubinemia and are good models for Dubin–Johnson syndrome, a human disease which is characterized by hyperbilirubinemia. This syndrome occurs in human with a hereditary MRP2 deficiency. Since bilirubin–glucuronides are endogenous substrates of MRP2 and excreted from the liver into bile by MRP2 (Gottesman and Ambudkar, 2001), inhibition of MRP2-regulated transport of bilirubin-glucuronides into the bile can potentially cause hyperbilirubinemia. Genetical analysis (Ieiri et al., 2004; Huang et al., 2004) and OATP-transfected cell studies (Campbell et al., 2004) also indicated that lack of OATP or inhibition of OATP could cause hyperbilirubinemia by the increase of bilirubin retention.

Pure cholestasis without hepatitis is observed most frequently with contraceptives and 17α-alkylated androgenic steroids, and the mechanism most likely involves interference with hepatocyte canalicular efflux systems for bile salts, organic anions, and phospholipids. The rate-limiting step in bile formation is considered to be the bile salt export pump (BSEP)-mediated translocation of bile salts across the canalicular hepatocyte membrane. Inhibition of BSEP function by metabolites of cyclosporine A, troglitazone, bosentan, rifampicin, and sex steroids is an important cause of drug induced cholestasis (Kullak-Ublick, 2004).

Many antiviral drugs (e.g., fialuridine; FIAU) produce clinically significant mitochondrial toxicity that limits their dosing or prevents their use in the clinic. Human ENT1 (hENT1) is expressed on the mitochondrial membrane and this expression may enhance the mitochondrial toxicity of nucleoside drugs such as FIAU (Lai et al., 2004). However, the lethal mitochondrial toxicity of fialuridine observed in the clinic was not predicted from preclinical toxicity studies in rodents (rats and mice), even at doses that were 1000-fold greater than that used in the human study. In fact, human ENT1 but not mouse Ent1 was expressed in the mitochondrial membrane, indicating that fialuridine can get into human but not mouse mitochondria via ENT1. This observation has been confirmed by hepatocyte studies. The mitochondrial uptake of fialuridine was higher in human hepatocytes than that in mouse hepatocytes, and this uptake could be reduced by an ENT inhibitor in human hepatocytes but not in mouse hepatocytes (Lee et al., 2006). Species difference in transporters may influence preclinical toxic species selection.

Adefovir and cidofovir are clinically important antiviral agents and have been shown to cause drug-associated nephrotoxicity in some patients. *In vitro* studies demonstrated that adefovir and cidofovir were about 500 and 400-fold more cytotoxic, respectively, in OAT1-transfected CHO cells compared to the vector control transfected CHO cells, suggesting that the drug associated nephrotoxicity could be caused by OAT1-mediated accumulation of adefovir and cidofovir in renal proximal tubules (Ho et al., 2000).

6.3   TRANSPORTERS IN DRUG RESISTANCE

To date, three major mechanisms of drug resistance in cells have been identified by using cellular and molecular biology techniques: first, decreased uptake of water-soluble drugs such as folate antagonists, nucleoside analogues, and cisplatin, which require transporters to enter cells; second, various changes in cells that affect the capacity of cytotoxic drugs to kill cells, including alterations in cell cycle, increased repair of DNA damage, reduced apoptosis and altered metabolism of drugs; and third, increased energy-dependent efflux of hydrophobic drugs that can easily enter the cells by diffusion through the plasma membrane (Szakacs et al., 2006). Among these mechanisms, both uptake (the first one) and efflux (the third one) processes are involved in transporters. Drug resistance due to MDR1/P-gp has not been overcome yet in the oncology arena to prevent remission of cancer. Overexpression of P-gp in tumors can confer two orders of magnitude of resistance for drugs that are P-gp substrates (Beketic-Oreskovic et al., 1995; Szakacs et al., 2006). Combination therapy with transporters inhibitors, like verapamil, PSC 833, GF120918 and cyclosporine, has offered some help against some refractory tumors (Beketic-Oreskovic et al., 1995; Szakacs et al., 2006). Such inhibitors in *in vitro* cell lines have markedly sensitized them to chemoagents. However, the clinical benefit from those P-gp modulators is still a question (Szakacs et al., 2006).

CsA, PSC833, and quinine showed some overall survival benefit in several anticancer drug treatments in P-gp positive patients with poor risk acute myelogenous leukemia (AML), untreated AML, and high myelodysplastic syndrome (MDS) respectively, but had no effect on the same anticancer drugs on the cancer type from different trial group (Table 6.2). All these controversial results could be partially due to the limitations of MDR inhibitors (such as potency, low specificity, potential toxicity, and nonoptimal PK profiles) and the inadequate clinical trial designs (Szakacs et al., 2006). Other efflux transporters like BCRP and MRP1/2 (Borst et al., 2000; Doyle et al., 1998) also have to be taken into account for development of drug resistance, and new chemoagents are needed for that. Except in the oncology area, efflux pumps have been also shown to confer resistance for the drugs targeting to the central nervous system (such as epilepsy) (Kwan and Brodie, 2005), central infections (such as HIV) (Kwan and Brodie, 2005), and T-cells (such as inflammation diseases) (Oerlemans et al., 2006; van der Heijden et al., 2004a, 2004b).

Other strategies for reversing the MDR are also being considered, some of these involve using coadministration of antisense oligonucleotides, hammerhead ribozymes and short-interfering RNA (iRNA) to suppress P-gp expression (Pichler et al., 2005; Xu et al., 2004); antagonism of xenobiotic nuclear receptor SXR involved in the induction of P-gp and CYP3A4 (Forman et al., 2002; Synold et al., 2001); and bolstering the P-gp expression in bone marrow stem cells, which are more prone than other cells to anticancer agents' toxicity and hence limiting their doses, by transfection with MDR1 cDNA and making these stem cells more resistant to chemoagents, thereby allowing higher doses of the drugs to be used for longer periods of time resulting in increased efficacy of the treatment (Gottesman et al., 1999). Although the transcriptional repression of MDR is promising and attractive strategy, it is still a challenging task to safely deliver the gene regulators to the cancer cells *in vivo* (Pichler et al., 2005; Szakacs et al., 2006; Xu et al., 2004). In contrast to normal stem cells, tumorigenic stem cells with high expression of drug transporters can also lead to drug resistance. In chemotherapy, normal cells are killed, but the tumor stem cells survive and proliferate, leading to recurring tumor composed of tumor stem cells and cells of variable, committed lineage (Dean et al., 2005). Mutations in the tumor stem cells and their descendents can further confer drug resistance phenotype, and resulting in tumor growth.

For some hydrophilic drug molecules, the rates of passive diffusion through cell membrane are low and transporter-mediated uptake is the major route for the drug getting into the target cells. *In vitro* cellular-based assays have demonstrated that inefficient cellular uptake is a potential mechanism of resistance to anticancer drugs such as the nucleoside drugs (Hoffman, 1991; Mackey et al., 1998a), in particular: cytarabine (Wiley et al., 1982, 1985), fludarabine (Gati et al., 1998), cladribine (Gati et al., 1998; Wright et al., 2002), 5-fluoro-2'-deoxyuridine (FdUrd) (Sobrero et al., 1985a, 1985b), 5-fluorouracil

TABLE 6.2 Outcome of some of the Phase III clinical trials with ABC transporter inhibitors.

Year	Trial group	Number of patients	Cancer type	Modulator	Anticancer drugs	Outcome	Reference
1995	MRC	235	Relapsed and refractory AML	Cyclosporin	Daunorubicin Cytarabine Etoposide	No benefit	Yin et al. (2001)
1998	SWOG	226	Poor-risk AML, RAEB-t	Cyclosporin	Daunorubicin Cytarabine	Improved OS in P-gp positive patients	List et al. (2001)
1996	Novartis	256	AML	PSC833	Daunorubicin Cytarabine Etoposide	No benefit	Sonneveld et al. (2001)
2000	CALGB	410	Untreated AML	PSC-833	Daunorubicin Etoposide Cytarabine	No OS advantage for those >45 years; OS benefit for those <45 years	Baer et al. (2002)
1996		315	Poor-risk acute leukemia	Quinine	Mitoxantrone Cytarabine	No benefit	Solary et al. (1996)
1996	GFM	131	High risk MDS	Quinine	Mitoxantrone Cytarabine	Improved OS in P-gp positive patients	Wattel et al. (1999)
1999	GEO-LAMS	425	De novo AML	Quinine	Idarubicin Cytarabine Mitoxantrone	Significant improvement in the CR rate in P-gp positive patients. No OS advantage	Solary et al. (2003)

AML, acute myelogenous leukemia; CR, complete response; MDS, myelodysplastic syndrome; OS, overall survival; RAEB-t, refractory anemia with excess of blasts in transformation; RR, response rate.

(Grem, 1992), gemcitabine (Mackey et al., 1998b), 6-mercaptopurine (Fotoohi et al., 2006), and 6-thiolguanine (Fotoohi et al., 2006). The NT-deficient murine T-cell lymphoma cells AE1 (Cohen et al., 1979) exhibit greatly reduced uptake of physiological nucleosides and high level resistance to cytotoxic nucleosides (Cohen et al., 1979). In FdUrd-resistant human HCT-8 colon cancer cells, there was no measurable uptake of FdUrd and no detectable ENT functions, resulting in 700-fold resistance to the cytotoxicity of FdUrd compared to naïve HCT-8 cell (Mackey et al., 1998b). However, the role of uptake transporter deficiency in clinical drug resistance is less clear partially due to the difficulties of performing transport studies on the malignant cells derived from clinical specimens and of quantifying transporter abundance in malignant clone mixed with normal cells. The efficiency of cytarabine uptake by leukemic blast cells has been related to clinical outcome in AML and ALL patients who received a standard dose of cytarabine (100–200 mg/m^2/day). The sensitivity to cytarabine therapy was highly correlated to nucleoside transporter content and a deficiency in NT may impart resistance to cytarabine (Wiley et al., 1982; Wright et al., 2002).

6.4 POLYMORPHISM OF TRANSPORTERS AND INTERINDIVIDUAL VARIATION

It has been estimated that genetics can account for 20–95% of variability in drug disposition and pharmacological effects although many other factors, such as age, organ function, concomitant therapy, drug-drug interactions, and the nature of disease, influence drug response (Kerb, 2006). Polymorphic genetic variations have been reported in human MDR1 (P-gp), MRP1, MRP2, BCRP, OATP (OATP1B1, OATP1B3, and OATP2B1), OAT1, OCT1, and OCT2 (Beketic-Oreskovic et al., 1995; Ho and Kim, 2005; Kerb, 2006 Sparreboom et al., 2003). Generally, the role of single nucleotide polymorphisms (SNPs) in drug disposition is confirmed *in vitro* by measuring the efflux or uptake activities of specific substrates in the cells or membranes expressing recombinant protein, and *in vivo* by measuring the expression of mRNA or protein in tissue samples, by assessing the intracellular accumulation of substrates, by evaluating the pharmacokinetic alterations of drug substrates, or by associating clinical outcome from drug substrates (Kerb, 2006).

Till date, genetic polymorphisms of human MDR1 has been extensively investigated. There are 29 SNPs and 13 of major haplotypes of MDR1 that have been identified. A lot of attentions have been focused on the silent mutation (not associated with any amino acid change) C3435T in exon 26 and nonsynonymous variants, G2677T (Ala893Ser) and G2677A (Ala893Thr). The SNPs of MDR1 was demonstrated to be associated to the changes in P-gp expression and function, and subsequent alteration in drug disposition. However, much of the clinical data has been contradictory or inconclusive (Sparreboom et al., 2003). The interindividual expression of P-gp varies manifold (two to eight fold) and seems to be related to MDR1 genotype (Lindell et al., 2003). The interindividual variation of P-gp in intestinal expression resulted in sevenfold range in the oral bioavailability of P-gp substrate digoxin (Greiner et al., 1999b) and 30% variability of cyclosporine A plasma concentration (Lown et al., 1997). The different P-gp expression causes much intersubject variability than intrasubject variability. For example, the intra and intersubject coefficients of variation for AUC_{0-inf} of orally administered P-gp substrate tanilolol were 14.0% and 20.4–29.5%, respectively (Siegmund et al., 2003).

SNPs in drug uptake transporters such as OATP1B1 (OATP2) have been shown to influence the exposure of HMG-CoA inhibitors and fexofenadine in healthy volunteers (Niemi et al., 2004, 2005, 2006b). Rifampicin, a potent inducer of CYP3A4 and a substrate of OATP2, has been demonstrated *in vitro* to show that OATP variants affect the uptake of rifampicin. However, a clinical study in 38 healthy volunteers suggested that OATP2 polymorphism did not affect the extent of induction of hepatic CYP3A4 by rifampicin (Niemi et al., 2006a).

The observed inconsistencies between the various drugs tested can be explained by various potentially confounding factors (Sparreboom et al., 2003). Numerous environmental factors affecting the phenotypical activity of drug transporters must be considered, which may include exogenous chemicals, food constituents, herbal preparations, and/or therapeutic drug use that may induce or inhibit the function or expression of the protein. Thus, these nongenetic factors might hide the potential genetic effects. Another explanation for the discrepancies is related to route of drug administration and drug-specific differences in metabolism and excretion for various substrates of drug transporters. For instance CsA is a substrate for P-gp and CYP3A4, but digoxin is only a substrate for P-gp. Rifampicin is not only a substrate for OATP2 but also for OATP8, which may compensate for reduced uptake of rifampicin by OATP2 variants (Niemi et al., 2006a). Finally, distribution of other unidentified variation(s) in the same gene and/or other genes relevant to drug disposition might be different among the different human populations studied.

6.5 TRANSPORTERS IN DRUG–DRUG OR DRUG–FOOD INTERACTIONS

In humans, the role of transporters in drug absorption has been indirectly shown by inhibition or induction studies. Transporter-related drug-drug interactions can occur during gastrointestinal absorption, hepatic excretion, renal excretion, blood–brain barrier penetration, and others due to the wide tissue distribution of transporters. Food and formulation effects on P-gp-mediated drug absorption process are also well studied. Most important P-gp-mediated drug-drug interactions observed in clinic are listed in Table 6.3.

TABLE 6.3 P-gp-mediated drug–drug interactions observed in clinic.

Substrate	Inhibitor	Inducer	Result	Possible interacting protein	Reference
Amitriptyline	—	St. John's wort	Decrease in AUC (-22%)	P-gp, CYP3A4	Johne et al. (2002)
Amprenavir	Saquinavir	—	Increase in AUC ($+32\%$)	P-gp, CYP3A4	Sadler et al. (2001)
Atorvastatin	Erythromycin	—	Increase in AUC (1.32-fold) and C_{max} (1.38-fold). No change in $t_{1/2}$	P-gp, CYP3A4	Siedlik et al. (1999)
Atorvastatin	Cyclosporin A	—	Increase in AUC (7.4-fold)	P-gp, CYP3A4	Asberg et al. (2000)
Azithromycin	Nelfinavir	—	Increase in AUC (2.06-fold) and C_{max} (2.0-fold)	P-gp	Amsden et al. (2000)
Cerivastatin	Cyclosporin A	—	Increase in AUC (3.8-fold)	OATPs	Muck et al. (1999)
Cyclosporin A	Grapefruit juice	—	Increase in AUC (1.5-fold)	P-gp and CYP3A4	Edwards et al. (1999)
Cyclosporin A	—	St. John's wort	Decrease in whole-blood trough concentration (mean dose-normalized trough concentration: $0.84 \rightarrow 0.48$ ng/ (mL $\times$ mg))	P-gp, CYP3A4	Mai et al. (2000)

(continued)

TABLE 6.3 *(Continued)*

Substrate	Inhibitor	Inducer	Result	Possible interacting protein	Reference
Cyclosporin A	Erythromycin	—	Increase in AUC (2.15-fold), increase in bioavailability (36% → 60%), decrease in total clearance (−13%)	P-gp, CYP3A4	Gupta et al. (1989)
Cyclosporine	Vitamin E	—	Increase in AUC (1.61-fold) and C_{max} (1.37-fold), no change in the ratio of AUC of metabolites to that of parent drugs	P-gp	Chang et al. (1996)
Dexamethasone	Erythromycin	—	Increase in AUC (1.24-fold)	P-gp	Kovarik et al. (1998)
Digoxin	Atorvastatin	—	Increase in AUC (1.15-fold) and C_{max} (1.2-fold). No change in renal clearance	P-gp	Boyd et al. (2000)
Digoxin	—	Rifampin	Decrease in AUC (−30%, po), decrease in bioavailability (−30%), Increase in P-gp expression level in duodenum (3.5-fold), no change in $t_{1/2}$, no change in renal clearance	P-gp	Greiner et al. (1999b)

Drug	Interacting agent	Effect	Mechanism	Reference
Digoxin	St. John's wort	Decrease in AUC (0.75-fold, Day 15) and C_{max} (0.74-fold)	P-gp	Johne et al. (1999)
Digoxin	PSC-833 (Valspodar)	Increase in AUC (+76%), decrease in renal clearance (−62%)	P-gp	Kovarik et al. (1999)
Digoxin	Quinidine	Increase in C_{max} (2-fold), Decrease in renal clearance (−50%)	P-gp	Yu (1999)
Digoxin	Talinolol	Increase in AUC (1.23-fold) and C_{max} (1.45), no change in renal clearance, no change in $t_{1/2}$	P-gp	Westphal et al. (2000a)
Digoxin	Grapefruit juice	Increase in AUC (1.1-fold), no change in C_{max}	P-gp, CYP3A4	Becquemont et al. (2001)
Digoxin	Verapamil	Increase in C_{max} (1.44-fold), decrease in biliary clearance (−43%), no change in renal clearance	P-gp	Hedman et al. (1991)

(continued)

TABLE 6.3 (*Continued*)

Substrate	Inhibitor	Inducer	Result	Possible interacting protein	Reference
Digoxin	Quinine	—	Increase in C_{max} (1.1-fold), decrease in biliary clearance (−34%), no change in renal clearance, no change in plasma concentration	P-gp	Hedman et al. (1990)
Digoxin	Quinidine	—	Increase in C_{max} (1.54-fold), decrease in biliary clearance (−45%) and renal clearance (−29%), increase in plasma concentration (+55%)	P-gp	Hedman et al. (1990)
Digoxin	Amiodarone	—	Increase in plasma level (C_{max}: 1.55–2.85 μg/L, AUC: 7.2–12.1 μg/L/h)	P-gp	Robinson et al. (1989)
Digoxin	Erythromycin	—	Increase in C_{max} (2.04-fold)	P-gp	Maxwell et al. (1989)
Digoxin	Propafenone	—	Decrease in total clearance (−31%), decrease in renal clearance (−32%)	P-gp	Calvo et al. (1989)
Digoxin	Diltiazem	—	The mean trough increased 1.38-fold.	P-gp	Andrejak et al. (1987)

Digoxin	Nifedipine	—	Increase in serum concentration (+15%)	P-gp	Kleinbloesem et al. (1985)
Digoxin	Tiapamil	—	Increase in plasma concentration (1.6-fold: 0.5–1.5 → 0.9–1.9 ng/mL)	P-gp	Lessem and Bellinetto (1983)
Digoxin	Spironolactone	—	Decrease in total clearance (−31%) and renal clearance (−18%)	P-gp	Fenster et al. (1984)
Digoxin	Itraconazole	—	Increase in AUC (+68%), decrease in renal clearance (−20%)	P-gp	Jalava et al. (1997)
Digoxin	Clarithromycin	—	Decrease in renal clearance (−48%)	P-gp	Wakasugi et al. (1998)
Digoxin	PSC-833 (Valspodar)	—	Increase in AUC (+76%), decrease in renal clearance (−62%)	P-gp	Kovarik et al. (1999)
Digoxin	Propafenone	—	Decrease in total clearance (−31%), decrease in renal clearance (−32%)	P-gp	Calvo et al. (1989)
Diltiazem	Grapefruit juice	—	Increase in AUC (+20%)	P-gp, CYP3A4	Christensen et al. (1989)

(continued)

163

TABLE 6.3 (*Continued*)

Substrate	Inhibitor	Inducer	Result	Possible interacting protein	Reference
Docetaxel	Cyclosporin A	—	Increase in AUC (7.3-fold), increase in bioavailability (8% to 90%)	P-gp, CYP3A4	Malingre et al. (2001b)
Docetaxel	R101933	—	Decrease in fecal excretion (8.47%–0.45%)	P-gp	van Zuylen et al. (2000)
Doxorubicin	GF120918	—	No change in AUC and renal clearance		Sparreboom et al. (1999)
Doxorubicin	Cyclosporine	—	Increase in AUC (1.8-fold), decrease in total clearance (−37%), no change in $t_{1/2}$, decrease in the ratio of AUC of metabolites to that of parent drugs (−75%)	P-gp, CYP3A4	Rushing et al. (1994)
Doxorubicin (Adriamycin)	Verapamil	—	Increase in AUC (2.0-fold)	P-gp, CYP3A4	Kerr et al. (1986)
Doxorubicin (Adriamycin)	Cremophor	—	Increase in AUC (+23%)	P-gp	Millward et al. (1998)
Etoposide	Cyclosporin A	—	Decrease in total clearance (−46%), renal clearance (−38%) and nonrenal clearance (−52%)	P-gp, CYP3A4	Lum et al. (1992)

Drug			Effect	Transporter/Enzyme	Reference
Etoposide	PSC-833 (Valspodar)	—	Decrease in total clearance (−41%), renal clearance (−32%) and nonrenal clearance (−48%)	P-gp, CYP3A4	Boote et al. (1996)
Erythromycin	—	Rifampin	Increase in erythromycin breath test (ERMBT value): (median ERMBT iv: twofold, oral: unchanged)	P-gp, CYP3A4	Paine et al. (2002)
Fexofenadine	Erythromycin	—	Increase in AUC (2.09-fold) and C_{max} (1.82-fold)	P-gp	Davit et al. (1999)
Fexofenadine	Erythromycin	—	Increase in AUC (+109%)	P-gp	Simpson and Jarvis (2000)
Fexofenadine	Azithromycin	—	Increase in AUC (+67%)	P-gp	Gupta et al. (2001)
Fexofenadine	—	Rifampin	Increase in oral clearance (1.9-fold) and C_{max} (0.59-fold), no change in renal clearance and $t_{1/2}$	P-gp	Hamman et al. (2001)
Fexofenadine	Azithromycin	—	Increase in AUC (+67%)	P-gp	Gupta et al. (2001)
Fexofenadine	—	St. John's wort	Decrease in AUC (−14%), after long-term administration	P-gp	Wang et al. (2002)
Fexofenadine	Ketoconazole	—	Increase in AUC (2.64-fold) and C_{max} (2.35-fold)	P-gp	Davit et al. (1999)

(continued)

TABLE 6.3 (*Continued*)

Substrate	Inhibitor	Inducer	Result	Possible interacting protein	Reference
Fluvastatin	Cyclosporin A	—	Increase in AUC (1.9-fold)		Goldberg and Roth (1996)
Glipizide	—	Rifampin	Decrease in AUC (-22%) and C_{max} ($+18\%$), Decrease in $t_{1/2}$ (-35%)	P-gp, CYP2C9	Niemi et al. (2001)
Glyburide	—	Rifampin	Decrease in AUC (-39%) and C_{max} (-22%), Decrease in $t_{1/2}$ (-17%)	P-gp, CYP2C9	Niemi et al. (2001)
Indinavir	—	St. John's wort	Decrease in AUC (-57%) and C_{max} ($12.3 \rightarrow 8.9$), no change in $t_{1/2}$	P-gp, CYP3A4	Piscitelli et al. (2000)
Irinotecan (CPT11)	Huangqin	—	significant improvement in diarrhea grades ($P = 0.044$) as well as a reduced frequency of diarrhea grades 3 and 4	Multiple factors, one of them could be due to modulation of P-gp	Mori et al. (2003)
Losartan	Grapefruit juice	—	Increase in AUC ($+17\%$)	P-gp, CYP3A4	Zaidenstein et al. (2001)
Loperamide	Quinidine (600 mg/m^2)	—	Increase in AUC (2.5-fold), expression of respiratory depression	P-pg	Sadeque et al. (2000b)

Lovastatin	Cyclosporin A	—	Increase in AUC (5-fold)	P-gp, CYP3A4	Kivisto et al. (1998)
Lovastatin	Diltiazem	—	Increase in AUC (3.6-fold)	P-gp, CYP3A4	Azie et al. (1998)
Lovastatin	Itraconazole	—	Increase in AUC (more than 15-fold)	P-gp, CYP3A4	Kivisto et al. (1998)
Phenytoin	Kava	—	piperine increased Ka, AUC(0–48), AUC (0-infinity), and delayed elimination of phenytoin.	P-gp, CYP3A4	Velpandian et al. (2001)
Quinacrine	Verapamil	—	Decrease in the frequency of myoclonus (side effect)	P-gp	Satoh et al. (2004)
Quinidine	Itraconazole	—	Increase in AUC (2.4-fold) and C_{max} (1.6-fold), decrease in renal clearance (-50%), increase in $t_{1/2}$ (1.6-fold), decrease in the ratio of AUC of metabolites to that of parent drugs (-50%)	P-gp, CYP3A4	Kaukonen et al. (1997)
Saquinavir	Erythromycin	—	Increase in AUC (1.99-fold)	P-gp, CYP3A4	Grub et al. (2001)
Saquinavir	Ketoconazole	—	Increase in AUC (2.9-fold) and C_{max} (2.71-fold)	P-gp, CYP3A4	Grub et al. (2001)
Saquinavir	Ritonavir	—	Increase in AUC (17-fold)	P-gp, CYP3A4	Buss et al. (2001)

(continued)

TABLE 6.3 (*Continued*)

Substrate	Inhibitor	Inducer	Result	Possible interacting protein	Reference
Saquinavir	—	Rifampin	Decrease in AUC (−70%)	P-gp, CYP3A4	Grub et al. (2001)
Saquinavir	—	Garlic	Decrease AUC by 51%, C_{max} by 54% and C_{8h} by 49%	P-gp, CYP3A4 (?)	Piscitelli et al. (2002)
Simvastatin	Cyclosporin A	—	Increase in AUC (2.6-fold)	P-gp, CYP3A4	Arnadottir et al. (1993)
Simvastatin	Diltiazem	—	Increase in AUC (5-fold)	P-gp, CYP3A4	Mousa et al. (2000)
Simvastatin	Erythromycin	—	Increase in AUC (6.2-fold) and C_{max} (3.4-fold)	P-gp, CYP3A4	Kantola et al. (1998)
Simvastatin	Grapefruit juice	—	Increase in AUC	P-gp, CYP3A4, OATP2?	Lilja et al. (1998) and Lilja et al. (2004)
Simvastatin	Itraconazole	—	Increase in AUC of total simvastatin (18.6-fold)	P-gp, CYP3A4	Neuvonen et al. (1998)
Simvastatin	Verapamil	—	Increase in AUC (4.6-fold) and C_{max} (2.6-fold)	P-gp, CYP3A4	Kantola et al. (1998)
Sirolimus	Cyclosporine	—	Increase in AUC (1.45-fold) and C_{max} (1.71-fold)	P-gp, CYP3A4	Kaplan et al. (1998)
Tacrolimus	Diltiazem	—	Increase in blood concentration (12.9–55 ng/mL)	P-gp, CYP3A4	Hebert and Lam (1999)

Drug	Inhibitor	Inducer	Effect	Transporter/Enzyme	Reference
Tacrolimus	Ketoconazole	—	Increase in AUC (two fold), increase in bioavailability (14%–30%), no change in total clearance and hepatic availability	P-gp, CYP3A4	Floren et al. (1997)
Tacrolimus	—	Rifampin	Increase in clearance (+47%), decrease in bioavailability (−51%)	P-gp, CYP3A4	Hebert et al. (1999)
Talinolol	Erythromycin	—	Increase in AUC (+52%)	P-gp	Schwarz et al. (2000)
Talinolol	Verapamil	—	Decrease in AUC (−24%)	P-gp	Schwarz et al. (1999)
Talinolol	Verapamil	—	Decrease in the secretion into small intestine (decrease of secretion rate: 29–59%)	P-gp	Gramatte and Oertel (1999)
Talinolol	—	Rifampin	Decrease in AUC (po: −21%; iv: −35%), increase in P-gp expression level (4.2-fold) in duodenum	P-gp	Westphal et al. (2000d)
Taxol (Paclitaxel)	Cyclosporin A	—	Increase in AUC (8.5-fold), increase in total clearance (eight fold)	P-gp, CYP3A4	Meerum Terwogt et al. (1999)
Taxol (Paclitaxel)	GF120918	—	Increase in AUC	P-gp	Malingre et al. (2001a)

(continued)

TABLE 6.3 (*Continued*)

Substrate	Inhibitor	Inducer	Result	Possible interacting protein	Reference
Taxol (Paclitaxel)	Verapamil	—	Decrease in clearance (−50%)	P-gp, CYP3A4	Berg et al. (1995)
Topotecan	GF120918	—	Increase in AUC (2.4-fold, po)	BCRP and P-gp	Kruijtzer et al. (2002)
Vincristine	Nifedipine	—	Increase in AUC (3.4-fold)	P-gp, CYP3A4	Fedeli et al. (1989)

6.5.1 Oral Absorption

In humans, the role of transporters in drug absorption has been indirectly shown by inhibition or induction studies. Drug-drug interactions due to P-gp-mediated absorption are generally limited to some biopharmaceutics classification system (BCS) class II and IV drugs, whereas there are minimal effects on class I drugs with high solubility and high permeability due to the saturation potential of P-gp at high therapeutic doses. Therefore, the importance of P-gp in oral drug bioavailability, drug disposition in the liver, drug efflux in the blood-brain barrier, and drug–drug interaction should be considered. This is especially important for drugs with narrow therapeutic windows.

The classic example of digoxin-quinidine interaction was observed in early 1980s. Coadministered quinidine increased the absorption rate constant of digoxin by 30%, C_{max} by 81%, and AUC by 77% in cardiac disease patients (Pedersen et al., 1983). Only recently, the underlining mechanism of digoxin–quinidine has been understood and has been attributed to intestinal P-gp (and also liver and renal P-gp). Moreover, in healthy volunteers, oral coadministration of 100 mg talinolol, a P-gp substrate, increased the AUC_{0-6h} and the AUC_{0-72h} of digoxin significantly by 18% and 23%, respectively, while infusion of talinolol with oral digoxin had no significant effects on digoxin pharmacokinetics. Digoxin did not affect the disposition of talinolol after both oral and intravenous administration (Westphal et al., 2000b). Another study showed that the talinolol (50 mg) AUC_{0-24h} and C_{max} were significantly increased after administration of oral erythromycin at 2 g compared to placebo, while the renal clearance of talinolol was unchanged in healthy volunteers. This suggests that the increase in oral bioavailability of talinolol after concomitant erythromycin is caused by increased intestinal net absorption due in turn to P-gp inhibition by erythromycin (Schwarz et al., 2000).

P-gp expression level in humans directly affected oral digoxin and talinolol absorption. Rifampin treatment (600 mg/day for 10 days) increased intestinal P-gp content 3.5-fold, which correlated with the decreased AUC after oral digoxin (1 mg) but not after intravenous digoxin (1 mg). Renal clearance and half-life of digoxin were not altered by rifampin (Greiner et al., 1999a). Similarly, the rifampin treatment was reported to result in increased expression of duodenal P-gp content 4.2-fold and decreased AUC of intravenous and oral talinolol (21% and 35%) in healthy volunteers, suggesting that rifampin induces P-gp-mediated excretion of talinolol predominantly in the gut wall (Westphal et al., 2000c). This implied that individual intestinal P-gp expression difference could contribute to the individual pharmacokinetics variation of digoxin and tanilolol.

For drugs with wide therapeutic index, the P-gp-mediated DDI may not be clinically significant. In healthy volunteers, dexamethasone (widely included in oncology antiemetic regimens) overall exposure was significantly increased by 24% by valspodar 400 mg, a P-gp modulator used as a chemotherapy adjunct. However, this AUC increase is unlikely to be considered in a clinical setting given dexamethasone's wide therapeutic index and the short duration of coadministration (Kovarik et al., 1998).

6.5.2 Brain Penetration

The high expressions of P-gp and BCRP at the luminal membrane of brain endothelia cells imply their roles in the blood–brain barrier. The function of P-gp in human brain penetration has been demonstrated by DDI studies. However, the impact of other transporters in human brain is not clear yet. Loperamide, a potent opiate, is used alone as an antidiarrheal drug without CNS effect due to P-gp restricted entry to the brain. When loperamide (16 mg) was given with quinidine at a dose of 600 mg in healthy volunteers, it elicited central opioid effects indicated by respiratory depression. This can be explained by P-gp inhibition in BBB and gut, resulting in increased brain penetration of loperamide and increased oral drug exposure (Sadeque et al., 2000a). In 12 human healthy volunteers, PET imaging studies have demonstrated that after IV infusion the brain concentration of [^{11}C] verapamil (a P-gp substrate) was significantly increased upon CsA (a potent P-gp inhibitor) coadministration (Sasongko et al., 2005), suggesting that P-gp limits its substrate, such as verapamil, across BBB in man. This PET imaging study appears to be the first report to demonstrate the function of P-gp in the human BBB.

Ivermectin, a neurotoxic compound in animals with low P-gp expressions, has been safely used in Africa for the prevention and treatment of river blindness. The lack of neurotoxicity in African might be due to the high P-gp expression in African population (Ameyaw et al., 2001). Gene analysis showed a high frequency of the C allele in the African group than British Caucasian, Portuguese, southwest Asian, Chinese, Filipino, and Saudi populations, which implies overexpression of P-gp in African population.

6.5.3 Renal Excretion and Hepatic Clearance

Renal excretion is a major elimination route of many antibiotics and antivirals, which is partially mediated via uptake by OATs into proximal tubular cells. With coadministration of probenecid, and nonspecific anion transporter inhibitor, many cephalosporins, including cephazedone, cefazolin, cefradine, cefoxitin, cefadroxil, exhibited increased peak concentration, half-life, and exposure due to inhibition of their renal excretion, enhancing the drug therapy (Brown, 1993). Nonsteroidal anti-inflammatory drugs (NSAIDs) including diflunisal, ketoprofen, flurbiprofen, indomethacin, naproxen, and ibuprofen can inhibit hOAT1-mediated transport of adefovir with IC$_{50}$s of 0.85–8 μM, which are at the clinically relevant concentrations. Therefore, this NSAIDs–adefovir interaction may reduce or delay the emergence of adefovir nephrotoxicity (Mulato et al., 2000).

Severe drug-drug interactions are known to occur between methotrexate and NSAIDs, probenecid, and penicillin G partially due to inhibition of renal

OAT-mediated secretion of methotrexate. By using mouse proximal tubule cells stably expressing human transporters, methotrexate has been demonstrated to be taken up via hOAT3 and hOAT1 at the basolateral side of the proximal tubule and effluxed or taken up at the apical side via hOAT4, where drug interactions occur between methotrexate and NSAIDs, probenecid, and penicillin G (Takeda et al., 2002).

Tsuruoka et al. recently reported the first case of severe arrhythmia as a result of the interaction of cetirizine and pilsicainide by competing renal excretion via P-gp and OCT. A patient with renal insufficiency who was taking oral pilsicainide was found to have a wide QRS wave with bradycardia 3 days after taking oral cetirizine. The plasma concentrations of both drugs were significantly increased during the coadministration. A follow-up pharmacokinetic study in healthy volunteers showed that the renal clearance of cetirizine (20 mg) or pilsicainide (50 mg) was significantly decreased by about two fold following coadministration of the two drugs. *In vitro* studies using Xenopus oocytes expressing OCT2 and renal cells transfected with P-gp revealed that the probe substrate transport was inhibited by either cetirizine or pilsicainide. These data explained that the elevated concentrations of these drugs are due to drug-drug interaction via either human P-gp or OCT2 in the renal tubular cells (Tsuruoka et al., 2006).

Cyclosporine increases the systemic exposure of all statins (lovastatin, simvastatin, pravastatin, cerivastatin, and rosuvastatin), due to drug-drug interaction (via either CYPs or transporters like P-gp and OATP) in the liver. Rosuvastatin has been shown to be a substrate for the human liver transporter OATP2 and BCRP, but not P-gp. It's metabolic clearance is low and mainly mediated by CYP2C9. CsA treatment in transplant recipients increased AUC_{0-2h} and C_{max} of rosuvastatin (10 mg) by 7.1 and 10.6-fold, respectively, compared with control values, due to CsA inhibition of OATP2-mediated rosuvastatin hepatic uptake (Simonson et al., 2004).

6.5.4 Food Effect

Drug–food interactions can often be caused by food or drug supplements such as St. John's Wort, an herbal medicine for the treatment of depression in humans. St. John's Wort products have shown remarkably decreased plasma exposure/concentration of certain comedicated drugs, such as CsA (Mai et al., 2000) and digoxin (Johne et al., 1999), which was attributed to an inducing effect of St. John's Wort on CYPs and P-gp activity. Potentially clinically significant drug interactions were observed with St. John's Wort (16/24 studies), garlic (2/5 studies), and American ginseng (1 study) (Mills et al., 2005).

Grapefruit juice at normal volume of 300 mL decreased AUC and C_{max} of fexofenadine (120 mg) to 58% and 53%, respectively, compared with dosing of corresponding volume of water in the healthy volunteers, and 1200 mL grapefruit juice reduced these parameters to 36% and 33%, respectively.

Grapefruit juice also diminished the AUC of fexofenadine variably among individuals. This decreased oral bioavailability is likely a contribution of direct inhibition of its uptake by intestinal OATP-A (Dresser et al., 2005).

6.5.5   Formulation Effect

Formulation strategies to overcome multidrug resistance have been evaluated for their enhancement of membrane permeability of a drug and inhibition of P-gp. Surfactants used in pharmaceutical formulations include nonionic detergents, polyoxyethylen(20)-sorbitanemonooleate (Tween 80), polyoxyethylen–polyoxypropylene block copolymers (e.g., Pluronic P85), and polyoxyethy-lenglycoltriticinoleate (Cremophor EL). These agents can modulate drug absorption by multiple mechanisms including inhibition of intestinal P-gp. Water-soluble vitamin E (D-alpha-tocopheryl poly(ethylene glycol) 1000 succinate (TPGS 1000), which is comprised of a hydrophilic polar head and a lipophilic alkyl tail, has been used as a solubilizer, an emulsifier, and an effective oral absorption enhancer.

In murine monocytic leukemia cells overexpressing P-gp, P-gp-mediated rhodamine123 transport was inhibited by five nonionic surfactants in a concentration-dependent manner and in the order TPGS > Pluronic PE8100 > Cremophor EL > Pluronic $\approx$ PE6100 Tween 80. In contrast, none of those surfactants showed a significant inhibition of MRP2-mediated efflux in MDCK-MRP2 cells (Chang et al., 1996). In nine healthy volunteers, talinolol solution, containing either talinolol alone (50 mg), talinolol and TPGS (0.04%), or talinolol and Poloxamer 188 (0.8%) was administered via nasogastrointestinal tube dosing. TPGS increased AUC of talinolol by 39% and C_{max} by 100%, whereas Poloxamer 188 did not significantly alter AUC or C_{max} of talinolol. This *in vivo* observation can be explained by Caco-2 data showing abolishment of P-gp-mediated talinolol efflux with TPGS (0.01%), but not Poloxamer 188 (Bogman et al., 2003). TPGS PEG chain length was demonstrated to influence on rhodamine123 transport in Caco-2 monolayers by using TPGS analogs containing different PEG chain length (TPGS 200/238/400/600/1000/2000/3400/3500/4000/6000 (Collnot et al., 2006)).

Pluronic P85 was also reported to cause a higher degree of alteration in the P-gp functions than MRP2 and MRP1 in ATPase assay of P-gp, MRP1, and MRP2 and inhibition assays with their substrates, vinblastine, and leucotriene-C_4 (Batrakova et al., 2004). Cremophor EL (cremophor) is a nonionic solubilizer and emulsifier used for some hydrophobic drugs and fat-soluble vitamins. In a Phase I trial of cremophor as a 6-h infusion every 3 weeks performed with bolus doxorubicin (50 mg/m^2), the AUC of doxorubicin increased from 1448 (CV of 24%) to 1786 h*ng/mL (CV of 15%) in the presence of cremophor, whereas the AUC of doxorubicinol increased from 252 (CV of 42%) to 486 (CV of 22%) h*ng/mL. Such interactions can be due to decreased clearance of doxorubicin 612 (CV of 29%) to 477 (CV of 15%)

mL/min by inhibiting P-gp in the liver and kidney and inhibiting metabolizing enzymes. In a dose escalation study, patients receiving $45 \, mL/m^2$ cremophor reached plasma levels $\geq 1.5 \, \mu L/mL$, a level well tolerated in humans. Cremophor $45 \, mL/m^2$ over 6 h with $35 \, mg/m^2$ doxorubicin are recommended for further studies (Millward et al., 1998). Similarly, in a study with 12 health volunteers, cremophor EL increased digoxin (0.5 mg) C_{max} by 22% and AUC by 22% (ten Tije et al., 2003).

Although most investigations have focused on formulation effect on P-gp, formulation may also affect other transporters in the gut and other organs like liver and kidney. This remains to be explored as more knowledge of other transporters becomes available.

6.5.6 *In vitro–In vivo* Correlation

Prediction of transporter-mediated drug-drug interactions in humans by using *in vitro* data is of great interest. Given the success of predicting CYP-based drug interactions, Endres et al. proposed to adopt the same approach using the following equations:

$$R = \frac{AUC(inh)}{AUC} = \frac{1}{\dfrac{f_{cl}}{1 + (f_u[I]/K_i)} + (1 - f_{cl})} \qquad (6.3)$$

where f_u and [I] are the free fraction and plasma concentration of the inhibitor, and f_{cl} is the fraction of the total clearance mediated by the affected transporter. When $f_{cl} = 1$ (transporter-mediated clearance contribution equals to the total clearance), the Equation 6.3 can be simplify to Equation 6.4 (Endres et al., 2006).

$$R = \frac{AUC(inh)}{AUC} = 1 + (f_u[I]/K_i) \qquad (6.4)$$

Unlike the progress in the predictions of CYP-based DDIs (Lu et al., 2007;35:79–85), transporter-based *in vitro–in vivo* correlations and predictions of DDI in humans are still in infancy stage. The predicted fold change in AUC based on Equation 6.3 or 6.4 was not close to every observed case yet because of multiple transporters and other mechanisms involved in the drug interactions and the difficulty to estimate the drug concentration at the site of interactions (Endres et al., 2006). Current FDA draft guidance on DDI studies recommends an *in vivo* drug interaction study with a P-gp substrate such as digoxin if a P-gp inhibitor shows $[I]/K_i$ or I/IC_{50} higher than 0.1. However, one shall also consider the overall clearance by transporters, therapeutic range, frequency of dosing, and therapeutic area for assessing the need for a DDI study. There is still a lack of understanding on evaluating quantitatively the role of transporters in overall clearance or uptake.

6.6 METHODS TO EVALUATE TRANSPORTER SUBSTRATE, INHIBITOR, OR INDUCER

Given the important role of transporters in ADMET and drug-drug interactions, the screening of drugs for their affinity as substrates or inhibitors of transporters should be a standard component of the ADME package in the discovery (Balani et al., 2005) and development stages.

6.6.1 *In vitro* Models

6.6.1.1 Membrane-Based Assays   Membranes prepared from cells expressing transporters have been widely used to study the function of ABC efflux pumps and to identify their substrates or inhibitors. Currently, there are two major membrane-based assays: the ATPase assay and the membrane vesicular transport (uptake) assay. Compared to the cell-based assay, the membrane-based assay has several advantages including: (1) the assay can be used to characterize the effect of a xenobiotic on one specific efflux transporter; (2) the assay can be easily employed in a high throughput mode; (3) membranes are easy to be maintained after preparation; and (4) the assay is easy to conduct.

The transport function of the ABC efflux pumps depends on the binding and the hydrolysis of cytoplasmic ATP within NBD. The ATPase activity of P-gp, MRP or BCRP expressing in insect cell or mammalian cell membranes is vanadate sensitive and can be stimulated or inhibited by substrates of these transporters (Chang et al., 1998; Ozvegy et al., 2001, 2002; Sarkadi et al., 1992; Senior and Urbatsch, 1995). The released phosphate can be determined by a sensitive colorimetric reaction under mild acidic conditions (Druekes and Palm, 1995) or the amount of ATP can be quantified by a luciferase-generated luminescent signal (Promega). Since the profile of altered ATPase activity by P-gp substrates has been shown to reflect the nature of the interaction between P-gp and a substrate (Ramachandra et al., 1996; Rao, 1995), ligand screening assays have been developed for P-gp and BCRP substrates/inhibitors using the ATPase assay (Polli et al., 2001; Scarborough, 1995; Xia et al., 2004). In addition to the stimulation of the basal ATPase activity of P-gp, it has been suggested to incorporate the inhibition of the stimulation of the ATPase activity by reference compounds, such as verapamil, progesterone, or vinblastine in screening studies (Garrigues et al., 2002; Scarborough, 1995). Coincubation of test agents with P-gp ATPase activators or inhibitors can provide information about the potential binding site to P-gp (Litman et al., 1997). ATPase can also be used to evaluate the interspecies difference of an ABC pump (Xia et al., 2006). The kinetic constant obtained from ATPase assay can be used to rank order the binding affinity between an ABC pump and its modulators (Boulton et al., 2002). The membrane ATPase assay is compound-independent, easy to perform and does not require the use of radiolabeled compounds. Therefore, it can easily be applied in a high throughput mode. However, the ATPase assay is not a functional transporter assay and can not distinguish substrates from inhibitors (Boulton et al., 2002; Polli et al., 2001; Ramachandra et al., 1996; Rao, 1995). Other drawbacks of this assay include large interday or intraday variations (Xia et al., 2006) and potential false negative results.

The membrane vesicular transport assay is another high throughput assay that can be used to identify substrates and/or inhibitors of efflux pumps, such as P-gp, BCRP, MRP or BSEP, and uptake transporters, such as NTCP. The membrane vesicles can be prepared not only from transporter-transfected or overexpressed cells but also from the brush-border membrane of intestine, kidney, and choroids plexus; hepatic sinusoidal and canalicular membranes; and luminal and albuminal membranes of the brain (Boyer and Meier, 1990; Meier and Boyer, 1990). The membrane vesicles obtained from tissues may contain many varied transporters that may limit their utility for evaluating the interaction of a given transporter with a given compound. The purity of these membrane vesicles from tissues can be evaluated by the enrichment of the relative activity of membrane marker enzymes, such as dipeptidyl peptidase IV for canulicular membrane of hepatocytes, for the target plasma membranes (Meier and Boyer, 1990). Most ABC transporters and some SLC transporters can be studied using membrane vesicles. For example, NTCP is a sodium dependent transporter. Thus, the NTCP activity can be assessed by measuring the difference of drug accumulation in the right-side-out membrane vesicles in the presence and absence of Na^+ buffer. Since ABC protein (efflux pump)-mediated transport is ATP-dependent, and ATP cannot traverse the lipid membrane due to its hydrophilic nature, only inside-out membrane vesicles can allow ABC transporters to bind ATP and subsequently pump a substrate into the vesicles. Utilizing a rapid filtration technique, the membrane vesicles can be collected on a filter membrane and the substrate trapped inside the vesicles can be quantitated by LC-MS/MS, fluorescence detector, or liquid scintillation counting (Tabas and Dantzig, 2002; Xia et al., 2004). The difference of the uptake of a substrate in the presence or absence of ATP or other driving force such as Na^+ and H^+ is attributed to ABC protein or uptake transporter-mediated transport, respectively. The membrane vesicular assay is an effective *in vitro* system without intracellular binding and metabolism to directly reflect transporter functions and to define the kinetic constants such as K_m for substrates (Keppler et al., 1998; Xia et al., 2004), and K_i or IC_{50} for inhibitors (Hirano et al., 2005; Horikawa et al., 2002). The assay can also be used to study the driving force of transporter of interest by manipulating the ionic composition of or ATP concentration in the assay buffer (Meier et al., 1984). This assay may give false negative results for highly lipophilic compounds due to high nonspecific binding to lipid membranes or high passive diffusion. The adsorption can be distinguished from active transporter-mediated intravesicular accumulation by measuring the uptake under different osmolarities since the intravesicular volume decreases with increasing osmolarity. Due to the consumption of ATP and subsequent drug release from vesicles, the drug uptake into membrane vesicles can reach a maximum level and then decrease ("overshoot" phenomenon). Therefore, the kinetic or inhibition assay must be used during the initial uptake phase, which is reflective of transporter activity. Sometimes, an ATP regeneration system is added in the buffer to ensure a sufficient energy supply for the duration of the incubation (Adachi et al., 1991).

As a functional assay, the membrane vesicle transport assay has been used to investigate binding sites (Shapiro et al., 1997, 1999), interspecies differences in transporter activity (Ishizuka et al., 1999; Ninomiya et al., 2005), polymorphisms in transporter activity (Center et al., 1998; Hirano et al., 2005), and substrate or inhibitor specificity for a given efflux pump (Keppler et al., 1998; Volk and Schneider, 2003).

6.6.1.2 Cell-Based Assays   Most cell-based studies are functional transporter assays. Due to the intact cell structure, cell-based assays may provide more definitive information about the interaction between drugs and transporters and can be employed to assess kinetic parameters, such as K_m and V_{max} for substrates as well as K_i or IC_{50} for inhibitors. These assays can also predict transporter-related DDI that may occur in the clinic. With automation and cell culture in a multiwell plate, cell-based assays can be adapted to a high throughput mode. Several drawbacks of cell-based assays include (1) expression of multiple transporters in a particular cell line including cell lines that have been engineered to express a given transporter; (2) the expression level of transporters changes with culture conditions and the number of cell passages; (3) cells need to be maintained under culture condition prior to use. This assay is more labor and time intensive than the ATPase assay and membrane vesicular transport studies.

Cell-based assays include uptake and transport studies. The transport assay is the most direct assay for evaluating a given transporter's function and measures the permeability of a test compound across cell monolayer. Cells can be seeded on the Transwell inserts and are ready to be used once they reach confluence and have completely differentiated. Transport experiment is initiated by the addition of a solution containing the test compound to either the apical (or upper chamber, for apical-to-basolateral (A-to-B) transport) or basolateral (or lower chamber, for basolateral-to-apical (B-to-A) transport) compartment and an aliquot of the solution is removed from the lower chamber (for A-to-B transport) or from the upper chamber (for B-to-A transport) at desired time points. The cumulative amount of drug (Q) in the receiving side is plotted as a function of time. The steady-state flux rate J is then estimated from the slope (dQ/dt). The apparent permeability coefficient (P_{app}) of unidirectional flux for the test compound is estimated by normalizing the flux rate J (mol/s) against the nominal surface area A and the initial drug concentration in the donor chamber C_0 (mol/mL), or $P_{app} = J/(A \times C_0)$. The B/A ratio is equal to the P_{app} value for B-to-A transport ($P_{app, \text{B-to-A}}$) divided by the P_{app} value for A-to-B transport ($P_{app, \text{A-to-B}}$). If an efflux transporter is expressed on the apical cell membrane, then $P_{app,\ A\text{-to-B}} < P_{app,\ B\text{-to-A}}$; if an uptake transporter is expressed on the apical cell membrane then $P_{app,\ A\text{-to-B}} > P_{app,\ B\text{-to-A}}$. These relationships will be reversed if transporter is localized on the basolateral cell membrane. For passively diffused compounds, the P_{app} value is concentration independent and the flux rate is correlated linearly with concentration. For actively transported compounds, the flux rate is saturable with increasing concentration of a test compound and the net transporter-mediated flux can be calculated by subtracting the passive diffusion flux from the total flux rate. The passive diffusion flux can be estimated either from the flux at 4°C (at this temperature all active transport processes cease to function) or in the presence of transporter inhibitors. Kinetic parameters such as K_m and V_{max} can be determined for net transporter-mediated flux. Polarized cells with the formation of tight junctions are suitable to be used in the transport assay. Currently, Caco-2, Madin–Darby canine kidney (MDCK), LLC-PK1, and certain primary cultured cells are commonly used. In the Draft FDA Guidance on DDI studies, transcellular transport assay is recommended as the efflux transporter functional assay and Caco-2, MDR1-MDCK, and MDR1-LLC-PK1 cells have been accepted as models to evaluate P-gp substrates or inhibitors.

Uptake assay is used to measure the amount of a compound accumulated in cells. Concentrations of a compound in cells can be quantified by different analytical tools based upon a given compound's properties. By using a fluorescent or a radiolabeled compound as a substrate, the uptake study can be set up as a high throughput assay for transporter inhibitory potential screening. In most cases a transporter substrate is also a competitive inhibitor although there are exceptions under certain circumstances (e.g., a substrate has low affinity for the transporter or a substrate is unable to reach sufficiently high concentrations due to cytotoxicity or poor solubility). Cell viability should be measured in the inhibitor screening studies in order to overcome the false negative for efflux pumps or positive results for uptake transporters.

The acetoxymethyl ester of calcein (Calcein-AM) is a highly lipid soluble and nonfluorescent compound that can rapidly penetrate cell plasma membranes. Once Calcein-AM is taken up into the cells, the ester bond can be quickly and irreversibly converted to hydrophilic, nonpermeable and intensively fluorescent free acid form, Calcein. When traversing the cell plasma membrane, Calcein-AM can be pumped out of the cells by P-gp. Inhibition of P-gp causes an increase in the cellular level of Calcein-AM and subsequently the appearance of fluorescent Calcein. Calcein AM uptake studies in P-gp expressing cells have been extensively used as a screening tool for P-gp inhibitors (Polli et al., 2001).

Daunorubicin and rhodamine123 (P-gp substrates), Calcein (a MRP substrate) (Sarkadi et al., 2001), LysoTracker (Xia et al., 2005b) (a BCRP substrate), and H_2FDA and BODIPY (BSEP substrates) have been used as fluorescent substrates for the screening of transporter inhibitors and their concentrations can be measured by fluorescence-activated cell sorter (FACS) flow cytometry (Wang et al., 2000, 2003) or any fluorescence detectors.

The cytotoxicity assay is an indirect measurement of the accumulation of cytotoxic compounds in cells and is a frequently used surrogate assay to determine whether a compound is a substrate or an inhibitor of a given transporter. Transporter substrates can be identified by comparing the IC_{50} (the concentration that inhibits the cell growth by 50%) in wildtype (naïve) and transporter expressing (drug resistant) cells, while inhibitors can be identified by their ability to potentiate (for efflux transporters) or attenuate (for uptake transporters) the cytotoxicity of a substrate in transporter expressing or drug resistant cells. The activity of the reversal agent is generally expressed as a fold reversion or MDR ratio. The MDR ratio is the ratio of the IC_{50} of a cytotoxic drug alone to the IC_{50} of a cytotoxic drug in the presence of a transporter modulator.

Freshly isolated or cryopreserved hepatocyte suspension can be used for hepatobiliary uptake transporter but not efflux transporter studies (Hirano et al., 2004; Hoffmaster et al., 2004a; Shitara et al., 2003) because efflux transporters are internalized after isolation (Hoffmaster et al., 2004a). The uptake of compounds into suspended hepatocytes was determined by a centrifugation filtration method (Hirano et al., 2004). Sandwich-cultured hepatocytes have become a valuable tool to evaluate both uptake and efflux hepatobiliary transporters since sandwich-cultured hepatocytes maintain the hepatocyte architecture, including tight junctions, canalicular biliary network, and the functional transporters. Because depletion of Ca^{2+} can open tight junctions, accumulation of compounds in hepatocytes, and bile ducts, or in hepatocytes is determined in Ca^{2+} and Ca^{2+} free buffer, respectively (Liu et al., 1999). Biliary elimination is then calculated as the difference of these two measurements. Sandwich cultured hepatocytes can be used to evaluate hepatobiliary transporter-mediated DDI and liver toxicities (Kemp et al., 2005). Other potential advantages of the hepatocyte model include the evaluation of the induction of transporters by xenobiotics and understanding interspecies difference in hepatobiliary transporters. Although cultured hepatocytes can provide information on an overall outcome of transporter-related DDI and hepatotoxicity, its utility is limited by the lack of the details of the individual interactions between a xenobiotic and an uptake or efflux transporter. This limitation could be overcome by using a specific transporter substrate or inhibitor or by using specific transporter expressed *in vitro* systems. The results obtained from one donor hepatocytes may not be reflective of the whole human population because of the possible polymorphisms of transporters.

6.6.1.3 *Oocytes and Yeast*

Both yeast and *Xenopus laevis* oocyte over-expressing transporters can be used to evaluate transporter functions. Yeast, which grows rapidly, can be used for the production of large quantities of transporter proteins and can be used for high throughput bioassay screening. Genetic manipulations in yeast are easy and cheap. Oocytes can be directly microinjected with mRNA into cytoplasm or cDNA into nucleus and they have low endogenous transporter background and short expression turnaround time (4–7 days). Additionally, the cells can perform the appropriate posttranslational modification; the transporters are appropriately localized in the membrane; and a high proportion of cells express the transferred genetic information (Sigel, 1990). The drawbacks of oocytes as an expression system include (Sigel, 1990) (1) transient expression of the transporter, which usually lasts no longer than 14 days; (2) seasonal variations in the quality of the oocytes; (3) low to medium throughput assays; (4) cDNA that has to be transcribed, capped, and polyadenylated *in vitro* unless it is directly injected into the nucleus of the oocytes and (5) requirement of microinjection skill for transfecting cDNA or mRNA and for efflux transporter studies.

6.6.1.4 Induction Models   Since Caco-2 cells lack steroid xenobiotic receptor (SXR), a human typical pregnane X receptor (PXR) (Li et al., 2003), it is a not a good model to evaluate P-gp inducers. SXR ligands, such as rifampicin, paclitaxel, nifedipine, and St. John's Wort, increase both P-gp and CYP3A4 expression in LS180 (Li et al., 2003; Perloff et al., 2001), a human intestinal carcinoma cell line, suggesting that LS180 cells can be used for the evaluation of P-gp inducers. Currently, all these models have shown an induction at the mRNA and protein levels but not in the transport function assay yet.

6.6.1.5 Mapping Transporter Substrates or Inhibitors   Estimation of a drug transporter substrate or inhibition properties can help to understand the drug disposition profile, pharmacological effects, and potential drug-drug interactions. Transporter substrate mapping can be done by using selective inhibitors or specific transporter transfected systems. Combined with efflux transporter inhibitor treatment, Caco-2 cells can be used to map efflux transporters to their substrates. This method is highly depended upon the selectivity of transporter inhibitors. The selectivity of several commonly used efflux transporter inhibitors is listed in Table 6.4. It is difficult to identify MRP2 substrate by using an inhibitor method because of the lack of the selective MRP2 inhibitors. It is noteworthy that not only the transfected transporters but also endogenous

TABLE 6.4   Selectivity of commonly used transporter inhibitors.

	IC_{50} (μM)			
	P-gp	BCRP	MRP2	OATP2
Test system	Caco-2 cells	Membrane vesicles	Membrane vesicles	Oocytes
Substrate (concentration)	Taxol (50 nM)	Estrone sulfate (2 μM)	Leukotriene C_4 (3.2 nM)	Estrone sulfate (2 μM)
GF120918	0.04	0.06	>100	>100
MK571	25	0.6	2.2	52.7
LY335979	0.05	12.0	>100	>100
Ko143	>50	0.03	>100	61.3

transporters are expressed in the transfected cell lines when using a transfected cell line to map transporter substrate. The vector control transfected cells and a known transporter inhibitor are always recommended to be used together with the transporter transfected cells to get more definitive results.

The IC_{50} and K_i of transporter inhibitors can be determined using selective transporter substrate by *in vitro* assays. For uptake transporters, hepatocyte, transporter transfected cells, oocytes, and yeast have been used as *in vitro* models; for efflux transporter, Caco-2 cells, and transporter transfected cells and membranes are commonly used models.

6.6.2   *In situ/Ex vivo* Models

Compared to *in vitro* assays, the isolated perfused intestine, liver, kidney, or brain allows a more accurate determination of the transporter functions in intestinal absorption, biliary elimination, renal excretion, or brain penetration and the interplay with drug metabolism enzymes. Liver perfusion studies have shown that transporter and metabolism enzymes acted in concert to govern drug metabolism in hepatocytes and elimination into the bile (Benet et al., 2004; Lau et al., 2004, 2006; Liu and Pang, 2005). In rat liver perfusion studies, the AUC of digoxin (a substrate of OATP2, P-gp, and CYP3A4) in perfusate was increased by rifampicin treatment (an OATP2 inhibitor) and decreased by quinidine treatment (a P-gp inhibitor). It is concluded that rifampicin limits the hepatic enzyme exposure to digoxin by inhibiting the sinusoidal Oatp2-mediated transport, whereas quinidine increased digoxin exposure to the hepatic enzymes by inhibiting the canalicular P-gp efflux. These data emphasize the importance of uptake and efflux transporters on hepatic drug metabolism (Lau et al., 2004). Perfused kidney studies proved the P-gp-mediated digoxin renal elimination and the interaction between digoxin and CsA (Hori et al., 1993; Okamura et al., 1993). By comparing the perfused tissue from transporter deficient animals with that from normal animals, the function of that specific deficient transporter can be evaluated. The disadvantage of the perfusion techniques is the need for specific surgical skills and sophisticated equipments.

6.6.3   *In vivo* Models

6.6.3.1   *Transporter Genetic Knockout Mice or Natural Mutant Animals*
Genetic knockout mice or natural mutant animal models have become important tools in understanding the physiological functions of drug transporters and in evaluating the effect of transporters on the pharmacokinetics and pharmacodynamics of drugs (Chen et al., 2003a; Loescher and Potschka, 2005; Xia et al., 2005b). Genetic knockout mice are generated by disrupting the endogenous transporter gene in order to investigate the specific targeted transporter. Naturally, mutant animals are subpopulations that have a spontaneous mutation in a transporter gene. Naturally occurring P-gp

(Mdr1) or Mrp2 mutations have been identified in mice (Lankas et al., 1997; Kwei et al., 1999), rats (Buechler et al., 1996; Elferink et al., 1989; Kurisu et al., 1991), or dogs (Paul et al., 1987). It is a useful tactic to use multiple gene knockout animals together to phenotype the impact of transporters on the pharmacokinetics of a given drug. By using Bcrp1(-/-) mice and EHBR rats, Hirano et al. (2005) demonstrated that biliary clearance of pitavastatin was accounted for to a large extent by Bcrp in mice and to a small extent by Mrp2 in rats. These findings differed for pravastatin, whose biliary clearance is largely mediated by Mrp2. A similar strategy has been applied for investigating the transporters involved in the transport of sulfasalazine. By comparing the PK profiles of sulfasalazine after IV and PO administration to wild type, Bcrp1(-/-) and Mdr1(-/-) mice, it was shown that Bcrp1, but not P-gp, was the important determinant for the oral bioavailability and the elimination of sulfasalazine in mice and that sulfasalazine can be potentially utilized as a specific *in vivo* probe for Bcrp1 (Zaher et al., 2006).

Caution needs to be taken when using these genetic animal models to elucidate transport mechanisms since deletion of one transporter can often cause alteration in the expression of other transporters or enzymes as well as change the physiology of the knockout animal. For example, the protein expression and activity of CYP3A2, 1A2, 2B1, and 2C11 in liver microsomes prepared from EHBR were lower than those in SD rats (Ohmori et al., 1991). The protein and mRNA expression levels of Mrp3 were significantly increased in the liver and kidney of EHBR compared to wild type SD rats. On the contrary, the protein and mRNA expressions of Oatp1 and Oatp2 were significantly decreased in the liver of EHBR.

6.6.3.2 Transporter Chemically Knockout Models   Mice or rats treated with a transporter inhibitor can provide useful information on the role of the transporter in absorption, elimination, and tissue distribution of a drug. The chemically knockout animal model is commonly used for the evaluation of the role of efflux transporters. The specificity of an inhibitor is critical when using chemically knockout animal models to evaluate the role of a specific transporter in the PK and toxicity of a drug. Whereas selective inhibitors are now available for P-gp and BCRP, no selective inhibitors exist for the MRP family of transporters. GF120918 (an inhibitor for P-gp and BCRP (Jonker et al., 2000)), LY335979 (a selective P-gp inhibitor (Shepard et al., 2003)), and Ku143 (a selective BCRP inhibitor (Allen et al., 2002)) can be safely used in rats or mice to chemically knockout the corresponding efflux transporter(s). However, no specific uptake transporter inhibitor for use in *in vivo* studies has been identified.

The combination of chemical and genetic knockout mice have been used to phenotype the role of efflux transporters in the absorption, disposition, or elimination of drugs. For example, etoposide has been demonstrated to be a P-gp and BCRP substrate using the cellular transport assay in P-gp or Bcrp- transfected cell lines. However, treatment of P-gp-deficient mice with

GF120918 did not increase the exposure to orally administered etoposide, suggesting that Bcrp1 activity is not a major limiting factor in the absorption of etoposide. In contrast, use of GF120918 to inhibit P-gp in wild type mice increased the plasma levels of etoposide to four fold (Allen et al., 2003). These results suggest that P-gp, but not BCRP, limited the absorption of etoposide.

For chemically knockout animals, the specificity of the transporter inhibitor is a major concern. For example, GF120918, a potent inhibitor of both P-gp and BCRP, has been extensively used to understand the role of P-gp and Bcrp *in vivo* (Allen et al., 2003; Jonker et al., 2000; Polli et al., 2004). However, Hoffmaster et al. recently indicated that GF120918 interacts with at least three transporters in the liver; two transporters at the canalicular membrane, P-gp and Bcrp, and an unknown efflux mechanism at the basolateral membrane (Hoffmaster et al., 2004b). Subsequently, Lee et al. have demonstrated that there could be one or more GF120918-sensitive efflux transporters distinct from BCRP or P-gp in the BBB that may contribute to the brain efflux of dehydroepiandrosterone sulfate (DHEAS) and mitoxantrone in mice. Understanding the specificity of transporter inhibitors can assist in more accurately determining the role of a given transporter *in vivo*.

The use of different *in vitro* and *in vivo* models depends largely on the objective of the studies. Multiple methodologies are often needed for better understanding of a transporter-mediated efflux or uptake of drug molecules. To address the potential DDI in humans, currently FDA suggests the use of the bidirectional transport assays in human transporter expressing cells in their draft guidance.

6.7 CONCLUSIONS AND PERSPECTIVES

With a large number of transporters cloned and their functions characterized in *in vitro* assays and in preclinical animals, the important roles of transporters in drug absorption, distribution, metabolism, elimination, as well as in efficacy and toxicity in humans have been well recognized. Although the in vivo transport mechanisms of most therapeutic drugs remain unknown, some clinical drug-drug interactions and toxicities in humans have been linked with the involvement of transporters. Therefore, both the regulatory agency and the pharmaceutical industry have recognized the need of evaluating the transporter substrate or inhibition potential of a drug candidate in causing drug–drug, drug–endobiotic or drug–food interactions in humans. Consequently, many pharmaceutical companies have started to standardize the *in vitro* screening and definitive transporter assays to help clinical trial design for the assessment of potential transporter interactions. Though the FDA has provided some guidance on the need to conduct DDI studies in humans based on empirical $[I]/K_i$ (or IC_{50}) ratio of >0.1, where $[I]$ is the inhibitor concentration in plasma at steady state, K_i is the inhibition constant for a compound for a transporter and IC_{50} is the concentration of the inhibitor that inhibits half of the maximal activity of transporters, this is open for interpretation and discussion. Once the comments from industry and academia are received by the FDA, sometime in future it is expected that the FDA will issue final guidance that will be agreeable to the industry. At this juncture, the authors of this chapter believe that in the absence of the knowledge-base on how to quantitatively assess the role of transporters on absorption and elimination of drugs, the current draft guidance on the trigger criteria of $[I]/K_i$ (or IC_{50}) >0.1 for the conduct of DDI studies in humans will face conflict from the experts.

The effort needs to be continually made to better predict transporter-mediated drug interactions by knowing the complexity of localization and function of transporters and their interplay with Phase I and Phase II enzymes. At this point, it is difficult to extrapolate the results from *in vitro* or animal studies to human. The challenge of understanding the quantitative significance of transporters in drug development remains. Research focusing on the development of methodology to delineate the contribution of major transporters to drug disposition, the establishment of *in vitro–in vivo* correlations, and understanding transporter polymorphisms will help us better use transporter information in drug development.

REFERENCES

Adachi Y, Kobayashi H, Kurumi Y, Shouji M, Kitano M, Yamamoto T. ATP-dependent taurocholate transport by rat liver canalicular membrane vesicles. Hepatology (Philadelphia, PA, United States) 1991;14:655–659.

Adachi Y, Suzuki H, Schinkel AH, Sugiyama Y. Role of breast cancer resistance protein (Bcrp1/Abcg2) in the extrusion of glucuronide and sulfate conjugates from enterocytes to intestinal lumen. Mol Pharmacol 2005;67:923–928.

Allen JD, Van Loevezijn A, Lakhai JM, Van der Valk M, Van Tellingen O, Reid G, Schellens JHM, Koomen G-J, Schinkel AH. Potent and specific inhibition of the breast cancer resistance protein multidrug transporter *in vitro* and in mouse intestine by a novel analogue of fumitremorgin C. Mol Cancer Ther 2002;1:417–425.

Allen JD, van Dort SC, Buitelaar M, van Tellingen O, Schinkel AH. Mouse breast cancer resistance protein (Bcrp1/Abcg2) mediates etoposide resistance and transport, but etoposide oral availability is limited primarily by P-glycoprotein. Cancer Res 2003;63:1339–1344.

Ameyaw M-M, Regateiro F, Li T, Liu X, Tariq M, Mobarek A, Thornton N, Folayan GO, Githang'a J, Indalo A, Ofori-Adjei D, Price-Evans DA, McLeod HL. MDR1 pharmacogenetics: frequency of the C3435T mutation in exon 26 is significantly influenced by ethnicity. Pharmacogenetics 2001;11:217–221.

Amsden G, Nafziger A, Foulds G, Cabelus L. A study of the pharmacokinetics of azithromycin and nelfinavir when coadministered in healthy volunteers. J Clin Pharmacol 2000;40:1522–1527.

Andrejak M, Hary L, Andrejak M, Lesbre J. Diltiazem increases steady state digoxin serum levels in patients with cardiac disease. J Clin Pharmacol 1987;27:967–970.

Arnadottir M, Eriksson LO, Thysell H, Karkas JD. Plasma concentration profiles of simvastatin 3-hydroxy-3-methyl-glutaryl-coenzyme A reductase inhibitory activity in kidney transplant recipients with and without ciclosporin. Nephron 1993;65:410–413.

Asberg A, Hartmann A, Fjeldsa E, Bergan S, Holdaas H. Bilateral pharmacokinetic interaction between cyclosporine A and atorvastatin in renal transplant recipients. Am J Transplant 2001;1:382–386. DOI:10.1034/j.1600-6143.2001.10415.x.

Azie NE, Brater DC, Becker PA, Jones DR, Hall SD. The interaction of diltiazem with lovastatin and pravastatin. Clin Pharmacol Ther 1998;64:369–377.

Baer MR, George SL, Dodge RK, O'Loughlin KL, Minderman H, Caligiuri MA, Anastasi J, Powell BL, Kolitz JE, Schiffer CA, Bloomfield CD, Larson RA. Phase 3 study of the multidrug resistance modulator PSC-833 in previously untreated patients 60 years of age and older with acute myeloid leukemia: cancer and leukemia group B study 9720. Blood 2002;100:1224–1232.

Balani SK, Miwa GT, Gan L-S, Wu J-T, Lee FW. Strategy of utilizing *in vitro* and *in vivo* ADME tools for lead optimization and drug candidate selection. Curr Top Med Chem 2005;5:1033–1038.

Balimane PV, Tamai I, Guo A, Nakanishi T, Kitada H, Leibach FH, Tsuji A, Sinko PJ. Direct evidence for peptide transporter (PepT1)-mediated uptake of a nonpeptide prodrug, valacyclovir. Biochem Biophys Res Commun 1998;250:246–251.

Batrakova EV, Li S, Li Y, Alakhov VY, Kabanov AV. Effect of pluronic P85 on ATPase activity of drug efflux transporters. Pharm Res 2004;21:2226–2233.

Becquemont L, Verstuyft C, Kerb R, Brinkmann U, Lebot M, Jaillon P, Funck-Brentano C. Effect of grapefruit juice on digoxin pharmacokinetics in humans. Clin Pharmacol Ther 2001;70:311–316.

Beketic-Oreskovic L, Duran GE, Chen G, Dumontet C, Sikic BI. Decreased mutation rate for cellular resistance to doxorubicin and suppression of mdr1 gene activation by the cyclosporin PSC 833. J Nat Cancer Inst 1995;87:1593–1602.

Benet LZ, Cummins CL, Wu CY. Transporter-enzyme interactions: Implications for predicting drug–drug interactions from *in vitro* data. Curr Drug Metab 2003;4:393–398.

Benet LZ, Cummins CL, Wu CY. Unmasking the dynamic interplay between efflux transporters and metabolic enzymes. Int J Pharm 2004;277:3–9.

Berg SL, Tolcher A, O'Shaughnessy JA, Denicoff AM, Noone M, Ognibene FP, Cowan KH, Balis FM. Effect of R-verapamil on the pharmacokinetics of paclitaxel in women with breast cancer. J Clin Oncol 1995;13:2039–2042.

Bogman K, Erne-Brand F, Alsenz J, Drewe J. The role of surfactants in the reversal of active transport mediated by multidrug resistance proteins. J Pharm Sci 2003;92:1250–1261.

Boote DJ, Dennis IF, Twentyman PR, Osborne RJ, Laburte C, Hensel S, Smyth JF, Brampton MH, Bleehen NM. Phase I study of etoposide with SDZ PSC 833 as a modulator of multidrug resistance in patients with cancer. J Clin Oncol 1996;14:610–618.

Borst P, Elferink RO. Mammalian ABC transporters in health and disease. Annu Rev Biochem 2002;71:537–592.

Borst P, Evers R, Kool M, Wijnholds J. A family of drug transporters: the multidrug resistance-associated proteins. J Nat Cancer Inst 2000;92:1295–1302.

Boulton DW, DeVane CL, Liston HL, Markowitz JS. *In vitro* P-glycoprotein affinity for atypical and conventional antipsychotics. Life Sci 2002;71:163–169.

Boyd R, Stern R, Stewart B, Wu X, Reyner E, Zegarac E, Randinitis E, Whitfield L. Atorvastatin coadministration may increase digoxin concentrations by inhibition of intestinal P-glycoprotein-mediated secretion. J Clin Pharmacol 2000;40:91–98.

Boyer JL, Meier PJ. Characterizing mechanisms of hepatic bile acid transport utilizing isolated membrane vesicles. Methods Enzymol 1990;192:517–533.

Breedveld P, Pluim D, Cipriani G, Wielinga P, Van Tellingen O, Schinkel AH, Schellens JHM. The effect of Bcrp1 (Abcg2) on the *in vivo* pharmacokinetics and brain penetration of imatinib mesylate (Gleevec): implications for the use of breast cancer resistance protein and P-glycoprotein inhibitors to enable the brain penetration of imatinib in patients. Cancer Res 2005;65:2577–2582.

Brown GR. Cephalosporin-probenecid drug interactions. Clin Pharmacokinet 1993; 24:289–300.

Buechler M, Koenig J, Brom M, Kartenbeck J, Spring H, Horie T, Keppler D. cDNA cloning of the hepatocyte canalicular isoform of the multidrug resistance protein, cMrp, reveals a novel conjugate export pump deficient in hyperbilirubinemic mutant rats. J Biol Chem 1996;271:15091–15098.

Buss N, Snell P, Bock J, Hsu A, Jorga K. Saquinavir and ritonavir pharmacokinetics following combined ritonavir and saquinavir (soft gelatin capsules) administration. Br J Clin Pharmacol 2001;52:255–264.

Calvo MV, Martin-Suarez A, Martin Luengo C, Avila C, Cascon M, Dominguez-Gil Hurle A. Interaction between digoxin and propafenone. Ther Drug Monit 1989;11:10–15.

Campbell SD, de Morais SM, Xu JJ. Inhibition of human organic anion transporting polypeptide OATP 1B1 as a mechanism of drug-induced hyperbilirubinemia. Chem Biol Interact 2004;150:179–187.

Center MS, Zhu Q, Sun H. Mutagenesis of the putative nucleotide-binding domains of the multidrug resistance associated protein (MRP). Analysis of the effect of these mutations on MRP mediated drug resistance and transport. Cytotechnology 1998;27:61–69.

Chang T, Benet LZ, Hebert MF. The effect of water-soluble vitamin E on cyclosporine pharmacokinetics in healthy volunteers. Clin Pharmacol Ther 1996;59:297–303.

Chang X-B, Hou Y-X, Riordan JR. Stimulation of ATPase activity of purified multidrug resistance-associated protein by nucleoside diphosphates. J Biol Chem 1998;273:23844–23848.

Chen C, Liu X, Smith BJ. Utility of Mdr1-gene deficient mice in assessing the impact of P-glycoprotein on pharmacokinetics and pharmacodynamics in drug discovery and development. Curr Drug Metab 2003a;4:272–291.

Chen C, Scott D, Hanson E, Franco J, Berryman E, Volberg M, Liu X. Impact of Mrp2 on the biliary excretion and intestinal absorption of furosemide, probenecid, and methotrexate using eisai hyperbilirubinemic rats. Pharm Res 2003b;20:31–37.

Christensen H, Asberg A, Holmboe AB, Berg KJ. Coadministration of grapefruit juice increases systemic exposure of diltiazem in healthy volunteers. Eur J Clin Pharmacol 2002;58:515–520.

Cohen A, Ullman B, Martin DW Jr. Characterization of a mutant mouse lymphoma cell with deficient transport of purine and pyrimidine nucleosides. J Biol Chem 1979;254:112–116.

Collnot EM, Baldes C, Wempe MF, Hyatt J, Navarro L, Edgar KJ, Schaefer UF, Lehr CM. Influence of vitamin E TPGS poly(ethylene glycol) chain length on apical efflux transporters in Caco-2 cell monolayers. J Control Release 2006;111:35–40.

Cui Y, Konig J, Leier I, Buchholz U, Keppler D. Hepatic uptake of bilirubin and its conjugates by the human organic anion transporter SLC21A6. J Biol Chem 2001;276:9626–9630.

Davit B, Reynolds K, Yuan R, Ajayi F, Conner D, Fadiran E, Gillespie B, Sahajwalla C, Huang S-M, Lesko LJ. FDA evaluations using *in vitro* metabolism to predict and interpret *in vivo* metabolic drug–drug interactions: impact on labeling. J Clin Pharmacol 1999;39:899–910.

Dean M, Fojo T, Bates S. Tumor stem cells and drug resistance. Nat Rev Cancer 2005;5:275–284.

Doyle LA, Yang W, Abruzzo LV, Krogmann T, Gao Y, Rishi AK, Ross DD. A multidrug resistance transporter from human MCF-7 breast cancer cells. Proc Nat Acad Sci USA 1998;95:15665–15670.

Dresser GK, Kim RB, Bailey DG. Effect of grapefruit juice volume on the reduction of fexofenadine bioavailability: possible role of organic anion transporting polypeptides. Clin Pharmacol Ther 2005;77:170–177.

Druekes P, Schinzel R, Palm D. Photometric microtiter assay of inorganic phosphate in the presence of acid-labile organic phosphates. Anal Biochem 1995;230:173–177.

Edwards DJ, Fitzsimmons ME, Schuetz EG, Yasuda K, Ducharme MP, Warbasse LH, Woster PM, Schuetz JD, Watkins P. 6′, 7′-Dihydroxybergamottin in grapefruit juice and Seville orange juice: effects on cyclosporine disposition, enterocyte CYP3A4, and P-glycoprotein. Clin Pharmacol Ther 1999;65:237–244.

Elferink RPJO, De Haan J, Lambert KJ, Hagey LR, Hofmann AF, Jansen PLM. Selective hepatobiliary transport of nordeoxycholate side chain conjugates in mutant rats with a canalicular transport defect. Hepatology (Philadelphia, PA, United States) 1989;9:861–865.

Endres CJ, Hsiao P, Chung FS, Unadkat JD. The role of transporters in drug interactions. Eur J Pharm Sci 2006;27:501–517.

Fedeli L, Colozza M, Boschetti E, Sabalich I, Aristei C, Guerciolini R, Del Favero A, Rossetti R, Tonato M, Rambotti P, Davis S. Pharmacokinetics of vincristine in cancer patients treated with nifedipine. Cancer 1989;64:1805–1811.

Fenster PE, Hager WD, Goodman MM. Digoxin–quinidine–spironolactone interaction. Clin Pharmacol Ther 1984;36:70–73.

Floren LC, Bekersky I, Benet LZ, Mekki Q, Dressler D, Lee JW, Roberts JP, Hebert MF. Tacrolimus oral bioavailability doubles with coadministration of ketoconazole. Clin Pharmacol Ther 1997;62:41–49.

Fojo AT, Ueda K, Slamon DJ, Poplack DG, Gottesman MM, Pastan I. Expression of a multidrug-resistance gene in human tumors and tissues. Proc Nat Acad Sci USA 1987;84:265–269.

Forman B, Dussault I, Synold TW. Methods for altering SXR activation using peptide mimetic HIV proteaseinhibitor SXR ligands to reduce drug resistance and drug clearance; US. Pat. Appl. Publ. 32 pp., Cont.-in-part of U.S. Ser. No. 815,300. USA,2002.

Fotoohi AK, Lindqvist M, Peterson C, Albertioni F. Involvement of the concentrative nucleoside transporter 3 and equilibrative nucleoside transporter 2 in the resistance of T-lymphoblastic cell lines to thiopurines. Biochem Biophys Res Commun 2006;343:208–215.

Garrigues A, Nugier J, Orlowski S, Ezan E. A high-throughput screening microplate test for the interaction of drugs with P-glycoprotein. Anal Biochem 2002;305:106–114.

Gati WP, Paterson ARP, Belch AR, Chlumecky V, Larratt LM, Mant MJ, Turner AR. Es nucleoside transporter content of acute leukemia cells: role in cell sensitivity to cytarabine (araC). Leuk Lymphoma 1998;32:45–54.

Goldberg R, Roth D. Evaluation of fluvastatin in the treatment of hypercholesterolemia in renal transplant recipients taking cyclosporine. Transplantation 1996;62:1559–1564.

Gottesman MM, Ambudkar SV. Overview: ABC transporters and human disease. J Bioenerg Biomembr 2001;33:453–458.

Gottesman MM, Pastan I, Ueda K.Methods of inducing multidrug resistance in mammalian cells using vectors carrying the human MDR1 gene; U.S. 11 pp., (United States Dept. of Health and Human Services, USA) 1999.

Gramatte T, Oertel R. Intestinal secretion of intravenous talinolol is inhibited by luminal R-verapamil. Clin Pharmacol Ther 1999;66:239–245.

Greiner B, Eichelbaum M, Fritz P, Kreichgauer HP, von Richter O, Zundler J, Kroemer HK. The role of intestinal P-glycoprotein in the interaction of digoxin and rifampin. [erratum appears in J Clin Invest 2002 Aug;110(4):571]. J Clin Invest 1999a;104:147–153.

Greiner B, Eichelbaum M, Fritz P, Kreichgauer H-P, von Richter O, Zundler J, Kroemer HK. The role of intestinal P-glycoprotein in the interaction of digoxin and rifampin. J Clin Invest 1999b;104:147–153.

Grem JL. Biochemical modulation of fluorouracil by dipyridamole: preclinical and clinical experience. Semin Oncol 1992;19:56–65.

Grub S, Bryson H, Goggin T, Ludin E, Jorga K. The interaction of saquinavir (soft gelatin capsule) with ketoconazole, erythromycin and rifampicin: comparison of the effect in healthy volunteers and in HIV-infected patients. Eur J Clin Pharmacol 2001;57:115–121.

Gupta SK, Bakran A, Johnson RW, Rowland M. Cyclosporin-erythromycin interaction in renal transplant patients. Br J Clin Pharmacol 1989;27:475–481.

Gupta S, Banfield C, Kantesaria B, Marino M, Clement R, Affrime M, Batra V. Pharmacokinetic and safety profile of desloratadine and fexofenadine when coadministered with azithromycin: a randomized, placebo-controlled, parallel-group study. Clin Ther 2001;23:451–466.

Gutmann H, Hruz P, Zimmermann C, Beglinger C, Drewe J. Distribution of breast cancer resistance protein (BCRP/ABCG2) mRNA expression along the human GI tract. Biochem Pharmacol 2005;70:695–699.

Haimeur A, Conseil G, Deeley RG, Cole SPC. The MRP-related and BCRP/ABCG2 multidrug resistance proteins: biology, substrate specificity and regulation. Curr Drug Metab 2004;5:21–53.

Hamman MA, Bruce MA, Haehner-Daniels BD, Hall SD. The effect of rifampin administration on the disposition of fexofenadine. Clin Pharmacol Ther (St. Louis, MO, United States) 2001;69:114–121.

Hebert MF, Fisher RM, Marsh CL, Dressler D, Bekersky I. Effects of rifampin on tacrolimus pharmacokinetics in healthy volunteers. J Clin Pharmacol 1999;39:91–96.

Hebert MF, Lam AY. Diltiazem increases tacrolimus concentrations. Ann Pharmacother 1999;33:680–682.

Hedman A, Angelin B, Arvidsson A, Dahlqvist R, Nilsson B. Interactions in the renal biliary elimination of digoxin: stereoselective difference between quinine and quinidine. Clin Pharmacol Ther (St. Louis, MO. United States) 1990;47:20–26.

Hedman A, Angelin B, Arvidsson A, Beck O, Dahlqvist R, Nilsson B, Olsson M, Schenck-Gustafsson K. Digoxin–verapamil interaction: reduction of biliary but not renal digoxin clearance in humans. Clin Pharmacol Ther 1991;49:256–262.

Hindmarch I. CNS effects of antihistamines: is there a third generation of non-sedative drugs? Clin Exp Allergy Rev 2002;2:26–31.

Hirano M, Maeda K, Shitara Y, Sugiyama Y. Contribution of OATP2 (OATP1B1) and OATP8 (OATP1B3) to the hepatic uptake of pitavastatin in humans. J Pharmacol Exp Ther 2004;311:139–146.

Hirano M, Maeda K, Matsushima S, Nozaki Y, Kusuhara H, Sugiyama Y. Involvement of BCRP (ABCG2) in the biliary excretion of pitavastatin. Mol Pharmacol 2005;68:800–807.

Ho ES, Lin DC, Mendel DB, Cihlar T. Cytotoxicity of antiviral nucleotides adefovir and cidofovir is induced by the expression of human renal organic anion transporter 1. J Am Soc Nephrol 2000;11:383–393.

Ho RH, Kim RB. Transporters and drug therapy: implications for drug disposition and disease. Clin Pharmacol Ther 2005;78:260–277.

Hoffman J. Murine erythroleukemia cells resistant to periodate-oxidized adenosine have lowered levels of nucleoside transporter. Adv Exp Med Biol 1991;309A: 443–446.

Hoffmaster KA, Turncliff RZ, LeCluyse EL, Kim RB, Meier PJ, Brouwer KLR. P-glycoprotein expression, localization, and function in sandwich-cultured primary rat and human hepatocytes: relevance to the hepatobiliary disposition of a model opioid peptide. Pharm Res 2004a;21:1294–1302.

Hoffmaster KA, Zamek-Gliszczynski MJ, Pollack GM, Brouwer KLR. Hepatobiliary disposition of the metabolically stable opioid peptide [D-Pen2, D-Pen5]-enkephalin (DPDPE): pharmacokinetic consequences of the interplay between multiple transport systems. J Pharmacol Exp Ther 2004b;311:1203–1210.

Hori R, Okamura N, Aiba T, Tanigawara Y. Role of P-glycoprotein in renal tubular secretion of digoxin in the isolated perfused rat kidney. J Pharmacol Exp Ther 1993;266:1620–1625.

Horikawa M, Kato Y, Tyson CA, Sugiyama Y. The potential for an interaction between MRP2 (ABCC2) and various therapeutic agents: probenecid as a candidate inhibitor of the biliary excretion of irinotecan metabolites. Drug Metab Pharmacokinet 2002;17:23–33.

Huang M-J, Kua K-E, Teng H-C, Tang K-S, Weng H-W, Huang C-S. Risk factors for severe hyperbilirubinemia in neonates. Pediatr Res 2004;56:682–689.

Ieiri I, Suzuki H, Kimura M, Takane H, Nishizato Y, Irie S, Urae A, Kawabata K, Higuchi S, Otsubo K, Sugiyama Y. Influence of common variants in the pharmacokinetic genes (OATP-C, UGT1A1, and MRP2) on serum bilirubin levels in healthy subjects. Hepatology Res 2004;30:91–95.

Ishizuka H, Konno K, Shiina T, Naganuma H, Nishimura K, Ito K, Suzuki H, Sugiyama Y. Species differences in the transport activity for organic anions across the bile canalicular membrane. J Pharmacol Exp Ther 1999;290:1324–1330.

Jalava KM, Partanen J, Neuvonen PJ. Itraconazole decreases renal clearance of digoxin. Ther Drug Monit 1997;19:609–613.

Johne A, Brockmoller J, Bauer S, Maurer A, Langheinrich M, Roots I. Pharmacokinetic interaction of digoxin with an herbal extract from St John's wort (*Hypericum perforatum*). Clin Pharmacol Ther 1999;66:338–345.

Johne A, Schmider J, Brockmoller J, Stadelmann AM, Stormer E, Bauer S, Scholler G, Langheinrich M, Roots I. Decreased plasma levels of amitriptyline and its metabolites on comedication with an extract from St. John's wort (*Hypericum perforatum*). J Clin Psychopharmacol 2002;22:46–54.

Jonker JW, Smit JW, Brinkhuis RF, Maliepaard M, Beijnen JH, Schellens JHM, Schinkel AH. Role of breast cancer resistance protein in the bioavailability and fetal penetration of topotecan. J Nat Cancer Inst 2000;92:1651–1656.

Jonker JW, Merino G, Musters S, van Herwaarden AE, Bolscher E, Wagenaar E, Mesman E, Dale TC, Schinkel AH. The breast cancer resistance protein BCRP (ABCG2) concentrates drugs and carcinogenic xenotoxins into milk. Nat Med 2005;11:127–129.

Juliano RL, Ling V. A surface glycoprotein modulating drug permeability in Chinese hamster ovary cell mutants. Biochim Biophys Acta Biomembr 1976; 455:152–162.

Kamisako T, Kobayashi Y, Takeuchi K, Ishihara T, Higuchi K, Tanaka Y, Gabazza EC, Adachi Y. Recent advances in bilirubin metabolism research: the molecular mechanism of hepatocyte bilirubin transport and its clinical relevance. J Gastroenterol 2000;35:659–664.

Kantola T, Kivisto KT, Neuvonen PJ. Erythromycin and verapamil considerably increase serum simvastatin and simvastatin acid concentrations. Clin Pharmacol Ther (St. Louis) 1998;64:177–182.

Kaplan B, Meier-Kriesche HU, Napoli KL, Kahan BD. The effects of relative timing of sirolimus and cyclosporine microemulsion formulation coadministration on the pharmacokinetics of each agent. Clin Pharmacol Ther 1998;63:48–53.

Kaukonen K-M, Olkkola KT, Neuvonen PJ. Itraconazole increases plasma concentrations of quinidine. Clin Pharmacol Ther (St. Louis) 1997;62:510–517.

Kemp DC, Zamek-Gliszczynski MJ, Brouwer KLR. Xenobiotics inhibit hepatic uptake and biliary excretion of taurocholate in rat hepatocytes. Toxicol Sci 2005;83:207–214.

Keppler D, Jedlitschky G, Leier I. Transport function and substrate specificity of multidrug resistance protein. Methods Enzymol 1998;292:607–616.

Kerb R. Implications of genetic polymorphisms in drug transporters for pharmacotherapy. Cancer Lett (Amsterdam, Netherlands) 2006;234:4–33.

Kerr DJ, Graham J, Cummings J, Morrison JG, Thompson GG, Brodie MJ, Kaye SB. The effect of verapamil on the pharmacokinetics of adriamycin. Cancer Chemother Pharmacol 1986;18:239–242.

Kilic T., Alberta JA, Zdunek PR, Acar M, Iannarelli P, O'Reilly T, Buchdunger E., Black PM, Stiles CD. Intracranial inhibition of platelet-derived growth factor-mediated glioblastoma cell growth by an orally active kinase inhibitor of the 2-phenylaminopyrimidine class. Cancer Res 2000;60:5143–5150.

Kivisto KT, Kantola T, Neuvonen PJ. Different effects of itraconazole on the pharmacokinetics of fluvastatin and lovastatin. Br J Clin Pharmacol 1998;46:49–53.

Kleinbloesem CH, van Brummelen P, Hillers J, Moolenaar AJ, Breimer DD. Interaction between digoxin and nifedipine at steady state in patients with atrial fibrillation. Ther Drug Monit 1985;7:372–376.

Kong W, Engel K, Wang J. Mammalian nucleoside transporters. Curr Drug Metab 2004;5:63–84.

Kovarik JM, Purba HS, Pongowski M, Gerbeau C, Humbert H, Mueller EA. Pharmacokinetics of dexamethasone and valspodar, a P-glycoprotein (mdr1) modulator: implications for coadministration. Pharmacotherapy 1998;18:1230–1236.

Kovarik JM, Rigaudy L, Guerret M, Gerbeau C, Rost KL. Longitudinal assessment of a P-glycoprotein-mediated drug interaction of valspodar on digoxin. Clin Pharmacol Ther 1999;66:391–400.

Kruijtzer CM, Beijnen JH, Rosing H, ten Bokkel Huinink WW, Schot M, Jewell RC, Paul EM, Schellens JH. Increased oral bioavailability of topotecan in combination with the breast cancer resistance protein and P-glycoprotein inhibitor GF120918. J Clin Oncol 2002;20:2943–2950.

Kullak-Ublick GA. Drug-induced cholestatic liver disease. Molecular Pathogenesis of Cholestasis. Landes Bioscience, Georgetown, Tx. 2004;256–265.

Kurisu H, Kamisaka K, Koyo T, Yamasuge S, Igarashi H, Maezawa H, Uesugi T, Tagaya O. Organic anion transport study in mutant rats with autosomal recessive conjugated hyperbilirubinemia. Life Sci 1991;49:1003–1011.

Kwan P, Brodie MJ. Potential role of drug transporters in the pathogenesis of medically intractable epilepsy. Epilepsia 2005;46:224–235.

Kwei GY, Alvaro RF, Chen Q, Jenkins HJ, Hop CEAC, Keohane CA, Ly VT, Strauss JR, Wang RW, Wang Z, Pippert TR, Umbenhauer DR. Disposition of ivermectin and cyclosporin A in CF-1 mice deficient in mdr1a P-glycoprotein. Drug Metab Dispos 1999;27:581–587.

Lai Y, Tse C-M, Unadkat JD. Mitochondrial expression of the human equilibrative nucleoside transporter 1 (hENT1) results in enhanced mitochondrial toxicity of antiviral drugs. J Biol Chem 2004;279:4490–4497.

Lankas GR, Cartwright ME, Umbenhauer D. P-glycoprotein deficiency in a subpopulation of CF-1 mice enhances avermectin-induced neurotoxicity. Toxicol Appl Pharmacol 1997;143:357–365.

Lau YY, Wu C-Y, Okochi H, Benet LZ. *Ex situ* inhibition of hepatic uptake and efflux significantly changes metabolism: hepatic enzyme-transporter interplay. J Pharmacol Exp Ther 2004;308:1040–1045.

Lau YY, Okochi H, Huang Y, Benet LZ. Multiple transporters affect the disposition of atorvastatin and its two active hydroxy metabolites: application of *in vitro* and *ex situ* systems. J Pharmacol Exp Ther 2006;316:762–771.

Lee E-W, Lai Y, Zhang H, Unadkat JD. Identification of the mitochondrial targeting signal of the human equilibrative nucleoside transporter 1 (hENT1): implications for interspecies differences in mitochondrial toxicity of fialuridine. J Biol Chem 2006.

Leis J, Stepan D, Curtin P, Ford J, Peng B, Schubach S, Druker B, Maziarz R. Central Nervous System Failure in Patients with Chronic Myelogenous Leukemia Lymphoid Blast Crisis and Philadelphia Chromosome Positive Acute Lymphoblastic Leukemia and Lymphoma 2004; Treated with Imatinib (STI-571). Leukemia– Lymphoma 2004;45(4):695–698.

Leggas M, Panetta JC, Zhuang Y, Schuetz JD, Johnston B, Bai F, Sorrentino B, Zhou S, Houghton PJ, Stewart CF. Gefitinib modulates the function of multiple ATP-binding cassette transporters in vivo. Cancer Res 2006;66:4802–4807.

Lessem J, Bellinetto A. Interaction between digoxin and the calcium antagonists nicardipine and tiapamil. Clin Ther 1983;5:595–602.

Li Q, Sai Y, Kato Y, Tamai I, Tsuji A. Influence of drugs and nutrients on transporter gene expression levels in Caco-2 and LS180 intestinal epithelial cell lines. Pharm Res 2003;20:1119–1124.

Lilja JJ, Kivisto KT, Neuvonen PJ. Grapefruit juice-simvastatin interaction: Effect on serum concentrations of simvastatin, simvastatin acid, and HMG-CoA reductase inhibitors. Clin Pharmacol Ther (St. Louis) 1998;64:477–483.

Lilja JJ, Neuvonen M, Neuvonen PJ. Effects of regular consumption of grapefruit juice on the pharmacokinetics of simvastatin. Br J Clin Pharmacol 2004;58:56–60.

Lindell M, Karlsson MO, Lennernas H, Pahlman L, Lang MA. Variable expression of CYP and Pgp genes in the human small intestine. Eur J Clin Invest 2003;33:493–499.

List AF, Kopecky KJ, Willman CL, Head DR, Persons DL, Slovak ML, Dorr R, Karanes C, Hynes HE, Doroshow JH, Shurafa M, Appelbaum FR. Benefit of cyclosporine modulation of drug resistance in patients with poor-risk acute myeloid leukemia: a Southwest Oncology Group study. Blood 2001;98:3212–3220.

Litman T, Zeuthen T, Skovsgaard T, Stein WD. Competitive, non-competitive and cooperative interactions between substrates of P-glycoprotein as measured by its ATPase activity. Biochim Biophys Acta (BBA) – Mol Basis of Disease 1997;1361:169–176.

Liu L, Pang KS. The roles of transporters and enzymes in hepatic drug processing. Drug Metab Dispos 2005;33:1–9.

Liu X, Lecluyse EL, Brouwer KR, Lightfoot RM, Lee JI, Brouwer KLR. Use of Ca^{2+} modulation to evaluate biliary excretion in sandwich-cultured rat hepatocytes. J Pharmacol Exp Ther 1999;289:1592–1599.

Loescher W, Potschka H. Role of drug efflux transporters in the brain for drug disposition and treatment of brain diseases. Prog Neurobiol (Amsterdam, Netherlands) 2005;76:22–76.

Lown KS, Mayo RR, Leichtman AB, Hsiao HL, Turgeon DK, Schmiedlin-Ren P, Brown MB, Guo W, Rossi SJ, Benet LZ, Watkins PB. Role of intestinal P-glycoprotein (mdr1) in interpatient variation in the oral bioavailability of cyclosporine. Clin Pharmacol Ther 1997;62:248–260.

Lu C, Miwa GT, Prakash SR, Gan L-S, Balani SK. A novel model for the prediction of drug–drug interactions in humans based on in vitro cyp phenotypic data. Drug Metab Dispos 2007;35:79–85.

Lum B, Kaubisch S, Yahanda A, Adler K, Jew L, Ehsan M, Brophy N, Halsey J, Gosland M, Sikic B. Alteration of etoposide pharmacokinetics and pharmacodynamics by cyclosporine in a phase I trial to modulate multidrug resistance. J Clin Oncol 1992;10:1635–1642.

Mackey JR, Baldwin SA, Young JD, Cass CE. Nucleoside transport and its significance for anticancer drug resistance. Drug Resis Updat 1998a;1:310–324.

Mackey JR, Mani RS, Selner M, Mowles D, Young JD, Belt JA, Crawford CR, Cass CE. Functional nucleoside transporters are required for gemcitabine influx and manifestation of toxicity in cancer cell lines. Cancer Res 1998b;58:4349–4357.

Mai I, Kruger H, Budde K, Johne A, Brockmoller J, Neumayer HH, Roots I. Hazardous pharmacokinetic interaction of Saint John's wort (*Hypericum perforatum*) with the immunosuppressant cyclosporin. Int J Clin Pharmacol Ther 2000;38:500–502.

Malingre MM, Beijnen JH, Rosing H, Koopman FJ, Jewell RC, Paul EM, Ten Bokkel Huinink WW, Schellens JH. Co-administration of GF120918 significantly increases the systemic exposure to oral paclitaxel in cancer patients. Br J Cancer 2001a;84:42–47.

Malingre MM, Richel DJ, Beijnen JH, Rosing H, Koopman FJ, Ten Bokkel Huinink WW, Schot ME, Schellens JHM. Coadministration of cyclosporine strongly enhances the oral bioavailability of docetaxel. J Clin Oncol 2001b;19:1160–1166.

Maxwell DL, Gilmour-White SK, Hall MR. Digoxin toxicity due to interaction of digoxin with erythromycin. BMJ 1989;298:572.

Meerum Terwogt JM, Malingre MM, Beijnen JH, ten Bokkel Huinink WW, Rosing H, Koopman FJ, van Tellingen O, Swart M, Schellens JHM. Coadministration of oral cyclosporin a enables oral therapy with paclitaxel. Clin Cancer Res 1999;5:3379–3384.

Meier PJ, Boyer JL. Preparation of basolateral (sinusoidal) and canalicular plasma membrane vesicles for the study of hepatic transport processes. Methods Enzymol 1990;192:534–545.

Meier PJ, St. Meier-Abt A, Barrett C, Boyer JL. Mechanisms of taurocholate transport in canalicular and basolateral rat liver plasma membrane vesicles. Evidence for an electrogenic canalicular organic anion carrier. J Biol Chem 1984;259:10614–10622.

Merino G, Jonker JW, Wagenaar E, van Herwaarden AE, Schinkel AH. The breast cancer resistance protein (BCRP/ABCG2) affects pharmacokinetics, hepatobiliary excretion, and milk secretion of the antibiotic nitrofurantoin. Mol Pharmacol 2005;67:1758–1764.

Mills E, Wu P, Johnston BC, Gallicano K, Clarke M, Guyatt G. Natural health product-drug interactions: a systematic review of clinical trials. Ther Drug Monit 2005;27:549–557.

Millward MJ, Webster LK, Rischin D, Stokes KH, Toner GC, Bishop JF, Olver IN, Linahan BM, Linsenmeyer ME, Woodcock DM. Phase I trial of cremophor EL with bolus doxorubicin. Clin Cancer Res 1998;4:2321–2329.

Mizuno N, Suzuki M, Kusuhara H, Suzuki H, Takeuchi K, Niwa T, Jonker JW, Sugiyama Y. Impaired renal excretion of 6-hydroxy-5,7-dimethyl-2-methylamino-4-(3-pyridylmethyl) benzothiazole (E3040) sulfate in breast cancer resistance protein (BCRP1/ABCG2) knockout mice. Drug Metab Dispos 2004;32:898–901.

Mori K, Kondo T, Kamiyama Y, Kano Y, Tominaga K. Preventive effect of Kampo medicine (Hangeshashin-to) against irinotecan-induced diarrhea in advanced non-small-cell lung cancer. Cancer Chemother Pharmacol 2003;51:403–406.

Mouly SJ, Paine MF, Watkins PB. Contributions of CYP3A4, P-glycoprotein, and serum protein binding to the intestinal first-pass extraction of saquinavir. J Pharmacol Exp Ther 2004;308:941–948.

Mousa O, Brater DC, Sunblad KJ, Hall SD. The interaction of diltiazem with simvastatin. Clin Pharmacol Ther 2000;67:267–274.

Muck W, Mai I, Fritsche L, Ochmann K, Rohde G, Unger S, Johne A, Bauer S, Budde K, Roots I, Neumayer HH, Kuhlmann J. Increase in cerivastatin systemic exposure after single and multiple dosing in cyclosporine-treated kidney transplant recipients. Clin Pharmacol Ther 1999;65:251–261.

Mulato AS, Ho ES, Cihlar T. Nonsteroidal anti-inflammatory drugs efficiently reduce the transport and cytotoxicity of adefovir mediated by the human renal organic anion transporter 1. J Pharmacol Exp Ther 2000;295:10–15.

Muller M, Jansen PLM. Molecular aspects of hepatobiliary transport. Am J Physiol 1997;272:G1285–G1303.

Neuvonen PJ, Kantola T, Kivisto KT. Simvastatin but not pravastatin is very susceptible to interaction with the CYP3A4 inhibitor itraconazole. Clin Pharmacol Ther (St. Louis) 1998;63:332–341.

Neville K, Parise RA, Thompson P, Aleksic A, Egorin MJ, Balis FM, McGuffey L, McCully C, Berg SL, Blaney SM. Plasma and Cerebrospinal Fluid Pharmacokinetics of Imatinib after Administration to Nonhuman Primates. Clin Cancer Res 2004;10:25257–2529.

Niemi M, Backman JT, Neuvonen M, Neuvonen PJ, Kivisto KT. Effects of rifampin on the pharmacokinetics and pharmacodynamics of glyburide and glipizide. Clin Pharmacol Ther (St. Louis, MO, United States) 2001;69:400–406.

Niemi M, Schaeffeler E, Lang T, Fromm MF, Neuvonen M, Kyrklund C, Backman JT, Kerb R, Schwab M, Neuvonen PJ, Eichelbaum M, Kivisto KT. High plasma pravastatin concentrations are associated with single nucleotide polymorphisms and haplotypes of organic anion transporting polypeptide-C (OATP-C. SLCO1B1). Pharmacogenetics 2004;14:429–440.

Niemi M, Kivisto KT, Hofmann U, Schwab M, Eichelbaum M, Fromm MF. Fexofenadine pharmacokinetics are associated with a polymorphism of the SLCO1B1 gene (encoding OATP1B1). Br J Clin Pharmacol 2005;59:602–604.

Niemi M, Kivisto KT, Diczfalusy U, Bodin K, Bertilsson L, Fromm MF, Eichelbaum M. Effect of SLCO1B1 polymorphism on induction of CYP3A4 by rifampicin. Pharmacogenet Genomics 2006a;16:565–568.

Niemi M, Pasanen MK, Neuvonen PJ. SLCO1B1 polymorphism and sex affect the pharmacokinetics of pravastatin but not fluvastatin. Clin Pharmacol Ther 2006b;80:356–366.

Ninomiya M, Ito K, Horie T. Functional analysis of dog multidrug resistance-associated protein 2 (MRP2) in comparison with rat MRP2. Drug Metab Dispos 2005;33:225–232.

Ocheltree SM, Shen H, Hu Y, Keep RF, Smith DE. Role and relevance of peptide transporter 2 (PEPT2) in the kidney and choroid plexus: in vivo studies with glycylsarcosine in wild type and PEPT2 knockout mice. J Pharm Exp Ther 2005;315:240–247.

Oerlemans R, van der Heijden J, Vink J, Dijkmans BAC, Kaspers GJL, Lems WF, Scheffer GL, Ifergan I, Scheper RJ, Cloos J, Assaraf YG, Jansen G. Acquired resistance to chloroquine in human CEM T cells is mediated by multidrug resistance-associated protein 1 and provokes high levels of cross-resistance to glucocorticoids. Arthritis Rheum 2006;54:557–568.

Ogihara T, Kamiya M, Ozawa M, Fujita T, Yamamoto A, Yamashita S, Ohnishi S, Isomura Y. What kinds of substrates show P-glycoprotein-dependent intestinal absorption? Comparison of verapamil with vinblastine. Drug Metab Pharmacokinet 2006;21:238–244.

Ohmori S, Kuriya S, Uesugi T, Horie T, Sagami F, Mikami T, Kawaguchi A, Rikihisa T, Kanakubo Y. Decrease in the specific forms of cytochrome P-450 in liver microsomes of a mutant strain of rat with hyperbilirubinuria. Res Commun Chem Pathol Pharmacol 1991;72:243–253.

Okamura N, Hirai M, Tanigawara Y, Tanaka K, Yasuhara M, Ueda K, Komano T, Hori R. Digoxin–cyclosporin A interaction: modulation of the multidrug transporter P-glycoprotein in the kidney. J Pharmacol Exp Ther 1993;266:1614–1619.

Ozvegy C, Litman T, Szakacs G, Nagy Z, Bates S, Varadi A, Sarkadi B. Functional characterization of the human multidrug transporter, ABCG2, expressed in insect cells. Biochem Biophys Res Commun 2001;285:111–117.

Ozvegy C, Varadi A, Sarkadi B. Characterization of drug transport, ATP hydrolysis, and nucleotide trapping by the human ABCG2 multidrug transporter. Modulation of substrate specificity by a point mutation. J Biol Chem 2002;277:47980–47990.

Paine MF, Wagner DA, Hoffmaster KA, Watkins PB. Cytochrome P450 3A4 and P-glycoprotein mediate the interaction between an oral erythromycin breath test and rifampin. Clin Pharmacol Ther (St. Louis, MO, United States) 2002;72:524–535.

Paul AJ, Tranquilli WJ, Seward RL, Todd KS, Jr, DiPietro JA. Clinical observations in collies given ivermectin orally. Am J Vet Res 1987;48:684–685.

Pedersen KE, Christiansen BD, Klitgaard NA, Nielsen-Kudsk F. Effect of quinidine on digoxin bioavailability. Eur J Clin Pharmacol 1983;24:41–47.

Perloff MD, Von Moltke LL, Stormer E, Shader RI, Greenblatt DJ. Saint John's wort: an *in vitro* analysis of P-glycoprotein induction due to extended exposure. Br J Pharmacol 2001;134:1601–1608.

Pfeifer H, Wassmann B, Hofmann WK, Komor M, Scheuring U, Brueck P, Binckebanck A, Schleyer E, Goekbuget N, Wolff T, Luebbert M, Leimer L, Gschaidmeier H, Hoelzer D, Ottmann OG. Risk and prognosis of central nervous system leukemia in patients with Philadelphia chromosome-positive acute leukemias treated with imatinib mesylate. Clin Cancer Res 2003;9:4674–4681.

Pichler A, Zelcer N, Prior JL, Kuil AJ, Piwnica-Worms D. *In vivo* RNA interference-mediated ablation of MDR1 P-Glycoprotein. Clin Cancer Res 2005;11:4487–4494.

Piscitelli SC, Burstein AH, Chaitt D, Alfaro RM, Falloon J. Indinavir concentrations and St John's wort. Lancet 2000;355:547–548.

Piscitelli SC, Burstein AH, Welden N, Gallicano KD, Falloon J. The effect of garlic supplements on the pharmacokinetics of saquinavir. Clin Infect Dis 2002;34:234–238.

Polli JW, Wring SA, Humphreys JE, Huang L, Morgan JB, Webster LO, Serabjit-Singh CS. Rational use of *in vitro* P-glycoprotein assays in drug discovery. J Pharmacol Exp Ther 2001;299:620–628, Promega. Available at http://www.promega.com/tbs/tb341/tb341.pdf.

Polli JW, Baughman TM, Humphreys JE, Jordan KH, Mote AL, Webster LO, Barnaby RJ, Vitulli G, Bertolotti L, Read KD, Serabjit-Singh CJ. The systemic exposure of an N-methyl-D-aspartate receptor antagonist is limited in mice by the P-glycoprotein and breast cancer resistance protein efflux transporters. Drug Metab Dispos 2004;32:722–726.

Ramachandra M, Ambudkar SV, Gottesman MM, Pastan I, Hrycyna CA. Functional characterization of a glycine 185-to-valine substitution in human P-glycoprotein by using a vaccinia-based transient expression system. Mol Biol Cell 1996;7:1485–1498.

Rao US. Mutation of glycine 185 to valine alters the ATPase function of the human P-glycoprotein expressed in Sf9 cells. J Biol Chem 1995;270:6686–6690.

Robinson K, Johnston A, Walker S, Mulrow JP, McKenna WJ, Holt DW. The digoxin–amiodarone interaction. Cardiovasc Drugs Ther 1989;3:25–28.

Rushing DA, Raber SR, Rodvold KA, Piscitelli SC, Plank GS, Tewksbury DA. The effects of cyclosporine on the pharmacokinetics of doxorubicin in patients with small cell lung cancer. Cancer 1994;74:834–841.

Sadeque AJ, Wandel C, He H, Shah S, Wood AJ. Increased drug delivery to the brain by P-glycoprotein inhibition. Clin Pharmacol Ther 2000a;68:231–237.

Sadeque AJM, Wandel C, He H, Shah S, Wood AJJ. Increased drug delivery to the brain by P-glycoprotein inhibition. Clin Pharmacol Ther (St. Louis) 2000b;68:231–237.

Sadler BM, Gillotin C, Lou Y, Eron JJ, Lang W, Haubrich R, Stein DS. Pharmacokinetic study of human immunodeficiency virus protease inhibitors used in combination with amprenavir. Antimicrob Agents Chemother 2001;45:3663–3668.

Sarkadi B, Price EM, Boucher RC, German UA, Scarborough GA. Expression of the human multidrug resistance cDNA insect cells generates a high activity drug-stimulated membrane ATPase. J Biol Chem 1992;267:4854–4858.

Sarkadi B, Homolya L, Hollo Z. Assay and reagent kit for evaluation of multi-drug resistance in cells; U.S. pp 17 pp., Cont.-in-part of U.S. 15,872,014, (Solvo Biotechnology, Hung). 2001.

Sasongko L, Link JM, Muzi M, Mankoff DA, Yang X, Collier AC, Shoner SC, Unadkat JD. Imaging P-glycoprotein transport activity at the human blood–brain barrier with positron emission tomography. Clin Pharmacol Ther (New York, NY, United States) 2005;77:503–514.

Satoh K, Shirabe S, Eguchi K, Yamauchi A, Kataoka Y, Niwa M, Nishida N, Katamine S. Toxicity of quinacrine can be reduced by co-administration of P-Glycoprotein inhibitor in sporadic creutzfeldt-jakob disease. Cell Mol Neurobiol 2004;24:873–875.

Scarborough GA. Drug-stimulated ATPase activity of the human P-glycoprotein. J Bioenerg Biomembr 1995;27:37–41.

Schwarz UI, Gramatte T, Krappweis J, Berndt A, Oertel R, von Richter O, Kirch W. Unexpected effect of verapamil on oral bioavailability of the [beta]-blocker talinolol in humans. Clin Pharmacol Ther 1999;65:283–290.

Schwarz UI, Gramatte T, Krappweis J, Oertel R, Kirch W. P-glycoprotein inhibitor erythromycin increases oral bioavailability of talinolol in humans. Int J Clin Pharmacol Ther 2000;38:161–167.

Senior AE, al-Shawi MK, Urbatsch IL. ATP hydrolysis by multidrug resistance protein from Chinese hamster ovary cells. J Bioenerg Biomembr 1995;27:31–36.

Shapiro AB, Corder AB, Ling V. P-glycoprotein-mediated Hoechst 33342 transport out of the lipid bilayer. Eur J Biochem 1997;250:115–121.

Shapiro AB, Fox K, Lam P, Ling V. Stimulation of P-glycoprotein-mediated drug transport by prazosin and progesterone. Evidence for a third drug-binding site. Eur J Biochem 1999;259:841–850.

Shepard RL, Cao J, Starling JJ, Dantzig AH. Modulation of P-glycoprotein but not MRP1- or BCRP-mediated drug resistance by LY335979. Int J Cancer 2003;103: 121–125.

Shitara Y, Li AP, Kato Y, Lu C, Ito K, Itoh T, Sugiyama Y. Function of uptake transporters for taurocholate and estradiol 17b-D-glucuronide in cryopreserved human hepatocytes. Drug Metab Pharmacokinet 2003;18:33–41.

Shitara Y, Sato H, Sugiyama Y. Evaluation of drug–drug interaction in the hepatobiliary and renal transport of drugs. Ann Rev Pharmacol Toxicol 2005;45:689–723.

Shitara Y, Horie T, Sugiyama Y. Transporters as a determinant of drug clearance and tissue distribution. Eur J Pharm Sci 2006;27:425–446.

Shitara Y, Sugiyama Y. Pharmacokinetic and pharmacodynamic alterations of 3-hydroxy-3-methylglutaryl coenzyme A (HMG-CoA) reductase inhibitors: drug–drug interactions and interindividual differences in transporter and metabolic enzyme functions. Pharmacol Ther 2006;112:71–105.

Siedlik P, Olson S, Yang B, Stern R. Erythromycin coadministration increases plasma atorvastatin concentrations. J Clin Pharmacol 1999;39:501–504.

Siegmund W, Ludwig K, Engel G, Zschiesche M, Franke G, Hoffmann A, Terhaag B, Weitschies W. Variability of intestinal expression of P-glycoprotein in healthy volunteers as described by absorption of talinolol from four bioequivalent tablets. J Pharm Sci 2003;92:604–610.

Sigel E. Use of Xenopus oocytes for the functional expression of plasma membrane proteins. J Membr Biol 1990;117:201–221.

Simonson SG, Raza A, Martin PD, Mitchell PD, Jarcho JA, Brown CD, Windass AS, Schneck DW. Rosuvastatin pharmacokinetics in heart transplant recipients administered an antirejection regimen including cyclosporine. Clin Pharmacol Ther 2004;76:167–177.

Simpson K, Jarvis B. Fexofenadine: a review of its use in the management of seasonal allergic rhinitis and chronic idiopathic urticaria. Drugs 2000;59:301–321.

Sobrero AF, Handschumacher RE, Bertino JR. Highly selective drug combinations for human colon cancer cells resistant *in vitro* to 5-fluoro-2'-deoxyuridine. Cancer Res 1985a;45:3161–3163.

Sobrero AF, Moir RD, Bertino JR, Handschumacher RE. Defective facilitated diffusion of nucleosides, a primary mechanism of resistance to 5-fluoro-2'-deoxyuridine in the HCT-8 human carcinoma line. Cancer Res 1985b;45:3155–3160.

Solary E, Witz B, Caillot D, Moreau P, Desablens B, Cahn J-Y, Sadoun A, Pignon B, Berthou C, Maloisel F, Guyotat D., Casassus P, Ifrah N, Lamy Y, Audhuy B, Colombat P, Harousseau JL. Combination of Quinine as a potential reversing agent with mitoxantrone and cytarabine for the treatment of acute leukemias a randomized multicenter study. Blood 1996;88:1198–1205.

Solary E, Drenou B, Campos L, De Cremoux P, Mugneret F, Moreau P, Lioure B, Falkenrodt A, Witz B, Bernard M, Hunault-Berger M, Delain M, Fernandes J,

Mounier C, Guilhot F, Garnache F, Berthou C, Kara-Slimane F, Harousseau J-L. Quinine as a multidrug resistance inhibitor: a phase 3 multicentric randomized study in adult *de novo* acute myelogenous leukemia. Blood 2003;102:1202–1210.

Sonneveld P, Suciu S, Weijermans P, Beksac M, Neuwirtova R, Solbu G, Lokhorst H, van der Lelie J, Dohner H, Gerhartz H, Segeren CM, Willemze R, Lowenberg B. Cyclosporin A combined with vincristine, doxorubicin and dexamethasone (VAD) compared with VAD alone; in patients with advanced refractory multiple myeloma: an EORTC-HOVON randomized phase III study (06914). Br J Haematol 2001;115:895–902.

Sparreboom A, van Asperen J, Mayer U, Schinkel AH, Smit JW, Meijer DKF, Borst P, Nooijen WJ, Beijnen JH, van Tellingen O, Limited oral bioavailability and active epithelial excretion of paclitaxel (Taxol) caused by P-glycoprotein in the intestine. PNAS 1997;94:2031–2035.

Sparreboom A, Planting AS, Jewell RC, van der Burg ME, van der Gaast A, de Bruijn P, Loos WJ, Nooter K, Chandler LH, Paul EM, Wissel PS, Verweij J. Clinical pharmacokinetics of doxorubicin in combination with GF120918, a potent inhibitor of MDR1 P-glycoprotein. Anticancer Drugs 1999;10:719–728.

Sparreboom A, Danesi R, Ando Y, Chan J, Figg WD. Pharmacogenomics of ABC transporters and its role in cancer chemotherapy. Drug Resist Updat 2003;6:71–84.

Synold TW, Dussault I, Forman BM. The orphan nuclear receptor SXR coordinately regulates drug metabolism and efflux. Nat Med (New York, NY, United States) 2001;7:584–590.

Szakacs G, Paterson JK, Ludwig JA, Booth-Genthe C, Gottesman MM. Targeting multidrug resistance in cancer. Nat Rev Drug Discov 2006;5:219–234.

Tabas LB, Dantzig AH. A high-throughput assay for measurement of multidrug resistance protein-mediated transport of leukotriene C4 into membrane vesicles. Anal Biochem 2002;310:61–66.

Tahara H, Kusuhara H, Fuse E, Sugiyama Y. P-glycoprotein plays a major role in the efflux of fexofenadine in the small intestine and blood–brain barrier, but only a limited role in its biliary excretion. Drug Metab Dispos 2005;33:963–968.

Takeda M, Khamdang S, Narikawa S, Kimura H, Hosoyamada M, Cha SH, Sekine T, Endou H. Characterization of methotrexate transport and its drug interactions with human organic anion transporters. J Pharmacol Exp Ther 2002;302: 666–671.

ten Tije AJ, Verweij J, Loos WJ, Sparreboom A. Pharmacological effects of formulation vehicles: implications for cancer chemotherapy. Clin Pharmacokinet 2003;42:665–685.

Thiebaut F, Tsuruo T, Hamada H, Gottesman MM, Pastan I, Willingham MC. Cellular localization of the multidrug-resistance gene product P-glycoprotein in normal human tissues. Proc Nat Acad Sci USA 1987;84:7735–7738.

Thoern M, Finnstroem N, Lundgren S, Rane A, Loeoef L. Cytochromes P450 and MDR1 mRNA expression along the human gastrointestinal tract. Br J Clin Pharmacol 2005;60:54–60.

Tokui T, Nakai D, Nakagomi R, Yawo H, Abe T, Sugiyama Y. Pravastatin, an HMG-CoA reductase inhibitor, is transported by rat organic anion transporting polypeptide, oatp2. Pharm Res 1999,16:904–908.

Tsuruoka S, Ioka T, Wakaumi M, Sakamoto K, Ookami H, Fujimura A. Severe arrhythmia as a result of the interaction of cetirizine and pilsicainide in a patient with renal insufficiency: first case presentation showing competition for excretion via renal multidrug resistance protein 1 and organic cation transporter 2. Clin Pharmacol Ther 2006;79:389–396.

van der Heijden J, de Jong MC, Dijkmans BAC, Lems WF, Oerlemans R, Kathmann I, Schalkwijk CG, Scheffer GI, Scheper RJ, Jansen G. Development of sulfasalazine resistance in human T cells induces expression of the multidrug resistance transporter ABCG2 (BCRP) and augmented production of TNFa. Ann Rheum Dis 2004a;63:138–143.

van der Heijden J, de Jong MC, Dijkmans BAC, Lems WF, Oerlemans R, Kathmann I, Scheffer GI, Scheper RJ, Assaraf YG, Jansen G. Acquired resistance of human T cells to sulfasalazine: stability of the resistant phenotype and sensitivity to non-related DMARDs. Ann Rheum Dis 2004b;63:131–137.

van Zuylen L, Verweij J, Nooter K, Brouwer E, Stoter G, Sparreboom A. Role of intestinal P-glycoprotein in the plasma and fecal disposition of docetaxel in humans. Clin Cancer Res 2000;6:2598–2603.

Velpandian T, Jasuja R, Bhardwaj RK, Jaiswal J, Gupta SK. Piperine in food: interference in the pharmacokinetics of phenytoin. Eur J Drug Metab Pharmacokine 2001;26:241–248.

Volk EL, Schneider E. Wild type breast cancer resistance protein (BCRP/ABCG2) is a methotrexate polyglutamate transporter. Cancer Res 2003;63:5538–5543.

Wakasugi H, Yano I, Ito T, Hashida T, Futami T, Nohara R, Sasayama S, Inui K. Effect of clarithromycin on renal excretion of digoxin: interaction with P-glycoprotein. Clin Pharmacol Ther 1998;64:123–128.

Wang E-J, Casciano CN, Clement RP, Johnson WW. In vitro flow cytometry method to quantitatively assess inhibitors of P-glycoprotein. Drug Metab Dispos 2000;28:522–528.

Wang E-j, Casciano CN, Clement RP, Johnson WW. Fluorescent substrates of sister-P-glycoprotein (BSEP) evaluated as markers of active transport and inhibition: evidence for contingent unequal binding sites. Pharm Res 2003;20:537–544.

Wang Z, Hamman MA, Huang S-M, Lesko LJ, Hall SD. Effect of St John's wort on the pharmacokinetics of fexofenadine. Clin Pharmacol Ther (St. Louis, MO, United States) 2002;71:414–420.

Wattel E, Solary E, Hecquet B, Caillot D, Ifrah N, Brion A, Mahe B, Milpied N, Janvier M, Guerci A, Rochant H, Cordonnier C, Dreyfus F, Buzyn A, Hoang-Ngoc L, Stoppa AM, Gratecos N, Sadoun A, Stamatoulas A, Tilly H, Brice P, Maloisel F, Lioure B, Desablens B, Pignon B, Abgrall JP, Leporrier M, Dupriez B, Guyotat D, Lepelley P, Fenaux P. Quinine improves the results of intensive chemotherapy in myelodysplastic syndromes expressing P glycoprotein: results of a randomized study. Br J Haematol 1998;102:1015–1024.

Wattel E, Solary E, Hecquet B, Caillot D, Ifrah N, Brion A, Milpied N, Janvier M, Guerci A, Rochant H, Cordonnier C, Dreyfus F, Veil A, Hoang-Ngoc L, Stoppa AM, Gratecos N, Sadoun A, Tilly H, Brice P, Lioure B, Desablens B, Pignon B, Abgrall JP, Leporrier M, Dupriez B, Guyotat D, Lepelley P, Fenaux P. Quinine improves results of intensive chemotherapy (IC) in myelodysplastic syndromes (MDS) expressing P-glycoprotein (PGP): updated results of a randomized study. Adv Exp Med Biol 1999;457:35–46.

Weller S, Blum MR, Doucette M, Burnette T, Cederberg DM, Miranda Pd, Smiley ML. Pharmacokinetics of the acyclovir pro-drug valaciclovir after escalating single- and multiple-dose administration to normal volunteers. Clin Pharmacol Ther (St. Louis) 1993;54:595–605.

Westphal K, Weinbrenner A, Giessmann T, Stuhr M, Franke G, Zschiesche M, Oertel R, Terhaag B, Kroemer HK, Siegmund W. Oral bioavailability of digoxin is enhanced by talinolol: evidence for involvement of intestinal P-glycoprotein. Clin Pharmacol Ther 2000a;68:6–12.

Westphal K, Weinbrenner A, Giessmann T, Stuhr M, Franke G, Zschiesche M, Oertel R, Terhaag B, Kroemer HK, Siegmund W. Oral bioavailability of digoxin is enhanced by talinolol: evidence for involvement of intestinal P-glycoprotein [see comment]. Clin Pharmacol Ther 2000b;68:6–12.

Westphal K, Weinbrenner A, Zschiesche M, Franke G, Knoke M, Oertel R, Fritz P, von Richter O, Warzok R, Hachenberg T, Kauffmann HM, Schrenk D, Terhaag B, Kroemer HK, Siegmund W. Induction of P-glycoprotein by rifampin increases intestinal secretion of talinolol in human beings: a new type of drug/drug interaction. Clin Pharmacol Ther 2000c;68:345–355.

Westphal K, Weinbrenner A, Zschiesche M, Franke G, Knoke M, Oertel R, Fritz P, von Richter O, Warzok R, Hachenberg T, Kauffmann H-M, Schrenk D, Terhaag B, Kroemer HK, Siegmund W. Induction of P-glycoprotein by rifampin increases intestinal secretion of talinolol in human beings: a new type of drug/drug interaction. Clin Pharmacol Ther 2000d;68:345–355.

Wiley JS, Jones SP, Sawyer WH, Paterson ARP. Cytosine arabinoside influx and nucleoside transport sites in acute leukemia. J Clin Invest 1982;69:479 489.

Wiley JS, Taupin J, Jamieson GP, Snook M, Sawyer WH, Finch LR. Cytosine arabinoside transport and metabolism in acute leukemias and T cell lymphoblastic lymphoma. J Clin Invest 1985;75:632–642.

Wolff NC, Richardson JA, Egorin M, Ilaria RL Jr. The CNS is a sanctuary for leukemic cells in mice receiving Imatinib mesylate for Bcr/Abl-induced leukemia. Blood 2003;101:5010–5013.

Wright AMP, Paterson ARP, Sowa B, Akabutu JJ, Grundy PE, Gati WP. Cytotoxicity of 2-chlorodeoxyadenosine and arabinosylcytosine in leukaemic lymphoblasts from paediatric patients: significance of cellular nucleoside transporter content. Br J Haematol 2002;116:528–537.

Wu C-Y, Benet LZ. Predicting drug disposition via application of BCS: transport/absorption/elimination interplay and development of a biopharmaceutics drug disposition classification system. Pharm Res 2005;22:11–23.

Xia CQ, Liu N, Miwa G, Gan L-S. Use of in vitro models to study interactions between the breast cancer resistant protein and its ligands. Drug Metab Rev 2004;36:53.

Xia CQ, Liu N, Yang D, Miwa G, Gan L-S. Expression, localization, and functional characteristics of breast cancer resistance protein in Caco-2 cells. Drug Metab Dispos 2005a;33:637–643.

Xia CQ, Yang JJ, Gan L-S. Breast cancer resistance protein in pharmacokinetics and drug–drug interactions. Expert Opin Drug Metab Toxicol 2005b;1:595–611.

Xia CQ, Xiao G, Liu N, Pimprale S, Fox L, Patten CJ, Crespi CL, Miwa G, Gan L-S. Comparison of species differences of P-Glycoproteins in beagle dog, rhesus monkey, and human using ATPase activity assays. Mol Pharm 2006;3:78–86.

Xu D, Kang H, Fisher M, Juliano RL. Strategies for inhibition of MDR1 gene expression. Mol Pharmacol 2004;66:268–275.

Yamazaki M, Akiyama S, Nishigaki R, Sugiyama Y. Uptake is the rate-limiting step in the overall hepatic elimination of pravastatin at steady-state in rats. Pharm Res 1996;13:1559–1564.

Yamazaki M, Akiyama S, Ni'inuma K, Nishigaki R, Sugiyama Y. Biliary excretion of pravastatin in rats: contribution of the excretion pathway mediated by canalicular multispecific organic anion transporter (cMOAT). Drug Metab Dispos 1997;25:1123–1129.

Yin JAL, Wheatley K, Rees JKH, Burnett AK. Comparison of \"sequential\" versus\ "standard\" chemotherapy as re-induction treatment, with or without cyclosporine, in refractory/relapsed acute myeloid leukemia (AML): results of the UK medical research council AML-R trial. Br J Haemat 2001;113:713–726.

Yu D. The contribution of P-glycoprotein to pharmacokinetic drug–drug interactions. J Clin Pharmacol 1999;39:1203–1211.

Zaher H, Khan AA, Palandra J, Brayman TG, Yu L, Ware JA. Breast cancer resistance protein (Bcrp/abcg2) is a major determinant of sulfasalazine absorption and elimination in the mouse. Mol Pharm 2006;3:55–61.

Zaidenstein R, Soback S, Gips M, Avni B, Dishi V, Weissgarten Y, Golik A, Scapa E. Effect of grapefruit juice on the pharmacokinetics of Losartan and its active metabolite E3174 in healthy volunteers. Ther Drug Monit 2001;23:369–373.

Zamek-Gliszczynski MJ, Hoffmaster KA, Tian X, Zhao R, Polli JW, Humphreys JE, Webster LO, Bridges AS, Kalvass JC, Brouwer KLR. Multiple mechanisms are involved in the biliary excretion of acetaminophen sulfate in the rat: role of Mrp2 and BCRP1. Drug Metab Dispos 2005;33:1158–1165.

7

REGULATORY CONSIDERATIONS OF DRUG METABOLISM AND DRUG INTERACTION STUDIES

XIAOXIONG WEI AND MINGSHE ZHU

7.1 INTRODUCTION

Historically, drug metabolism research in the pharmaceutical industry concentrated on radiolabeled absorption, distribution, metabolism, and excretion (ADME) studies in humans and preclinical species. These studies were conducted in the late phase of drug development mainly for supporting new drug regulatory registration. In the late 1990s, regulatory guidances on drug metabolism and drug–drug interactions (DDI) were issued by the regulatory authorities in the United States (FDA, 1997, 1999) and Europe, which define the role and practice of the contemporary drug metabolism in drug development and registration. Since then, drug metabolism research in the industry has focused on to early *in vitro* assessment of potential DDI of drug candidates, including screening of CYP enzyme inhibition and induction, and reaction phenotyping of metabolizing enzymes. Consequently, the drug metabolism organization or function within industry was rapidly extended to drug discovery, and ultimately drug metabolism has become a vital and integrated part of the drug discovery and development process (Kola and Landis, 2004; Smith et al., 2002). In 2005, the US Food and Drug Administration (FDA) published a draft guidance entitled "Safety Testing of Drug Metabolites" (FDA, 2005b). This guidance emphasizes the critical role of animal and human ADME studies in the selection of metabolites for

Drug Metabolism in Drug Design and Development, Edited by Donglu Zhang, Mingshe Zhu and W. Griffith Humphreys Copyright © 2008 John Wiley & Sons, Inc.

bioanalysis in preclinical and clinical studies and for safety evaluation *in vitro* and in toxicological animal species. To avoid delays in the drug development process due to the discovery of unique human metabolites in the late stages of clinical trials, some pharmaceutical companies have moved radiolabeled human ADME studies to early stages of clinical trials.

Regulatory guidances not only direct trends in drug metabolism research in the pharmaceutical industry but also provide detailed recommendations on experimental design, study execution, and data interpretation. For example, the newly published FDA draft guidance on DDI (FDA, 2006a) includes specific CYP probe substrates, inhibitors and inducers for *in vitro* and *in vivo* drug interaction studies. To provide high quality and complete drug metabolism data for drug development and registration, it is essential to follow regulatory guidances when designing, conducting, and reporting drug metabolism studies.

The FDA, the Ministry of Health & Welfare in Japan, and the European Medicinal Evaluation Agency (EMEA) in Europe are three major regional regulatory agencies. Each of the agencies publishes their own regulatory guidelines. In addition, to achieve greater harmonization in the interpretation and application of regulatory guidelines for drug product registration and to reduce or obviate the need to duplicate the testing carried out during the research and development of new medicines, the International Conference on Harmonization of Technical Requirements for Registration of Pharmaceuticals for Human Use (ICH) publishes guidances. EMEA publishes and distributes the Step 2 ICH guidelines for comments and endorses the Step 4 guidelines (http://www.emea.eu.int). The FDA publishes a notice with the full text of the Step 2 or Step 4 ICH guidances in the Federal Register. Step 4 guidances are available for use on the date they are published in the Federal Register (CDER: http://www.fda.gov/cder/guidance/index.htm). In this chapter, we briefly review the FDA and the ICH guidances relevant to drug metabolism. In particular, from the regulatory perspective, the overall strategies, study designs, data interpretation, and special considerations in the assessment of metabolite safety and drug–drug interaction potential are discussed.

7.2 REGULATORY GUIDANCES RELEVANT TO DRUG METABOLISM

Table 7.1 lists selected ICH and the FDA guidances that are associated with drug metabolism, such as the guidances on drug interaction (FDA, 1999, 1997, 2006a) and safety testing of metabolites (FDA, 2005b); or focus on other subjects with more limited relevance to drug metabolism, such as guidances on toxicokinetics (ICH, 1995a) and regulatory submissions (FDA, 1995, 2003a). In the following sections, we attempt to give a brief description of these regulatory guidances.

TABLE 7.1   Selected ICH and the US FDA guidances relevant to drug metabolism and drug–drug interaction studies.

Regulatory guidance	ICH	FDA
Toxicokinetics	S3A—Toxicokinetics: the assessment of systemic exposure in toxicity studies, (ICH, 1995A)	
Use of radioactive drugs		CFR Title 21, Part 361.1— Radioactive drugs for certain research Uses (FDA, 2005a)
Safety assessment of metabolites	S1C—Dose selection for carcinogenicity studies of pharmaceuticals (ICH, 1995b)	Carcinogenicity study protocol submissions (FDA, 2002) Safety testing of drug metabolites (FDA, 2005b)
Drug interactions		Drug metabolism/drug interaction studies in the drug development process: studies *in vitro* (FDA, 1997) *In vivo* metabolism/drug interaction studies— study design, data analysis, and recommendations for dosing and labeling (FDA, 1999) Drug interaction studies—study design, data analysis, and implications for dosing and labeling (FDA, 2006a)
Analytical method	Q3B(R) - Impurities in new drug substances (ICH, 2006) Q2(R1) - Validation of analytical procedures: text and methodology (ICH, 2005)	Bioanalytical method validation (ICH, 1997b)
Quality and compliance	E6—Good clinical practice: consolidated guidance (ICH, 1997)	Good laboratory practice for nonclinical laboratory. CFR 21, Part 58 (FDA, 2003d) Part 11, Electronic records; electronic signatures—scope and application (FDA, 2003c)

TABLE 7.1   (*Continued*)

Regulatory guidance	ICH	FDA
Regulatory submission format and contents	M4—Common technical document for the registration of pharmaceuticals for human use	IND meetings for human drugs and biologics (FDA, 2001b) Content and format of investigational new drug applications (INDs) for phase I studies of drugs (FDA, 1995) CFR Title 21, Part 312—Investigational new drug application (FDA, 2003a) CFR Title 21, 314—Applications for FDA approval to market a new drug (FDA, 2003b) Labeling for human prescription drug and biological products— implementing the new content and format requirements (FDA, 2006b)

7.2.1   Toxicokinetic Studies

The ICH published a guidance for toxicokinetics in 1995 entitled "the Assessment of Systemic Exposure in Toxicity Studies" (ICH, S3A, Table 7.1) (ICH, 1995a). In order to aid in the interpretation of toxicology findings and promote rational study design development, this document gives guidance on developing testing strategies in toxicokinetics and the need to integrate pharmacokinetics into toxicity testing. The primary objective of toxicokinetics is to evaluate systemic exposure achieved in animals and its relationship to dose levels in toxicity studies (Cayen, 1995a, 1995b). Assessment of the systemic exposure in toxicokinetic studies is mainly carried out by measurement of drug concentrations in plasma or other body fluids using a validated bioanalytical assay. However, in some cases, measurement of metabolite concentrations in plasma or excreta is important in the conduct of toxicokinetics when (1) a metabolite is the primary active entity of a prodrug, (2) a metabolite is pharmacologically or toxicologically active and makes a significant contribution to tissue/organ responses, and (3) the measurement of a metabolite in plasma or tissues is the only practical means of estimating exposure of drug-related components because the parent drug is very extensively metabolized. The decision to monitor a circulating metabolite by a validated bioanalytical assay in toxicity studies often relies on metabolite profiles of a radiolabeled drug in the toxicology species as well as knowledge of pharmacological activity or toxicity of the metabolite.

7.2.2 Use of Radiolabeled Materials

The FDA regulation of the use of radiolabeled drugs in drug development was published in 21 CFR 361.1 Radioactive Drug for Certain Research Uses (FDA, 2005a). It defines the objectives of the use of radioactive drugs in human subjects to obtain basic information regarding metabolism, human physiology, pathophysiology, and biochemistry, but not intended for immediate therapeutic or diagnostic purposes or to determine the safety and effectiveness of the drug. To ensure the safe and effective use of radioactive drugs, the document requires that studies with a radiolabeled drug should be approved by Radioactive Drug Research Committee. Doses used in these studies should not exceed the dose limitations applicable to the separate administration of the active ingredients with nonradiolabeled materials, and the recommended maximal dose of radioactivity. Tissue distribution studies in rats can provide the data to calculate dosimetry for the use of the radioactive drugs in human subjects. The details of the rat tissue distribution study design and radioactivity dosimetry calculation are described in Chapter 18.

7.2.3 Metabolite Safety Assessment

Traditionally, plasma concentrations of a parent drug have been used as an index of systemic exposure to drug-related materials in animals and humans. When drug blood or plasma concentrations in toxicological species are higher than those in humans, it is assumed that potential clinical safety of the drug has been appropriately evaluated during standard nonclinical toxicity studies. However, in some cases, the toxicity of drugs is mediated by their metabolites. Therefore, if a metabolite is present in humans but absent in toxicological species or has higher exposure levels in humans than in the toxicological species, the safety of the metabolite may not be adequately assessed in toxicity studies with the animal models. To address the concern of metabolite safety assessment, the FDA guidance "Carcinogenicity Study Protocol Submissions" (FDA, 2002) (Table 7.1) recommends that the following drug metabolism information be included in the carcinogenicity study protocol:

(1) Metabolite profiles in humans and in the species employed in carcinogenicity study. In cases where *in vivo* metabolism data are unavailable, *in vitro* data can be used.

(2) Toxicokinetic data that are sufficient to estimate steady state AUC (0–24) for the drug and each major human metabolites at doses employed in the range finding study.

(3) Exposure [steady state AUC (0–24) data] for the drug and major metabolites in humans.

Similar suggestions are also included in the ICH Guideline, "Dose Selection for Carcinogenicity Studies of Pharmaceuticals" (ICH, 1995).

The "Metabolite in Safety Testing" (MIST) document was published by a committee of the Pharmaceutical Research and Manufactures of America (PhRMA) in May 2002 (Baillie et al., 2002). It outlines the thinking of the pharmaceutical industry on how to generate and use metabolite data in support of drug safety programs, and aimed to promote continued discussions among industrial scientists and representatives of the regulatory agencies. Three years later, "Safety Testing of Drug Metabolites" (FDA 2005b, Table 7.1), a FDA draft guidance was issued. It is the first regulatory document worldwide to provide recommendations on "when and how to identify, characterize, and evaluate the safety of unique human metabolites and major metabolites of small molecule (nonbiologic) drug products." This guidance recommends a battery of nonclinical toxicity studies for directly assessing the safety of a metabolite that is produced only in humans or formed to a much greater extent in humans than toxicological animal species. Metabolite safety testing is a very complex subject. It involves quantitative and qualitative metabolite profiling in humans and animal species with radiolabeled drugs, pharmacological activity testing of the major metabolites in human circulation, monitoring of metabolites in selected clinical and nonclinical studies using a validated bioanalytical assay, and nonclinical toxicity evaluation of selected metabolites. The entire process requires a great deal of resources and can take several years. As the FDA guidance indicates, "the discovery of unique or major human metabolites late in drug development can cause development delays and could have possible implications for marketing approval." On the contrary, decisions to select metabolites for pharmacological activity testing, monitoring in humans and animals, and direct safety testing should be considered on a case-by-case basis. The FDA has recommended that drug sponsors contact the agency early to discuss such situations. The MIST document and the FDA draft guidance have opened another important chapter of drug metabolism research in the pharmaceutical industry, and the discussions on how to conduct metabolite safety testing will continue. (Guengerich, 2006; Hastings, 2003; Humphreys and Unger, 2006; Prueksaritanont, 2006; Smith and Obach, 2005, 2006; Davis-Bruno and Atrakchi, 2006) In Section 7.3, we discuss basic concepts, general approaches, and special considerations involved in safety evaluation of metabolites.

7.2.4 Drug–Drug Interaction Studies

The FDA guidance "Drug Metabolism/Drug Interaction Studies in the Drug Development Process: Studies *in Vitro*" (FDA, 1997), and the European Agency for the Evaluation of Medicinal Products (EMEA) guidance "Note for Guidance on the Investigation of Drug Interactions" were issued in 1997. The FDA guidance "*in Vivo* Drug Metabolism/Drug Interaction Studies" was issued two years later (FDA, 1999). These regulatory documents, for the first time, clearly define the role and practice of drug metabolism in the drug development and registration process by stating that "the metabolism of an investigational new drug should be defined during drug development and that its interactions with other drugs should be explored as part of an adequate assessment of its safety and effectiveness." More specifically, "the primary metabolic pathway(s), the primary site of metabolism for the drug, and the proportion of the total clearance that each primary metabolic pathway constitutes should be determined at an early stage. Furthermore, the enzymes responsible for the metabolism, potential effects of their inhibition or induction, and possible polymorphism in the metabolism should be investigated." These documents and similar drug interaction guidances issued by the Ministry of Health, Labor and Welfare in Japan (MHLW) describe strategies, general methodologies and points for considerations in designing and assessing *in vitro* and *in vivo* drug interaction studies. To address specific designs of the drug interaction studies that are not presented in these guidances, and to harmonize the study designs and experimental approaches, the recommendations of a workshop attended by experts from academia, industry, and regulatory bodies were published (Tucker, 2001). Furthermore, the PhRMA published a white paper representing their opinion on the conduct of DDI studies (Bjornsson et al., 2003). Recently, the FDA published a new draft guidance on drug interaction studies (Table 7.1) (FDA, 2006a). It covers both *in vitro* and *in vivo* drug interaction studies with a much more detailed description of study design and data interpretation compared to the two previously published guidances (Table 7.1). In addition, transporter-mediated drug interactions are included in the regulatory guidance for the first time. In Section 7.4, special considerations in drug interaction studies are discussed.

7.2.5 Analytical Method Validation and Compliance

The FDA guidance for industry "Bioanalytical Method Validation" (FDA, 2001a) (Table 7.1) and the ICH guidance for "Validation of Analytical Procedures: Text and Methodology" (ICH, 2005) provide general recommendations for validation of bioanalytical assays using LC/MS and other analytical techniques. These assays are mainly applied to the quantitative determination of drugs and their metabolites in biological matrices, including blood, plasma, urine or bile, in clinical and nonclinical studies that require pharmacokinetic evaluation. The FDA also recommends consulting the ICH Q3A guidance "Impurities in New Drug Substances" (ICH, 2006) (Table 7.1) with regard to the development of analytical methods for measuring metabolites in selected matrices. Maintaining records and submitting information to the FDA electronically from bioanalytical studies should follow FDA guidance, Part 11, Electronic Records: Electronic Signatures— Scope and Applications (FDA, 2003c) (Table 7.1). However, there are no specific FDA or ICH requirements or regulations with respect to analytical methods used in drug metabolism studies. However, following the spirit of Good Laboratory Practice (GLP) or GLP-like procedures has been recommended by representatives from the pharmaceutical industry, regulatory agencies, and academia (Bjornsson et al., 2003; Tucker et al., 2001). A detailed description of GLP can be found on the FDA guidance website, Good Laboratory Practice for Nonclinical Laboratory (FDA, 2003d) (Table 7.1). On the contrary, like any clinical study, human ADME studies should follow the ICH guidance "Good Clinical Practices" (GCP) (ICH, 1997) (Table 7.1). This document describes the responsibilities and expectations of all participants in the conduct of clinical trials and covers aspects of monitoring, reporting, and archiving of clinical trials and incorporating addenda into the essential documents and into the investigator's brochure.

7.2.6   Regulatory Submission Format and Content

7.2.6.1 Investigational New Drug Application (IND)   The FDA regulations require a sponsor to submit an IND to the agency before the initiation of human studies in the United States for an investigational drug. The general principles underlying the IND submission and the general requirements for IND content and format are published in the CFR Title 21, Part 312 Investigational New Drug Application (FDA, 2003a). These documents suggest including a summary of the pharmacological effects and mechanism(s) of action of the drug in animals, and information on the absorption, distribution, metabolism, and excretion of the drug, if known, in the section of pharmacology and drug disposition of IND submission. The FDA guidance "Content and Format on Investigational New Drug Applications for Phase I Studies of Drugs, Including Well-Characterized, Therapeutic, Biotechnology-derived Products" (Table 7.1) clarifies requirements for data and data presentation outlined in 21 CFR 312.22 and 312.23.

7.2.6.2 Investigator's Brochure (IB)   The IB is a collection of the clinical and nonclinical data on the investigational product that support the rationale for the proposed clinical trial and the safe use of the product. The IB is provided to investigators and others involved in clinical trials. As a part of the IND submission and the clinical trials applications in the European Commission, the IB is also submitted to regulatory authorities. The IB document should be updated annually to include new clinical and nonclinical information and should be presented in the format of summaries. The ICH guidance (E6, Good Clinical Practice, Table 7.1) (ICH, 1997) describes the content and format of IB documents. It suggests including a summary of the pharmacokinetics and product metabolism in animals in the section of Nonclincal Studies, and a summary of the pharmacokinetics and product metabolism in humans in the section of Effects in Humans.

7.2.6.3 New Drug Application (NDA) (FDA, 2003b)   The FDA document "Applications for FDA Approval to Market a New Drug" (CFR Title 21, 314), sets forth procedures and requirements for the submission to, and the review by, the FDA of applications to market a new drug. The content and format of an NDA is described in 21 CFR 314.50 section. A summary of the application, technical sections, samples and labeling, and other parts must be included in the NDA. The technical sections consist of (1) chemistry, manufacturing, and controls, (2) nonclinical pharmacology and toxicology, (3) human pharmacokinetics and bioavailability, (4) microbiology, (5) clinical data, (6) statistical section, and (7) pediatric use. Any studies of the absorption, distribution, metabolism, and excretion of the drug in animals should be included in the technical section (2). A summary of analysis of the pharmacokinetics and metabolism of the active ingredients should be included in the technical section (3).

7.2.6.4 Common Technical Document for the Registration of Pharmaceuticals for Human Use (CTD) ICH, 2001)   The CTD provides for a harmonized structure, content, and format for new product applications. The document is divided into four separate sections. The four sections address the application organization (M4 organize), the quality section (M4Q), the safety section (M4S) and the efficacy section (M4E) of the harmonized application. Drug metabolism and pharmacokinetic data should be summarized in the safety section that includes the nonclinical overview, nonclinical written summaries, and nonclinical tabulated summaries. More specifically, in Section 2.6.4.5 metabolism (interspecies comparison) of the written summaries, the following data should be included:

- Chemical structures and quantities of metabolites in biological samples
- Possible metabolic pathways
- Presystemic metabolism (GI/hepatic first-pass effects)
- *in vitro* metabolism including P450 studies
- Enzyme induction and inhibition

The CTD also provides templates for the preparation of tabulated summaries. Similar templates are often used for IND filing by some of pharmaceutical companies.

7.3   METABOLISM STUDIES RELEVANT TO METABOLITE SAFETY ASSESSMENT

7.3.1   Goals and General Strategies

One of major tasks of drug metabolism in drug development is to determine the exposure of drug metabolites in humans and toxicological species, especially the assessment of whether human circulating metabolites are present in the plasma of toxicology species at appropriate levels. The task usually is accomplished by qualitative and quantitative profiling in human and animal

ADME studies with radiolabeled material. Comparative metabolic profiling data, especially metabolite profiles in plasma across species, have very important implications for drug development, including:

(1) Selection and/or validation of the most appropriate toxicological animal species to be used for drug safety testing (Guengerich, 2006), such as carcinogenicity studies (ICH, 1995b);
(2) Selection of metabolites for pharmacological and/or biological activity testing (Baillie et al., 2002);
(3) Selection of metabolites for monitoring by a validated bioanalytical method in selected nonclinical toxicology studies and clinical studies (Baillie et al., 2002; Humphreys and Unger, 2006);
(4) Dtermination of whether the safety of a particular human metabolite need to be directly evaluated in toxicology studies (FDA, 2005b).

In general, metabolite profiles of a nonradiolabeled therapeutic agent in liver tissues from humans and animals are obtained in drug discovery, which are used for selection of animal species for pre-IND safety evaluation. Later on, *in vitro* metabolite profiling of a radiolabeled drug is often conducted to gain better quantitative and qualitative analysis evaluation of metabolites. Radiolabeled ADME studies provide the ultimate assessment of the similarity of biotrans formation between humans and the toxicology species. Animal ADME studies are usually conducted from late drug discovery to early stages of drug development. Human ADME studies, which require far more resources and follow more rigorous regulatory requirements, are carried out from late Phase I trials to the early Phase III studies. In some cases, the exposure to a specific metabolite is further analyzed using a validated bioanalytical assay in selected clinical and toxicological studies in which radiolabeled drugs are not suitable (FDA, 2001a).

Finding a unique and/or major metabolite in human circulation late in drug development could slow down the drug development process. Therefore, the FDA has recommended "the *in vivo* metabolic evaluation in humans be performed as early as feasible" (FDA, 2005b).

7.3.2 *In vitro* Metabolite Profiling Studies

In vitro metabolite profiles are often obtained for early assessment of the similarity of drug metabolism across species (FDA 2005b and 2006a). Liver microsomes and hepatocytes are the most used *in vitro* models for this purpose although several other *in vitro* models, including liver slits, liver S9 fractions, and recombinant metabolizing enzymes, are available. Liver microsomes have higher CYP enzyme activities, while hepatocytes contain a more comprehensive set of metabolizing enzymes, transporters, and cofactors. If analytical sensitivity is not an issue, hepatocytes should be the first choice for comparative metabolite profiling. On the contrary, if the metabolism rate of a test drug in hepatocytes is relatively low, resulting in difficulty in metabolite detection, or if metabolite profiles between human liver microsomes and hepaotcytes are qualitatively similar, the liver microsome model is a good choice for the profiling of *in vitro* metabolites across species.

For the purpose of metabolite profiling, *in vitro* models have several advantages over *in vivo* ADME studies (1) *In vitro* metabolism studies need much less resources. (2) Due to lower amounts of endogenous components present in *in vitro* systems than those in plasma, urine, and feces, identification of *in vitro* metabolites by LC/MS and quantitative analysis by LC/UV can be readily accomplished without using radiolabeled compounds. (3) *In vitro* metabolism models allow for determining the enzymes involved in a bioactivation reaction. (4) *In vitro* metabolism experiments can provide meaningful assessment of reactive metabolite formation by measuring covalent protein binding or using trapping agents. However, *in vitro* metabolite profiles are not consistent with those in plasma in many cases. Additionally, *in vitro* metabolism models often underestimated the secondary metabolites or metabolites formed from non-CYP-mediated biotransformation reactions.

7.3.3   ADME Studies

In vivo metabolite profiling data are routinely obtained from ADME studies using radiolabeled material (Marathe, 2004; Dalvie, 2000). An ADME study consists of two parts: (1) mass balance measurement by analyzing the total radioactivity in excreta, such as urine and feces, and/or bile, and (2) quantitative and qualitative metabolite profiling of biological matrices, including plasma, urine, bile, and/or feces (see Chapters 9 and 18). Radiochromatograms in plasma reveal the relative abundance (percentage of the total radioactivity) of circulating radiolabeled components. Consequentially, concentrations of individual radioactive components in the plasma, which are utilized for calculating exposure to metabolites, can be derived from the relative abundance of radioactive peaks and the concentration of total radioactivity in the plasma. The total amount of a metabolite as a percent of dose can be calculated based on the total radioactivity excreted into urine and feces (or bile), and the relative abundance of this metabolite in urine and feces (or bile). Results from ADME studies, together with pharmacological and/or toxicological activity data of the metabolite, are essential for the determination of whether a specific metabolite needs to be monitored in toxicological and clinical studies.

7.3.4   Analytical Methods for Metabolite Profiling

Metabolite profiling involves detection, quantitative analysis, and structural elucidation of drug metabolites present in a biological matrix using HPLC coupled with a UV detector, radioactivity detector, and/or mass spectrometer.

Information obtained from metabolite profiling includes the number, relative concentrations, and structures of metabolites present in a biological sample. LC/MS/MS is the most used analytical technique for sensitive metabolite detection and structural elucidation (see Chapter 11). However, the sensitivity and response of an LC/MS instrument depend on metabolite structure and biological matrix. Therefore, metabolite quantification by LC/MS without metabolite standards may not be reliable. On-line UV detection is useful for semiquantification of high concentrations of metabolites when the parent drug has a good UV chromophore. In the drug discovery stage when radiolabeled compounds are not available, LC/UV/MS is the most often used analytical technique for profiling *in vitro* and *in vivo* metabolites. Quantitative metabolite profiling in the late discovery and development stages, especially *in vivo* metabolite profiling, is mainly accomplished using radiolabeled materials. Liquid radiochromatographic techniques, including online radioflow detection, and off-line microplate scintillation counting (Zhu et al., 2005a, 2005b), provide sensitive and accurate measurement of the radiolabeled drug and its metabolites. LC/radiodetection/MS is the choice for metabolite profiling in radiolabeled ADME studies. Application of radiochromatographic techniques to metabolite profiling is described in Chapter 10.

7.3.5 Special Considerations

7.3.5.1 Terms Involved in Metabolite Safety Testing

Exposure   This term refers to the systemic exposure to drug-related material. It can be expressed as the plasma AUC of administered drug or metabolite within a dosage interval, typically assessed at steady state. The FDA metabolite safety draft guidance also uses the percent of the dose as the measurement of exposure to a metabolite.

Major Metabolites   The MIST document defines a major metabolite "as one that accounts for 25% or more of the exposure to circulating drug-related material." The FDA draft guidance defines it as "a metabolite in humans that accounts for plasma levels greater than 10% percent of the administered dose or systemic exposure, whichever is less." The "major metabolite" concept used in the FDA draft guidance refers to two types of metabolites: (1) a circulating metabolite that accounts for 10% or more of the total drug-related components in the plasma, and (2) a metabolite that accounts for 10% or more of the administered dose.

Unique Human Metabolites   The FDA draft guidance document defines the term as "a metabolite only in humans." The MIST document refers to a unique human metabolite as a metabolite identified in the plasma of humans but not in animals. However, it is very rare to find a truly unique human metabolite that is not formed in any animal species (Davis-Bruno and Atrakchi, 2006).

Active Metabolites or Pharmacologically Active Metabolites   The term refers to a metabolite that has target pharmacological activity greater than, equal to, or less than the parent drug.

Important Active Metabolites   The MIST document defines an important active metabolite "as one which is deemed to make a significant contribution to the total known pharmacological activity of a given dose of the parent compound." The FDA draft guidance document does not use this term.

Structural Alerts   The term refers to a metabolite that belongs to a chemical class with known toxicity, a metabolite that contains positive structural alerts for certain toxicity, or a GSH adduct/mercapturic acid conjugate indicative of a trapped reactive intermediate.

7.3.5.2 Selection of Metabolite(s) for Pharmacological Activity Testing

Understanding whether a circulating metabolite in humans makes a significant contribution to the drug-related target pharmacological effect has important implications for selection of the metabolite for monitoring using a bioanalytical method and for safety testing. Therefore, it is a common practice to test the pharmacological activity of the major human plasma metabolites. It should be also considered for pharmacological activity testing if a unique metabolite is identified in humans or a significant nonlinear PK/PD relationship is observed in humans or preclinical species. Pharmacological activity testing of a metabolite, especially a plasma metabolite, often requires time- and resource-consuming processes, namely definitive metabolite identification by LC/MS and NMR followed by chemical synthesis. In some cases large-scale metabolite isolation from urine or bile may provide sufficient amounts of metabolite(s) for activity testing. Because of resource restrictions, it is often not feasible to test the pharmacological activity for all major metabolites in humans or animals. Both the MIST document and the FDA draft guidance have suggested that the pharmacological activity of metabolites is an important factor in the consideration of metabolite monitoring and safety testing. However, these documents do not elaborate what types of metabolites should be tested for receptor-mediated pharmacological activity. We believe that the issue should be handled on a case-by-case basis.

7.3.5.3 Selection of Metabolite(s) for Monitoring in Clinical Studies and Nonclinical Toxicological Studies

The objective of the metabolite measurement using a bioanalytical assay, such as an LC/MS method, is to determine its pharmacokinetic parameters (such as AUC and C_{max} values) of a specific metabolite. The MIST document recommends quantifying certain human plasma metabolites, including "major metabolites," "important active metabolites," and "metabolites associated with structural alerts" using a validated bioanalytical assay in selected clinical and nonclinical studies. However, metabolite monitoring in clinical and nonclinical studies for the purpose of metabolite safety testing is not discussed in detail in the FDA draft guidance. Smith and Obach have proposed a decision-making process for selecting human plasma metabolites for monitoring in toxicological and clinical studies based on their absolute concentrations in the circulation rather than relative abundance and pharmacological activity (Smith and Obach, 2006). The question of what metabolites to be monitored in what studies will continue to be debated (Humphreys and Unger, 2006; Prueksaritanont et al., 2006). In general, radiolabeled human and animal ADME studies after a single dose of radiolabeled drug can provide the necessary information on what are the major or unique human metabolites. This consideration is consistent with the MIST document and the FDA metabolite safety draft guidance, in which radioactive metabolite profiling data are considered to be the key information resource for the assessment of metabolite exposure. If the exposure of metabolites in humans and toxicological species is evaluated properly in radiolabeled ADME studies, metabolite monitoring by a validated analytical assay in additional clinical and nonclinical studies may be not necessary. Quantification of selected metabolites by a validated LC/MS assay can provide more accurate or confirmatory assessment of metabolite exposure, which is especially useful in the measurement of the exposure to metabolites in special populations or after multiple dosing.

7.3.5.4 *Selection of Metabolite(s) for Monitoring in Toxicokinetic Studies*   The primary objective of toxicokinetic studies is to evaluate system exposure achieved in animals and its relationship to the dose level in the toxicity studies (Table 7.1). The assessment of the systemic exposure in the toxicity studies is mainly carried out by measurement of drug concentrations in plasma or other body fluids using a validated bioanalytical assay. However, in some cases the measurement of metabolite concentrations in plasma or excreta is important, particularly when the metabolite is the primary active entity of a prodrug or is the pharmacologically or toxicologically active compound (ICH S3A). In addition, significant plasma metabolites that are present at the levels equivalent to or higher than those of the parent drugs in circulation are often monitored in the toxicokinetic studies. Metabolic profiling data from animal ADME studies are essential in selecting appropriate analytes to be measured in certain matrixes. For example, a unique toxic metabolite in animal species or a metabolite that is a major in animal plasma and minor in human plasma may be considered for monitoring for toxicokinetic evaluation, but not necessarily for formal metabolite safety testing.

7.3.5.5 *Recommended Studies for Metabolite Safety Testing*   The FDA metabolite safety testing draft guidance outlines four kinds of safety studies for assessing the safety of unique or major human metabolites. These include (1) general toxicity studies with multiple dosing of the metabolite to species used for toxicological testing, (2) genotoxicity studies that consist of one *in vitro* assay to detect point mutations and another to detect chromosomal aberrations, (3) embryo–fetal development studies for a drug candidate that is interned for use in women of childbearing potential, and (4) carcinogenicity studies. The guidance also suggests these safety studies be completed and the study reports be submitted to the agency before beginning large-scale Phase III trials if toxicity studies of a human metabolite are warranted. However, since safety testing of drug metabolites is very complex, there is no clear cut-off criteria for if direct safety testing of a metabolite is necessary. Both members from regulatory agencies and scientists from the pharmaceutical industry have emphasized that these complex issues should be treated on a case-by-case basis (Davis-Bruno and Atrakchi, 2006).

Unique Human Metabolites   The FDA metabolite safety testing guidance indicates that the unique human metabolite(s) should be considered for safety assessment in animals, especially those that have pharmacological activity or associated with a structural alert. However, there are some situations where a full set of safety studies for unique human metabolites may not be necessary, such as if the metabolite is very minor in human plasma or it is only found in human feces.

Major Metabolites in Human Circulation   The FDA draft guidance suggests a major metabolite in human circulation (>10% of drug-related material), which is not present at comparable or higher levels in animals, to be considered for direct safety evaluation. On the contrary, direct safety assessment is not necessary for a major human plasma metabolite that is present at sufficient levels in animal circulation.

Major Metabolites in Human Excreta   The FDA draft guidance suggests considering direct safety evaluation for a major human excreted metabolite (>10% of dose) if it is not formed at comparable or higher levels in toxicological species. The concept of the percentage of dose reflects the total body burden of a metabolite. For example, an excretory metabolite that represents 15% of dose (200 mg/subject) has a total body burden 30 mg. The percentage of dose is a useful measurement of exposure to a metabolite that represents a major metabolic pathway, but may not be present at significant levels in plasma.

Conjugated Metabolites   Glucuronide and sulfate conjugates are usually pharmacologically inactive, and they are rarely associated with a toxic response. These conjugates are more water soluble and more readily eliminated than oxidative metabolites. Therefore, the treatment of conjugated metabolites with respect to metabolite safety testing may warrant a less stringent safety testing paradigm than that of oxidative metabolites (Davis-Bruno and Atrakchi, 2006).

Reactive Metabolites or Intermediates   Reactive metabolites are chemically unstable and often rapidly bind to proteins or glutathione (GSH). In general, GSH adducts are not present at significant levels in circulation, because they undergo rapid elimination via direct biliary excretion or further metabolism to mercapturic acids that are excreted into urine. Reactive metabolites can be monitored via quantitative analysis of their GSH adducts or mercapturic acids in excreta. However, it is not practical to test the toxicity of reactive metabolites on animals directly since the synthesis and dosing preparation of reactive metabolites are extremely difficult or impossible due to their chemical instability (Davis-Bruno and Atrakchi, 2006).

7.4   DRUG–DRUG INTERACTION STUDIES

Drug interactions are part of the assessment of drug safety and effectiveness. In the past, adverse pharmacokinetic drug interactions were one of the most frequent reasons that the drugs failed in drug development or needed to be withdrawn from markets (Prentis et al., 1988). The failure related to drug interactions/drug metabolism in drug development has been greatly reduced in the United States since the FDA issued the guidance for *in vitro* drug metabolism studies in 1997 and the guidance for *in vivo* drug interaction studies in 1999 (Kola and Landis, 2004). Now industry sponsors address issues of drug metabolism and drug interactions in the early phase drug development. Interactive dialogues between the regulatory agencies and sponsors have generated a positive impact on new drug development programs. The lessons learned from past experiences have helped in both regulatory agencies and industry sponsors understand the potential consequences of incomplete understanding and the necessary measures needed to minimize risk resulting from severely harmful drug interactions. A risk management plan has to be considered when a new, strong CYP inhibitory drug is being developed for oral administration, especially when CYP3A4 is the isoform that is affected. The understanding of the potential for drug–drug interaction continues to advance with more knowledge about drug transporters (Kim, 2006), emergence of the role of CYP2C8 (Totah and Rettie, 2005), the classification of CYP inhibitors, among others. Recently, the FDA issued its latest draft guidance for drug interaction studies (FDA, 2006a). In addition, the agency has listed on its Web site more detailed information regarding the selection of CYP substrates and inhibitors and the experimental design for studying effects on transporters (http://www.fda.gov/cder/drug/drugInteractions/default.htm-overview). In this section, the main focus is on special considerations in the conduct of drug interaction studies.

7.4.1   General Strategies

Drug interaction studies should be planned in the early exploratory phase of drug development. The *in vivo* studies should be conducted according to the findings of *in vitro* studies. Negative findings from *in vitro* work can help eliminate the need for related *in vivo* studies. Proper use of suitable substrates and inhibitors can help interpret the outcome of drug interaction studies. Experience with a wide range of new drugs and behavior with regard to drug–drug interaction has allowed the regulatory authorities to accept the results of representative *in vitro* and *in vivo* studies involving relevant CYP enzymes. The results can be applied to all drugs that are metabolized via the same CYP enzymes in order to predict potential outcome of metabolic drug interactions.

7.4.1.1 In vitro Studies *In vitro* drug metabolism and drug interaction studies should be conducted during an early phase of drug development. The purposes of these early phase *in vitro* studies are to identify (1) the drug-metabolizing enzymes responsible for the metabolic pathway of the new drug and the potential for other drugs to modify the metabolism of the new drug, and (2) the potential for metabolic drug interactions including inhibition and induction between the new drug and concomitant drugs.

In vitro studies can frequently serve as a screening tool to rule out the involvement of a particular metabolic pathway and the drug–drug interactions that occur through that pathway, so that subsequent *in vivo* studies are not needed. The 2006 FDA draft guidance for drug interactions states that since CYP3A, CYP2C8, CYP2C9, CYP2C19 enzymes, and the P-glycoprotein (P-gp) are all generally induced through the same mechanism, *in vivo* induction studies of the new drug on CYP3A may be sufficient to indicate that the new drug does or does not induce these CYP enzymes and P-gp. However, this judgment should be based on appropriately validated experimental methods and rational selection of substrate/interacting drug concentrations.

Appropriate *in vitro* methods are essential in order to draw reliable conclusions and to make a decision about the need for subsequent *in vivo* studies. A FDA survey of 194 new drugs approved in the United States from 1992 to 1997 indicated that industry investigators used different probe reactions to represent the same CYP enzyme activities for evaluating the modulatory potential of a new drug (Yuan et al., 2002). Some of those are not selective enough for the CYP enzymes of interest and are not generally preferred substrates. It is important to use "preferred substrates," to which the regulatory agencies and the scientific community have agreed (Bjornsson et al., 2003). The Office of Clinical Pharmacology of the FDA has listed on its Web site the preferred substrates and inhibitors useful for *in vitro* drug metabolism and drug interaction studies.

The 2006 draft FDA guidance for drug interactions suggests that *in vitro* studies should be conducted to identify drug metabolizing enzymes if the metabolic pathway contributes more than 25% of a drug's clearance based on human *in vivo* data. Whenever possible, these *in vitro* experiments should be conducted with drug concentrations deemed appropriate by kinetic experiments. Enzyme identification experiments should be conducted under initial rate conditions, where the linearity of metabolite production rate is assumed

TABLE 7.2 The *in vitro* prediction and actual *in vivo* drug interaction between erythromycin and midazolam.

$K_i{}^a$	$C_{max}{}^b$	$[I]/K_i$	Actual $AUC_I/AUC_{CON}{}^c$
8.8 μM	2.9 μM	0.33	>fourfold

[a]Atkinson et al. (2005).
[b]Yu et al. (2001).
[c]Olkkola et al. (1993).

with respect to time and enzyme concentrations. The 2006 draft guidance for drug interaction studies adds CYP2B6 and CYP2C8 as required enzymes for *in vitro* and *in vivo* studies.

7.4.1.2 Time-Dependent Inhibition Time-dependent inhibition of CYP enzymes should be conducted as a routine part of *in vitro* screening studies, because the prediction of the extent of drug–drug interactions by the new drug may be underestimated significantly by using competitive models if the new drug is a mechanism based inhibitor. For instance, the K_i value for inhibiting midazolam metabolism by erythromycin is quite large, and the prediction using $[I]/K_i$ would far underestimate the extent of inhibitory drug interaction between these two drugs (Table 7.2). In general, a various time of preincubation with a new drug before the addition of substrate is recommended. Any time-dependent loss of initial product formation rate indicates mechanism-based inhibition. Detection of time-dependent inhibition kinetics *in vitro* may suggest the need for follow-up with *in vivo* studies in humans after multiple dose administration.

7.4.1.3 [I]/K_i In the past few years, NDA and IND submissions increasingly contain prediction of *in vivo* drug interactions using *in vitro* drug metabolism data, although a complete understanding of the relationship between the *in vitro* findings and *in vivo* results of metabolism/drug–drug interaction studies is still emerging. The most popular and simplest approach is to directly apply $[I]/K_i$ to the prediction of fold increase in exposure when an inhibitor is present. With this approach, [I] represents the mean steady-state C_{max} value for total drug (bound plus unbound) following administration of the highest therapeutic clinical dose and the K_i is the inhibition constant (Table 7.3). The consensus cut-off value for the $[I]/K_i$ ratio is 0.1. When $[I]/K_i$ is greater than 0.1, a clinically relevant drug interaction may occur, and a follow-up *in vivo* drug interaction study is recommended. When $[I]/K_i$ is <0.1, an

TABLE 7.3 Prediction of CYP inhibition.

$[I]/K_i$	Prediction
$[I]/K_i > 1$	Likely
$1 > [I]/K_i > 0.1$	Possible
$0.1 > [I]/K_i$	Remote

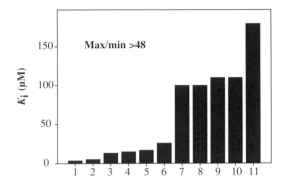

FIGURE 7.1 K_i in human liver microsomes for ketoconazole to inhibit the formation of 1-OH-midazolam from *in vitro* studies (data from 11 published articles from University of Washington Drug Metabolism database).

in vivo drug interaction study may not be needed. The ratio of $[I]/K_i$ only predicts the likelihood of an *in vivo* interaction, not the exact magnitude of the interaction. As the ratio increases, the likelihood of an interaction increases. However, both $[I]$ and K_i can be quite variable depending on the methods used for each individual drug. A short survey shows that the K_i values can be up to 48-fold apart in experiments using the same substrate and the same inhibitor toward the same CYP enzyme (Fig. 7.1). The variation in K_i values using the same inhibitor but different substrates is far greater across various sponsors and laboratories (Table 7.4). Differential sensitivity of CYP3A4 substrates toward inhibition by the same inhibitor indicates that enzyme–substrate interactions at the CYP3A4 active site(s) are substrate dependent (Wang et al., 2000). The importance of estimating inhibitor potency using the drug of interest as the substrate has been emphasized. Use of alternate substrate drugs for the same enzyme can potentially yield inaccurate predictions of *in vivo* drug interactions.

The regulatory agencies recommend the use of total drug concentrations of the mean steady-state C_{max} (bound plus unbound) of the highest proposed

TABLE 7.4 Ketoconazole K_i values determined with different CYP3A4 substrates.

Substrate	$K_i(\mu M)$	Reference
Midazolam	0.011	Li et al. (2004)
Triazolam	0.026	von Moltke et al. (1996)
Alprazolam	0.046	von Moltke et al. (1994)
Nifedipine	0.05	Wandel et al. (2000)
Trazodone	0.12	Zalma et al. (2000)
Lovastatin	0.4	Jacobsen et al. (1999)
Cisapride	1.5	Desta et al. (2000)
Terfenadine	3	Jurima-Romet et al. (1994)
Tacrolimus	11	Christians et al. (1996)

TABLE 7.5   Effect of ketoconazole on oral triazolam AUC using mean reported K_i value, total concentrations (bound + unbound fractions) of C_{max} and unbound fraction of C_{max} (I_u) as [I] for prediction compared with actually observed *in vivo* study results.

Dose	C_{max} (μM)	f_u	I_u	K_i^a (μM)	$1 + C_{max}/K_i$	$1 + I_u/K_i$	Observed
200 mg S.D.	9.3	0.01	0.093	0.0165	565	6.6	9^b
200 mg, Q12 × 2.5 days	10	0.01	0.10	0.0165	607	7.1	13.7^c
400 mg, QD × 4 days	15	0.01	0.15	0.0165	910	9.1	22.4^d

The AUCi/AUCc header spans the last three columns ($1 + C_{max}/K_i$, $1 + I_u/K_i$, Observed).

[a]Mean value from published reports.
[b]von Moltke et al. (1996).
[c]Greenblatt et al. (1998).
[d]Varhe et al. (1994).

clinical dose as inhibitor concentrations [I] for prediction. This is a conservative way to calculate the ratio. Table 7.5 lists examples of predicted and observed magnitude of drug interactions between ketoconazole and triazolam. Using both total and unbound fractions to predict the fold increase in AUC indicates that there is huge discrepancy between predicted and observed studies using total and unbound fractions. Although most drugs are better predicted using the unbound fraction rather than total concentrations, it is a conservative measure to use the total concentrations for regulatory use because there are a few exceptions, i.e. itraconazole, where the prediction using unbound fractions is much underestimated.

Alternatively, an IC_{50} can be used for prediction since IC_{50} and K_i are correlated (Fig. 7.2). In general, the IC_{50} value is greater than or equal to the

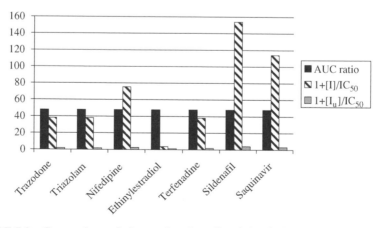

FIGURE 7.2 Comparison of observed and predicted AUC changes of saquinavir with coadministration of ritonavir using different IC_{50} values from different substrates (data from University of Washington Drug Metabolism database).

K_i. If the inhibition process is strictly noncompetitive, the IC_{50} should be equal to the K_i. For competitive inhibition, IC_{50} values are substrate concentration dependent, and the IC_{50} determined at low substrate concentrations will be very close to K_i since $IC_{50} = (1 + [S]/K_m) \times K_i$. The predicted magnitude of the interaction between ritonavir and saquinavir based on the IC_{50} values obtained with various substrates is shown in Figure 7.2. The in vivo study has shown that ritonavir increased saquinavir AUC by 48-fold. The prediction using total ritonavir peak concentrations as [I] and IC_{50} values from different substrates is closer to the actual ratio compared to the ratio that is obtained with when unbound fraction is used as [I].

7.4.1.4 CYP2C8 CYP2C8 has emerged recently as a key enzyme for clearance of several drugs, especially oral antidiabetic drugs (Totah and Rettie, 2005). Therefore, CYP2C8 is now recommended for in vitro and in vivo drug interaction studies by the FDA (FDA, 2006). Based on current knowledge, gemfibrozil is considered to be a strong, specific CYP2C8 inhibitor (Ogilvie et al., 2006). Gemfibrozil increases the plasma concentrations of rosiglitazone in vivo (Niemi et al., 2003a). Moreover, cerivastatin metabolism is believed to be partially inhibited by gemfibrozil via a CYP2C8 pathway. Inhibition of this pathway is possibly contributed to serious adverse events including rhabdomyolysis (Shitara et al., 2004).

The study of gemfibrozil, itraconazole, and their combination on the pharmacokinetics of repaglinide has further confirmed that gemfibrozil is a strong inhibitor of CYP2C8 (Fig. 7.3). When both the CYP2C8 (major)- and CYP3A4 (minor)- mediated metabolic pathways of repaglinide are blocked, the synergistic accumulation of repaglinide exposure is observed (Table 7.6) (Niemi et al., 2003b). Repaglinide is now recommended as a sensitive CYP2C8 substrate for human in vivo drug interaction studies.

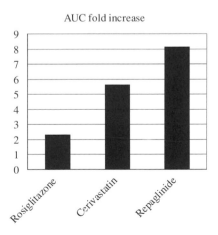

FIGURE 7.3 Gemfibrozil 600 mg for 3 days increased the exposures of CYP2C8 substrate drugs (Backman et al., 2002; Niemi et al., 2003).

TABLE 7.6   Fold increase in repaglinide exposure by gemfibrozil and itraconazole.

Drug	C_{max}	AUC
Gemfibrozil	2.4	8.1
Itraconazole	1.5	1.4
Gem + Itra	2.8	19.3

7.4.1.5 Induction   The knowledge of mechanisms and experimental conditions for drug induction studies has greatly advanced in recent years. Since CYP3A4 and CYP2C family (2C8, 2C9, and 2C19) share the same induction mechanism, if the induction studies with a new drug confirm that it is not an inducer of CYP3A4, then it can be concluded that the new drug is also not an inducer of the CYP2C family. The experiments of *in vitro* induction studies should include an acceptable enzyme inducer as a positive control. The inducer used as a positive control should increase enzyme activity by more than twofold at the inducer concentrations less than 500 μM. Although screening of PXR ligands has become a useful tool in drug development in order to select molecules with a lesser capacity to induce drug-metabolizing enzymes and *P-gp*, the most reliable method to study induction is to use the enzyme activity index obtained from primary hepatocyte cultures. Either freshly isolated human hepatocytes or cryopreserved hepatocytes are acceptable. The concentrations of a new drug for *in vitro* induction studies should encompass the actual plasma drug concentrations obscured in humans. At least one concentration should be a full order of magnitude greater than the average plasma concentration for the new drug. If the drug produces a change that is equal to or greater than 40% of the positive control, the induction study is considered positive and a further *in vivo* evaluation is warranted.

7.4.1.6 Conjugation Enzymes   In general, probe substrates and inhibitors for conjugation enzymes are less selective. In most cases, these enzymes play less important roles in metabolism-based drug interactions. However, the requirement by regulatory agencies for study of conjugation enzymes is on a case-by-case basis. It may be important to determine the impact of conjugation of a new drug if these pathways are involved in the detoxification of the drug and/or its active metabolites. One example is that metabolic conversion of irinotecan to the active metabolite SN-38, which is mediated by carboxylesterase enzymes in the liver. SN-38 is subsequently conjugated predominantly by the enzyme UDP-glucuronosyltransferase 1A1 (UGT1A1) to form a glucuronide metabolite (Iyer et al., 1998). The UGT1A1 activity is reduced in individuals with certain genetic polymorphisms such as the UGT1A1*28 polymorphism. In a prospective study, in which irinotecan was administered as a single-agent on a once-every-3-week schedule, patients who were homozygous for UGT1A1*28 had a higher exposure to SN-38 than patients with the wild-type UGT1A1 allele (Soepenberg et al., 2005). Also, a case-controlled study of Japanese cancer patients revealed that those with variant UGT1A1 alleles were at significantly higher risk of severe adverse reactions to irinotecan (Ando et al., 2005).

Another example is the major metabolite of gemfibrozil, gemfibrozil glucuronide, which is produced by a conjugation reaction, namely glucuronidation of the parent molecule. Gemfibrozil glucuronide was found to be a mechanism-based inhibitor of CYP2C8. The IC_{50} for inhibition of CYP2C8 by gemfibrozil glucuronide decreased from 24 to 1.8 μM after a 30-min incubation with human liver microsomes (Ogilvie et al., 2006). The drug interaction between cerivastatin and gemfibrozil was observed clinically, since the frequency of rhabdomyolysis was much higher when these two drugs were administered concurrently (Pogson et al., 1999). The mechanism of this drug–drug interaction remained unexplained for an extended period at least partially due to the emphasis put on cytochrome P450 enzymes (Shitara et al., 2004).

7.4.1.7 Transporters

7.4.1.7 Transporters   Transporter-based drug interactions have been increasingly reported. P–glycoprotein (P–gp), organic anion transporter (OAT), organic anion transporting polypeptide (OATP), organic cation transporter (OCT), multidrug resistance-associated proteins (MRP), and breast cancer resistant protein (BCRP) are among those relatively well studied. These drug transporters are related to drug bioavailability and cellular uptake, which impacts drug clearance, adverse drug interactions, and the enzyme induction and inhibition. For example, the large inter-subject and inter-racial variations in the potency of rifampin induction of CYP enzymes now are known to be related to OATP-C uptake modulation, which results in variations in hepatic cellular levels of rifampin (Tirona et al., 2003). The needs for studies of various transporters in the new drug development are dependent on individual programs and the characteristics of drugs being investigated.

P-gp is located in the apical domain of the enterocyte of the lower gastrointestinal tract (jejunum, duodenum, ileum, and colon), thereby limiting the absorption of some drug substrates from the gastrointestinal tract. In other organs such as the liver and kidney, expression of this transporter at the apical membrane of hepatocytes and kidney proximal tubular cells results in enhanced excretion of drug substrates into bile and urine, respectively. P-gp is also an important component in the blood-brain barrier, limiting the CNS entry of a variety of drug substrates. P-gp is as well found in other tissues known to have tissue–blood barriers, such as placenta and testis. Therefore, P-gp plays an important role in the absorption, distribution, metabolism, and excretion of test drugs. Since P-gp is the most well understood among drug transporters, it is generally recommended to evaluate whether the new drug is a substrate, inhibitor, and/or inducer of P-gp during drug development.

In evaluating whether a new drug is a substrate for P-gp, the net flux ratios across Caco-2 cells or MDR1-overexpressed cell lines are normally determined. Net flux ratios ≥ 2 are considered as positive for P-gp involvement and an *in vivo* drug interaction study with a P-gp inhibitor may be warranted. A net flux ratio <2 is considered as negative for P-gp involvement. Additional *in vitro* transporter studies will be dependent on the validity of other available methods to evaluate those transporters. Whether a new drug is an inhibitor of P-gp is also dependent on the net flux ratio measured for a model substrate in the presence of the new drug. If K_i or IC_{50} is $<10\,\mu M$, the new drug is likely a potent P-gp inhibitor and an *in vivo* drug interaction study is required. If K_i or IC_{50} is $>10\,\mu M$, the test drug may not be a potent P-gp inhibitor and further *in vivo* drug interaction study generally may not be needed, except when K_i or IC_{50} is within the range of therapeutic concentrations (Zhang et al., 2006).

7.4.2 *In vivo* Studies

Appropriately designed pharmacokinetic drug interaction studies can provide important information about metabolic pathways and their contribution to overall elimination. Together with information from *in vitro* studies, these *in vivo* studies can be a primary basis of labeling statements and often can help avoid the need for further unnecessary investigations.

7.4.2.1 Study Design   *In vivo* drug–drug interaction studies can be designed in many different ways. The fundamental requirement of such a study design is to explore the worst case scenario for the potential drug interactions from a regulatory perspective. In general, the study is designed to compare substrate exposures with and without an interacting drug. A study can use a randomized crossover, a one-sequence crossover, or a parallel design depending on the characteristics of the substrates and the interacting drugs. The dose selection for an interacting drug should be the highest recommended therapeutic dose. The time to give a concurrent inhibitory drug should be dependent on the mechanism of inhibition. For a rapidly reversible inhibitor, administration of the interacting drug either just before or simultaneously with the substrate on the test day is preferred to increase interaction potential. For a mechanism-based inhibitor, administration of the inhibitor prior to the administration of the substrate drug is required to exhibit metabolism-based inhibitory effects.

Ketoconazole, a strong inhibitor of CYP3A4, is often used in prototypical CYP3A4 drug interaction studies. The dose regimen of ketoconazole can dramatically change the outcome. To evaluate the worst case scenario of inhibitory drug interactions with ketoconazole, the regimen of 400 mg, once a day for 4 days has been demonstrated to produce maximal drug interactions (Table 7.5).

7.4.2.2 Substrates and Inhibitors   The regulatory agency has adopted the definition of sensitive substrates and the classification of potency of inhibitors for CYP3A4. The probe substrate is defined as a sensitive substrate if the exposure of substrate can be increased by fivefold or more by inhibition of its main metabolic pathways. The inhibitors are classified as strong, moderate, and weak, depending on the increase in AUC of sensitive substrates or decrease in clearance of sensitive substrates when the inhibitor is given at the highest

TABLE 7.7   Classification of CYP inhibitors.

Classification	Sensitive substrate	
	AUC increase	CL decrease
Strong inhibitor	>5fold	80%↓
Moderate inhibitor	>2but <5fold	50–80%↓
Weak inhibitor	>1.25 - but <2fold	20–50%↓

approved dose and the shortest dosing interval in human *in vivo* studies (Table 7.7).

This classification is quite useful since information on the clinical evaluation of substrates and inhibitors is readily available from public domains. The application of the inhibitor classification system allows generalizations such that a strong inhibitor of CYP3A does not need to be tested with multiple CYP3A substrates in order to lead to a warning in the label. A newly identified strong inhibitor of CYP3A4 can have a class labeling such that it may be contraindicated for CYP3A4 substrate drugs after a small number of studies. However, caution must be exercised, because the actual magnitude of interactions by different strong inhibitors can vary greatly. From the case study below, we know that the same oral dose and duration of conivaptan increased simvastatin AUC by 20-fold, but midazolam AUC only by sixfold.

On the contrary, although both ketoconazole and ritonavir are strong inhibitors of CYP3A4, ritonavir actually produces more severe inhibitory drug interactions. In a series of CYP3A4 drug interaction studies described in the vardenafil (Levitra™) package insert (http://www.fda.gov/cder/foi/label/ 2005/021400s004lbl.pdf), ritonavir increased vardenafil AUC by 49-fold compared to10-fold by ketoconazole (Fig. 7.4). Another case is the interaction

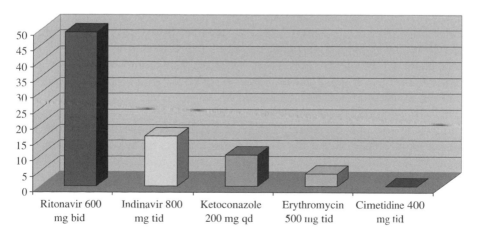

FIGURE 7.4   AUC fold increase of 5 mg vardenafil after coadministration of various CYP3A4 inhibitors.

TABLE 7.8   Pharmacokinetic parameters of fluticasone nasal spray (200 μg once daily) +7 days with ritonavir (100 mg twice daily).

PK parameter	Fluticasone alone	Fluticasone + ritonavir	Ratio
C_{max} (pg/mL)	11.9	318	28
AUC 0–t (pg•h/mL)	8.43	3102.6	368

between ritonavir and fluticasone (http://www.fda.gov/cder/foi/label/2004/20549slr016,20548slr020,20121slr030_flonase_lbl.pdf). In a multiple-dose, crossover drug interaction study in 18 healthy subjects fluticasone nasal spray (200 mcg once daily) was coadministered for 7 days with ritonavir (100 mg twice daily). The fluticasone AUC was elevated by 368-fold after coadministration of ritonavir with fluticasone nasal spray (Table 7.8). Clinical cases have been reported for patients with AIDS who suffered from asthma and developed an iatrogenic Cushing's syndrome subsequent to receiving both fluticasone and ritonavir at the same time (Hillebrand-Haverkort et al., 1999).

7.4.2.3   Cocktail Approaches   Simultaneous administration of a mixture of two or more probe substrates of CYP enzymes in one study in human volunteers is referred to as a "cocktail approach."

A number of drug metabolism cocktails have been proposed and evaluated in clinical studies. Frye et al. reported a five-drug cocktail (the so-called "Pittsburgh cocktail") containing caffeine, chlorzoxazone, dapsone, debrisoquine, and mephenytoin for the simultaneous phenotyping of CYP1A2, CYP2E1, CYP3A, CYP2D6, and CYP2C19 and N-acetyltransferase activities (Frye et al., 1997). However, the study of CYP2E1 has been less emphasized and dapsone is a controversial substrate for CYP3A4 due to its lack of specificity/selectivity. Streetman et al. reported a four-drug cocktail (the so-called "Cooperstown cocktail") of oral caffeine, omeprazole, dextromethorphan, and intravenous midazolam for simultaneous phenotyping of CYP1A2, CYP2C19, CYP2D6, CYP3A, N-acetyltransferase-2, and xanthine oxidase (Streetman et al., 2000). The Cooperstown cocktail focuses on more important CYP enzymes and the substrate specificity and availability. Zhu et al. reported a five-drug cocktail, caffeine, chlorzoxazone, mephenytoin, metoprolol, and midazolam for phenotyping CYP1A2, CYP2E1, CYP2C19, CYP2D6, and CYP3A4, in which the oral administration of midazolam increases convenience (Zhu et al., 2001). Blakey et al. also reported a five-drug cocktail of oral caffeine (CYP1A2), tolbutamide (CYP2C9), debrisoquine (CYP2D6), chlorzoxazone (CYP2E1), and midazolam (CYP3A4) for simultaneous phenotyping of CYP1A2, CYP2C9, CYP2C19, CYP2D6, and CYP3A4 (Blakey et al., 2004). All of the four cocktail approaches mentioned above were validated to rule out drug interactions among the probe substrates. Wang et al. also reported a cocktail approach to evaluate the short-term and long-term effects of St John's wort on human CYP enzymes, in which caffeine, tolbutamide, dextromethorphan, oral and intravenous midazolam were used for simultaneous phenotyping of CYP1A2, CYP2C9, CYP2D6, and intestinal and hepatic CYP3A4. The study focused on the induction effect of St. John's wort on a cocktail of substrates, but it did not validate whether there were any drug interactions among these substrates after adding midazolam to the previously validated three substrates cocktail (Bruce et al., 2001; Wang et al., 2001).

The cocktail approach has been proposed as a screening tool for potential *in vivo* drug–drug interactions since in many cases a cocktail approach uses only metabolic ratios to determine the interactions. The same equivalence criterion of 80–125% confidence intervals should be applied to the changes in ratios for plasma or urine to conclude positive or negative outcomes. Negative results from a cocktail study can eliminate the need for further evaluation of particular CYP enzymes. However, positive results can indicate the need for further *in vivo* evaluation to provide quantitative exposure changes (such as AUC, C_{max}), if the initial evaluation only assessed the changes in metabolic ratios. One must be aware that any modifications from a validated cocktail protocol, such as dose increase, replacement of different substrates, and/or adding additional substrates will require additional validation to rule out any potential drug interactions. For example, Palmer et al. reported that chlorzoxazone significantly increased the exposure of oral midazolam (Palmer et al., 2001). In general, the data collected from a properly validated cocktail study can supplement data from other *in vitro* and *in vivo* studies to help judge a drug's potential to inhibit or induce CYP enzymes.

7.4.2.4 Population Pharmacokinetics

Population pharmacokinetic analysis from clinical trials with sparse or intensive blood sampling can be valuable to evaluate the clinical impact of drug interactions, and to suggest dose adjustments. The results from population pharmacokinetic studies are informative, but not conclusive in most cases. They can be conclusive if the population pharmacokinetic components in clinical trials are adequately designed to detect significant changes in drug exposure due to drug–drug interactions. Population pharmacokinetic evaluations may detect signals of unsuspected drug–drug interactions. When the results from a population pharmacokinetic analysis are different from that which a conventional pharmacokinetic drug interaction study have strongly suggested, the results from the conventional pharmacokinetic study overwrite the population pharmacokinetic analysis. Because there are many confounding factors in large-scale clinical trials, population pharmacokinetic studies must have carefully designed study protocols and sample collections. It is unlikely for the drug label to claim there are no drug interactions for comedications based on population pharmacokinetic analysis, especially when a conventional pharmacokinetic drug interaction study indicates the presence of an interaction. To be acceptable for drug labeling, population pharmacokinetic studies should be properly designed and be prospective in nature.

7.4.2.5   Outcomes   The FDA 2006 draft guidance for drug interactions states that the results of drug–drug interaction studies should be reported as 90% confidence intervals about the geometric mean ratio of the observed pharmacokinetic parameters with (substrate + inhibitor) and without the interacting drug (substrate alone). Tests of statistical significance are not appropriate, because small, consistent systemic exposure differences can be statistically significant ($P < 0.05$) but not clinically relevant. When the 90% confidence intervals for systemic exposure ratios fall entirely within the equivalence range of 80–125%, the standard practice of regulatory authorities is to conclude that no clinically significant differences are present unless the sponsor has alternative predefined no effect boundaries. No effect boundaries can also be defined based on average dose and/or concentration–response relationships, PK/PD models, and others. No effect boundaries define the degree of difference caused by the interaction that is of no clinical consequence. For example, a sponsor has determined that 100 mg for Drug X is the highest recommended dose, however, has sufficient data from the Phase III clinical trials with 200 mg dose and the safety profiles between 100 and 200 mg are very similar. Therefore, the sponsor could define a twofold increase in the exposure as a no effect boundary and any pharmacokinetic drug interaction is considered to be negative if the increase in the ratio with and without an interacting drug is less than twofold.

7.4.3   Case Study

Conivaptan (Vaprisol®), a nonpeptide, dual antagonist of arginine vasopressin (AVP) V_{1A} and V_2 receptors, was approved for marketing under NDA21-697 in 2005 (http://www.fda.gov/cder/foi/nda/2005/021697s000_VaprisolTOC.htm). Conivaptan was originally developed as a tablet intended for oral administration. Conivaptan has nonlinear pharmacokinetics following intravenous infusion and oral administration. CYP3A4 was identified as the sole cytochrome P450 isozyme responsible for its metabolism. Coadministration of oral conivaptan hydrochloride (10 mg) with ketoconazole (200 mg) resulted in a 4- and 11-fold increase in C_{max} and AUC of conivaptan, respectively. It also is a potent inhibitor of multiple CYP enzymes from *in vitro* studies. In one drug interaction study, 15 mg conivaptan was intravenously administered twice a day (8 am and 8 pm) over 120 min for 3 days (Day 3 to Day 5), and 60 mg simvastatin was given orally on Day 1 and Day 5. The pharmacokinetic parameters of simvastatin were increased significantly (Table 7.9). However, in

TABLE 7.9   Effect of 15 mg iv conivaptan bid on simvastatin and simvastatin acid: mean ratio of Day 1 versus Day 5.

Parameter	Simvastatin	Simvastatin acid
C_{max}	3.12	2.87
AUC $(0-\infty)$	4.22	2.64

TABLE 7.10 **Effect of oral conivaptan bid on simvastatin and simvastatin acid: mean ratio of Day 1 versus Day 6.**

Parameter	Conivaptan 20 mg q12 h × 6 days		Conivaptan 40 mg q12 h × 6 days	
	Simvastatin	Simvastatin acid	Simvastatin	Simvastatin acid
C_{max}	7.9	14.4	3.12	18.1
AUC (0–∞)	6.0	13.0	4.22	19.8

another study, 20 or 40 mg conivaptan was given orally twice a day from Day 2 to Day 6 and simvastatin 60 mg was given orally on Day 1 and Day 6. The pharmacokinetic parameters of simvastatin and simvastatin acid were increased to a much greater extent (Table 7.10). It is worth noting that the treatment of oral conivaptan 20 and 40 mg bid for 5 days only increase oral midazolam (2 mg) AUC by 3.5- and 5.8- fold, respectively.

Because demonstrated above, oral administration will carry a high risk for severe adverse drug interactions. Conivaptan was eventually reformulated for intravenous administration late in the drug development program, which greatly reduced the risk for adverse drug interactions, especially for marketed CYP3A4 sensitive substrate drugs and strong inhibitory drugs. The intravenous administration of conivaptan makes the risk management program much easier to limit the drug to hospitalized patients only. The regulatory agency concluded that the drug can be administered effectively and safely if limited to the intravenous route.

7.5 CONCLUSIONS

Regulatory agencies have been advocating a science-based review process, have been closely monitoring the emergence of new science and have been updating their regulatory guidances in a timely manner, especially in the field of drug metabolism and drug interactions. The FDA 2006 draft guidance for "Drug Interaction Studies—Study Design, Data Analysis, and Implications for Dosing and Labeling" has collected a consensus among scientists from regulatory, pharmaceutical industry, and academic fields. Its contents and scope are substantial and sophisticated. When the new draft guidance is finalized, it will eventually replace the 1997 *in vitro* metabolism and the 1999 *in vivo* metabolism/drug interaction guidances. The 2006 draft guidance for drug interactions has incorporated many new developments and can be used as an excellent reference tool for all scientists in the field. The new guidance on drug interactions also will increase acceptance for including study results in the drug product labeling using preferred substrates, inhibitors, and study design from *in vitro* and *in vivo* drug metabolism and drug interaction studies.

The FDA draft guidance "Safety Testing of Drug Metabolites," together with the MIST document, has opened a new chapter of drug metabolism research in the pharmaceutical industry although the discussions on how to conduct metabolite safety testing will continue. Metabolite safety testing is a very complex subject. It involves metabolite profiling in humans and toxicology species with radiolabeled drugs, pharmacological activity testing of the major metabolites in human circulation, monitoring of metabolites in selected clinical and nonclinical studies using a validated bioanalytical assay, and nonclinical toxicity evaluation of selected metabolites. The FDA has suggested that decisions to select metabolites for pharmacological activity testing, monitoring in humans and animals, and direct safety testing should be considered on a case-by-case basis.

Regulatory guidances help pharmaceutical scientists design optimal drug development programs, weigh the risk over benefit ratio in the early phase, increase assurance of safety, reduce failure and cost in drug development, and eliminate unnecessary studies. Therefore, pharmaceutical scientists should be familiar with relevant regulatory guidances in order to assure best practices in drug development.

ACKNOWLEDGMENT

The authors greatly appreciate Dr. Robert E. Vestal, in critical reading and editing this chapter.

REFERENCES

Ando M, Hasegawa Y, Ando Y. Pharmacogenetics of irinotecan: a promoter polymorphism of UGT1A1 gene and severe adverse reactions to irinotecan. Invest New Drugs 2005;23:539–549.

Atkinson A, Kenny JR, Grime K. Automated assessment of time-dependent inhibition of human cytochrome P450 enzymes using liquid chromatography-tandem mass spectrometry analysis. Drug Metab Dispos 2005;33:1637–1647.

Backman JT, Kyrklund C, Neuvonen M, Neuvonen PJ. Gemfibrozil greatly increases plasma concentrations of cerivastatin. Clin Pharmacol Ther 2002;72:685–691.

Baillie TA, Cayen MN, Fouda H, Gerson RJ, Green JD, Grossman SJ, Klunk LJ, LeBlanc B, Perkins DG, Shipley LA. Drug metabolites in safety testing. Toxicol Appl Pharmacol 2002;182:188–196.

Blakey GE, Lockton JA, Perrett J, Norwood P, Russell M, Aherne Z, Plume J. Pharmacokinetic and pharmacodynamic assessment of a five-probe metabolic cocktail for CYPs 1A2, 3A4, 2C9, 2D6 and 2E1. Br J Clin Pharmacol 2004;57:162–169.

Bjornsson TD, Callaghan JT, Einolf HJ, Fischer V, Gan L, Grimm S, Kao J, King SP, Miwa G, Ni L, Kumar G, McLeod J, Obach SR, Roberts S, Roe A, Shah A, Snikeris F, Sullivan JT, Tweedie D, Vega JM, Walsh J, Wrighton SA. The conduct of *in vitro* and *in vivo* drug–drug interaction studies: a PhRMA perspective. J Clin Pharmacol 2003;43:443–469.

Bruce MA, Stephen DH, Haehner-Daniels DB, Gorski JC. *In vivo* effect of clarithromycin on multiple cytochrome P450s. Drug Metab Dispos 2001;29:1023–1028.

Cayen MN. Considerations in the design of toxicokinetic programs. Toxicol Pathol 1995a;23 (2):148–157.

Cayen MN. Toxicokinetic challenges in the pharmaceutical industry. Toxicol Pathol 1995;23 (2):217–219.

Christians U, Schmidt G, Bader A, Lampen A, Schottmann R, Linck A, Sewing KF. Identification of drugs inhibiting the *in vitro* metabolism of tacrolimus by human liver microsomes. Br J Clin Pharmacol 1996;41:187–190.

Dalvie D. Recent advances in the applications of radioisotopes in drug metabolism, toxicology and pharmacokinetics. Curr Pharm Des 2000;6:1009–1028.

Davis-Bruno K and Atrakchi A. A regulatory perspective on issues and approaches in characterizing human metabolites. Chem Res Toxicol 2006;19:1561–1563.

Desta Z, Soukhova N, Mahal SK, Flockhart DA. Interaction of cisapride with the human cytochrome P450 system: metabolism and inhibition studies. Drug Metab Dispos 2000;28:789–800.

FDA. Content and format of investigational new drug Applications (INDs) for phase 1 studies of drugs;1995.

FDA. Drug metabolism/drug interaction studies in the drug development process: studies *in vitro*;1997.

FDA. *In Vivo* metabolism/drug interaction studies — study design, data analysis, and recommendations for dosing and labeling;1999.

FDA. Bioanalytical method validation;2001a.

FDA. IND meetings for human drugs and biologics; chemistry, manufacturing, and controls information;2001b.

FDA. Carcinogenicity study protocol submissions;2002.

FDA. CFR title 21, part 312—investigational new drug application;2003a.

FDA. CFR title 21, part 314—applications for FDA approval to market a new drug;2003b.

FDA. Part 11, electronic records, electronic signatures—scope and application;2003c.

FDA. CFR title 21, part 58—good laboratory practice for nonclinical laboratory studies;2003d.

FDA. CFR title 21, part 361.1—radioactive drugs for certain research uses;2005a.

FDA. Safety testing of drug metabolites (draft);2005b.

FDA. Drug interaction studies—study design, data analysis, and implications for dosing and labeling (Draft guidance);2006a.

FDA. Labeling for human prescription drug and biological products—implementing the new content and format requirements (draft);2006b.

Frye RF, Matzke GR, Adedoyin A, Porter JA, Branch RA. Validation of the five-drug "Pittsburgh cocktail" approach for assessment of selective regulation of drug-metabolizing enzymes. Clin Pharmacol Ther 1997;62:365–376.

Greenblatt DJ, Wright CE, vonMoltke LL, Harmatz JS, Ehrenberg BL, Harrel LM, Corbett K, Counihan M, Tobias S, Shader RI. Ketoconazole inhibition of triazolam and alprazolam clearance: differential kinetic and dynamic consequences. Clin Pharmacol Ther 1998;64:237–247.

Guengerich FP. Safety assessment of stable drug metabolites. Chem Res Toxicol 2006;19:1559–1560.

Hastings KL, El-Hage J, Jacobs A, Leighton J, Morse D, Osterberg RE. Drug metabolites in safety testing. Toxicol Appl Pharmacol 2003;190:93–94.

Hillebrand-Haverkort ME, Prummel MF, TenVeen JH. Ritonavir-induced Cushing's syndrome in a patient treated with nasal fluticasone. AIDS 1999;13:1803.

Humphreys WG, Unger SE. Safety assessment of drug metabolites: characterization of chemically stable metabolites. Chem Res Toxicol 2006;19:1564–1569.

ICH. S3A—Toxicokinetics: the assessment of systemic exposure in toxicity studies; 1995a.

ICH. S1C—Dose selection for carcinogenicity studies of pharmaceuticals;1995b.

ICH. E6—Good clinical practice: consolidated guideline;1997.

ICH. Q2B—Validation of analytical procedures: methodology;1997.

ICH. M4: The CTD—Safety;2001.

ICH. Validation of analytical procedures: text and methodology;2005.

ICH. Q3B(R) impurities in new drug products (Revision 2);2006.

Iyer L, King CD, Whitington PF, Green MD, Roy SK, Tephly TR, Coffman BL, Ratain MJ. Genetic predisposition to the metabolism of irinotecan (CPT-11): role of uridine diphosphate glucuronosyltransferase isoform 1A1 in the glucuronidation of its active metabolite (SN-38) in human liver microsomes. J Clin Invest 1998;101:847–854.

Jacobsen W, Kirchner G, Hallensleben K, Mancinelli L, Deters M, Hackbarth I, Baner K, Benet LZ, Sewing KF, Christians U. Small intestinal metabolism of the 3-hydroxy-3-methylglutaryl-coenzyme A reductase inhibitor lovastatin and comparison with pravastatin. J Pharmacol Exp Ther 1999;291:131–139.

Jurima-Romet M, Crawford K, Cyr T, Inaba T. Terfenadine metabolism in human liver. *In vitro* inhibition by macrolide antibiotics and azole antifungals. Drug Metab Dispos 1994;22:849–857.

Kim RB. Transporters and drug discovery: why, when, and how. Mol Pharm 2006;3:26–32.

Kola I, Landis J. Can the pharmaceutical industry reduce attrition rates? Nat Rev Drug Discov 2004;3:711–715.

Li XQ, Andersson TB, Ahlstrom M, Weidolf L. Comparison of inhibitory effects of the proton pump-inhibiting drugs omeprazole, esomeprazole, lansoprazole, pantoprazole, and rabeprazole on human cytochrome P450 activities. Drug Metab Dispos 2004;32:821–827.

Marathe PH, Shyu WC, Humphreys WG. The use of radiolabeled compounds for ADME studies in discovery and exploratory development. Curr Pharm Des 2004;10:2991–3008.

Niemi M, Backman JT, Granfors M, Laitila J, Neuvonen M, Neuvonen PJ. Gemfibrozil considerably increases the plasma concentrations of rosiglitazone. Diabetologia 2003a;46:1319–1323.

Niemi M, Backman JT, Neuvonen M, Neuvonen PJ. Effects of gemfibrozil, itraconazole, and their combination on the pharmacokinetics and pharmacodynamics of repaglinide: potentially hazardous interaction between gemfibrozil and repaglinide. Diabetologia 2003b;46:347–351.

Ogilvie BW, Zhang D, Li W, Rodrigues AD, Gipson AE, Holsapple J, Toren P, Parkinson A. Glucuronidation converts gemfibrozil to a potent, metabolism-dependent inhibitor of CYP2C8: implications for drug–drug interactions. Drug Metab Dispos 2006;34:191–197.

Olkkola KT, Aranko K, Luurila H, Hiller A, Saarnivaara L, Himberg JJ, Neuvonen PJ. A potentially hazardous interaction between erythromycin and midazolam. Clin Pharmacol Ther 1993;53:298–305.

Palmer JL, Scott RJ, Gibson A, Dickins M, Pleasance S. An interaction between the cytochrome P450 probe substrates chlorzoxazone (CYP2E1) and midazolam (CYP3A). Br J Clin Pharmacol 2001;52:555–561.

Pogson GW, Kindred LH, Carper BG. Rhabdomyolysis and renal failure associated with cerivastatin-gemfibrozil combination therapy. Am J Cardiol 1999;83:1146.

Prentis RA, Lis Y, Walker SR. Pharmaceutical innovation by the seven UK-owned pharmaceutical companies (1964–1985). Br J Clin Pharmacol 1988;25:387–396.

Prueksaritanont T, Lin JH, Baillie TA. Complicating factors in safety testing of drug metabolites: kinetic differences between generated and preformed metabolites. Toxicol Appl Pharmacol 2006;217:143–152.

Shitara Y, Hirano M, Sato H, Sugiyama Y. Gemfibrozil and its glucuronide inhibit the organic anion transporting polypeptide 2 (OATP2/OATP1B1:SLC21A6)-mediated hepatic uptake and CYP2C8-mediated metabolism of cerivastatin: analysis of the mechanism of the clinically relevant drug-drug interaction between cerivastatin and gemfibrozil. J Pharmacol Exp Ther 2004;311:228–236.

Smith D, Schmid E, Jones B. Do drug metabolism and pharmacokinetic departments make any contribution to drug discovery?Clin Pharmacokinet 2002;41:1005–1019.

Smith DA, Obach RS. Metabolites and safety: what are the concerns, and how should we address them? Chem Res Toxicol 2006;19:1570–1579.

Smith DA, Obach RS. Seeing through the mist: abundance versus percentage. Commentary on metabolites in safety testing. Drug Metab Dispos 2005;33:1409–1417.

Soepenberg O, Dumez H, Verweij J, deJong FA, deJonge MJ, Thomas J, Eskens FA, van Schaik RH, Selleslach J, Ter Steeg J, Lefebvre P, Assadourian S, Sanderink GJ, Sparreboom A, van Oosterom AT. Phase I pharmacokinetic, food effect, and pharmacogenetic study of oral irinotecan given as semisolid matrix capsules in patients with solid tumors. Clin Cancer Res 2005;11:1504–1511.

Streetman DS, Bleakley JF, Kim JS, Nafziger AN, Leeder JS, Gaedigk A, Gotschall R, Kearns GL, Bertino JS. Jr. Combined phenotypic assessment of CYP1A2, CYP2C19, CYP2D6, CYP3A, N-acetyltransferase-2, and xanthine oxidase with the "Cooperstown cocktail". Clin Pharmacol Ther 2000;68:375–383.

Tirona RG, Leake BF, Wolkoff AW, Kim RB. Human organic anion transporting polypeptide-C (SLC21A6) is a major determinant of rifampin-mediated pregnane X receptor activation. J Pharmacol Exp Ther 2003;304:223–228.

Totah RA, Rettie AE. Cytochrome P450 2C8: substrates, inhibitors, pharmacogenetics, and clinical relevance. Clin Pharmacol Ther 2005;77:341–352.

Tucker GT, Houston JB, Huang SM. Optimizing drug development: strategies to assess drug metabolism/transporter interaction potential—toward a consensus. Pharm Res 2001;18:1071–1080.

UW database. available at http://www.druginteractioninfo.org.

Varhe A, Olkkola KT, Neuvonen PJ. Oral triazolam is potentially hazardous to patients receiving systemic antimycotics ketoconazole or itraconazole. Clin Pharmacol Ther 1994;56 (6 Pt 1):601–607.

vonMoltke LL, Greenblatt DJ, Cotreau-Bibbo MM, Harmatz JS, Shader RI. Inhibitors of alprazolam metabolism *in vitro*: effect of serotonin-reuptake-inhibitor antidepressants, ketoconazole and quinidine. Br J Clin Pharmacol 1994;38:23–31.

vonMoltke LL, Greenblatt DJ, Harmatz JS, Duan SX, Harrel LM, Cotreau-Bibbo MM, Pritchard GA, Wright CE, Shader RI. Triazolam biotransformation by human liver microsomes *in vitro*: effects of metabolic inhibitors and clinical confirmation of a predicted interaction with ketoconazole. J Pharmacol Exp Ther 1996;276:370–379.

Wandel C, Kim RB, Guengerich FP, Wood AJ. Mibefradil is a P-glycoprotein substrate and a potent inhibitor of both P-glycoprotein and CYP3A *in vitro*. Drug Metab Dispos 2000;28:895–898.

Wang RW, Newton DJ, Liu N, Atkins WM, Lu AY. Human cytochrome P450 3A4: *in vitro* drug–drug interaction patterns are substrate-dependent. Drug Metab Dispos 2000;28:360–366.

Wang Z, Gorski JC, Hamman MA, Huang SM, Lesko LJ, Hall SD. The effects of St John's wort (*Hypericum perforatum*) on human cytochrome P450 activity. Clin Pharmacol Ther 2001;70:317–326.

Yu KS, Cho JY, Shon JH, Bae KS, Yi SY, Lim HS, Jang IJ, Shin SG. Ethnic differences and relationships in the oral pharmacokinetics of nifedipine and erythromycin. Clin Pharmacol Ther 2001;70:228–236.

Yuan R, Madani S, Wei XX, Reynolds K, Huang SM. Evaluation of cytochrome P450 probe substrates commonly used by the pharmaceutical industry to study *in vitro* drug interactions. Drug Metab Dispos 2002;30:1311–1319.

Zalma A, von Moltke LL, Granda BW, Harmatz JS, Shader RI, Greenblatt DJ. *In vitro* metabolism of trazodone by CYP3A: inhibition by ketoconazole and human immunodeficiency viral protease inhibitors. Biol Psychiatry 2000;47:655–661.

Zhang L, Strong JM, Qiu W, Lesko LJ, Huang SM. Scientific perspectives on drug transporters and their role in drug interactionst. Mol Pharm 2006;3:62–69.

Zhu B, Ou-Yang DS, Chen XP, Huang SL, Tan ZR, He N, Zhou HH. Assessment of cytochrome P450 activity by a five-drug cocktail approach. Clin Pharmacol Ther 2001;70:455–461.

Zhu M, Zhao W, Vazquez N, Mitroka JG. Analysis of low level radioactive metabolites in biological fluids using high-performance liquid chromatography with microplate scintillation counting: method validation and application. J Pharm Biomed Anal 2005;39:233–245.

Zhu M, Zhang D, Skills G.Quantification and structural elucidation of low quantities for radiolabeled metabolites using microplate scintillation counting (MSC) techniques in conjunction with LC/MS. Chowdhury S.K, editor. Identification and Quantification of Drugs, Metabolites and Metabolizing Enzymes by LC-MS. Amsterdam: Elsevier; 2005. p:195–223.

PART II

ROLE OF DRUG METABOLISM IN THE PHARMACEUTICAL INDUSTRY

8

DRUG METABOLISM RESEARCH AS AN INTEGRAL PART OF THE DRUG DISCOVERY PROCESS

W. GRIFFITH HUMPHREYS

8.1 INTRODUCTION

Drug metabolism research has become an integral part of the lead optimization phase of drug discovery. The goal of this work is to attempt to optimize the metabolism properties of clinical candidates in parallel with the optimization of the potency and efficacy. The major reason to optimize the metabolism of a new chemical entity is so the clearance properties can be matched to the indication. For most indications involving chronic, oral therapy that means that compounds must be found with preclinical clearance properties that predict a human clearance consistent with qd administration (this is not true in all cases, some indications may require a short duration of action and thus rapid metabolism). Drugs that have rapid metabolic clearance and low percentage of absolute oral bioavailability (%F) due to high preabsorptive and first-pass metabolism are likely to have a high degree of inter and intrapatient variability (Hellriegel et al., 1996) and be more likely to suffer from drug–drug interactions. Also, drugs with rapid clearance and low %F are likely to require suboptimal dosing regimes, that is, bid or tid, and/or relatively high doses. The former leads to poor patient compliance, suboptimal efficacy, and marketing issues and the later can lead to unanticipated toxicities due to a large flux of drug and drug metabolites. Other benefits to understanding and optimizing the metabolism of new chemical entity (NCEs) are

Drug Metabolism in Drug Design and Development,   Edited by Donglu Zhang, Mingshe Zhu and W. Griffith Humphreys

(1) a decreased clearance may translate into a lower overall dose;

(2) lower rates of formation and overall amounts of reactive intermediates that may mediate acute or idiosyncratic toxicities;

(3) increased pharmacokinetic half-life that will hopefully translate into a longer duration of action, less frequent dosing, and better patient compliance;

(4) better understanding of the extrapolation of animal data to humans, making human dose projections more reliable, and reducing risk upon entry into clinical development;

(5) lower risk of drug–drug interactions, even if the compound is still dependent on metabolism for the bulk of its clearance, low clearance drugs are less susceptible in drug–drug interaction caused by the coadministration of inhibitor;

(6) lower risk of drug–food interactions due to the reduced dose;

(7) decreased formation of metabolites that may have pharmacological activity against the target or may have significant off-target activity.

These considerations make it important to understand the metabolism characteristics of candidate drug molecules and to optimize these character- istics preclinically when possible. Drug metabolism plays a central role in modern drug discovery and candidate optimization, and recent reviews have detailed how metabolism has impacted the discovery process and challenges that the field faces in the future (Cox et al., 2002; Smith et al., 2002; Thompson, 2001; White, 2000).

This chapter will give an overview of metabolism-related topics that are often incorporated into drug discovery efforts in an effort to advance the optimal drug candidate into clinical investigations. Many of the topics covered briefly in this chapter will also be covered in more depth in other chapters of this book.

8.2   METABOLIC CLEARANCE

8.2.1   General

The metabolic clearance of NCEs is most often studied with a combination of *in vitro* and *in vivo* approaches. There are several *in vivo* approaches that can be used to study metabolism in preclinical species and these along with *in vitro* results can often shed mechanistic insight into the problems associated with rapid metabolic clearance and incomplete oral bioavailability due to first-pass metabolism. Modern LC–MS/MS measurement of plasma drug concentrations provide a rapid tool to assess oral bioavailability of new candidate compounds and allows for early definition of bioavailability problems. When bioavailability concerns do arise, there are several avenues discussed below that can be followed to isolate the factor(s) limiting the oral delivery of the compound.

In vitro experiments often provide valuable insight into the clearance mechanisms for NCEs. The experiments that are most often employed in tandem to understand bioavailablilty are determinations of compound solubility, membrane permeability, and stability in subcellular fractions. The subcellular fractions most often employed are plasma (for ester containing compounds) and liver subcellular fractions with the addition of either NADPH or UDPGA as cofactor. Hepatocytes are also a very useful *in vitro* tool that provides a more complete system for studying metabolism.

All of the assays mentioned can be set up in an automated, medium throughput system, however, all require a specific assay to measure compound concentration at the end of the assay that places significant limitations on throughput. There have been recent attempts to solve the throughput problems inherent with this type of assay by developing generic endpoint assays for metabolic stability, but there has been no clear solution to date.

8.2.2 Prediction of Human Clearance

The rapid determination of pharmacokinetic parameters, solubility, permeability, and *in vitro* stability in plasma or liver tissue can often provide a reasonable explanation of the mechanisms limiting oral bioavailability. An approach that is often used is to extrapolate the *in vitro* rate of metabolism to estimate the hepatic clearance using *in vitro–in vivo* correlation methodology (Houston and Carlile, 1997; Ito et al., 1998; Lave et al., 1997, 1999, 2002; Lin, 1998; Miners et al., 2006; Obach, 1999, 2001; Obach et al., 1997; Shiran et al., 2006). These methods use *in vitro* kinetic parameters, usually V_{max}/K_m or *in vitro* $t_{1/2}$, to determine an intrinsic clearance, which is then scaled to hepatic clearance using amount of tissue in the *in vitro* incubation, the weight of the liver and the well-stirred model for hepatic clearance.

Care must be used when determining *in vitro* parameters (Grime and Riley, 2006; Margolis and Obach, 2003; Miners et al., 2006; Obach, 1999; Obach et al., 1997; Tran et al., 2002; Ziegler, 2002). The methods used to extrapolate to the *in vivo* situation all assume that the drug concentration at the active site of the CYP enzymes will be much less than the K_m value. This is a reasonable assumption for most drugs used clinically under typical *in vitro* experimental conditions; however, experimental design, especially *in vitro* $t_{1/2}$ determinations, should take this into account. The incorporation of protein binding corrections in these calculations has been somewhat controversial. Clearance of "free drug" was included in the first theoretical models for predicting hepatic extraction from *in vitro* data (Rane et al., 1997) and was the favored method in early *in vitro–in vivo* correlation attempts. However, the inclusion of protein binding into the scaling equations tends to underpredict actual *in vivo* values and generally better results are achieved when protein binding is left out of the equation (Obach, 1997, 1999). The reasons for this could be (1) the intrahepatic concentration of most drugs is closer to the total plasma concentration than the free plasma concentration due to efficient uptake by the liver, (2) protein binding to microsomal proteins may serve to "cancel out" protein binding in plasma, that is, the free concentration in microsomal incubations is closer to the free concentration in plasma so the effect protein binding is already accounted for in the *in vitro* parameters. Methods to account for protein binding or lessen its effect are (1) to determine free concentration in microsomal incubations and then calculate *in vitro* parameters based on free drug (this would then be compared to free plasma clearance values), or (2) to use as little protein in microsomal incubations as possible to lessen the effect of binding. The scale-up from *in vitro* parameters to predict *in vivo* human clearance of new chemical entities is still a difficult task to do prospectively, especially in cases where the CYP kinetic data does not follow classical Michaelis–Menten behavior (Houston and Kenworthy, 2000). The uncertainty of this calculation can be somewhat mitigated if the same scale-up methods provide accurate results when applied to preclinical animal data.

This method has most often been applied in conjunction to rates of CYP-mediated oxidative metabolism in microsomal systems but can also be applied to conjugation (Lin and Wong, 2002; Miners et al., 2004, 2006; Soars et al., 2002) or FMO-catalyzed reactions (Fisher et al., 2002) or data derived from hepatocytes (Lau et al., 2002) or other *in vitro* systems (Yamamoto et al., 2005). The scaled clearance values can be compared to the values determined *in vivo* for a single compound or a set of analogs to give some idea of the predictive capacity of the *in vitro* systems.

The use of microsomes along with UDPGA as cofactor assay to measure UGT enzyme activity has been hampered historically by the fact that this enzymatic activity in microsomes is often in a "latent" form and requires activation by physical or detergent-induced disruption of the membrane matrices. Recently, a generic method involving the addition of the pore-forming peptide alamethicin to overcome the latency exhibited by this enzyme system has been described (Fisher et al., 2000). The inclusion of alamethicin seems to provide a more consistent method of assessing UGT enzyme activity.

8.2.3   *In vivo* Methods to Study Metabolism

The methods outlined above may not lead to a satisfactory determination of mechanism of incomplete bioavailability and additional methods may be necessary in some cases to fully characterize the factors responsible. Bile duct cannulated animals provide a powerful model to examine incomplete bioavailability issues, and the rat provides the most flexibility because terminal studies can routinely be done. For compounds that are thought to have dissolution limitations or instability in the GI tract, the GI tract can be removed at the end of the experiment and the contents assayed for drug and metabolites. The amount of parent in bile and urine can be quantitated by LC/MS/MS and this method can also be used for any metabolites where authentic standards have been prepared.

Alternatively, bile and urine metabolites can be estimated with HPLC–UV chromatography using the extinction coefficient of the parent. These measurements will begin to define the total absorption of the compound and how that relates to the systemic bioavailabilty. The availability of radiolabeled compound at this point makes the bile duct cannulated rat experiment especially powerful.

A second *in vivo* model system that is very useful in sorting through problem of low oral bioavailability is portal vein cannulated animals. There are two ways this experiment can be conducted to determine hepatic extraction: (1) measure systemic plasma concentration after oral, portal vein and systemic administration and (2) measure portal vein and hepatic vein concentrations after an oral dose. Both methods yield information on hepatic extraction and the percentage of dose reaching the portal circulation (the product of the fraction absorbed and the fraction metabolized by the gut wall).

Both the bile duct cannulated model and the portal vein cannulated model can be combined with a number of methods of modulating absorption or metabolism to ask specific mechanistic questions regarding stability, permeability, and metabolism. The following are several methods that can be used to modulate metabolism *in vivo*. The most commonly used method to modulate oxidative metabolism *in vivo* is to coadminister either ketoconazole or 1-aminobenzotriazole (Balani et al., 2002) to inhibit CYP enzymes. Both of these compounds will inhibit intestinal and liver CYP enzymes after oral administration. Care must be exercised when using ketoconazole for this purpose because the compound also has effects on transporters and this effect may make interpretation of results ambiguous. Alternatively, a CYP inducer can be coadministered to determine the effect on the clearance of the compound of interest. This can be accomplished with one of many known inducers of CYP enzymes. For inhibition of esterase enzyme activity, bis-[*p*-nitrophenyl]phosphate (BNPP) can be coadministered because this compound does not inhibit cholinesterase activity (Buch et al., 1969). Methods for inhibiting conjugation enzymes *in vivo* are not as well defined, since in most cases good inhibitors of the enzymes have not been identified. It is possible to deplete cofactor stores by coadministering large amounts of substrate to ask mechanistic questions, although this method is most easily applied to sulfotransferase (SULT) enzymes (Kim et al., 1995). Many conjugation enzymes are subject to induction, which may be an avenue available for modulating enzyme activities *in vivo*.

8.2.4 Screening Strategies

The approach often taken in candidate optimization is to try to isolate bioavailability problems for a member of a chemotype of interest with a combination of *in vitro* and *in vivo* experiments and then to try to devise rapid techniques to screen for the liability (Fig. 8.1). This approach relies on a good deal of up-front work to fully understand the bioavailability limitations and periodic checking of the property to ensure that the screen is providing reliable

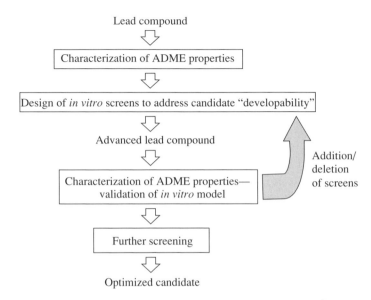

Lead compound

Characterization of ADME properties

Design of *in vitro* screens to address candidate "developability"

Advanced lead compound

Characterization of ADME properties—
validation of *in vitro* model

Addition/
deletion
of screens

Further screening

Optimized candidate

FIGURE 8.1   Scheme for incorporation of ADME-based developability screen during the candidate optimization phase of the drug discovery process.

results (Fig. 8.1). Systematic decision trees can be employed to allow clear pathways for evaluation of new compounds. Alternatively, screens can be run in parallel so that all information is generated for all compounds. This strategy makes workflows simpler and increases efficiency but can lead to information overload and complex decision making.

For metabolic stability screens to be most effective, the screens must be tightly linked to some means of gathering information on metabolite structure. Methods for rapidly determining metabolite molecular weight and limited structural information have improved dramatically and allow this approach to be routinely employed (Anari and Baillie, 2005; Watt et al., 2003). The goal of this type of approach is to allow the identification of metabolic "soft spots" which can then be altered to produce compounds with improved metabolic stability. The literature of successful structural modification to increase stability has recently been reviewed (Thompson, 2001).

8.2.5   *In silico* Methods to Study Metabolism

Metabolism prediction through *in silico* methods may be possible in the future, although there is still a great deal of technical development needed before it is a viable alternative (van de Waterbeemd and Gifford, 2003). Predictive metabolism programs may allow metabolism scientists and medicinal chemists to obtain metabolic rate or site of metabolism information prior to first synthesis of compounds. The majority of the methods in this area that have been published to date fall into four categories: (1) expert systems designed to identify probable sites of metabolism based on comparison to database information (Erhardt, 2003; Hawkins, 1999), (2) programs to predict the site of metabolism based on molecular orbital theory (Jones et al., 2002), (3) programs to predict the rate of metabolism based on physical chemical attributes of the molecule and QSAR development (Andrews et al., 2000; Balakin et al., 2005; de Groot and Ekins, 2002; Ekins and Obach, 2000; Ekins et al., 2005; Lewis, 2003, 2004; Long and Walker, 2003; Taskinen et al., 2003; Yoshida and Topliss, 2000) and (4) the prediction of site of metabolism based on knowledge of the CYP active site (Lewis, 2005; Lewis et al., 2006).

8.3 METABOLITE PROFILING

Although advances in LC/MS/MS technology have made the determination of metabolic profiles without radiolabeled material more reliable, methods for obtaining quantitative, complete profiles using cold material are still lacking. The use of radiolabeled material for the rigorous determination of metabolic profiles remains the "gold standard" at all stages of drug discovery and development. However, as mentioned above, new HPLC technology and techniques have made qualitative metabolite profiling of *in vitro* and *in vivo* samples much more robust. New highly sensitive triple quadrapole and ion-trap mass spectrometers along with techniques such as neutral loss and product ion scanning allow routine detection of metabolites even from complex matrices. Future advances in technology, such as accurate mass–mass spectrometers coupled with mass filters and nanospray technologies will likely lead to significant improvements in both qualitative and quantitative metabolite profiling data and will continue to make metabolite profiling at the early stage more definitive.

The majority of *in vitro* metabolite profiling experiments are designed to evaluate interspecies differences in profiles and are generally done to provide background context for toxicology studies. These studies generally involve HPLC separation of metabolites followed by quantitation by flow or standard scintillation counting. There have been recent advances in technology that have made the coupling of mass spectrometers in-line with radioactive flow detectors possible so that metabolites can be identified in the same run that they are being quantified (Egnash and Ramanathan, 2002; Nassar et al., 2003, 2004). Also, new scintillation counters and technologies are available that allow much greater sensitivity and thus reduce the amount of radioactivity necessary to detect low level metabolites (Nassar et al., 2003, 2004).

A major use of radiolabeled drugs in the *in vivo* situation is for metabolic profiling. These studies can be utilized to investigate interspecies comparisons of metabolic profiles in plasma urine and feces (or bile) in conjunction with toxicology or carcinogenicity studies or for performing definitive metabolite profiling in human ADME studies.

8.4 REACTION PHENOTYPING

It is extremely important to gain an early understanding of the enzymes predicted to be important for metabolic clearance of an NCE so that some level of predictions can be made as to the potential for drug–drug interactions and polymorphic clearance in humans can be made. It should be noted that the true contribution of individual enzymes to the clearance of a compound will in most cases be difficult to predict until the human ADME study is complete and the contribution of metabolic clearance to overall clearance of the drug has been determined.

The major goal of phenotyping at the discovery stage is to make predictions regarding the potential for the dependence of a single enzyme for a major fraction (usually defined as >20%) of the clearance of a NCE. The type of information that can reasonably generated at this stage is the contribution of individual CYP enzymes to the intrinsic clearance in human liver microsomes using well-established experimental techniques. The *in vitro* information can be augmented with *in vivo* information in animal models establishing the contribution of oxidative biotransformation to the overall clearance, but this preclinical information cannot be directly extrapolated to humans and does not add to the prediction of contribution of individual CYP enzymes. The estimation of fractional clearance by transferase enzymes producing direct conjugates, esterases, or other drug metabolizing enzymes remains problematic as does the prediction of transporters to the direct clearance via biliary or urinary excretion. This information is certainly important for the design of early clinical drug–drug interaction studies but in some cases may be used in the discovery stage for decision-making purposes.

8.5 ASSESSMENT OF POTENTIAL TOXICOLOGY OF METABOLITES

8.5.1 Reactive Metabolite Studies—*In vitro*

The metabolism of drugs to reactive intermediates followed by covalent binding to cellular components is hypothesized to be the basis for the acute or idiosyncratic toxicities caused by some drugs (Guengerich, 2001; Ju and Uetrecht, 2002; Kaplowitz, 2005; Liebler and Guengerich, 2005; Park et al., 2005; Walgren et al., 2005). The testing of new drug candidates for their potential to form reactive metabolites has received much recent attention (Baillie, 2006; Guengerich, 2005; Uetrecht, 2003;). Reactive intermediates are most commonly thought to arise through the generation of high energy intermediates during the oxidation of drugs by CYP enzymes (Amacher, 2006; Guengerich, 2003, 2006). Examples of these intermediates are epoxides, oxirenes, arene oxides, and quinoid species. Myeloperoxidase is another human oxidative enzyme that is known to catalyze the formation of reactive intermediates. Reactive esters formed by the conjugation of carboxylic acids with glucuronic acid or acyl coenzyme A are also thought to be a source of reactive metabolites.

The study of the interaction of drugs with cellular components can be broken down into two types of experiments, either measurement of covalent binding after reaction with nucleophilic sites on cellular macromolecules or studies with small molecule nucleophiles. These two types of experiments provide complementary information and are probably both important determinants in the overall toxicology associated with reactive intermediates.

Covalent binding studies usually require radiolabeled drug and are thus usually carried out late in the discovery phase or in early development. A typical experiment is done either with labeled drug in microsomes or by administering the labeled drug to rodents. Both types of studies involve isolation of cellular proteins through precipitation, followed by extensive washing steps to remove noncovalently bound drug. Residual radioactivity bound to proteins is then determined by scintillation counting. The use of this type of data has received much recent attention as a potential predictor of toxicity, especially of idiosyncratic toxicities, although this remains a controversial subject.

A recent example published by Samuel et al. illustrated how ^{3}H-compounds could be used in the discovery phase to reduce the extent of covalent adducts produced by candidates from a specific chemotype (Samuel et al., 2003). These authors found high levels of covalent binding to rat and human liver microsomes when a lead compound was tested. The compound was then incubated in the presence of GSH and the structure of the resultant GSH adduct elucidated. The knowledge of the adduct structure allowed new analog synthesis aimed at modulating metabolism of the site of reactive intermediate formation. An analog was synthesized that reduced the amount of covalent binding to human liver microsomes by 44-fold compared to the initial structure. The authors propose that this approach could be generally applied and provides a viable method of decreasing the amount of covalent binding associated with metabolic activation during the lead optimization phase.

There are many literature examples of covalent binding studies completed with toxicologically interesting molecules (Lecoeur et al., 1994; Masubuchi et al., 1996; Mays et al., 1990, 1995; Naisbitt et al., 2002; Pirmohamed et al., 1992; Singh et al., 2003; Ward et al., 1982; White et al., 1995). These studies clearly show that there is a wide range of values for the extent of covalent binding for different compounds. This is at least partially due to the different experimental conditions employed for each experiment, and likely also reflects a wide spectrum of both the efficiency of activation by microsomal enzymes and the efficiency of trapping by cellular components displayed by these compounds. The multifactorial nature of hepatotoxicity and blood toxicities displayed by compounds that form reactive intermediates makes the absolute translation of the extent of covalent binding number into a prediction of severity of toxicity difficult to accomplish. However, many compounds that show hepatotoxicity and blood toxicities do form reactive intermediates that bind cellular macromolecules. Although there are many literature values for covalent binding of compounds associated with toxicity, there are relatively few reports of compounds that are clinically safe and thus the lower limit of covalent binding values has not been well established.

Trapping experiments are most often done with unlabeled drug and glutathione with detection by mass spectrometry. Because unlabeled drug is used, these experiments can be done earlier in the discovery cycle than the covalent binding experiments. The detection of glutathione adducts is aided by the characteristic fragmentation pattern of glutathione. The use of this fragmentation has recently been adapted into a relatively high throughput assay based on the neutral loss of the lysine fragment of glutathione (Chen et al., 2001). Other analytical strategies have been developed to provide qualitative and/or quantitative determination of reactive metabolite formation at the screening stage (Argoti et al., 2005; Dieckhaus et al., 2005; Gan et al., 2005; Mutlib et al., 2005; Yan et al., 2005). These assays provide only qualitative information or utilize a nonphysiological trapping agent and are thus somewhat limited in utility. The approach most often employed to circumvent this problem is through the use of radiolabeled trapping agents, typically tritiated glutathione. The information generated from this approach can be utilized in a number of ways: (1) after determination of the formation of adduct and quantitation of adduct level, the adduct structure can be characterized. The adduct structure can then lead to medicinal chemistry approaches with the goal of limiting the amount of adduct formed. (2) as a trigger to perform more advanced covalent binding studies. (3) as a trigger to do advanced toxicology studies in animals or in hepatocytes. The link between the absolute amount of glutathione adduct formed upon reaction of a compound with liver microsomes in the presence of glutathione and prediction of toxicological endpoints has not been made and is likely to remain elusive. For this reason, the information gained in trapping experiments is usually used to modify structure to minimize the problem or to trigger more definitive studies.

8.5.2 Reactive Metabolite Studies—*In vivo*

Questions regarding the formation of metabolites that covalently bind proteins can be addressed *in vivo* with experiments similar in nature to the experiments described in the *in vitro* section. Animal studies are often focused on the liver proteins as a target for drug binding, while human studies are usually focused on blood components due to obvious ethical limitations. Baillie and colleagues have proposed the use of extent of covalent modification as a measure of risk of adverse events for new chemical entities. In this proposal, a level of greater than 50 pmol compound bound/mg microsomal protein has been put forth as a level of binding that would provide a safety margin (approximately 20-fold) over compounds that have shown liver toxicities in the clinic and are thought to produce toxicity through formation of reactive intermediate (Evans et al., 2004).

8.5.3   Toxicology Mediated Through Metabolite Interaction with Off-Target Receptors

The biotransformation of parent molecules often produces metabolites with relatively minor structural alterations (addition of a hydroxyl group, demethylation, etc.). As these changes may not be dramatic enough to depart from the parent structure activity relationship, these metabolites may have potent interactions with the same pharmacological target as the parent. Although it is rare for metabolites to have greater potency than parent, a loss of less that one order of magnitude is fairly common and the metabolites may be potent enough to contribute to the overall pharmacological response.

Although it is fairly common for metabolites to potently interact with the same pharmacological target as the parent, it is fairly uncommon for metabolites to have potent interactions with other "off-target" receptors that are not already affected by parent. Of course, this relationship will likely break down as the dose and plasma concentration of drug and metabolites increase. This relationship between dose and consideration of metabolite toxicity forms the basis of arguments put forth by Smith and Obach (Smith and Obach, 2005) that the metabolites of low dose drugs do not need to be characterized as extensively as high dose drugs. The chances for new pharmacology would be expected to increase for metabolites that result from major structural alterations, but even in these cases there are relatively few examples of this type of behavior. The chances for off-target pharmacology of metabolites would also be expected to increase when the target is a member of a receptor family with multiple related members (e.g., PPAR receptors, kinase families).

8.6   ASSESSMENT OF POTENTIAL FOR ACTIVE METABOLITES

In most cases, the metabolism of drugs leads to pharmacological inactivation through biotransformation to therapeutically inactive molecules. However, drug metabolism can also result in pharmacological activation, where pharmacologically active metabolites are generated. Although formation of pharmacologically active metabolites (bioactivation) can be mediated by both oxidation and conjugation reactions, bioactivation resulting from oxidative metabolism mediated by CYP enzymes is the more common pathway.

Active metabolites may have superior pharmacological, pharmacokinetic, and safety profiles compared to their respective parent molecules (Fura, 2006; Fura et al., 2004). As a result, a number of active metabolites have been developed and marketed as drugs with improved profiles relative to their parent molecules. Examples of active metabolites of marketed drugs that have been developed as drugs include acetaminophen, oxyphenbutazone, oxazepam, cetirizine (Zyrtec), fexofenadine (Allegra), and desloratadine (Clarinex). Each of these drugs provides a specific benefit over the parent molecule and is superior in one or more of the categories described above.

During lead optimization, drug candidates are routinely screened for metabolic stability or *in vivo* systemic exposure and rank ordered according to the rate and extent of metabolism or systemic exposure level (White, 2000). In the case of metabolic screening, this is usually performed *in vitro* after incubations of the drug candidates with subcellular fractions such as liver microsomes or intact cellular systems (e.g., hepatocytes) containing full complement of drug-metabolizing enzymes. Compounds with low metabolic stability are then excluded from further consideration because most therapeutic targets require compounds with an extended pharmacokinetic half-life. The same is true with *in vivo* exposure studies, where high clearance compounds are discarded. In these early screens, the concentration of the parent compound is typically the only measurement made. Consequently, there is no information on the number, identity and pharmacological significance of metabolites that may have been formed. Even when metabolic profiling is completed and metabolites are identified, the information is typically used to direct synthesis of analogs with improved metabolic stability through the modification of metabolic soft/hot spots. Thus, the information is rarely used for the purpose of searching for pharmacologically active products as new analogs. However, rapid metabolism of parent compounds could lead to the formation of pharmacologically active metabolites that may have comparatively superior developability characteristics. As a result, metabolic instability that otherwise may be considered a liability can become advantageous as a method of drug design.

There are a number of advantages for screening drug candidates for active metabolites during drug discovery. The primary reason is that the process could lead to the discovery of a drug candidate with superior drug developability attributes such as

(1) improved pharmacodynamics (PD);
(2) improved pharmacokinetics (PK);
(3) lower probability for drug–drug interactions;
(4) less variable pharmacokinetics and/or pharmacodynamics;
(5) improved overall safety profile;
(6) improved physicochemical properties (e.g., solubility).

Other advantages of early screening for active metabolites include the potential for modifications of the entire chemical class (chemotype) to improve overall characteristics (Clader, 2004; Fura et al., 2004). Further, early discovery of active metabolites will allow for more complete patent protection of the parent molecule. Additionally, tracking active metabolites at the drug discovery stage will allow for the correct interpretation of the pharmacological effect observed in preclinical species in relation to a predicted effect in humans. In other words, if an active metabolite is responsible for significant activity in a species used for preclinical efficacy determination, there is a significant risk that the effect will be dramatically different in humans unless similar levels of metabolite can be expected in humans.

An active metabolite may have a low potential for off-target toxicity as it, in most cases, leads to the formation of fewer number of metabolites compared to the parent compound. Moreover, most active metabolites are products of functionalization reactions, and as such are more susceptible to conjugation reactions. Conjugation reactions result in the formation of secondary metabolites that, in general, are safely cleared from the body. For example, phenacetin, which is no longer in use in humans, is metabolized to a number of metabolites. Of the many phenacetin metabolic pathways, the O-deethylation pathway leads to the formation of acetaminophen, a more analgesic agent, whereas N-hydroxylation of phenacetin leads to the formation of a toxic metabolite. On the contrary, the corresponding active metabolite, acetaminophen, is predominantly cleared via Phase II conjugation reactions (sulfation and glucuronidation) and has a greater margin of safety relative to phenacetin.

In general, drug metabolism reactions convert lipophilic compounds to more hydrophilic, more water-soluble products. An improvement in the solubility profile is an added advantage, particularly in the current drug discovery paradigm where many drug candidates generated during the lead optimization have poor aqueous solubility.

The discovery of an active metabolite can serve as a modified lead compound around which new structure–activity relationships can be examined during the lead optimization stage of drug discovery. For example, this approach was used in the discovery of ezetimibe, a cholesterol absorption inhibitor (Clader, 2004; van Heek et al., 1997). In these studies, a lead candidate (SCH48461) gave rise to a pharmacologically active biliary metabolite upon oral administration to rats that was approximately 30-fold more potent than the parent molecule. Further optimization of the metabolite through structural modification led to the discovery of ezetimibe, a molecule that was approximately 400-fold more potent than the initial lead candidate.

In summary, tracking active metabolites at the drug discovery stage is not only important to correctly interpret the pharmacological effects in preclinical species but may also lead to the discovery of a lead candidate with superior drug developability characteristics.

8.6.1 Detection of Active Metabolites During Drug Discovery

The exploration of the potential for formation of active metabolites can be carried out with varying degrees of direction from information gathered through, metabolism, pharmacokinetics, and biological/pharmacological assays. An example of undirected screening of active metabolites would be the modification of chemical libraries by subjecting them to metabolizing systems and subsequently using these modified libraries for high-throughput screens, either against the intended target or more broadly. This example is a way to generate increased molecular diversity from a given chemical library.

However, this approach requires significant "deconvolution" efforts when activity is found in mixtures. To increase the success rate and decrease the number of compounds screened to a manageable size, the search for active metabolites could be limited to those compounds/chemotypes showing a high clearance rates in *in vitro* metabolic stability or *in vivo* exposure screens.

However, activity assays may serve as a more rationale approach to the exploration of active metabolites. This is most often and most effectively done in the setting of an *in vivo* efficacy experiment that allows for both pharmacodynamic and pharmacokinetic information to be gathered. Analysis of the relationship between the PD endpoint and the PK profile will sometimes demonstrate an apparent disconnect between the two data sets and point to the possibility that an active metabolite is responsible for some of the activity. These disconnects can serve as clear trigger points for the initiation of active metabolite searches.

For example, van Heek and coworkers observed a lead candidate that underwent extensive first-pass metabolism and yet elicited a significant level of pharmacological activity (van Heek et al., 1997). To evaluate the biological activity of the *in vivo* biotransformation products, they collected samples of bile from rats dosed with a lead compound and directly administered the samples to a bile duct cannulated rats via an intraduodenal cannula. As a control study, the parent compound prepared in a blank bile was dosed in a similar fashion to the recipient rats. The results indicated that the *in vivo* activity elicited by the bile samples was higher than the parent control sample, clearly indicating the presence of an active metabolite(s) that was more potent than the parent compound. To identify the active component, the bile sample was then fractionated and each fraction tested for biological activity. The structure of the metabolite was then established following the detection of the active fraction. As mentioned before, further modification of the active metabolite led to the discovery of ezetimibe.

Although a lack of correlation between PK and PD data is the clearest trigger point for pursuing the possibility of metabolite contributions to the observed pharmacology, there are several other potential triggers that can be used that include (1) the observation of a greater pharmacological effect upon extravascular administration of a compound relative to parenteral administration. (2) a reduced pharmacological effect upon coadministration *in vivo/in vitro* with compounds that inhibit metabolism (e.g., aminobenzotriazole, ketoconazole, (3) a prolonged PD effect despite rapid *in vitro* metabolism. Examples of the utilization of metabolite structural information in drug design can be found in recent reviews (Fura, 2006; Fura et al., 2004).

8.6.2 Methods for Assessing and Evaluating the Biological Activity of Metabolites

In order to assess the biological activity and hence usefulness of metabolic products, several approaches can be used. The most straightforward approach is to take samples of interest (i.e, microsomal incubations, plasma samples) and isolate and purify any metabolites present, after which the structure of the metabolites is determined and each is tested for biological activity. An alternative approach is to use bioassay-guided methods where biological samples containing biotransformation products are first evaluated for their pharmacological activity without any effort to isolate or structurally characterize the metabolites. The bioassay methods may be based on the assessment of the pharmacological activity using *in vitro*-ligand binding (Lim et al., 1999; Soldner et al., 1998), cell-based assays or *in vivo* pharmacological assays (van Heek et al., 1997). Metabolites can be generated by any of the *in vitro* and *in vivo* methods discussed in the following sections. Biological activity in the sample mixture can then be evaluated as is or after fractionation of the sample mixture by using chromatographic techniques. The structural identity of the active metabolite can then be determined and its *in vitro* and *in vivo* activity confirmed after isolation and/or after further biological or chemical synthesis.

A systematic approach to profiling active metabolites using a 96-well plate format was recently described (Shu et al., 2002). The approach is based on a rapid bioassay-guided metabolite detection and characterization. Drug metabolite mixtures (generated by various methods described below) are separated and fractions collected into microtiter plates such as 96-well plates. The fractions are then subjected to one or more relevant activity (e.g., receptor ligand binding) assays.

8.6.3   Methods for Generation of Metabolites

There are a number of *in vitro* and *in vivo* biotransformation techniques available to generate metabolites. The *in vitro* techniques include the use of subcellular fractions prepared from cells that mediate drug metabolism, intact cell-based systems, intact organs, and isolated expressed enzymes. *In vivo* methods involve the use of biological fluids (plasma, bile, urine, etc.) obtained from laboratory animals or humans dosed with the parent molecule. Microbial methods and biomimetic systems based on metalloporphyrin chemistry can also be used as bioreactors to produce metabolites.

8.7   SUMMARY

The past several decades have witnessed an explosion in our knowledge of drug metabolizing enzymes. It is now possible to fully characterize and predict the metabolic fate of new chemical entities in humans with reasonable certainty. The *in vitro* methods used to do this, that is, human liver microsomes, expressed enzymes, cryopreserved, or freshly isolated hepatocytes, can be adapted to be run in medium-throughput fashion and allow the metabolic properties of new compounds to be optimized during the discovery phase.

Also, medium-throughput screens for bioavailability are now possible with rapid quantitation by LC–MS/MS techniques. These techniques will need to continue to evolve to allow metabolism scientists the ability to keep pace with the increasing speed of drug discovery. The increasing complexity of the process will place an ever greater dependence on information management and the ability to turn that information into knowledge will be tantamount. This is especially true when one considers the advances in other fields of ADME-related research, for example, transporters, interactions with nuclear hormone receptors, that need to be characterized during the discovery of a drug. This complexity ensures that the field of drug metabolism will continue to play an increasing role in the endeavor to provide high quality drug candidates that will become safe, highly efficacious medicines.

REFERENCES

Amacher DE. Reactive intermediates and the pathogenesis of adverse drug reactions: the toxicology perspective. Curr Drug Metab 2006;7:219–229.

Anari MR, Baillie, TA. Bridging cheminformatic metabolite prediction and tandem mass spectrometry. Drug Discov Today 2005;10:711–717.

Andrews CW, Bennett L, Yu LX. Predicting human oral bioavailability of a compound: development of a novel quantitative structure-bioavailability relationship. Pharm Res 2000;17:639–644.

Argoti D, Liang L, Conteh A, Chen L, Bershas D, Yu CP, Vouros P, Yang E. Cyanide trapping of iminium ion reactive intermediates followed by detection and structure identification using liquid chromatography–tandem mass spectrometry (LC–MS/MS). Chem Res Toxicol 2005;18:1537–1544.

Baillie, TA. Future of toxicology-metabolic activation and drug design: challenges and opportunities in chemical toxicology. Chem Res Toxicol 2006;19:889–893.

Balakin KV, Ivanenkov YA, Savchuk NP, Ivashchenko AA, Ekins S. Comprehensive computational assessment of ADME properties using mapping techniques. Curr Drug Discov Technol 2005;2:99–113.

Balani SK, Zhu T, Yang TJ, Liu Z, He B, Lee FW. Effective dosing regimen of 1-aminobenzotriazole for inhibition of antipyrine clearance in rats, dogs, and monkeys. Drug Metab Dispos 2002;30:1059–1062.

Buch H, Buzello W, Heymann E, Krisch, K. Inhibition of phenacetin- and acetanilide-induced methemoglobinemia in the rat by the carboxylesterase inhibitor bis-[p-nitrophenyl] phosphate. Biochem Pharmacol 1969;18:801–811.

Chen WG, Zhang C, Avery MJ, Fouda HG. Reactive metabolite screen for reducing candidate attrition in drug discovery. Adv Exp Med Biol 2001;500:521–524.

Clader JW. The discovery of ezetimibe: a view from outside the receptor. J Med Chem 2004;47:1–9.

Cox KA, White RE, Korfmacher WA. Rapid determination of pharmacokinetic properties of new chemical entities: *in vivo* approaches. Comb Chem High Throughput Screen 2002;5:29–37.

de Groot MJ, Ekins S. Pharmacophore modeling of cytochromes P450. Adv Drug Deliv Rev 2002;54:367–383.

Dieckhaus CM, Fernandez-Metzler CL, King R, Krolikowski PH, Baillie TA. Negative ion tandem mass spectrometry for the detection of glutathione conjugates. Chem Res Toxicol 2005;18:630–638.

Egnash LA, Ramanathan R. Comparison of heterogeneous and homogeneous radioactivity flow detectors for simultaneous profiling and LC-MS/MS characterization of metabolites. J Pharm Biomed Anal 2002;27:271–284.

Ekins S, Andreyev S, Ryabov A, Kirillov E, Rakhmatulin EA, Bugrim A, Nikolskaya T. Computational prediction of human drug metabolism. Expert Opin Drug Metab Toxicol 2005;1:303–324.

Ekins S, Obach RS. Three-dimensional quantitative structure activity relationship computational approaches for prediction of human *in vitro* intrinsic clearance. J Pharmacol Exp Ther 2000;295:463–473.

Erhardt PW. A human drug metabolism database: potential roles in the quantitative predictions of drug metabolism and metabolism-related drug–drug interactions. Curr Drug Metab 2003;4:411–422.

Evans DC, Watt AP, Nicoll-Griffith DA, Baillie TA. Drug-protein adducts: an industry perspective on minimizing the potential for drug bioactivation in drug discovery and development. Chem Res Toxicol 2004;17:3–16.

Fisher MB, Campanale K, Ackermann BL, VandenBranden M, Wrighton SA. *In vitro* glucuronidation using human liver microsomes and the pore-forming peptide alamethicin. Drug Metab Dispos 2000;28:560–566.

Fisher MB, Yoon K, Vaughn ML, Strelevitz TJ, Foti RS. Flavin-containing monooxygenase activity in hepatocytes and microsomes: *in vitro* characterization and *in vivo* scaling of benzydamine clearance. Drug Metab Dispos 2002;30:1087–1093.

Fura A. Role of pharmacologically-active metabolites in drug discovery and development. Drug Discov Today 2006;11:133–142.

Fura A, Shu YZ, Zhu M, Hanson RL, Roongta V, Humphreys WG. Discovering drugs through biological transformation: role of pharmacologically active metabolites in drug discovery. J Med Chem 2004;47:4339–4351.

Gan J, Harper TW, Hsueh MM, Qu Q, Humphreys WG. Dansyl glutathione as a trapping agent for the quantitative estimation and identification of reactive metabolites. Chem Res Toxicol 2005;18:896–903.

Grime K, Riley RJ. The impact of *in vitro* binding on *in vitro–in vivo* extrapolations, projections of metabolic clearance and clinical drug-drug interactions. Curr Drug Metab 2006;7:251–264.

Guengerich FP. Common and uncommon cytochrome P450 reactions related to metabolism and chemical toxicity. Chem Res Toxicol 2001;14:611–650.

Guengerich FP. Cytochrome P450 oxidations in the generation of reactive electrophiles: epoxidation and related reactions. Arch Biochem Biophys 2003;409:59–71.

Guengerich FP. Principles of covalent binding of reactive metabolites and examples of activation of bis-electrophiles by conjugation. Arch Biochem Biophys 2005;433:369–378.

Guengerich FP. Cytochrome P450S and other enzymes in drug metabolism and toxicity. AAPS J 2006;8:E101–E111.

Hawkins DR. Use and value of metabolism databases. Drug Discov Today 1999;4:466–471.

Hellriegel ET, Bjornsson TD, Hauck WW. Interpatient variability in bioavailability is related to the extent of absorption: implications for bioavailability and bioequivalence studies. Clin Pharmacol Ther 1996;60:601–607.

Houston JB, Carlile DJ. Prediction of hepatic clearance from microsomes, hepatocytes, and liver slices. Drug Metab Rev 1997;29:891–922.

Houston JB, Kenworthy KE. In vitro–in vivo scaling of CYP kinetic data not consistent with the classical Michaelis–Menten model. Drug Metab Dispos 2000;28:246–254.

Ito K, Iwatsubo T, Kanamitsu S, Nakajima Y, Sugiyama Y. Quantitative prediction of in vivo drug clearance and drug interactions from in vitro data on metabolism, together with binding and transport. Annu Rev Pharmacol Toxicol 1998; 38:461–499.

Jones JP, Mysinger M, Korzekwa, KR. Computational models for cytochrome P450:a predictive electronic model for aromatic oxidation and hydrogen atom abstraction. Drug Metab Dispos 2002;30:7–12.

Ju C, Uetrecht JP. Mechanism of idiosyncratic drug reactions: reactive metabolite formation, protein binding and the regulation of the immune system. Curr Drug Metab 2002;3:367–377.

Kaplowitz N. Idiosyncratic drug hepatotoxicity. Nat Rev Drug Discov 2005;4:489–499.

Kim HJ, Cho JH, Klaassen CD. Depletion of hepatic 3'-phosphoadenosine 5'-phosphosulfate (PAPS) and sulfate in rats by xenobiotics that are sulfated. J Pharmacol Exp Ther 1995;275:654–658.

Lau YY, Sapidou E, Cui X, White RE, Cheng, KC. Development of a novel in vitro model to predict hepatic clearance using fresh, cryopreserved, and sandwich-cultured hepatocytes. Drug Metab Dispos 2002;30:1446–1454.

Lave T, Coassolo P, Reigner B. Prediction of hepatic metabolic clearance based on interspecies allometric scaling techniques and in vitro–in vivo correlations. Clin Pharmacokinet 1999;36:211–231.

Lave T, Dupin S, Schmitt C, Chou RC, Jaeck D, Coassolo P. Integration of in vitro data into allometric scaling to predict hepatic metabolic clearance in man: application to 10 extensively metabolized drugs. J Pharm Sci 1997;86:584–590.

Lave T, Luttringer O, Zuegge J, Schneider G, Coassolo P, Theil FP. Prediction of human pharmacokinetics based on preclinical in vitro and in vivo data. Ernst Schering Res Found Workshop 2002;81–104

Lecoeur S, Bonierbale E, Challine D, Gautier JC, Valadon P, Dansette PM, Catinot R, Ballet F, Mansuy D, Beaune PH. Specificity of in vitro covalent binding of tienilic acid metabolites to human liver microsomes in relationship to the type of hepatotoxicity: comparison with two directly hepatotoxic drugs. Chem Res Toxicol 1994;7:434–442.

Lewis DF. Quantitative structure–activity relationships (QSARs) within the cytochrome P450 system: QSARs describing substrate binding, inhibition and induction of P450s. Inflammopharmacology 2003;11:43–73.

Lewis DF. Quantitative structure-activity relationships (QSARs) for substrates of human cytochromes P450 CYP2 family enzymes. Toxicol In Vitro 2004;18:89–97.

Lewis DF. Human P450s in the metabolism of drugs: molecular modelling of enzyme–substrate interactions. Expert Opin Drug Metab Toxicol 2005;1:5–8.

Lewis DF, Ito Y, Goldfarb PS. Investigating human P450s involved in drug metabolism via homology with high-resolution P450 crystal structures of the CYP2C subfamily. Curr Drug Metab 2006;7:589–598.

Liebler DC, Guengerich FP. Elucidating mechanisms of drug-induced toxicity. Nat Rev Drug Discov 2005;4:410–420.

Lim HK, Stellingweif S, Sisenwine S, Chan KW. Rapid drug metabolite profiling using fast liquid chromatography, automated multiple-stage mass spectrometry and receptor-binding. J Chromatogr A 1999;831:227–241.

Lin JH. Applications and limitations of interspecies scaling and *in vitro* extrapolation in pharmacokinetics. Drug Metab Dispos 1998;26:1202–1212.

Lin JH, Wong BK. Complexities of glucuronidation affecting *in vitro–in vivo* extrapolation. Curr Drug Metab 2002;3:623–646.

Long A, Walker JD. Quantitative structure-activity relationships for predicting metabolism and modeling cytochrome p450 enzyme activities. Environ Toxicol Chem 2003;22:1894–1899.

Margolis JM, Obach RS. Impact of nonspecific binding to microsomes and phospholipid on the inhibition of cytochrome P4502 D6: implications for relating *in vitro* inhibition data to *in vivo* drug interactions. Drug Metab Dispos 2003;31:606–611.

Masubuchi Y, Igarashi S, Suzuki T, Horie T, Narimatsu S. Imipramine-induced inactivation of a cytochrome P450 2D enzyme in rat liver microsomes: in relation to covalent binding of its reactive intermediate. J Pharmacol Exp Ther 1996;279:724–731.

Mays DC, Hilliard JB, Wong DD, Chambers MA, Park SS, Gelboin HV, Gerber N. Bioactivation of 8-methoxypsoralen and irreversible inactivation of cytochrome P-450 in mouse liver microsomes: modification by monoclonal antibodies, inhibition of drug metabolism and distribution of covalent adducts. J Pharmacol Exp Ther 1990;254:720–731.

Mays DC, Pawluk LJ, Apseloff G, Davis WB, She ZW, Sagone AL, Gerber N. Metabolism of phenytoin and covalent binding of reactive intermediates in activated human neutrophils. Biochem Pharmacol 1995;50:367–380.

Miners JO, Knights KM, Houston JB, Mackenzie PI. *In vitro–in vivo* correlation for drugs and other compounds eliminated by glucuronidation in humans: pitfalls and promises. Biochem Pharmacol 2006;71:1531–1539.

Miners JO, Smith PA, Sorich MJ, McKinnon RA, Mackenzie PI. Predicting human drug glucuronidation parameters: application of *in vitro* and *in silico* modeling approaches. Annu Rev Pharmacol Toxicol 2004;44:1–25.

Mutlib A, Lam W, Atherton J, Chen H, Galatsis P, Stolle W. Application of stable isotope labeled glutathione and rapid scanning mass spectrometers in detecting and characterizing reactive metabolites. Rapid Commun Mass Spectrom 2005;19:3482–3492.

Naisbitt DJ, Farrell J, Gordon SF, Maggs JL, Burkhart C, Pichler WJ, Pirmohamed M, Park BK. Covalent binding of the nitroso metabolite of sulfamethoxazole leads to toxicity and major histocompatibility complex-restricted antigen presentation. Mol Pharmacol 2002;62:628–637.

Nassar AE, Bjorge SM, Lee DY. On-line liquid chromatography-accurate radioisotope counting coupled with a radioactivity detector and mass spectrometer for metabolite identification in drug discovery and development. Anal Chem 2003;75:785–790.

Nassar AE, Parmentier Y, Martinet M, Lee DY. Liquid chromatography-accurate radioisotope counting and microplate scintillation counter technologies in drug metabolism studies. J Chromatogr Sci 2004;42:348–353.

Obach RS. Nonspecific binding to microsomes: impact on scale-up of in vitro intrinsic clearance to hepatic clearance as assessed through examination of warfarin, imipramine, and propranolol. Drug Metab Dispos 1997;25:1359–1369.

Obach RS. Prediction of human clearance of twenty-nine drugs from hepatic microsomal intrinsic clearance data: an examination of in vitro half-life approach and nonspecific binding to microsomes. Drug Metab Dispos 1999;27:1350–1359.

Obach RS. The prediction of human clearance from hepatic microsomal metabolism data. Curr Opin Drug Discov Devel 2001;4:36–44.

Obach RS, Baxter JG, Liston TE, Silber BM, Jones BC, MacIntyre F, Rance DJ, Wastall P. The prediction of human pharmacokinetic parameters from preclinical and in vitro metabolism data. J Pharmacol Exp Ther 1997;283:46–58.

Park BK, Kitteringham NR, Maggs JL, Pirmohamed M, Williams DP. The role of metabolic activation in drug-induced hepatotoxicity. Annu Rev Pharmacol Toxicol 2005;45:177–202.

Pirmohamed M, Kitteringham NR, Guenthner TM, Breckenridge AM, Park BK. An investigation of the formation of cytotoxic, protein-reactive and stable metabolites from carbamazepine in vitro. Biochem Pharmacol 1992;43:1675–1682.

Rane A, Wilkinson GR, Shand DG. Prediction of hepatic extraction ratio from in vitro measurement of intrinsic clearance. J Pharmacol Exp Ther 1997;200:420–424.

Samuel K, Yin W, Stearns RA, Tang YS, Chaudhary AG, Jewell JP, Lanza T Jr,Lin LS, Hagmann WK, Evans DC, Kumar S. Addressing the metabolic activation potential of new leads in drug discovery: a case study using ion-trap mass spectrometry and tritium labeling techniques. J Mass Spectrom 2003;38:211–221.

Shiran MR, Proctor NJ, Howgate EM, Rowland-Yeo K, Tucker GT, Rostami-Hodjegan A. Prediction of metabolic drug clearance in humans: in vitro–in vivo extrapolation vs allometric scaling. Xenobiotica 2006;36:567–580.

Shu YZ, Li W, Leet JE, Alberts J, Arora VK, Yeola S, Philip T, Qian-Cutrone J, Zhao N, Santone K, Humphreys WG, Dickinson K. Biogram enabled evaluation of active metabolites: an exploratory approach for detecting and characterizing active/toxic drug metabolites. Drug Metab Rev 2002;34 (1 Suppl):75.

Singh R, Silva Elipe MV, Pearson PG, Arison BH, Wong BK, White R, Yu X, Burgey CS, Lin JH, Baillie TA. Metabolic activation of a pyrazinone-containing thrombin inhibitor. Evidence for novel biotransformation involving pyrazinone ring oxidation, rearrangement, and covalent binding to proteins. Chem Res Toxicol 2003;16:198–207.

✓Smith D, Schmid E, Jones B. Do drug metabolism and pharmacokinetic departments make any contribution to drug discovery? Clin Pharmacokinet 2002;41:1005–1019.

Smith DA, Obach RS. Seeing through the mist: abundance versus percentage. Commentary on metabolites in safety testing. Drug Metab Dispos 2005;33:1409–1417.

Soars MG, Burchell B, Riley RJ. In vitro analysis of human drug glucuronidation and prediction of *in vivo* metabolic clearance. J Pharmacol Exp Ther 2002;301:382–390.

Soldner A, Spahn-Langguth H, Palm D, Mutschler E. A radioreceptor assay for the analysis of AT1-receptor antagonists. Correlation with complementary LC data reveals a potential contribution of active metabolites. J Pharm Biomed Anal 1998;17:111–124.

Taskinen J, Ethell BT, Pihlavisto P, Hood AM, Burchell B, Coughtrie MW. Conjugation of catechols by recombinant human sulfotransferases, UDP-glucuronosyltransferases, and soluble catechol O-methyltransferase: structure-conjugation relationships and predictive models. Drug Metab Dispos 2003;31:1187–1197.

✓Thompson TN. Optimization of metabolic stability as a goal of modern drug design. Med Res Rev 2001;21:412–449.

Tran TH, von Moltke LL, Venkatakrishnan K, Granda BW, Gibbs MA, Obach RS, Harmatz JS, Greenblatt DJ. Microsomal protein concentration modifies the apparent inhibitory potency of CYP3A inhibitors. Drug Metab Dispos 2002;30:1441–1445.

Uetrecht J. Screening for the potential of a drug candidate to cause idiosyncratic drug reactions. Drug Discov Today 2003;8:832–837.

van de Waterbeemd H, Gifford E. ADMET *in silico* modelling: towards prediction paradise? Nat Rev Drug Discov 2003;2:192–204.

van Heek M, France CF, Compton DS, McLeod RL, Yumibe NP, Alton KB, Sybertz EJ, Davis, HR Jr. *In vivo* metabolism-based discovery of a potent cholesterol absorption inhibitor, SCH58235, in the rat and rhesus monkey through the identification of the active metabolites of S CH48461. J Pharmacol Exp Ther 1997;283:157–163.

Walgren JL, Mitchell MD, Thompson DC. Role of metabolism in drug-induced idiosyncratic hepatotoxicity. Crit Rev Toxicol 2005;35:325–361.

Ward DP, Trevor AJ, Kalir A, Adams JD, Baillie TA, Castagnoli N Jr. Metabolism of phencyclidine. The role of iminium ion formation in covalent binding to rabbit microsomal protein. Drug Metab Dispos 1982;10:690–695.

Watt AP, Mortishire-Smith RJ, Gerhard U, Thomas SR. Metabolite identification in drug discovery. Curr Opin Drug Discov Devel 2003;6:57–65.

White IN, De Matteis F, Gibbs AH, Lim CK, Wolf CR, Henderson C, Smith LL. Species differences in the covalent binding of [14C]tamoxifen to liver microsomes and the forms of cytochrome P450 involved. Biochem Pharmacol 1995;49:1035–1042.

White, RE. High-throughput screening in drug metabolism and pharmacokinetic support of drug discovery. Annu Rev Pharmacol Toxicol 2000;40:133–157.

Yamamoto T, Itoga H, Kohno Y, Nagata K, Yamazoe Y. Prediction of oral clearance from *in vitro* metabolic data using recombinant CYPs: comparison among well-stirred, parallel-tube, distributed and dispersion models. Xenobiotica 2005;35: 627–646.

Yan Z, Maher N, Torres R, Caldwell GW, Huebert N. Rapid detection and characterization of minor reactive metabolites using stable-isotope trapping in combination with tandem mass spectrometry. Rapid Commun Mass Spectrom 2005;19:3322–3330.

Yoshida F, Topliss JG. QSAR model for drug human oral bioavailability. J Med Chem 2000;43:2575–2585.

Ziegler DM. An overview of the mechanism, substrate specificities, and structure of FMOs. Drug Metab Rev 2002;34:503–511.

9

ROLE OF DRUG METABOLISM IN DRUG DEVELOPMENT

RAMASWAMY IYER AND DONGLU ZHANG

9.1 INTRODUCTION TO THE ROLE OF DRUG METABOLISM IN DRUG DEVELOPMENT

In drug discovery, metabolism studies are conducted to select compounds with superior absorption, distribution, metabolism, and excretion (ADME) properties and to eliminate compounds with potential liabilities. In drug development, metabolism studies are conducted to provide detailed data about the disposition of the selected compounds in animals and humans. Metabolism studies in the early drug discovery stages often involve evaluating a series of compounds with the main purpose of selecting a compound with high metabolic stability; multiple clearance pathway; low potential for enzyme inhibition or induction, and, finally, low potential for forming reactive intermediates. Once the compound is selected for clinical development, then detailed studies, often with radiolabeled compound, are conducted during the drug development process in animals (toxicology species) and humans to predict potential drug–drug interactions, help address safety issues and provide data for registration. The focus of this chapter is on the role of drug metabolism in drug development.

9.1.1 Drug Metabolism and Clinical Interactions

Absorption, distribution, metabolism, and excretion are four key processes that determines exposure and duration of a drug's activity. Understanding

Drug Metabolism in Drug Design and Development, Edited by Donglu Zhang, Mingshe Zhu and W. Griffith Humphreys

261

these processes and the role they play in the disposition of the drug are important since they help us understand and predict potential for various types of interactions, particularly drug–drug interactions. Drugs that are currently being developed are mainly targeted toward receptors and enzymes and, therefore, tend to be lipophilic with a molecular weight ranging from 300 to 700 Da. One consequence of selecting compounds with these physicochemical properties is that most of these drugs are primarily eliminated through metabolism. Oxidative, hydrolytic, reductive, or conjugative (glucuronidation, sulfation, etc.) metabolic pathways are often responsible for clearance of modern drugs (Parkinson, 2001). Oxidative metabolism, mainly catalyzed by cytochrome P450s, is probably responsible for the clearance of >75% of the drugs on the market (Williams et al., 2004). Activities of CYPs are affected by numerous factors like genetics, age, sex, disease, inhibitors, and inducers. Genetic polymorphisms affect the activity of a number of drug-metabolizing enzymes such as CYP2C19, CYP2D6, CYP1A2. For example, genetic polymorphism in the CYP2D6, which is found in 5–10% of Caucasians, gives rise to poor metabolizers that have low or lack of activity (Parkinson, 2001). Therefore, for drugs metabolized by CYP2D6, dose adjustment may be needed in patients that are poor metabolizers. Similarly, age, sex, and disease state could also have an effect on the activity of the drug–metabolizing enzyme. Thus, during drug development, where appropriate, subjects that show genetic polymorphisms toward the enzyme(s) that is involved in the drugs metabolism (e.g., CYP2D6 poor metabolizer), subjects of different age (old and young), both sexes, and special populations (e.g., hepatically impaired) are included in the clinical trial process to understand the pharmacokinetic variability. The results from these studies are included in the label to help prescribing physicians adjust the dose or, if necessary, avoid giving the drug to patients. The other two factors that have an effect on the disposition of a drug are coadministered medicines that are inhibitors or inducers of drug metabolism enzymes. These types of drug–drug interactions are increasingly being observed and have resulted in the withdrawal of numerous drugs from the market or have restricted their use (Huang and Lesko, 2004). Because of the implication that these types of interactions have toward safety and efficacy, the FDA and European regulatory agencies have provided guidelines to address them during the drug development process (Products; US Food and Drug Administration). Understanding the ADME properties of the drug helps to address some of the drug–drug interaction potential for a drug (Kumar and Surapaneni, 2001).

9.1.2 Drug Metabolism and Issue Resolution

Issue-driven metabolism studies, conducted when either animal toxicity or clinical safety findings are observed, are an important part of the drug development process. These studies help to define the role of drug and drug-derived metabolites in the observed toxicity or safety findings. Although issues regarding clinical safety or toxicity findings in animals are frequent in drug development, there are no common approaches and systems one can use to address these issues. The approaches used depend on the nature of the issue. For example, if it is a toxicology finding in long-term toxicity studies in animals then having an understanding about the quantitative and qualitative differences in metabolism across species, along with exposure to drug and drug-derived metabolites, helps to further define the issue and assists in the planning of key follow-up metabolism-related studies. There are three examples at the end of this chapter that illustrate the role of metabolism-related studies to address issues in the drug development process. These three examples and numerous examples in the literature further highlight the need for a thorough understanding of the *in vitro* and *in vivo* metabolism profile of a drug.

9.1.3 Drug Metabolism and Regulatory Requirements

Ultimately, the metabolism-related data that is generated during the drug development process is to satisfy the regulatory agencies requirements. Table 9.1 shows the table of contents for the drug metabolism sections in the common technical document (CTD) for a drug application. Table 9.2 shows the template used to summarize drug metabolism data for an *in vivo* study that is included in the summary tables of the CTD. There are a number of regulatory guidelines published by the FDA and EMEA regarding the type of *in vitro* and *in vivo* metabolism data needed for registration of a drug. Chapter 7

TABLE 9.1   The sections of the CTD where metabolism studies are summarized.

2.6.4 Pharmacokinetics written summary
The sequence of the pharmacokinetics written summary:
 Brief summary
 Methods of analysis
 Absorption
 Distribution
 Metabolism
 Excretion
 Pharmacokinetic drug interactions
 Other pharmacokinetic studie
 Discussion and conclusions

2.6.4.5 Metabolism (interspecies comparison)
The following data should be summarized in this section:
 Chemical structures and quantities of metabolites in biological samples
 Possible metabolic pathways
 Presystemic metabolism (GI/hepatic first-pass effects)
 In vitro metabolism including P450 studies
 Enzyme induction and inhibition

TABLE 9.2 Example of a CTD table for summarizing *in vivo* metabolism data.

2.6.5.9 Pharmacokinetics: metabolism *in vivo*

Test article:

Gender (M/F)/
Number of animals:
Feeding condition:
Vehicle/formulation:
Method of administration:
Dose (mg/kg):
Radionuclide:
Specific activity:

Species	Sample	Sampling time or period	Percentage of dose in sample	Percentage of compound in sample			Study No.	Location in CTD	
				Parent	M1	M2		Vol	Section
	Sample								
	Plasma								
	Urine								
	Bile								
	Feces								
	Plasma								
	Urine								
	Bile								
	Feces								

of this book specifically addresses regulatory guidelines for drug metabolism studies.

In addition to the CTD, metabolism-related data are also incorporated into the various sections of the product label to provide prescribing physicians with the relevant information to either recommend adjusting the dose or avoid prescribing the medicine when patients are taking concomitant interacting medications. These are discussed in more detail in the later half of the chapter (Section 9.7).

9.2 STAGING AND TYPES OF DRUG METABOLISM STUDIES IN DRUG DEVELOPMENT

Drug metabolism studies are conducted throughout the discovery and development lifetime of a drug. The type of metabolism studies performed depends on the stage of the drug development process (Fig. 9.1). As mentioned before, the focus of the drug metabolism studies in early drug discovery is to aid in selection and identify potential liabilities, whereas, studies conducted after the compound is selected for development are for registrational filing and to provide information for the product label. The timing of the metabolism-related studies depends on how the information generated will be used in the compound's development. Since the focus of this chapter is on the role of metabolism studies after the compound is selected to be taken into clinical

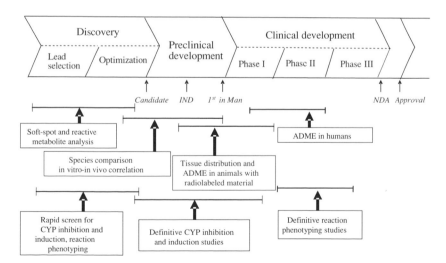

FIGURE 9.1 Staging of metabolism-related studies in the drug discovery and development process. The timing and the sequence of the studies presented here are based on author's experience in conducting these studies in Bristol Myers Squibb. The studies mentioned here include the core set of metabolism-related work that is currently conducted for registrational filing.

trials, the staging of the studies that support clinical development and long-term toxicology studies will be discussed here.

One major goal for conducting metabolism-related studies in both nonclinical and human subjects is to make sure that the metabolites detected in human circulation show exposure in the toxicology species. Therefore, metabolism studies are conducted both in humans and in the toxicology species used in the long-term safety studies. Prior to IND filing, in the absence of *in vivo* human data, *in vitro* systems like human liver microsomes and hepatocytes are heavily relied on to get an early read on the potential human metabolites. These studies along with *in vitro* and *in vivo* animal studies provide a qualitative comparison of metabolism across species.

The timing of ADME studies in animals and humans, which are typically conducted with radiolabeled material, depends on the drug development strategy. Given the regulatory climate, and that *in vivo* data are preferred (see FDA and ICH guidelines) over *in vitro* data, the current strategy employed in much of the pharmaceutical industry is to do detailed ADME studies much earlier during the lifetime of drug's development. Often these studies are done during the preclinical development to Phase 1 timeframe. The main objectives of these studies are to answer the following questions: What is the major route of elimination (urinary or biliary) in animals and humans? Will renal or hepatic impairment affect the drug's clearance? Does metabolism play a significant role prior to elimination and, if so, what are the primary pathways of metabolism? What are major metabolites in human plasma? Are major drug–drug interaction anticipated? Will polymorphisms have a significant effect on clearance? Are the toxicology species exposed to human metabolites? and finally, will metabolites contribute to the pharmacological effects?

Data generated from the ADME studies could trigger more studies like detailed reaction phenotyping studies to identify the enzymes involved in the primary pathways of metabolism that in turn can guide some key clinical interaction studies. Furthermore, the comparative data generated from the ADME studies get incorporated into the carcinogenicity protocol to guide in the dose selection for these studies. Decisions about monitoring for major circulating metabolites, as per the MIST and FDA (draft) guidelines, can also be made and appropriate assays developed for measuring the metabolites in the upcoming studies.

9.3 *IN VIVO* ADME STUDIES

The following section will focus on the detailed ADME studies conducted during the drug development process to help understand the disposition of a drug in nonclinical species and in humans. Since these studies are typically conducted with radiolabeled compound, a short section on the use of radiolabeled compound is included here along with a section on tissue distribution studies conducted to support the human ADME study.

9.3.1 Use of Radiolabeled Compound in ADME Studies

During the process of drug development there are numerous studies conducted with radiolabeled material (C-14 or tritium), because these studies provide quantitative information about parent and metabolites (Marathe et al., 2004). To provide reliable information, it is critical that the position of the C-14 or tritium label on the molecule be such that there is no loss of the label either due to exchange or due to a small fragment that gets incorporated or through expiration. The position of the label on the molecule is based on prior knowledge about the metabolism of the compound conducted during lead optimization. In cases where metabolism of a compound leads to splitting of the molecule into large fragments, it is useful to label both sides of the molecule. This strategy has been successfully used in ADME studies with omapatrilat and gemopatrilat. Amide hydrolysis, resulting in the formation of (S)-2-thio-3-phenylpropionic acid, was shown to be one of the major metabolic pathways for both of the structurally similar ACE and NEP inhibitors. Therefore, the radioactivity studies with a single [^{14}C] label would not elucidate the metabolic fate of both sides of the molecule. Based on this consideration, ADME studies were conducted with a mixture of [^{14}C]-labeled derivatives: one [^{14}C] label at the exocyclic carbonyl carbon, and the other [^{14}C] label on the amine side of the molecule that is generated after hydrolysis. This approach helped to track metabolites that arose from both sides of the molecule (Iyer et al., 2001; Wait et al., 2006).

In addition to quantitation of parent and metabolites, radiolabeled studies also aid in the detailed structural characterization of metabolites since radioactivity can be used as a selective marker for isolation and identification purposes. LC/MS/MS analysis can then be performed on the metabolite of interest and the structure of the metabolite assigned based on mass spec fragmentation patterns. One of the challenges faced during the analysis of a complex matrix like bile or urine, where there is interference from endogenous compounds, is to identify the base peak for MS/MS analysis. One approach to circumvent this issue is to administer a stable-labeled drug along with radiolabeled drug in a fixed proportion during the ADME studies. The use of stable-labeled compound in toxicology studies with small molecules and chemicals is well reported in the literature (Fennell et al., 1991; Fennell and Summer, 1994; Sumner et al., 1992). This strategy was successfully used in ADME studies with omapatrilat. The stable-labeled $^{13}C_2$ omapatrilat was included in the dose to help in the identification of the metabolites by LC/MS. The metabolites that retained the C-13 labels showed a characteristic $^{12}C - ^{13}C$ isotope cluster in the full mass spectrum. LC/MS/MS analysis of the molecular ions (m/z and $^{13}C_2 m/z$) further confirmed the identity of the metabolites. The observation that the daughter ions either retained or lost the $^{13}C_2$-label helped establish the fragmentation pattern for the metabolites (Iyer et al., 2003). Recently, mass defect filtering techniques were developed and effectively applied to identify molecular ions of drug metabolites in complex biological matrices such as plasma, urine, bile, and feces (Zhang et al., 2003).

9.3.2 Tissue Distribution Study to Support the Human ADME Study

One of the key studies that the FDA and investigative review boards (IRB) require the sponsors to do to support a human ADME protocol with radiolabeled compound is to conduct tissue distribution studies in pigmented animals, to provide dosimetry to various tissues and organs. The Long–Evans rat, since they are pigmented, is the most common species used for this purpose. Typically, these studies are limited to single dose by the intended route of administration (PO, IV, SC, etc.). Various tissues are collected at different time points and measured for radioactivity. The common techniques used for counting radioactivity in tissues are (1) liquid scintillation counting (LSC) of the tissue homogenate after appropriate processing or combustion of the tissues to CO_2 that is trapped and counted for radioactivity and (2) quantitative whole body autoradiography (QWBA), where thin sections of the whole animals are prepared and then exposed to a phosphorimaging screen that is then scanned with a phosphor imager. Figure 9.2 shows a typical QWBA image of rat after administration of a C-14-labeled compound. Readers are encouraged to refer to the review articles referenced here that describes in detail the use of tissue distribution studies in the pharmaceutical industry (Solon and Kraus, 2001; Solon et al., 2002). Based on body surface area and body weight, exposure of radioactivity to the tissues in rat is extrapolated to human tissues and used to estimate a whole body exposure (WHO,1977). For most compounds, administration of a 100-μCi radioactive single dose typically exposes the human subjects to an effective dose equivalent of <20 mrem, well below the radiation limit set by the Nuclear Radiation Committee of 3000 mrem (Hall, 1994; Valentin, 2002).

In addition to the tissue distribution study mentioned above for human dosimetry, other tissue distribution studies, namely, maternal–fetal and milk excretion studies are also conducted during the drug development process. These studies are required for the registrational filing and are typically conducted in rat species used in the toxicological evaluation. Data from pregnant rat studies provide information about the drug's potential to cross the placental barrier and expose the fetus. Data from lactating animals provide information about the drug's potential to be excreted into the milk. Based on the outcome of these studies, appropriate cautionary statements are made in the product label for treating pregnant or lactating female patients.

9.3.3 Nonclinical and Clinical ADME Studies

ADME studies provide information on absorption, distribution, metabolism, and excretion for the compound of interest in animals and humans. In drug development, these studies are performed with either C-14- or tritium-labeled material to provide detailed quantitative information on the circulating metabolites, the extent of metabolism and routes of excretion for drug and its metabolites. Readers are encouraged to refer Chapter 18 of this book for more detailed discussion on ADME study design and data presentation.

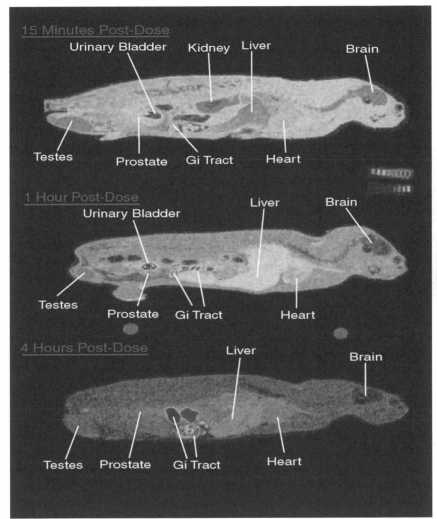

FIGURE 9.2 An example of quantitiative whole body autoradiography (QWBA) in rat after administration of C-14-labeled material. The figure shows the distribution of radioactivity at various time points postdose. The image was generated by taking a sliced section of the whole body of rat and exposing it to a phosphorimaging film.

The nonclinical species used in the ADME studies are based on toxicology species used in the long-term safety evaluation of the compound. These are typically rodent (rat and mouse) and nonrodent (rabbit, dog or monkey) species. The study is designed to closely mimic the toxicology studies keeping the dose selected, the route of administration, and the dosing vehicle as close as possible to those used in the actual studies. The amount of radioactive dose administered in animal species is usually determined by the pharmacokinetic properties of the parent. The typical administered radioactivity range is 60–200 μCi/kg. In our experience, this is sufficient to generate metabolic profiles in plasma at multiple time points with adequate radioactive sensitivity. The duration of the study is determined by the terminal half-life of the parent (in that species) and should be conducted for ≥5- plasma half-lives. For compounds with extremely long half-lives, the duration of the study can be based on the criteria that the study can be terminated when ≤1% of the total dose is excreted in a given 24-h interval in both urine and feces. Plasma, urine, and fecal samples are collected during the duration of the study to analyze for radioactivity and for metabolite profiling. If needed, additional tissue samples are collected to determine the concentration and accumulation of drug and drug-related component in a particular tissue.

The first set of information generated from the ADME studies are the overall plasma profile of total radioactivity (TRA) versus time, which is compared to the plasma profile of parent versus time measured by a validated LC/MS/MS assay. For those drugs where the parent is the major component at all time points in plasma, the total radioactivity profile usually parallels the profile of the parent. Metabolic profiles of plasma samples generated at different time points by HPLC analysis, followed by radioactivity and mass spectrometric detection, provides exposure-related information for parent and metabolites in humans and animal species. In addition, it provides information about metabolites that humans are exposed to and how this compares to exposures in animal species. This is illustrated in Fig. 9.3, which shows the plasma profiles in rats, dogs, and humans after an oral dose of a C-14-labeled drug. The plasma profiles clearly show that the metabolic profiles are similar across species. The Fig. 9.3 table insert summarizes the percent AUC of parent and two circulating metabolites as compared to AUC of total radioactivity. Based on this data, it can be clearly concluded that the monohydroxylated and N-demethylated metabolites are major circulating metabolites in humans that are also represented in the animal species. These comparative radioactivity profiles could be used to generate the absolute concentration of the metabolites present in human and animal plasma samples at the relevant doses and gives an estimate of the exposure margin for the metabolites in these species.

Metabolic profiles of urine and fecal samples generated in ADME studies provide information about the extent of metabolism and routes of excretion for parent and metabolites. The excretion of radioactivity after administration of [14C] gemopatrilat to healthy human volunteers shows that the compound is excreted in both urine and feces with most of the radioactive dose excreted by 48 h postdose. Based on urinary excretion, at least 55% of the dose is absorbed when administered orally. Figure 9.4 shows the comparative urinary and fecal profiles of orally administered [14C] gemopatrilat in rat, dog, and human (Wait et al., 2006). These types of data representations help understand the quantitative and qualitative differences in metabolism across species. As can be clearly seen in Fig. 9.4, the drug is extensively metabolized in all species,

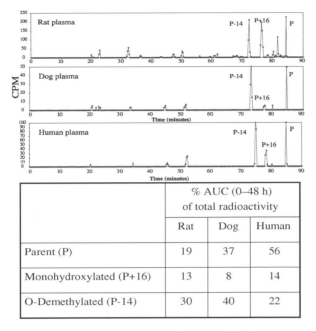

	% AUC (0–48 h) of total radioactivity		
	Rat	Dog	Human
Parent (P)	19	37	56
Monohydroxylated (P+16)	13	8	14
O-Demethylated (P-14)	30	40	22

FIGURE 9.3 Comparative plasma profiles of radioactivity in rat, dog, and humans showing parent and metabolites after a single oral dose of C-14-labeled compound. The table in the figure shows the percent AUC of total radioactivity for parent (P) and two major circulating metabolites in plasma samples. The AUCs for parent and metabolite were generated from radioactivity profiles generated at several time points.

since no parent peak is observed in the radioactivity profiles in both urine and feces. All the metabolites that the humans generated are also observed in animals suggesting that the primary pathways of metabolism is similar across species. Taken together the metabolic profiles in plasma, urine, and feces help us understand the complete disposition of drug.

For highly lipophilic drugs in the 300–700 Da range, metabolism and excretion through the bile often plays a major role in their disposition. For these types of molecules, conducting ADME studies in bile duct cannulated (BDC) animals, where bile is collected during the duration of the study, is extremely useful. BDC studies are particularly helpful if conjugative pathways like glucuronidation or sulfation are involved in the metabolic clearance of the drug. Glucuronide and sulfate conjugates, when excreted through the bile into the gastrointestinal (GI) tract, can be hydrolyzed during their passage through the intestine (Marier et al., 2002; Parker et al., 1980). Therefore, in the absence of a bile profile the role of conjugation in the overall metabolic clearance of the drug would be missed. BDC studies are usually run for a shorter duration, ≤24 h in rat and between 24 and 72 h in dog or monkey. These studies often do not achieve full mass balance studies. Where nonclinical data shows that most of the drug is excreted in the feces through bile as conjugative metabolites, it is

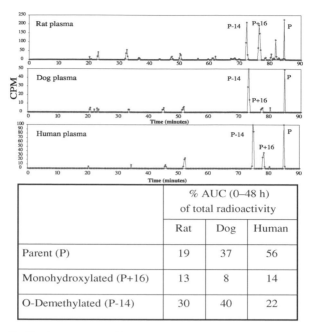

	% AUC (0–48 h) of total radioactivity		
	Rat	Dog	Human
Parent (P)	19	37	56
Monohydroxylated (P+16)	13	8	14
O-Demethylated (P-14)	30	40	22

FIGURE 9.4 Radiochromatographic profiles of pooled urine and fecal samples in rat, dog, and human administered a single oral dose of [^{14}C] gemopatrilat. The profiles are a background subtracted reconstructed radiochromatogram of 15-s fractions collected from an HPLC run.

useful to include a panel in the human ADME study where bile can be collected. There are several methods for this that has been recently reviewed (Ghibellini et al., 2006). This is exemplified in Fig. 9.5 for an ADME study conducted with [^{14}C] muraglitazar, where bile was collected for a short duration (3–8 h) after dosing. When the bile profile is compared against fecal profile it can be clearly seen that the dose in the bile is excreted as conjugates that is hydrolyzed during their passage through the GI tract (Wang et al., 2006).

The data from the human ADME study provides information about the primary pathways of metabolism for the compound. This is based on the identification of metabolites in plasma, urine, and feces/bile. This in turn can lead to detailed reaction phenotyping studies that is performed to identify the enzymes that generate the primary metabolites. Furthermore, metabolism data from human ADME studies in conjunction with reaction phenotyping can drive decisions regarding the conduct of key drug–drug interaction studies. For the example illustrated in Fig. 9.3, reaction phenotyping studies showed that the monohydroxylated and O-demethylated metabolites were generated mainly by CYP3A4. Moreover, metabolism through these pathways, based on the

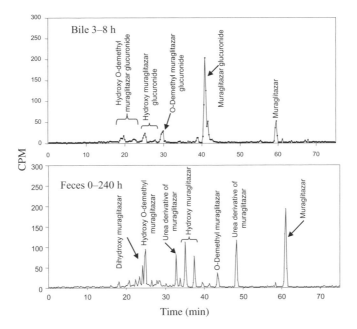

FIGURE 9.5 Radiochromatographic profiles of human bile and feces after oral administration of [¹⁴C]muraglitazar. The profiles are a background subtracted reconstructed radiochromatogram of 15-s fractions collected from an HPLC run.

metabolites identified in urine, accounted for clearance of ≥25% of the parent dose. Therefore, a decision was made to conduct a ketoconazole drug–drug interaction study to assess the effect of inhibiting CYP3A4 on the exposure of the parent.

9.4 METABOLITES IN SAFETY TESTING (MIST)

A recent PhRMA perspective published by Baillie et al. outlines a consensus of pharmaceutical industry representatives on monitoring metabolites in safety studies (Baillie et al., 2002). This report discusses in detail the role of metabolites in safety testing and decisions that are made based on data from various *in vitro* and *in vivo* metabolism studies. Basically, the decision to monitor metabolite(s) in humans and animals needs to be considered on a case-by-case basis. A general guideline put forth by the PhRMA committee for a metabolite that contributes to ≥25% AUC of all the drug-related component is to monitor this metabolite in human and animal studies. The FDA is planning to provide their guidelines in the near future, however, based on their comments in response to the PhRMA proposal and presentations at different workshops over the past 2–3 years, it seems that they would like to see a more stringent criteria (≥10% AUC) used to trigger considerations with regard to monitoring specific metabolites. In Fig. 9.3, two metabolites reach the threshold of $\geq 10\%$ as per FDA draft guidelines, and since the monohydroxylated metabolite had pharmacological activity and the N-demethylated metabolite was relatively abundant in humans, a decision was made to monitor both the metabolites in future studies.

To make decisions regarding monitoring of metabolites in humans and animal species, it is critical to generate quantitative exposure information regarding parent and metabolites. The quantitative information is generated by dosing animals and humans with radiolabeled drug and following the fate of the label in plasma and the excreta (urine, feces, or bile). In cases where synthetic standards are available for metabolites, similar quantitative information can be generated by monitoring for the metabolites by a specific LC/MS/MS assay. Alternatively, when synthetic metabolite standards are not available, radioactive metabolites generated in animals or *in vitro* sources could be used as reference standards to estimate metabolite exposures in the first-in-man and early toxicology studies (Zhang et al., 2007a).

9.5 *IN VITRO* DRUG METABOLISM STUDIES IN DRUG DEVELOPMENT

In vitro drug metabolism studies are conducted, both at the discovery and development stages, to help understand the potential safety and efficacy liabilities of a drug. In drug discovery/early drug development where the drug has not been tested in humans, *in vitro* metabolism studies provide an early read on the metabolic profiles likely to be observed in humans. Generally, the *in vitro* studies conducted are (1) metabolite profiling studies with liver microsomes, S-9 fraction or hepatocytes to understand the metabolic profile across different toxicology species and humans, (2) detailed reaction phenotyping studies to identify the enzymes responsible for metabolism of the drug, (3) inhibition studies to see if the drug is an inhibitor of CYP enzymes, and (4) induction studies to see if the drug induces CYP enzymes. The last three types of studies are particularly tailored toward identifying the CYP enzymes involved, since most of the drugs in the market are metabolized by this class of enzymes. The information generated from reaction phenotyping, inhibition, and induction studies helps understand the potential for drug–drug interaction and drives decisions regarding the clinical studies that should be done to address these potential interactions (Davit et al., 1999). Described below is a brief summary of the various *in vitro* tools used in drug discovery and development. Readers are encouraged to refer to the PhRMA position paper by Bjornsson et al., recent FDA guidances and Chapters 5 and 7 of this book for additional information on drug–drug interaction studies. (Bjornsson et al., 2003). The PhRMA position paper also outlines the optimal substrates, inhibitors or inducers of CYP enzymes that should be used in assessing the potential of drug to cause drug–drug interaction.

9.5.1 *In vitro* Metabolite Profiling

In vitro metabolite profiling work is typically done in liver fractions (microsomes or S-9) or hepatocytes (freshly isolated or cryopreserved). The latter is often preferred because hepatocytes provide cellular integrity with respect to enzyme architecture and contain the full complement of drug–metabolizing enzymes and cofactors (both oxidative and conjugative) (Gomez-Lechon et al., 2003; Hewitt and Utesch, 2004; McGinnity et al., 2004). Alternatively, freshly isolated liver slices can be used (Ekins et al., 1996; Martignoni et al., 2004). These systems tend to complement each other. When oxidative metabolism is involved, the turnover in hepatocytes are usually lower compared to subcellular tissue fractions or recombinant enzymes in the presence of cofactors (de Graaf et al., 2002). Incubation of a drug with hepatocytes, liver slices, or subcellular tissue fractions followed by chromatographic analysis of the incubation medium by radioactivity detection and LC/MS/MS is typically conducted for metabolite profiling and identification. The comparative biotransformation profiles provide an analysis of the similarities and differences in metabolism across species; information that could be helpful in the selection of the toxicology species to be evaluated in long-term safety studies and potentially to help understand toxicity in animal species mediated through metabolism.

Figure 9.6 shows a comparative profile generated from incubations with hepatocytes and liver microsomes of rat, monkey, and human for a C-14-labeled compound. The data shows that metabolism of this compound is qualitatively similar across species in both microsomes and hepatocyte systems. Profiles in microsomes showed the oxidative pathways of metabolism (see Fig. 9.6) whereas hepatocyte profiles, in addition to the oxidative metabolites, also showed conjugative metabolites (Fig. 9.6, metabolites designated M21, M30, and M31). This again emphasizes the need to use complementary *in vitro* systems to understand the overall metabolism of a compound.

9.5.2 Identification of Drug-Metabolizing Enzyme(s)

Reaction phenotyping refers to a set of experiments that is conducted to identify the specific enzymes responsible for the metabolism of a drug (Zhang et al., 2007). For studies with CYP enzymes, various tools like expressed enzymes, specific chemical inhibitors, antibodies, and liver banks are available (Lu et al., 2003; Williams et al., 2003). Identification of drug-metabolizing CYP enzymes is crucial for predicting the potential for drug–drug interactions in the clinic. It is well known for drugs that are metabolized mainly by CYP3A4, there is potential for significant drug–drug interaction when coadministered with a strong inhibitor like ketoconazole, where changes in AUCs could be $\geq$5–10-fold (Dresser et al., 2000; Greenblatt and von Moltke, 2002). It is obvious that the potential for drug–drug interaction decreases for compounds that are not primarily eliminated through metabolism, are low clearance compounds, or are metabolized by multiple enzymes.

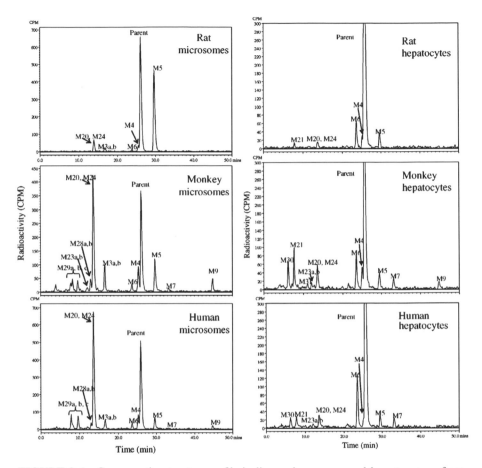

FIGURE 9.6 Comparative *in vitro* profile in liver microsomes and hepatocytes of rat, monkey, and human incubated with [^{14}C]dasatinib. The profiles are radiochrotogram of HPLC eluant flowing through an online IN/US β-Ram radioactivity detector.

Reaction phenotyping studies are conducted at multiple stages in the drug development process. Earlier studies performed during the drug discovery stages measure loss of the parent to understand the potential liability associated with clearance by a single polymorphic enzyme. Detailed reaction phenotying studies are often conducted after results from the human ADME become available. As indicated in the ADME section of this chapter, the human study with radiolabeled drug is an information rich study that helps understand: (1) if metabolism is responsible for clearance of a drug, (2) if a major metabolic pathway contributes to >25% of clearance, and (3) if there are major circulating metabolites. Once the primary pathways that contribute to ≥25% of the clearance have been identified, then detailed reaction phenotyping studies should be performed to identify the enzyme(s) responsible for clearance through that pathway. In addition, attempts should be made to identify the enzyme(s) responsible for formation of major circulating metabolite(s) even though that pathway may be minor (<25%) in the overall clearance of the compound. This is especially true for metabolites that may have pharmacological or toxicological consequences.

There are numerous articles and reviews written on the conduct of the reaction phenotyping studies using tools outlined above (Bjornsson et al., 2003; Lu et al., 2003; Mei et al., 1999; Rodrigues, 1999; Williams et al., 2003; Zhang et al., 2007). Chapter 15 provides additional information on the conduct of the reaction phenotyping studies. In general, three methods for identifying the individual CYP enzymes responsible for a metabolic pathway are typically used: (1) incubation with individual human cDNA-expressed CYP enzymes, (2) incubation with specific chemical inhibitors or monoclonal antibodies in pooled human liver microsomes, and (3) incubation with a bank of human liver microsomes that are prepared from individual donors and are well characterized for CYP activities. At least two of the methods should be performed to identify the specific enzyme(s) responsible for a metabolic pathway. If biotransformation through a particular pathway is inhibited by a chemicals/monoclonal antibody inhibitor and catalyzed by the expressed enzyme, then there is strong evidence for the role of that enzyme in metabolism of the drug.

Human CYPs involved in metabolism can be easily characterized because of the tools that are commercially available. However, similar tools are not available for other drug-metabolizing enzymes like FMOs, UGTs, and sulfotransferases. There are no specific inhibitors or monoclonal anitibodies identified for any of the commercially available expressed UGTs or sulfo-transferases. Due to the lack of necessary tools it is difficult to quantitatively assess their contribution to the overall clearance of drugs that are metabolized by these enzymes.

9.5.3 Evaluation of CYP Inhibition

A key set of studies that is conducted during discovery and development is to understand the potential of a drug to inhibit different CYP enzymes. Potent inhibition of CYPs may affect further development, specifically if they inhibit key drug-metabolizing enzymes such as CYP1A2, CYP2A6, CYP3A4, CYP2B6, CYP2C8, CYP2C9, CYP2C19, and CYP2D6. Evaluation of CYP inhibition is described in detail in Chapters 5 and 16. A potential for CYP inhibition can be evaluated either with human liver microsomes or with expressed enzymes. A number of fluorescent substrates have been developed for various CYPs (Cohen et al., 2003; Ghosal et al., 2003; Yamamoto et al., 2002). These fluorescent substrates makes the assays amenable to high throughput analysis. Such screens are usually used in the drug discovery stage. Once a lead compound is identified for development then the inhibition experiments are repeated in human liver microsomes to confirm the results obtained from screening assays. FDA and EMEA usually consider inhibition studies in human liver microsomes as a better representative of the *in vivo* situation.

Clinical inhibition potential can be predicted based on inhibition type and *in vitro* inhibition data with various degree of success as described in Chapter 5 (Bjornsson et al., 2003; Zhang and Wong, 2005; Zhou et al., 2004, 2005).

9.5.4 Evaluation of CYP induction

Induction of enzyme refers to increase in an enzyme activity through over expression of the protein. Induction of drug-metabolizing enzyme by administration of a xenobiotic occurs through multiple mechanisms. Increasing gene transcription by modulating the activity of the transcription factors is the most common mechanism by which enzyme inducers cause induction. Induction of CYP enzymes has potential for clinical drug–drug interaction (Luo et al., 2004) and FDA guidelines require that drugs in development be tested for their induction potential of CYP P450 enzymes (US Food and Drug Administration). There are number of different ways in which enzyme induction potential can be tested. Since induction of CYP3A4 is considered most relevant and common for drug–drug interaction, in the drug discovery stage high throughput reporter-based assays are used for screening compounds for their induction potential of CYP3A4 (see Chapter 6) (Sinz et al., 2006). FDA considers data from cell-based system more relevant to *in vivo* situation. An FDA accepted method is to evaluate the enzyme induction potential of a compound in primary hepatocytes from three individual donors after treatment with the drug of interest for 3 days (see Chapter 17 for details). This is compared against known inducers as a positive control and with solvent as a negative control. A greater than twofold increase in probe enzyme activity compared to solvent control or ≥40% of activity compared to positive control is considered an enzyme inducer. The rationale for selection of various concentrations of a drug in the induction assay is similar to the one used for CYP inhibition. Typically, at least two indicators of enzyme induction are looked at: namely, enzyme activity toward known substrate and mRNA levels or protein expression by Western immunoblotting. It is important to look at two indicators of enzyme induction since an inducer could also be an inhibitor of enzyme activity and in the absence of mRNA or protein expression data this could be misleading. Ritanovir, for example, is both an inhibitor and an inducer of CYP3A4, and in primary cultured human hepatocyte measured for only CYP3A4 activity, incubation with this compound would lead to an erroneous conclusion, that is, no induction (Luo et al., 2002).

9.6   EXAMPLES OF ROLE OF DRUG METABOLISM
TO ADDRESS SAFETY ISSUES

When metabolites are involved in eliciting toxicity in animal species it is important to establish the relevance of these findings to humans. If a metabolite-mediated toxicity pathway is indeed thought to be animal specific then it is necessary to demonstrate this specificity in animals and humans or show that there is enough of a safety margin established in animals such that there is minimal concern in humans. There are numerous examples in the literature where bioactivation of a xenobiotic to a reactive intermediates is proposed to lead to toxicity (Baillie and Kassahun, 2001; Evans et al., 2004). However, there are very few examples for marketed drugs where detailed studies have been conducted to explain toxicity findings in animals that are not relevant to humans due to lack of that pathway in humans. The example of rat-specific toxicity of efavirenz is discussed below. Similarly, investigative approaches can be used when unexpected drug–drug interactions or toxicities occur in humans. There are two examples provided below were inhibition of a drug-metabolizing enzyme could play a possible role in the observed drug–drug interaction or toxicity in humans.

9.6.1 Rat-Specific Toxicity of Efavirenz Caused by Species-Specific Bioactivation

Efavirenz (Sustiva), an anti-HIV drug was found to cause necrosis of the renal proximal tubular epithelium in rats after an oral dose of 700 mg/kg. Similar toxicity was not observed in cynomolgus monkey or human. The nephrotoxicity in rats was attributed to the formation of a glutathione adduct that was further processed to a mixture of cysteine–glycine adducts and cysteine adducts (Mutlib et al., 1999). Mechanistic studies with stable-labeled efavirenz and coadministration with acivicin, a γ-glutamyltranspeptidase inhibitor, confirmed that the formation of the glutathione conjugate and its processing finally to the cysteine conjugate was necessary for eliciting the nephrotoxicity in rats (Mutlib et al., 2000). This glutathione conjugation pathway was not observed in cynomolgus monkey or in humans. This demonstration of species-specific pathway and its absence in humans, followed by detailed mechanistic studies, helped to place this issue in the proper perspective in terms of risk to patients. It was concluded that the nephrotoxicity in rat observed at high doses was not a safety concern in humans.

9.6.2 UGT1A1 Inhibition-Mediated Hyperbilirubinemia by HIV Protease Inhibitors

In this example, investigative studies were conducted to address an issue of unconjugated hyperbilirubinemia in human subjects observed after administration of HIV protease inhibitors indinavir and atazanavir. The proposed hypothesis for hyperbilirubinemia was inhibition of glucuronidation of bilirubin by these two drugs, a key step in the excretion of bilirubin into bile. To test this hypothesis, a panel of HIV protease inhibitors, including atazanavir, indinavir, lopinavir, nelfinavir, ritonavir, and saquinavir, were tested as inhibitors of UGT1A1, 1A3, 1A4, 1A6, and 1A9 in human liver microsomes (Zhang et al., 2005). All the protease inhibitors inhibited UGT1A1, UGT1A3, and UGT1A4 with IC_{50} values that ranged from 2 to 87 μM. No inhibition of UGT1A6, 1A9, and 2B7 was observed (IC_{50} >100 μM). Further analysis of UGT1A1 inhibition by atazanavir and indinavir demonstrated a linear-mixed type inhibition with K_i values of 1.9 and 47.9 μM, respectively. Furthermore, an *in vitro–in vivo* scaling with $[I]/K_i$, where [I] was the free C_{max} drug concentration, predicted that atazanavir and indinavir were more likely to induce hyperbilirubinemia than other HIV protease inhibitors (Zhang et al., 2005).

9.6.3 Mechanism-Based Inactivation of CYP2C8 by Gemfibrozil Glucuronide

In vitro experiments demonstrated that gemfibrozil, a lipid-lowering drug, is a more potent inhibitor of CYP2C9 than of CYP2C8 (Wen et al., 2001). Coadministration of gemfibrozil with the CYP2C9 substrate warfarin did not increase the plasma concentrations of either *R*- or *S*-warfarin (in fact, it actually decreases them) (Lilja et al., 2005). However, coadministration of gemfibrozil with CYP2C8 substrates such as cerivastatin, repaglinide, rosiglitazone, and pioglitazone led to increased exposures of these drugs (Backman et al., 2002; Jaakkola et al., 2005; Niemi et al., 2003a, 2003b). Furthermore, Shitara et al. demonstrated that gemfibrozil 1-*O*-β-glucuronide inhibits *in vitro* the CYP2C8-mediated metabolism of cerivastatin as well as the OATP2-mediated uptake of cerivastatin (Shitara et al., 2004). Given this apparent differences between *in vitro* and *in vivo* data a study was conducted were both gemfibrozil and its major metabolite, an acyl-glucuronide (gemfibrozil 1-*O*-β-glucuronide) were evaluated as direct-acting and metabolism-dependent inhibitors of the major drug-metabolizing CYP enzymes (CYP1A2, 2B6, 2C8, 2C9, 2C19, 2D6, and 3A4) in human liver microsomes. Gemfibrozil most potently inhibited CYP2C9 (IC_{50} of 30 μM), whereas gemfibrozil glucuronide most potently inhibited CYP2C8 (IC_{50} of 24 μM). Unexpectedly, gemfibrozil glucuronide, but not gemfibrozil, was found to be a metabolism-dependent inhibitor of CYP2C8. The IC_{50} for inhibition of CYP2C8 by gemfibrozil glucuronide decreased from 24 to 1.8 μM after a 30-min incubation with human liver microsomes and NADPH. Inactivation of CYP2C8 by gemfibrozil glucuronide required NADPH, and proceeded with a K_i (inhibitor concentration that supports half the maximal rate of enzyme inactivation) of 20–52 μM and a k_{inact} (maximal rate of inactivation) of 0.21 min^{-1} (Ogilvie et al., 2006). The results described seems to suggest that the mechanism of the clinical interaction reported between gemfibrozil and CYP2C8 substrates, at least in part, could probably be due to inhibition of CYP2C8 by gemfibrozil glucuronide.

9.7 IMPACT OF METABOLISM INFORMATION ON NDA FILING AND LABELING

FDA and ICH guidelines (see Chapter 7 for specific guidelines) require that detailed studies be conducted to completely characterize the metabolic profile of a drug. The list of *in vivo* and *in vitro* studies summarized in the sections above aim to fulfill that purpose. The results generated from these studies are included in the new drug application (NDA) where the data generated in animal species and humans are integrated in the metabolism sections of the nonclinical and the clinical pharmacology summaries. This integration of data helps the reviewers in the regulatory agencies understand the comparative metabolism across species and, specifically, review the data to see if animals are adequately exposed to human metabolites. Reaction phenotyping studies in conjunction with human ADME data, and CYP inhibition and induction studies, provide information to the regulatory agencies about the potential for drug–drug interaction and the metabolic liabilities, if any, associated with the drug.

The results generated from the studies outlined above are also summarized in the various sections of the product label. The key sections where drug metabolism information and the recommendations based on the data appear in the label are (1) Pharmacokinetics—where the overall ADME properties of the drug in humans is summarized based on pharmacokinetic and radiolabeled human studies, (2) Drug—drug interactions—where the potential interactions of the drug as a substrate, inhibitor or inducer of the CYP enzymes are summarized based on key clinical drug–drug interaction (DDI) data as well as *in vitro* reaction phenotyping and CYP inhibition and induction data and, (3) Precautions—where information is provided for interactions with different classes of drugs and how it can be clinically managed. Both the drug–drug interaction and precaution sections of the label provide the prescribing physicians with the relevant information to either adjust the dose or avoid prescribing the medicine when the patients are taking concomitant interacting medications.

REFERENCES

Backman JT, Kyrklund C, Neuvonen M, Neuvonen PJ. Gemfibrozil greatly increases plasma concentrations of cerivastatin. Clin Pharmacol Ther 2002;72:685–691.

Baillie TA, Cayen MN, Fouda H, Gerson RJ, Green JD, Grossman SJ, Klunk LJ, LeBlanc B, Perkins DG, Shipley LA. Drug metabolites in safety testing. Toxicol Appl Pharmacol 2002;182:188–196.

Baillie TA, Kassahun K. Biological reactive intermediates in drug discovery and development: a perspective from the pharmaceutical industry. Adv Exp Med Biol 2001;500:45–51.

Bjornsson TD, Callaghan JT, Einolf HJ, Fischer V, Gan L, Grimm S, Kao J, King SP, Miwa G, Ni L, Kumar G, McLeod J, Obach SR, Roberts S, Roe A, Shah A, Snikeris F, Sullivan JT, Tweedie D, Vega JM, Walsh J, Wrighton SA. The conduct of *in vitro* and *in vivo* drug–drug interaction studies: a PhRMA perspective. J Clin Pharmacol 2003;43:443–469.

Cohen LH, Remley MJ, Raunig D, Vaz AD. *In vitro* drug interactions of cytochrome P450: an evaluation of fluorogenic to conventional substrates. Drug Metab Dispos 2003;31:1005–1015.

Davit B, Reynolds K, Yuan R, Ajayi F, Conner D, Fadiran E, Gillespie B, Sahajwalla C, Huang SM, Lesko LJ. FDA evaluations using *in vitro* metabolism to predict and interpret *in vivo* metabolic drug–drug interactions: impact on labeling. J Clin Pharmacol 1999;39:899–910.

de Graaf IAM, van Meijeren CE, Pektas F, Koster HJ. Comparison of *in vitro* preparations for semiquantitative prediction of *in vivo* drug metabolism. Drug Metab Dispos 2002;30:1129–1136. DOI:10.1124/dmd.30.10.1129.

Dresser GK, Spence JD, Bailey DG. Pharmacokinetic–pharmacodynamic consequences and clinical relevance of cytochrome P450 3A4 inhibition. Clin Pharmacokinet 2000;38:41–57.

Ekins S, Williams JA, Murray GI, Burke MD, Marchant NC, Engeset J, Hawksworth GM. Xenobiotic metabolism in rat, dog, and human precision-cut liver slices, freshly isolated hepatocytes, and vitrified precision-cut liver slices. Drug Metab Dispos 1996;24:990–995.

Evans DC, Watt AP, Nicoll-Griffith DA, Baillie TA. Drug–protein adducts: an industry perspective on minimizing the potential for drug bioactivation in drug discovery and development. Chem Res Toxicol 2004;17:3–16.

Fennell TR, Kedderis GL, Sumner SCJ. Urinary metabolites of [1,2,3-^{13}C]acrylonitrile in rats and mice detected by ^{13}C nuclear magnetic resonance spectroscopy. Chem Res Toxicol 1991;4:678–687.

Fennell TR, Sumner SCJ. Identification of metabolites of carcinogens by ^{13}C NMR spectroscopy. Drug Metab Rev 1994;26:469–481.

Ghibellini G, Leslie EM, Brouwer KL. Methods to evaluate biliary excretion of drugs in humans: an updated review. Mol Pharmacol 2006;3:198–211.

Ghosal A, Hapangama N, Yuan Y, Lu X, Horne D, Patrick JE, Zbaida S. Rapid determination of enzyme activities of recombinant human cytochromes P450, human liver microsomes and hepatocytes. Biopharm Drug Dispos 2003;24:375–384.

Gomez-Lechon MJ, Donato MT, Castell JV, Jover R. Human hepatocytes as a tool for studying toxicity and drug metabolism. Curr Drug Metab 2003;4:292–312.

Greenblatt DJ, von Moltke LL. Drug–drug interactions: clinical perspective. In: Rodrigues AD, editor. Drug–Drug Interactions. New York: Marcel Dekker; 2002. p 565–584.

Hall EJ. Radiobiology for the Radiologist. JB Lippincott Company; 1994.

Hewitt NJ, Utesch D. Cryopreserved rat, dog and monkey hepatocytes: measurement of drug metabolizing enzymes in suspensions and cultures. Hum Exp Toxicol 2004;23:307–316.

Huang SM, Lesko LJ. Drug–drug, drug–dietary supplement, and drug–citrus fruit and other food interactions: what have we learned? J Clin Pharmacol 2004;44: 559–569.

Iyer RA, Malhotra B, Sanaullah K, Mitroka J, Bonacorsi SJ, Waller SC, Rinehart JK, Kripalani K. Comparative biotransformation of radiolabeled [^{14}C]omapatrilat and stable-labeled [^{13}C$_2$]omapatrilat after oral administration to rats, dogs, and humans. Drug Metab Dispos 2003;31:67–75.

Iyer RA, Mitroka J, Malhotra B, Bonacorsi SJ, Waller SC, Rinehart JK, Roongta VA, Kripalani K. Metabolism of [^{14}C]omapatrilat, a sulfhydryl-containing vasopeptidase inhibitor in humans. Drug Metab Dispos 2001;29:60–69.

Jaakkola T, Backman JT, Neuvonen M, Neuvonen PJ. Effects of gemfibrozil, itraconazole, and their combination on the pharmacokinetics of pioglitazone. Clin Pharmacol Ther 2005;77:404–414.

Kumar GN, Surapaneni S. Role of drug metabolism in drug discovery and development. Med Res Rev 2001;21:397–411.

Lilja JJ, Backman JT, Neuvonen PJ. Effect of gemfibrozil on the pharmacokinetics and pharmacodynamics of racemic warfarin in healthy subjects. Br J Clin Pharmacol 2005;59:433–439.

Lu AY, Wang RW, Lin JH. Cytochrome P450 *in vitro* reaction phenotyping: a re-evaluation of approaches used for P450 isoform identification. Drug Metab Dispos 2003;31:345–350.

Luo G, Cunningham M, Kim S, Burn T, Lin J, Sinz M, Hamilton G, Rizzo C, Jolley S, Gilbert D, Downey A, Mudra D, Graham R, Carroll K, Xie J, Madan A, Parkinson A, Christ D, Selling B, LeCluyse E, Gan LS. CYP3A4 induction by drugs: correlation between a pregnane X receptor reporter gene assay and CYP3A4 expression in human hepatocytes. Drug Metab Dispos 2002;30:795–804.

Luo G, Guenthner T, Gan LS, Humphreys WG. CYP3A4 induction by xenobiotics: biochemistry, experimental methods and impact on drug discovery and development. Curr Drug Metab 2004;5:483–505.

Marathe PH, Shyu WC, Humphreys WG. The use of radiolabeled compounds for ADME studies in discovery and exploratory development. Curr Pharm Des 2004;10:2991–3008.

Marier JF, Vachon P, Gritsas A, Zhang J, Moreau JP, Ducharme MP. Metabolism and disposition of resveratrol in rats: extent of absorption, glucuronidation, and enterohepatic recirculation evidenced by a linked-rat model. J Pharmacol Exp Ther 2002;302:369–373.

Martignoni M, Monshouwer M, de Kanter R, Pezzetta D, Moscone A, Grossi P. Phase I and phase II metabolic activities are retained in liver slices from mouse, rat, dog, monkey and human after cryopreservation. Toxicol In Vitro 2004;18:121–128.

McGinnity DF, Soars MG, Urbanowicz RA, Riley RJ. Evaluation of fresh and cryopreserved hepatocytes as *in vitro* drug metabolism tools for the prediction of metabolic clearance. Drug Metab Dispos 2004;32:1247–1253.

Mei Q, Tang C, Assang C, Lin Y, Slaughter D, Rodrigues AD, Baillie TA, Rushmore TH, Shou M. Role of a potent inhibitory monoclonal antibody to cytochrome P-450 3A4 in assessment of human drug metabolism. J Pharmacol Exp Ther 1999;291:749–759.

Mutlib AE, Chen H, Nemeth GA, Markwalder JA, Seitz SP, Gan LS, Christ DD. Identification and characterization of efavirenz metabolites by liquid chromatography/mass spectrometry and high field NMR: species differences in the metabolism of efavirenz. Drug Metab Dispos 1999;27:1319–1333.

Mutlib AE, Gerson RJ, Meunier PC, Haley PJ, Chen H, Gan LS, Davies MH, Gemzik B, Christ DD, Krahn DF, Markwalder JA, Seitz SP, Robertson RT, Miwa GT. The species-dependent metabolism of efavirenz produces a nephrotoxic glutathione conjugate in rats. Toxicol Appl Pharmacol 2000;169:102–113.

Niemi M, Backman JT, Granfors M, Laitila J, Neuvonen M, Neuvonen PJ. Gemfibrozil considerably increases the plasma concentrations of rosiglitazone. Diabetologia 2003a;46:1319–1323.

Niemi M, Backman JT, Neuvonen M, Neuvonen PJ. Effects of gemfibrozil, itraconazole, and their combination on the pharmacokinetics and pharmacodynamics of repaglinide: potentially hazardous interaction between gemfibrozil and repaglinide. Diabetologia 2003b;46:347–351.

Ogilvie BW, Zhang D, Li W, Rodrigues AD, Gipson AE, Holsapple J, Toren P, Parkinson A. Glucuronidation converts gemfibrozil to a potent, metabolism-dependent inhibitor of CYP2C8: implications for drug–drug interactions. Drug Metab Dispos 2006;34:191–197.

Parker RJ, Hirom PC, Millburn P. Enterohepatic recycling of phenolphthalein, morphine, lysergic acid diethylamide (LSD) and diphenylacetic acid in the rat. Hydrolysis of glucuronic acid conjugates in the gut lumen. Xenobiotica 1980;10:689–703.

Parkinson A. Biotransformation of xenobiotics. In: Klassen CD, editor. Casarett & Doull's Toxicology: The Basic Science of Poisons McGraw-Hill; 2001. p 133–224.

Product Note for guidance on the investigation of drug interactions (CPMP/EWP/560/95). Available atwww.eudra.org/humandocs/humans/EWP.html.

Rodrigues AD. Integrated cytochrome P450 reaction phenotyping: attempting to bridge the gap between cDNA-expressed cytochromes P450 and native human liver microsomes. Biochem Pharmacol 1999;57:465–480.

Shitara Y, Hirano M, Sato H, Sugiyama Y. Gemfibrozil and its glucuronide inhibit the organic anion transporting polypeptide 2 (OATP2/OATP1B1:SLC21A6)-mediated hepatic uptake and CYP2C8-mediated metabolism of cerivastatin: analysis of the mechanism of the clinically relevant drug–drug interaction between cerivastatin and gemfibrozil. J Pharmacol Exp Ther 2004;311:228–236.

Sinz M, Kim S, Zhu Z, Chen T, Anthony M, Dickinson K, Rodrigues AD. Evaluation of 170 xenobiotics as transactivators of human pregnane X receptor (hPXR) and correlation to known CYP3A4 drug interactions. Curr Drug Metab 2006;7:375–388.

Solon EG, Balani SK, Lee FW. Whole-body autoradiography in drug discovery. Curr Drug Metab 2002;3:451–462.

Solon EG, Kraus L. Quantitative whole-body autoradiography in the pharmaceutical industry. Survey results on study design, methods, and regulatory compliance. J Pharmacol Toxicol Methods 2001;46:73–81.

Sumner SCJ, Stedman DB, Clarke DO, Welsch F, Fennell TR. Characterization of urinary metabolites from [1,2,methoxy-^{13}C]-2-methoxyethanol in mice using ^{13}C nuclear magnetic resonance spectroscopy. Chem Res Toxicol 1992;5:553–560.

US Food and Drug Administration Guidance for industry: drug metabolism/drug interaction studies in the drug development process: studies in vitro. Available at www.fda.gov/cder/guidance.html.

Valentin J. Basic anatomical and physiological data for use in radiological protection: reference values: ICRP Publication 89. Annals of the ICRP 2002;32:1–277.

Wait JCM, Vaccharajani N, Mitroka J, Jemal M, Khan S, Bonacorsi SJ, Rinehart JK, Iyer RA. Metabolism of [^{14}C]gemopatrilat after oral administration to rats, dogs, and humans. Drug Metab Dispos 2006;34:961–970.

Wang L, Zhang D, Swaminathan A, Xue Y, Cheng PT, Wu S, Mosqueda-Garcia R, Aurang C, Everett DW, Humphreys WG. Glucuronidation as a major metabolic clearance pathway of ^{14}C-labeled muraglitazar in humans: metabolic profiles in subjects with or without bile collection. Drug Metab Dispos 2006;34:427–439. Epub 2005 Dec 20–28.

Wen X, Wang J-S, Backman JT, Kivisto KT, Neuvonen PJ. Gemfibrozil is a potent inhibitor of human cytochrome P450 2C9. Drug Metab Dispos 2001;29:1359–1361.

WHO.Use of ionizing radiation and radionuclides on human beings for medical research, training and nonmedical purpose. *WHO Publication* 611;1977.

Williams JA, Hurst SI, Bauman J, Jones BC, Hyland R, Gibbs JP, Obach RS, Ball SE. Reaction phenotyping in drug discovery: moving forward with confidence? Curr Drug Metab 2003;4:527–534.

Williams JA, Hyland R, Jones BC, Smith DA, Hurst S, Goosen TC, Peterkin V, Koup JR, Ball SE. Drug-drug interactions for UDP-glucuronosyltransferase substrates: a pharmacokinetic explanation for typically observed low exposure (AUCi/AUC) ratios. Drug Metab Dispos 2004;32:1201–1208.

Yamamoto T, Suzuki A, Kohno Y. Application of microtiter plate assay to evaluate inhibitory effects of various compounds on nine cytochrome P450 isoforms and to estimate their inhibition patterns. Drug Metab Pharmacokinet 2002;17:437–448.

Zhang D, Chando TJ, Everett DW, Patten CJ, Dehal SS, Humphreys WG. *In vitro* inhibition of UDP glucuronosyltransferases by atazanavir and other HIV protease inhibitors and the relationship of this property to *in vivo* bilirubin glucuronidation. Drug Metab Dispos 2005;33:1729–1739.

Zhang D, Wang L, Chandrasena G, Ma L, Zhu M, Zhang H, Davis CD, Humphreys WG. Involvement of multiple cytochrome P450 and UDP-glucuronosyltransferase enzymes in the *in vitro* metabolism of muraglitazar. Drug Metab Dispos 2007;35:139–149.

Zhang D, Raghavan N, Chando T, Gambardella J, Fu Y, Zhang DX, Unger S, Humphreys WG. LC-MS/MS-based approach for obtaining exposure estimates of metabolites in early clinical trials by using radioactive metabolites as reference standards. Drug Metab Lett 2007a; in print.

Zhang H, Zhang D, Ray K. A software filter to remove interference ions from drug metabolites in accurate mass liquid chromatography/mass spectrometric analysis. J Mass Spectrom 2003;38:1110–1112.

Zhang ZY, Wong YN. Enzyme kinetics for clinically relevant CYP inhibition. Curr Drug Metab 2005;6:241–257.

Zhou S, Chan E, Lim LY, Boelsterli UA, Li SC, Wang J, Zhang Q, Huang M, Xu A. Therapeutic drugs that behave as mechanism-based inhibitors of cytochrome P450 3A4. Curr Drug Metab 2004;5:415–442.

Zhou S, Yung Chan S, Cher Goh B, Chan E, Duan W, Huang M, McLeod HL. Mechanism-based inhibition of cytochrome P450 3A4 by therapeutic drugs. Clin Pharmacokinet 2005;44:279–304.

PART III

ANALYTICAL TECHNIQUES IN DRUG METABOLISM

10

APPLICATIONS OF LIQUID RADIOCHROMATOGRAPHY TECHNIQUES IN DRUG METABOLISM STUDIES

Mingshe Zhu, Weiping Zhao, and W. Griffith Humphreys

10.1 INTRODUCTION

Radioactive isotopes are routinely used to trace the fate of drugs in disposition and metabolism studies conducted during drug discovery and development (Dalvie, 2000; Marathe et al., 2004; Veltkamp, 1990). Most studies involving radioactive compounds use either ^{3}H or ^{14}C to replace an H or C atom of the original drug molecule. ^{3}H-labeled compounds that can be readily prepared in most cases are primarily used for investigative metabolism and selected biotransformation studies to address specific drug metabolism-related issues in drug discovery. Due to the risk of ^{3}H exchange with water and loss of the tritium label via biotransformation reactions, radiolabeled drug metabolism and disposition studies in support of drug development and registrations mainly rely on ^{14}C-radiolabeled analogs even though their synthesis requires a resource-intensive effort. These studies include: (1) comparative *in vitro* metabolism; (2) absorption, distribution, metabolism and elimination (ADME) studies in toxicological species (Fouda et al., 1997; Graf et al., 2005; Li et al., 2006; Zhang et al., 2007b); (3) tissue distribution study in rats or other species (Hazai et al., 1999; Magyar et al., 1998); (4) ADME study in humans (Zhang et al., 2005, 2007b); (5) enzymatic studies to identify the enzymes involved in the metabolism of the drug, and the kinetics of the metabolic reactions (Ilett et al., 2002; Lautala et al., 1999; Zhang et al., 2004, 2007a; Zhu et al., 2005a); and (6) investigative drug metabolism studies to address issues related to drug metabolism and disposition (Evans et al., 2004; Li et al., 2006; Meneses-Lorente et al., 2006; Zhang et al., 2003). All these radiolabeled studies, except for the tissue distribution study, require metabolite profiling to determine the number and relative concentrations of individual metabolites in various biological matrices using a radiochromatographic technique.

Applications of radioisotopes to studies of ADME have been comprehensively reviewed recently (Dalvie, 2000; Marathe et al., 2004; Veltkamp, 1990). This chapter focuses on the recent advance in liquid radiochromatographic techniques and their application to metabolite detection and quantification in drug metabolism studies. In addition, the use of liquid radiochromatography coupled with mass spectrometry in metabolite profiling and structural elucidation is described.

10.2 TRADITIONAL RADIOCHROMATOGRAPHY TECHNIQUES

Planar radiochromatography was the first widely used technique for separation and detection of radiolabeled metabolites (Veltkamp, 1990). The technique is very sensitive, fast, and has relatively low equipment and operating costs, however, due to low separation resolution and poor precision in quantification, it has largely been replaced by HPLC-based radiochromatographic techniques in the pharmaceutical industry. Online radio flow detection (RFD) and off-line liquid scintillation counting (LSC) have been routinely used for analysis of radiolabeled metabolites for the last 15 years (Veltkamp, 1990).

10.2.1 HPLC-RFD

Figure 10.1a and b illustrate general setups of HPLC–RFD for radioactivity profiling, in which a mass spectrometric detection can be coupled for simultaneous metabolite characterization with or without splitting of the HPLC effluent (Egnash and Ramanathan, 2002; Heath et al., 1997; Hyllbrant et al., 1999; Vlasakova et al., 1998). Online HPLC–RFD has several advantages including rapid analysis, high separation resolution (Fig. 10.2c), and good accuracy and precision for radioactivity measurement. Currently, HPLC–RFD is the most common liquid radiochromatographic technique used in drug metabolism studies, and is especially suitable for radiochromatographic method development, analysis of a large of number of samples necessary for completing studies such as enzyme kinetic determination, and metabolite characterization through coupling with a mass spectrometer (Morovjan et al., 2002). The single major disadvantage of HPLC–RFD is its relatively poor sensitivity due to short residence times (5–15 s) of the radioactive peaks in the RFD detection cell (Zhu et al., 2005b). As listed in Table 10.1, the detection of

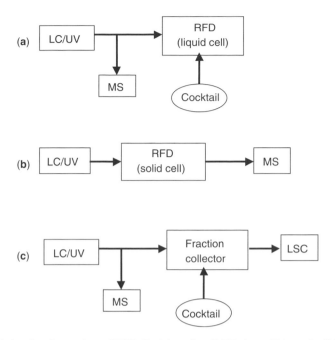

FIGURE 10.1 Configuration of HPLC with online RFD (**a** and **b**); and off-line of LSC (**c**) with or without coupling with a mass spectrometer.

limit of RFD is 250–500 disintigrations per minutes (DPM) and the limit of quantification of RFD ranges from 750 to 1500 DPM, which greatly limits its use in the analysis of low levels of radiolabeled metabolites such as plasma metabolites.

10.2.2   HPLC-LSC

Figure 10.1c illustrates a general setup of an HPLC coupled with an LSC instrument for radioactivity profiling. Off-line LSC radioactivity counting is at least 25-fold more sensitive than HPLC–RFD (Figs. 10.2b and 10.2c and Table 10.1) because individually collected effluent fractions can be counted for a longer time (usually 10 min or more). LSC has been traditionally employed for quantification of low levels of radiolabeled metabolites (Chando et al., 1998; Everett et al., 1991). Radioactivity analysis by off-line LSC consists of four steps: HPLC separation, fraction collection into individual vessels, mixing with scintillation cocktail, and radioactivity counting one fraction at a time. The entire process is time consuming and relatively labor intensive. In addition, due to the labor and reagent consumption of the process HPLC fractions are often collected at time intervals, which result in suboptimal peak resolution compared to other techniques (Fig. 10.2b and 10.2c). Because of these reasons, the use of HPLC–LSC for profiling of low level radiolabeled metabolites has generally been replaced by HPLC–microplate scintillation

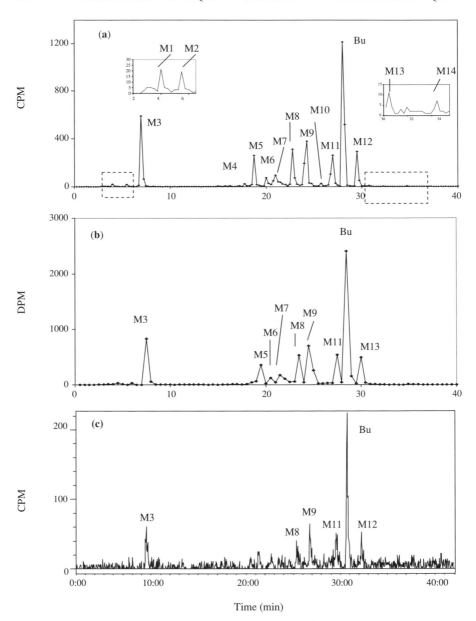

FIGURE 10.2 Metabolite profiles of buspirone determined by HPLC with MSC, LSC, and RFD (Zhu et al., 2005b). Buspirone metabolites from human liver microsome incubations were separated by HPLC (1 mL/min). Radioactivity in HPLC effluent was determined by; **(a)** TopCount: 8000 DPM injected, four fractions per min and 10 min counting time; **(b)** LSC: 8000 DPM injected, two fractions per min, 10 min counting time; and **(c)** RFD: 32,000 DPM injected.

TABLE 10.1   Sensitivity of various liquid radiochromatographic techniques.

Radio-detection	Background (CPM)	Counting efficiency (%)	Counting time (min)	Limit of detection[a] (DPM)	Limit of quantification[b] (DPM)
LC–LSC	25	90	10	10	31
LC–RFD	15	70	5–10 s	250–500	750–1500
LC–MSC[c]	2	70	10	5	15
Stop-flow RFD[d]	15	70	1	25–50	75–150
LC–AMS[e]				0.0001	

[a]Limit of detection (LOD) for LSC, RFD, MSC and stop-flow RFD was calculated based on Equation 10.1. The parameters used for the calculation are listed Table 10.1.
[b]Limit of quantification (LOQ) was calculated based on an equation presented in reference (Zhu et al., 2005b).
[c]TopCount instrument was used.
[d]Stop by-fraction mode was utilized.
[e]See references (Brown et al., 2005, 2006).

counting (MSC) and other recently developed liquid radiochromatographic techniques.

10.3   NEW RADIOCHROMATOGRAPHY TECHNIQUES

10.3.1   HPLC-MSC

MSC was recently introduced as an off-line liquid radiochromatographic detection technique for radioactive metabolite profiling (Boernsen et al., 2000; Bruin et al., 2006; Kiffe et al., 2003; Wallace et al., 2004; Zhu et al., 2005b, 2005c). Compared to HPLC–LSC, HPLC–MSC not only increases analytical throughput and sensitivity but also reduces radioactive waste and manual operations involved in LSC vial handling. The HPLC–MSC has become the method of choice for analysis of low level radioactive metabolites in some metabolism laboratories (Boernsen et al., 2000; Bruin et al., 2006; Kiffe et al., 2003; Wallace et al., 2004; Zhu et al., 2005b). In HPLC–MSC analysis (Fig. 10.3a), HPLC effluent is collected into 96-well microplates, and then evaporated using a speed vacuum system. The radioactivity of the residue in the 96-well plates is determined by counting up to 12 wells at a time with a microplate scintillation counter. Two types of MSC instruments, TopCount and MicroBeta counter (Nedderman et al., 2004; Zhu et al., 2005b), are commercially available. The TopCount instrument uses deep-well LumaPlates, in which yttrium silicate scintillators are deposited at the bottom of each well. The MicroBeta counter uses 96-well Scintiplates® that consist of white frames with clear wells embedded with solid scintillators. The radiodetection sensitivity of TopCount is slightly better than MicroBeta counter, while radiolabeled components can be readily recovered from Scintiplates but not from LumaPlates. In HPLC–MSC analysis, counts per minutes (CPM) values

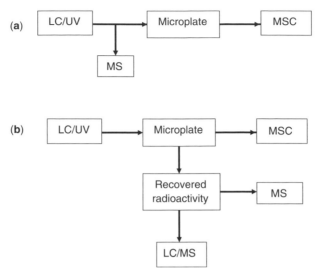

FIGURE 10.3 Configuration of HPLC with an off-line MSC with or without coupling with a mass spectrometer.

are generally utilized for the calculation of relative radioactivity abundance. If counting efficiencies of HPLC effluent fractions across the entire HPLC run are consistent, or vary within an accepted range, the relative abundance of radioactivity peaks calculated from CPM values will not be significantly different from those of DPM values. In the following sections sensitivity, precision, and accuracy of radioactivity profiling by HPLC–TopCount are described. In addition, biological matrix effects on HPLC–MSC performance are discussed.

10.3.1.1 Sensitivity   The limit of detection of radioactivity counting can be calculated using the following Equation 10.1 (Zhu et al., 2005b), where LD is the limit of detection expressed in DPM, B is the background radioactivity expressed in DPM, T is counting time (min) and E is counting efficiency.

$$LD = 2.71/TE + 4.65\sqrt{B/TE} \qquad (10.1)$$

A primary factor contributing to the superior sensitivity of off-line scintillation counting techniques is the greater counting time available. In HPLC–MSC analysis, the counting time is typically around 10 min, which is approximately 30-to 60-fold longer than that in HPLC–RFD analysis (Table 10.1). The low background radioactivity (1–2 DPM) of MSC is another factor contributing to its superior sensitivity (Zhu et al., 2005b). The limit of detection for ^{14}C on a TopCount instrument with 10 min counting was favored to be approximately 5 DPM (Table 10.1). This was approximately twofold better than LSC, and 50–100-fold better than RFD (Zhu et al., 2005b). A comparison of the radiodetection sensitivity of MSC, LSC, and RFD is illustrated in Fig.

10.2. Minor metabolite peaks M1, M2, M4, M10, M13 and M14 were detected in all analyses with TopCount (Fig. 10.2a), but not detected by RFD even when four times as much sample was injected (Fig. 10.2c). Most minor metabolites were detected by LSC, but a few minor peaks detected by MSC were not observed with LSC analysis. For example, M14 (15 CPM) was not observed in the LSC radiochromatogram shown in Fig. 10.2b.

10.3.1.2 Precision and Accuracy The precision and accuracy of the TopCount instrument for metabolite profiling were recently determined and compared to those from RFD and/or LSC analysis (Bruin et al., 2006; Zhu et al., 2005b). The precision for determining the relative radioactivity abundance of buspirone metabolites by HPLC–TopCount (Fig. 10.2a) ranged from 2 to 11 %, which was comparable to that determined by the LSC and RFD methods (Table 10.2). The accuracy values obtained on a TopCount for the buspirone metabolites (Fig. 10.2a) were ±1 % to 12 % of nominal values obtained by LSC and were similar to those determined by radio-flow detection (Table 10.2). Excellent accuracy for metabolite profiling by HPLC–MicroBeta counter was also demonstrated in the analysis of a plasma sample (Wallace et al., 2004).

TABLE 10.2 Precision of metabolite profiling by LSC, MSC, and RFD (Zhu et al., 2005b).

Radioactivity peak	Percentage radioactivity[a]					
	LSC		RFD		MSC	
	Mean[b]	RSD[c]	Mean	RSD	Mean	RSD
M3	11.24	7.0	11.73	8.8	12.56	4.5
M5	5.96	10.8	5.40	12.7	6.00	3.1
M6	6.88[d]	10.0[d]	2.17	11.5	2.18	10.5
M7			4.77	11.4	4.92	4.9
M8	8.22	8.4	8.05	8.6	7.68	3.4
M9	13.20	5.0	13.42	3.5	12.92	4.2
M11	7.98	4.6	7.89	10.4	7.90	1.6
Buspirone	32.80	4.3	34.96	7.5	33.00	2.9
M12	7.26	3.7	6.84	8.7	6.60	2.1

[a]Profiles of buspirone metabolites were determined by HPLC with three radiodetection techniques (see Fig. 10.2). Percent distribution of radioactive peaks was calculated by dividing the radioactivity of a metabolite peak by the total radioactivity determined in the HPLC run. (~8000 DPM per injection for HPLC–LSC and HPLC–MSC, and ~32,000 DPM per an injection for HPLC–RCD).
[b]Mean of % radioactivity ($n = 5$), which was calculated based on five HPLC injections.
[c]RSD is the relative standard deviation.
[d]The value represents the sum of M6 and M7 radioactivity since the two metabolites were not separated in HPLC–LSC analysis.

10.3.1.3   Matrix Effect of Biological Samples   MSC uses solid scintillators to convert radioactivity decay energy into photons. Biological samples, such as plasma, urine, and feces, which contain significant amounts of proteins, salts, and small organic chemicals, may coelute with radiolabeled analytes, and be present on the bottom of the wells after solvent is removed. These components could affect the performance of MSC through color quenching, or other mechanisms. The matrix effect of biological samples on the performance of a TopCount was recently evaluated (Bruin et al., 2006; Zhu et al., 2005b). Extracts or concentrates of plasma (Fig. 10.4d), urine (Fig. 10.4e), and HLM incubation solution (Fig. 10.4c) showed no effect on the measurement of radioactivity in any fractions except for those corresponding to the retention time of the HPLC void volume in the chromatograms of the HLM incubation solution and urine. Unlike the plasma sample, the urine and HLM samples were not treated with solid phase extraction and, therefore, proteins, salts, and other polar components were not removed from these samples. Elution of these components in the void volume and deposition in the 96-well plates likely acted as a physical barrier between the radiolabeled compound and the solid scintillators, resulting in lower counting efficiency. Lower CPM values were also observed in radiochromatograms of fecal extract (fractions 84–88, Fig. 10.4f). In a manner similar to that caused by salts eluting at the solvent front, this might be due to a physical interaction of endogenous components with the scintillators rather than color quenching because these fractions did not display intense color. Compared to the MicroBeta counter (Wallace et al., 2004), the matrix effects of urine and fecal samples on TopCount performance appeared to be minimal, suggesting that TopCount is better for analyzing these types of samples.

The quenching observed in the fecal sample analysis was proportional to the amount of fecal extract injected. Therefore, it could be minimized to an accepted level while not sacrificing sensitivity by limiting the amount of sample injected. Based on data from a study with ^{14}C it was recommended that injection of human samples (equivalent to original volume or mass) should be ≤1 mL for liver microsomal incubations and plasma, ≤2 mL for urine, and ≤50 mg for feces. In addition, conditions must be set such that polar

---

FIGURE 10.4   The effect of biological matrices on TopCount performance (Zhu et al., 2005b). Extracted or concentrated biological samples were injected onto HPLC and eluted with solvents A (water) and B (acetonitrile) up to 80 % solvent B. A [^{14}C] labeled drug was infused into the HPLC effluent using a post column Tee. The mixed effluent was collected into 96-well plates for TopCount analyses or test tubes for LSC analysis at four fractions per min for 10 min counting. (**a**) TopCount control, no injection of a biological sample; (**b**) LSC control, no injection of a biological sample; (**c**) HLM (equivalent to 1 mL HLM incubation); (**d**) human plasma (equivalent to 1 mL plasma); (**e**) human urine (equivalent to 2 mL urine); (**f**) human feces (equivalent to 100 mg feces). The up and down lines in each figure represent ±15 % of the mean values.

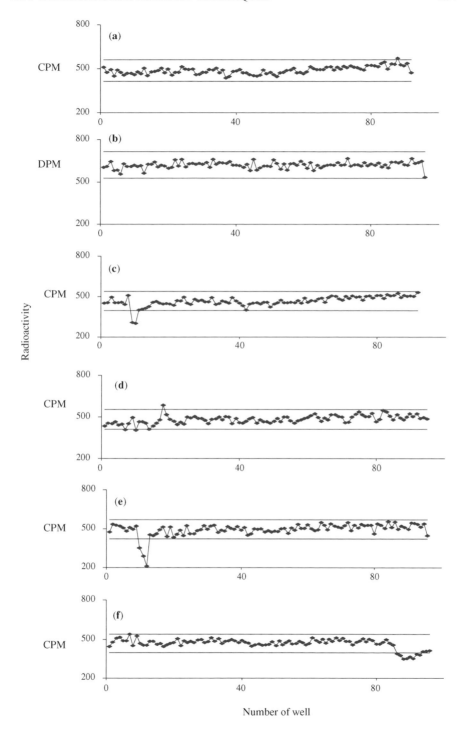

metabolites are retained somewhat and not eluted in the HPLC solvent front. Since feces displayed the most significant matrix effect among the samples tested, it is recommended that injections contain the smallest possible amount of fecal sample. Furthermore, when dealing with radioisotopes other than the ^{14}C or biological matrices different from those tested in that study, potential matrix effects of the samples on TopCount performance should be evaluated.

10.3.1.4 Radioactivity Recovery Determination   Determination of the radioactivity recovery from an HPLC analysis is a necessary procedure to ensure accurate determination of all radioactive components in the original sample. In addition, because metabolite profiling by MSC requires the additional step of in vacuo removal of HPLC solvents, it is also important to determine whether volatile metabolites are lost in the process. We have developed a simple method to determine HPLC column recovery, plate recovery, and the total recovery for an HPLC–MSC analysis (Zhu et al., 2005b). In general, aliquots of a sample are injected onto an HPLC with and without an HPLC column. All effluent from each HPLC run are separately collected and aliquots are analyzed for total radioactivity by LSC with and without solvent evaporation. The DPM values obtained are used to calculate the recoveries.

10.3.2 Stop-Flow HPLC-RFD

To improve the detection sensitivity of RFD, a stop-flow liquid radiochromatographic detection technique (Accurate radioisotope counting, ARC) was recently developed by AIM Research Company. The stop-flow RFD system offers three operation modes: (1) by-fraction; (2) by-level; and (3) nonstop (Nassar et al., 2003, 2004). The by-fraction mode performs stop flow in preset count zones or an entire HPLC run. The by-level mode performs stop flow when radioactivity is above a preset minimal value. The nonstop mode operates the same way as the regular RFD. Figure 10.5 displays typical radiochromatograms acquired by the three operation modes on a stop-flow RFD instrument after a radioactivity sample was repeatedly injected. The data showed that the by-fraction analysis at the 1 min counting time was approximately 10-fold more sensitive than traditional RFD. On the contrary, the time to complete a radiochromatographic run using by fraction mode was increased to 4 h from 45 min in the nonstop flow analysis. The by level mode provided a shorter total run time (52 min) than the by-fraction mode and significantly better sensitivity than the nonstop mode.

The stop-flow technique greatly extends the capability of RFD by improving its sensitivity, while it retains the advantages of online radiodetection. The stop-flow RFD method has the flexibility to allow it to operate in different modes suitable for high speed analysis, automation, and higher sensitivity. The new online radiochromatographic technique has been employed in profiling of *in vitro* and *in vivo* metabolites and metabolite structural characterization by coupling with a mass spectrometer (Nassar et al., 2003, 2004). Stop-flow RFD

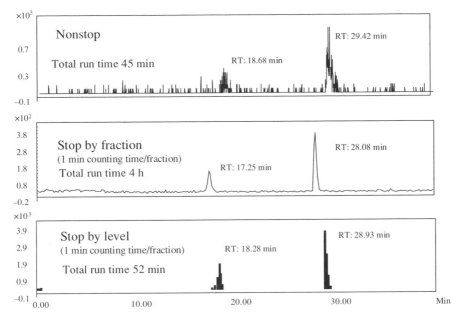

FIGURE 10.5   Analysis of two radioactive compounds using stop flow HPLC-RFD; **(a)** nonstop mode; **(b)** stop by fraction mode; and **(c)** stop by level mode. Radioactivity in stop by fraction and stop by level analyses were counted for 1 min each fraction.

is especially useful in HPLC method development and rapid profiling of metabolites with medium-to-low levels of radioactivity. Another attractive application of stop-flow RFD is in the quantitative measurement of the metabolite formation of radiolabeled substrates in enzyme kinetic studies such as CYP reaction phenotyping and determination of K_m and V_{max} values, since these studies often deal with a large number of *in vitro* incubation samples (Zhao, 2004). One of the major limitations of stop-flow RFD is the significant variability of the analyte retention times when different operation modes are used. As showed in Fig. 10.5a, the major radioactivity peak eluted at 29.42 min in the analysis using the nonstop mode, while the same peak eluted earlier in the analyses using the by-level mode (28.93 min, Fig. 10.5c), and the by-fraction mode (28.08 min, Fig. 10.5b). The shifts of metabolite HPLC retention times seem to depend on the total HPLC run times: the longer total run time, the shorter retention times of analytes. This observation is consistent with those reported in the literature (Nassar et al., 2003, 2004). Most likely, the decrease of the HPLC retention times in the stop by-fraction or stop by-level analyses is the result of some diffusion of analytes when the HPLC flows are stopped. The significant shifts in analyte retention times among different operation modes or of different HPLC run times make the stop-flow HPLC–RFD very difficult to determine identities of radioactive metabolites on the basis of their HPLC retention times.

10.3.3 Dynamic Flow HPLC-RFD

A dynamic flow radiodetection technology has been recently developed to enhance the performance of the stop-flow RFD (Lee, 2006). The new Dynamic Flow device (AIM Research Inc,) employs a dynamic flow mode added to the original stop-flow RFD system. The dynamic flow mode operates at a consistent HPLC flow rate, while the ratio of liquid scintillation cocktail to the HPLC flow rate varies. As a result, the total flow rate in the radiodetection cell varies. A decrease of the total flow rate in the RFD detection cell increases the counting time of a radioactivity peak. Therefore, radiodetection sensitivity of the dynamic flow RFD is improved. Figure 10.6 shows comparison of radiochromatograms from the analyses of the conventional HPLC–RFD versus the dynamic flow HPLC–RFD. A sample with a total radioactivity of 130 DPM was analyzed separately using the nonstop mode or the dynamic flow mode with the same run time (26 min). The dynamic flow RFD analysis clearly detected two major radioactivity peaks (Fig. 10.6b), while the RFD failed to detect any radioactivity components (Fig. 10.6a). Dynamic flow HPLC–RFD

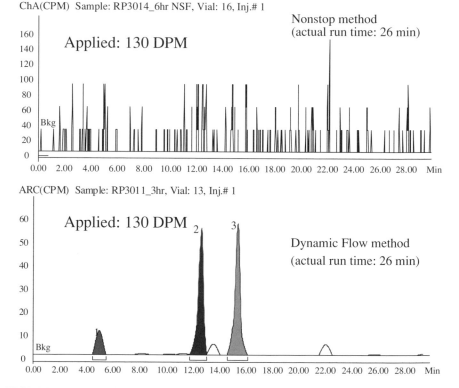

FIGURE 10.6 Analysis of radiolabeled metabolites using HPLC-RFD (top panel, nonstop mode) and dynamic HPLC-RFD (bottom panel). The same amount of sample was injected, and the total HPLC run times were the same in both analyses.

seems to have no problem in maintaining the consistency of the analyte retention times since the total HPLC run time and the HPLC flow rate in the dynamic flow analysis are the same as those in conventional HPLC–RFD analysis (Fig. 10.6) (Lee, 2006). Although further evaluation and detailed validation of the use of dynamic flow LC–RFD for metabolite profiling are required, the preliminary data have shown that the new radiochromatographic technique enables a significant increase in the sensitivity of RFD without a stop or interruption of the HPLC flow.

10.3.4 UPLC-Radiodetection

Ultra-performance liquid chromatography (UPLC) coupled with UV and MS has been applied to metabolite profiling with improved sensitivity and shorter HPLC run times (Johnson and Plumb, 2005; Wang et al., 2006a, 2006b). The width of metabolite peaks in UPLC is usually 2–6 s, equivalent to peak volumes of approximately 10–20 μL. Even with low volume detection cells the residence times of peaks in such a UPLC–RFD setting would result in a very significant decrease in radiodetection sensitivity. Therefore, RFD is not practically suitable for UPLC. Alternatively, MSC has been applied as an off-line radioactivity detector of UPLC for radioactive metabolite profiling (Dear et al., 2006). The configuration of UPLC–MSC is the same as that of HPLC–MSC (Fig. 10.3a). In the analysis, the UPLC effluent was collected into 96-well plates at a rate of 2 s/well using a modified rapid fraction collection system, and then the 96-well plates were counted on TopCount. Compared to HPLC–MSC, UPLC–MSC provided better radiodetection sensitivity (2–5 fold), faster analysis, and comparable separation resolution (Dear et al., 2006). Additionally, since smaller volume fractions are collected in the UPLC–MSC analysis, liquid scintillation cocktail can be directly added into 96-well plates without a drying process, which eliminates the possibility of the loss of volatile metabolites and reduces the matrix effect. Ultra-performance liquid chromatography coupled with MSC provides an alternative tool for high sensitivity radiolabeled metabolite profiling.

10.3.5 HPLC-AMS

Accelerator mass spectrometry (AMS) is an ultrasensitive analytical method for radioactivity analysis. AMS offers 10^3–10^9-fold increases in sensitivity over LSC or other decay counting methods so that levels as low as 0.0001 DPM can be detected (Brown et al., 2005, 2006). AMS has been applied to mass balance determination, pharmacokinetic studies of total radioactivity, and measurement of chemically modified DNA and proteins in humans after the administration of a low radioisotope dose (approximately 10 nCi/person for mass balance and drug metabolism studies) (Buchholz et al., 1999; Garner, 2000; Garner et al., 2002; Liberman et al., 2004; White and Brown, 2004). In addition, off-line HPLC–AMS has been explored for metabolite profiling after injection of 0.25–5 DPM of radioactive materials (Brown et al., 2005, 2006). For example, metabolite profiles of ^{14}C-R115777, a farnesyl transferase inhibitor, in human urine, faces, and plasma have been determined by HPLC–AMS after 50 mg ^{14}C-R115777 (687 nCi/mg) was dosed to subjects (Brown et al., 2005, 2006). HPLC–AMS is especially useful for human ADME studies when it is necessary to administer a very low radioisotope dose (either due to safety considerations or because the total dose of the drug is low, i.e., <1 mg) or for the determination of long-term kinetics and metabolism in humans. HPLC–AMS could be envisioned as a tool to allow metabolite profiles to be determined very early in clinical development. However, the use of HPLC–AMS in drug metabolism has not been widespread due to the relatively high cost of AMS analysis and the challenge as inherent in sample collection and preparation for HPLC–AMS analysis.

10.4   RADIOCHROMATOGRAPHY IN CONJUNCTION WITH MASS SPECTROMETRY

The use of radiolabeled drugs is not only crucial for the quantification of unknown metabolites but has also long played a critical role in metabolite identification studies. The utility of radiolabeled substrates include: (1) their use as tracers of drug-related components during sample clean-up, concentration, profiling, and isolation from complex biological matrixes; and (2) facilitation of LC-MS/MS detection of radiolabeled metabolites based on their HPLC retention times, peak shapes, and in some cases, isotopic ratios.

10.4.1   LC-RFD-MS

General configurations of LC–RFD–MS are shown in Figs. 10.1a and 10.1b. The configuration (A) is designed for conventional HPLC at flow rates of approximately 1 mL/min. After splitting, a small portion ($\approx$20 %) of the HPLC effluent is introduced into a mass spectrometer, and the rest of the effluent goes to the liquid detection cell of the RFD. This setup provides flow rates suitable for both mass spectrometry and RFD and a large sample loading capacity suitable for *in vivo* sample analysis (Egnash and Ramanathan, 2002; Hyllbrant et al., 1999; Mullen et al., 2002, 2003). Configuration (B) (Fig. 10.1b) is more comparable with liquid radiochromatographic analysis at a flow rate of approximately 0.2 mL/min. This setup is more sensitive in MS analysis, since there is no splitting of HPLC effluent. Use of solid radioactivity detection cell greatly reduces volume of radioactivity waste; however, the solid cell has lower counting efficiency. Also, in some cases, radiolabeled drug or metabolites adhere to the solid cell, resulting in peak broadening. LC–RFD–MS is the most popular technique for simultaneous radioactivity profiling and metabolite characterization because LC–RFD can provide results in the time frame comparable to LC/MS (Mullen et al., 2002, 2003). In addition, metabolite retention times and peak shapes in a radiochromatogram and its corresponding ion chromatograms recorded from LC–RFD–MS analysis are the same, which can greatly enhance the identification of metabolite peaks. However, the relatively poorer radioactivity detection of RFD limits the use of LC–RFD–MS in the analysis of the samples with low levels of radioactivity.

10.4.2   Stop-Flow and Dynamic Flow LC–RFD–MS

Stop-flow LC–RFD coupled with a mass spectrometer has been applied to radioactivity profiling and metabolite characterization (Nassar et al., 2003). The configuration of this system was similar to that displayed in Fig. 10.1a. In the analysis, the HPLC flow was stopped for both radioactivity counting by RFD and mass spectral data acquisition by the mass spectrometer. Although mass spectral data of metabolites were obtained from these studies, no ion chromatographs from mass spectrometric analysis were reported. When the HPLC flow is completely stopped, the HPLC effluent is not introduced into the mass spectrometer. As a result, the retention times and shapes of radiolabeled metabolite parks in reconstituted ion chromatograms may be different than those in the corresponding radiochromatograms. Another disadvantage of the stop-flow HPLC–RFD–MS is that it takes a much longer run time (up to 4 h, Fig. 10.5). The newly developed dynamic flow LC–RFD would be an improved online radiodetection technology regarding coupling with a mass spectrometer, since the HPLC flow is not stopped or interrupted in the dynamic flow analysis (Lee, 2006).

10.4.3   LC-MSC-MS

The combination of HPLC–MSC with mass spectrometry (LC–MSC–MS) is often employed for metabolite analysis and is especially useful for metabolites with a low level of radioactivity (Zhang., et al 2007a and 2007b). Two configurations of LC–MSC–MS are shown in Fig. 10.3. In configuration (A) the HPLC effluent is split so that a small portion flows to the mass spectrometer and the remaining portion flows to 96-well microplates. Radioactivity in the 96-well plates is counted by MSC after either removing solvent under vacuum or direct addition of scintillation cocktail. Conventional HPLC columns (3.9–4.6 mm ID) are typically employed in LC–MSC–MS analyses because a large column provides consistent retention times and high separation resolution of analytes in complex biological samples. It also provides better MSC and mass spectrometric sensitivity because larger sample volumes can be injected without sacrificing chromatographic performance. In general, MSC provides a level of detection comparable to or better than LC–MS/MS when 40–100 μCi radioactivity is dosed to each animal or human subject in ADME studies. Figure 10.7 shows an example of using LC–MSC–MS for quantification and structure elucidation of metabolites in a rat bile

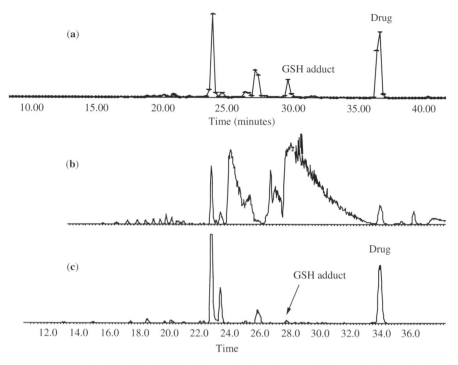

FIGURE 10.7 Profiles of radiolabeled metabolites in a rat bile sample determined by HPLC-MSC-MS. A majority of HPLC effluent (1 mL/min) was collected into 96-well microplates (four fractions per min) followed by radioactivity counting using TopCount. The top panel is the radiochromatogram of this sample. A potion of the effluent was analyzed by LTQ FTMS. The middle panel is the total ion chromatogram (TIC) from accurate mass full scan MS analysis. The bottom panel is a mass defect filter processed TIC (Zhu et al., 2006). The peak indicated is a GSH adduct that was not presented in the unprocessed TIC.

sample. MSC was able to detect and quantify both minor and major metabolites (Fig. 10.7a), which is consistent with the metabolite profile determined by high resolution LC/MS (Fig. 10.7c).

   To enhance the sensitivity of mass spectrometry in the analysis of radiolabeled metabolites, an alternative method using a combination of LC–MSC and MS has been recently developed as shown in Fig. 10.3b (Gedamke, 2003; Zhu, 2002, 2003, 2005c). First, a radioactivity profile is determined by MSC without splitting HPLC effluent. Based on the resultant metabolite profile, radioactive metabolites of interest are then recovered from the microplates and structurally characterized by capillary and nano LC–MS or nanospray mass spectrometry via direct infusion. A combination of LC–MSC and nano LC/MS for analyzing rat plasma metabolites showed several advantages to this approach (Gedamke, 2003). (1) The superior radiodetection sensitivity was maintained due to the high loading capacity of the large HPLC column. (2) Suppression of metabolite ionization by coeluting components can be significantly reduced or eliminated by the dual stages of chromatographic separation. (3) Higher sensitivity of mass spectrometric analysis is achieved by the use of very low flow rates. (4) Use of the second HPLC allows LC–MS conditions to be optimized for the analysis of each metabolite. The method using an infusion nanoelectrospray mass spectrometer (Nanomate) to analyze radiolabeled metabolites recovered from microplates (Fig. 10.3b) is also employed for metabolite profiling. The approach increases acquisition time up to 30–60 min for each recovered radioactivity peak so that multiple MS experiments such as MS^n analyses can be conducted with improved sensitivity (Staack et al., 2007).

10.4.4 An Integrated Radiochromatography–Mass Spectrometry Approach

Recently, an integrated approach using a combination of RFD, MSC and mass spectrometers has been developed (Fig. 10.8) for quantification and structural characterization of both high and low level radiolabeled metabolites (Zhao, 2006). In that study, a sample from a rat hepatocyte incubation of [^{14}C]nefazodone (NEF) was injected onto an HPLC (4.6 × 250 mm). A portion (20 %) of the HPLC effluent was directed into a linear ion-trap MS (LTQ, Thermo) and the remainder was passed through a solid scintillation cell and then collected into 96-well plates. Radioactive peaks of interest were recovered and infused using chip-based nanoelectrospray (Nanomate, Advion) into a high resolution MS (LTQ FT) with its mass resolution set to 25,000 for

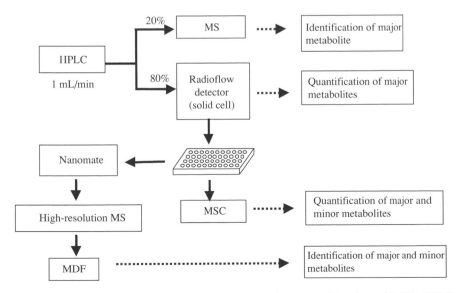

FIGURE 10.8 A set up of integrated approach using a combination of RFD, MSC, and mass spectrometers for quantification and identification of both high and low levels of radiolabeled metabolites.

acquisition of accurate mass spectral data. The online RFD and linear ion-trap MS analyses detected multiple oxidative and conjugated metabolites with high levels of radioactivity (>250 DPM). The off-line MSC radiochromatogram from the same HPLC run displayed nine additional metabolites with low level of radioactivity. Nanoelectrospray FTMS analyses of a number of the metabolites recovered from microplates were carried out under optimized analytical conditions with continuous infusion for up to 30 min. As a result, high quality accurate mass full scan MS and MS^n spectra were obtained. In some cases, metabolite ions could not be recognized in the full scan MS spectra due to interference from coeluting biological components and background ions. Mass defect fiter (Zhu et al., 2006) removed the majority of these intense interfering ions so that the metabolite ions became predominant and were easily recognizable. As a result, a single LC/RFD/MS run followed by MSC radioactivity profiling and nanoelectrospray FTMS analysis enabled the detection, quantitative determination, and structural characterization of more than 20 NEF-related components, including several very minor metabolites. This example demonstrates that the integrated approach can provide both quantitative analysis and structural characterization of radiolabeled metabolites at both high and low levels.

10.5 APPLICATION OF NEW RADIOCHROMATOGRAPHY TECHNIQUES IN DRUG METABOLISM STUDIES

10.5.1 Profiling of Radiolabeled Metabolites in Plasma

One of important objectives of drug metabolism studies in drug development is to assess whether human circulating metabolites are present in plasma of toxicology species at appropriate levels, which is usually accomplished by profiling plasma metabolites in radiolabeled human and animal ADME studies. However, radioactivity analysis of plasma metabolites is a challenging task because concentrations of metabolites in plasma are much lower compared to those in urine, bile or feces, and volumes of plasma sample obtained from ADME studies are very limited, especially when multiple plasma samples were collected for AUC determination. In most cases, HPLC–RFD is not sensitive enough for this application. In addition, high throughput is not required in plasma metabolite profiling since a limited of number of pooled plasma samples are subjected to analysis in ADME studies (Zhang et al., 2007b). Therefore, off-line LC–MSC technique well suits for analysis of plasma and other samples from ADME studies. Furthermore, unlike stop-flow or dynamic flow LC–RFD methods, the use of LC–MSC technique for metabolite profiling has been fully validated with respect to accuracy, precision, sensitivity, matrix effect and radioactivity recovery (Bruin et al., 2006; Zhu et al., 2005b). Thus, high quality results of metabolite profiling by MSC can be obtained as long as proper procedures are followed.

An example of using off-line HPLC–MSC for profiling plasma metabolites is displayed in Fig. 10.9. Plasma samples containing a total of 400–900 CPM of radioactivity were injected. A minor metabolite (M3, 9 CPM, Fig. 10.9c), approximately 3 % of the total radioactivity injected was clearly detected. The percent distribution of all drug-related components in the plasma sample was determined in the same analysis (Fig. 10.9). Furthermore, drug and metabolite concentrations can be calculated by multiplying the percent distributions to the concentrations of the total radioactivity in the corresponding plasma samples. As a result, AUC_{0-8h} values of the drug, M3 and M6 were obtained by calculating their concentrations in multiple plasma samples collected from 0.5 to 8 h.

Interestingly, although the relative concentration (7.7 %) of M6 in the 8 h dog plasma was much less than that (16.3 %) in the 1 h human plasma, the AUC_{0-8h} values of M6 in dogs and humans were comparable. Based on the sensitivity provided by TopCount analysis (Table 10.1), samples containing

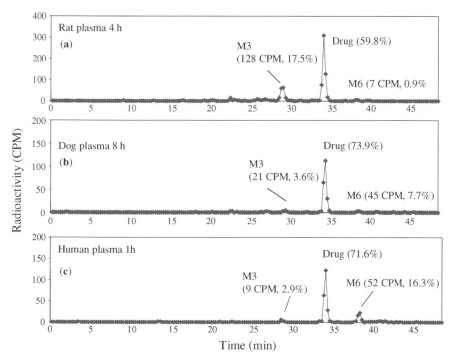

FIGURE 10.9 Representative plasma metabolite profiles of a radiolabeled drug candidate in rats (**a**), dogs (**b**), and humans (**c**). The pooled plasma samples from radiolabeled ADME studies after an oral administration were collected around the maximum concentration of the drug. The radioactivity profiles were determined by off-line HPLC–TopCount (four functions per min and 10 min counting time). Approximately, 400–9000 DPM of radioactivity was injected.

approximately 500 DPM of radioactivity or more are required for detection of minor metabolites (10 DPM, 10 min counting time) corresponding to 2 % of the total radioactivity on column. For the detection of the same minor radioactive components, approximately 25,000–50,000 DPM of radioactivity is required in the LC–RFD analysis. MSC is also applicable in the detection and profiling of *in vitro* metabolites when the specific radioactivity of the radiolabeled drug is low, metabolic turnover is low, or quantification of minor radiolabeled metabolites is required.

10.5.2 Analysis of Metabolites of Nonradiolabeled Drugs Using Radiolabeled Cofactors or Trapping Agents

Another application of liquid radiochromatographic techniques in drug metabolism is the detection and quantification of conjugated metabolites or reactive metabolites of nonradiolabeled drugs formed in incubations with radiolabeled cofactors such as [^{14}C]uridine diphosphate glucuroinic acid (UDPGA) (Ethell et al., 1998) or reactive metabolite trapping agents such as [^{3}H]- or [^{35}S]-glutathione (GSH) (Thompson et al., 1995; Zhu et al., 2005b). For analyzing a large number of samples in enzyme kinetic experiments or screening reactive metabolites, the use of online radiochromatographic techniques such as HPLC–RFD and stop-flow LC–RFD should be the first choice of the methods. HPLC–MSC is an alternative tool for sensitive analysis of a limited number of samples. Figure 10.10 shows radioactivity profiles of a nonradiolabeled drug incubated with human liver microsomes in the presence of 1 mM [^{3}H]GSH with and without the addition of NADPH. GSH trapped reactive metabolites M1 and M2 were detected and quantified by HPLC–MSC (Fig. 10.10a). These same conjugates were not detected in the incubation in which NADPH was omitted (Fig. 10.10b). The sensitivity of MSC allows for detection of low quantities of GSH adducts even when the specific activity and total concentration of [^{3}H]GSH in the incubations were kept low. If the same experiment was conducted using a radio-flow detector, 50–100-fold more [^{3}H]GSH would be required to achieve the same level of detection.

10.5.3 Determination of Structures and Formation Pathways of Sequential Metabolites

Many drugs, such as buspirone, undergo multiple oxidative biotransformation reactions. In cases such as buspirone the secondary metabolites, which are minor metabolites *in vitro*, are major metabolites in the circulation or excreta in humans and animals because primary metabolites are rapidly converted to secondary metabolites. To determine the formation pathways and structures of secondary or sequential metabolites, a method using "metabolite incubation" and HPLC–MSC–MS analysis was developed and demonstrated using buspirone as an example (Zhu, 2002). [^{14}C]buspirone was incubated with HLM, and metabolic profiling was carried out using HPLC–MSC. The

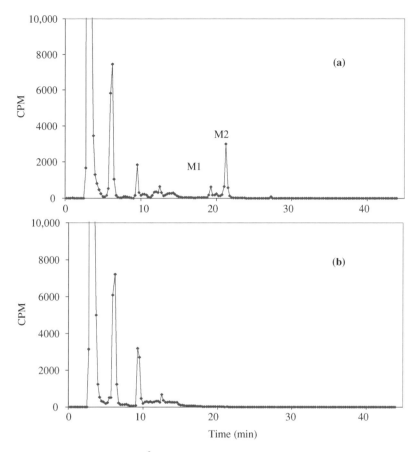

FIGURE 10.10   Analysis of [³H]GSH trapped reactive metabolites by HPLC with TopCount (Zhu et al., 2005b). A nonlabeled drug (50 mM) was incubated with a mixture of GSH (1 mM) and trace [³H]GSH (1–2 μCi/mL) in human liver microsomes. After precipitating proteins, the samples were analyzed by HPLC (1 mL/min) with TopCount (4 wlls/min, 10-min counting time); **(a)** Radioactivity profile of the incubation. M1 and M2 were GSH-trapped reactive metabolites; and **(b)** Radioactivity profile of a control incubation sample (without NADPH).

radioactive peaks corresponding to the major buspirone metabolites 3′-OH, 5-OH and *N*-oxide were recovered from the 96-well microplates and reincubated with HLM. Secondary metabolites derived from the incubations of buspirone and these primary metabolites were analyzed by HPLC–MSC in conjunction with ion-trap and accurate mass LC/MS.

Based on HPLC retention times, MSⁿ spectra and diagnostic fragment ions from accurate mass analysis, four secondary metabolites 5,3′-di-OH, *N*,3′-di-OH, 5,6′-di-OH and *N*,6′-di-OH derivatives were identified. The major precursors of these secondary metabolites were also elucidated based on formation rates of the secondary metabolites in the incubations of the individual primary metabolites as determined by radiochromatographic analysis. The sequential metabolites identified in the metabolite incubation approach were further used as metabolite standards for successfully determining the identities of buspirone metabolites in rat plasma. Results from this example illustrate that this *in vitro* approach, using a combination of metabolite incubation and LC–MSC–MS, is a simple and effective way to determine the formation pathways and structures of sequential metabolites, such as dihydroxy metabolites. A similar approach was applied to the quantitative determination of formation pathways of capravirine in humans using stop-flow HPLC–RFD analysis (Bu et al., 2005). The approach includes three steps: (1) 30-min primary incubation of [^{14}C]capravirine; (2) isolation of radiolabeled metabolites from the primary incubation; and (3) 30-min reincubation of the isolated metabolites supplemented with an ongoing (30-min) microsomal incubation with nonlabeled capravirine. The formation pathways of sequential metabolites were assigned based on the extent of disappearance of the isolated primary metabolites and the formation of sequential metabolites.

10.5.4 Enzyme Kinetic Studies

Enzymology studies in drug metabolism include the determination of kinetic parameters, such as K_m and V_{max}, and reaction phenotyping of enzymes that catalyze formation of metabolites. These studies require quantification of metabolites formed in a large number of incubation samples. Two analytical approaches are often employed for these studies: (1) LC–MS/MS using synthetic standards of metabolites; and (2) liquid radiochromatography using radiolabeled substrates. Typically in the pharmaceutical industry, radiolabeled drugs are available when a single compound is selected for development. The online RFD method is the primary radiochromatographic technique used in enzyme kinetic studies since it provides fast results and high separation resolution (Zhu et al., 2005a). In some cases, metabolite quantitation is challenging because the experiment is conducted with a drug at a low specific activity or low substrate concentrations, or with slow turnover. Thus, use of more sensitive liquid radiochromatographic techniques such as HPLC–MSC or stop-flow RFD is required. Although HPLC–MSC provides excellent sensitivity, due to relatively slow throughput (usually 6–8 HPLC injections per day) and manual operation, it is not well suited for enzyme kinetic studies. Stop-flow HPLC–RFD appears more attractive for enzyme kinetic experiments since it can be performed in an automated fashion (Zhao, 2004). In addition, the stop-flow RFD instrument has the conventional RFD function suitable for high level radioactivity and the stop-flow function for analysis of relatively low levels of radiolabeled metabolites, which is especially useful in the determination of enzyme kinetic parameters.

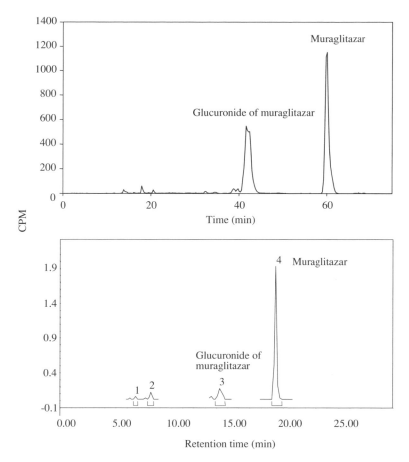

FIGURE 10.11   Quantification of muraglitazar glucuronide in the incubation with UGT1A3 by off-line HPLC-LSC (four fractions per min) (top panel) and online stop-flow HPLC–RFD (operated with the stop by-fraction mode) (bottom panel).

An example of using stop-flow HPLC–RFD to determine the enzyme kinetic parameters of the glucuronidation of muraglitazar in human UGT1A3 was recently presented (Zhao, 2004). Muraglitazar was incubated with recombinant human UGT2B7 enzyme and UDPGA. These samples were analyzed using the stop by-fraction mode so that the HPLC flow rate was stopped and counted for 1 min in the selected time zones when significant drug-related components passed through the RFD cell (Fig. 10.11). The relative distribution of the muraglitazar glucuronide in the sample was determined and its concentrations in the incubations were calculated by multiplying the relative distribution to the initial drug concentration in the incubations. The precision and accuracy of the stop by fraction analysis for determining the percentage distribution of the drug and its glucuronide were found to be quite good (Table 10.3). Accordingly, the relationship of muraglitazar glucuronidation

TABLE 10.3 Precision and accuracy of stop-flow HPLC/RFD for determining relative radioactivity of muraglitazar and its metabolites in the incubation of [^{14}C] muraglitazar with UGT1A3.

Radioactivity peak		1	2	3 (muraglitazar glucuronide)	4 (muraglitazar)
Precision (N = 5)	Stop-flow	2.6 ± 0.3	4.7 ± 0.6	12 ± 0.6	70 ± 1.4
	LSC	2.3 ± 0.5	5.1 ± 0.6	13 ± 2	72 ± 1.5
Accuracy (%)		16	8.2	8.2	3.0

Stop flow LS/RFD (1 mL/min) was carried using the by-fraction mode (Fig. 10.11b) and counting time was 1 min. LC–LSC was carried by counting HPLC fractions (four fractions per min) for 10 min by LSC.

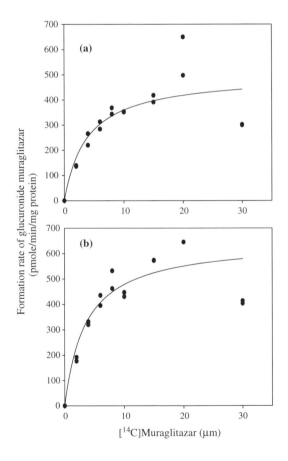

FIGURE 10.12 Formation rates of muraglitazar glucuronide catalized by UGT1A3 determined by stop-flow HPLC–RFD (a) and HPLC–TopCount (b).

rates with substrate concentrations was obtained by the stop-flow HPLC–RFD method (Fig. 10.12a), which was very similar to that determined using off-line HPLC–MSC (Fig. 10.12b). The results suggest that both stop-flow LC–RFD and off-line LC–MSC can provide high quality data in the analysis of low level radioactivity, however, stop-flow LC–RFD is 3–4 fold faster than HPLC–MSC.

10.6 SUMMARY

Metabolite profiling by liquid radiochromatographic techniques determines the number and relative concentrations (% distribution of radioactivity) of metabolites presented in a biological sample. Conventional online HPLC–RFD is currently the primary liquid radiochromatographic tool employed in drug metabolism laboratories for profiling radiolabeled metabolites, enzyme kinetic determination of radiolabeled substrates, and metabolite structural characterization by coupling with mass spectrometry. The recently introduced stop-flow and dynamic flow techniques have significantly improved the sensitivity of RFD. Ideally any new RFD devices would include nonstop, stop-flow and dynamic flow functions that are completely interchangeable. Off-line HPLC–MSC is a new and important tool for analysis of low level radioactivity, which is especially useful in radiolabeled *in vivo* ADME studies that require very high radiodetection sensitivity for analysis of plasma metabolites, and validated procedures for supporting drug development and registration. In addition, the combination of HPLC–MSC and mass spectrometry allows for the detection and structural characterization of minor radiolabeled metabolites that may not be detected by LC–RFD–MS. HPLC coupled with AMS is an ultrasensitive liquid radiochromatographic technique that has been applied in a few clinical studies to meet special needs. However, the cost of the instrument and operation, and the difficulty in dealing with radioactivity contamination limit its application in drug metabolism studies.

REFERENCES

Boernsen KO, Floeckher JM, Bruin GJ. Use of a microplate scintillation counter as a radioactivity detector for miniaturized separation techniques in drug metabolism. Anal Chem 2000;72:3956–3959.

Brown K, Dingley KH, Turteltaub KW. Accelerator mass spectrometry for biomedical research. Methods Enzymol 2005;402:423–443.

Brown K, Tompkins EM, White IN. Applications of accelerator mass spectrometry for pharmacological and toxicological research. Mass Spectrom Rev 2006;25:127–145.

Bruin GJ, Waldmeier F, Boernsen KO, Pfaar U, Gross G, Zollinger M. A microplate solid scintillation counter as a radioactivity detector for high performance liquid chromatography in drug metabolism: validation and applications. J Chromatogr A 2006;1133:184–194.

Bu HZ, Kang P, Zhao P, Pool WF, Wu EY. A simple sequential incubation method for deconvoluting the complicated sequential metabolism of capravirine in humans. Drug Metab Dispos 2005;33:1438–1445.

Buchholz BA, Fultz E, Haack KW, Vogel JS, Gilman SD, Gee SJ, Hammock BD, Hui X, Wester RC, Maibach HI. HPLC–accelerator MS measurement of atrazine metabolites in human urine after dermal exposure. Anal Chem 1999;71:3519–3525.

Chando TJ, Everett DW, Kahle AD, Starrett AM, Vachharajani N, Shyu WC, Kripalani KJ, Barbhaiya RH. Biotransformation of irbesartan in man. Drug Metab Dispos 1998;26:408–417.

Dalvie D. Recent advances in the applications of radioisotopes in drug metabolism, toxicology, and pharmacokinetics. Curr Pharm Des 2000;6:1009–1028.

Dear GJ, Patel N, Kelly PJ, Webber L, Yung M. TopCount coupled to ultra-performance liquid chromatography for the profiling of radiolabeled drug metabolites in complex biological samples. J Chromatogr B Analyt Technol Biomed Life Sci 2006;844:96–103.

Egnash LA, Ramanathan R. Comparison of heterogeneous and homogeneous radioactivity flow detectors for simultaneous profiling and LC–MS/MS characterization of metabolites. J Pharm Biomed Anal 2002;27:271–284.

Ethell BT, Anderson GD, Beaumont K, Rance DJ, Burchell B. A universal radiochemical high-performance liquid chromatographic assay for the determination of UDP-glucuronosyltransferase activity. Anal Biochem 1998;255:142–147.

Evans DC, Watt AP, Nicoll-Griffith DA, Baillie TA. Drug-protein adducts: an industry perspective on minimizing the potential for drug bioactivation in drug discovery and development. Chem Res Toxicol 2004;17:3–16.

Everett DW, Chando TJ, Didonato GC, Singhvi SM, Pan HY, Weinstein SH. Biotransformation of pravastatin sodium in humans. Drug Metab Dispos 1991;19:740–748.

Fouda HG, Avery MJ, Dalvie D, Falkner FC, Melvin LS, Ronfeld RA. Disposition and metabolism of tenidap in the rat. Drug Metab Dispos 1997;25:140–148.

Garner RC. Accelerator mass spectrometry in pharmaceutical research and development a new ultrasensitive analytical method for isotope measurement. Curr Drug Metab 2000;1:205–213.

Garner RC, Goris I, Laenen AA, Vanhoutte E, Meuldermans W, Gregory S, Garner JV, Leong D, Whattam M, Calam A, Snel CA. Evaluation of accelerator mass spectrometry in a human mass balance and pharmacokinetic study-experience with 14C-labeled (R)-6-[amino(4- chlorophenyl)(1-methyl-1H-imidazol-5-yl)methyl]-4-(3-chlorophenyl)-1- methyl-2(1H)-quinolinone (R115777), a farnesyl transferase inhibitor. Drug Metab Dispos 2002;30:823–830.

Gedamke R, Zhao W, Gozo S, Mitroka J, Zhu M. A sensitive method for plasma metabolite identification using nano LC/ion trap MS in conjunction with a microplate scintillation counter. ASMS Archive Abstract 2003.

Graf BA, Mullen W, Caldwell ST, Hartley RC, Duthie GG, Lean ME, Crozier A, Edwards CA. Disposition and metabolism of [2-14C]quercetin-4′-glucoside in rats. Drug Metab Dispos 2005;33:1036–1043.

Hazai I, Patfalusi M, Klebovich I, Urmos I. Whole-body autoradiography and quantitative organ-level distribution study of deramciclane in rats. J Pharm Pharmacol 1999;51:165–174.

Heath TG, Mooney JP, Broersma R. Narrow-bore liquid chromatography-tandem mass spectrometry with simultaneous radioactivity monitoring for partially characterizing the biliary metabolites of an arginine fluoroalkyl ketone analog of D-MePhe-Pro-Arg, a potent thrombin inhibitor. J Chromatogr B Biomed Sci Appl 1997;688:281–289.

Hyllbrant B, Tyrefors N, Markides KE, Langstrom B. On the use of liquid chromatography with radio- and ultraviolet absorbance detection coupled to mass spectrometry for improved sensitivity and selectivity in determination of specific radioactivity of radiopharmaceuticals. J Pharm Biomed Anal 1999;20:493–501.

Hett KF, Ethell BT, Maggs JL, Davis TM, Batty KT, Burchell B, Binh TQ, Thu le TA, Hung NC, Pirmohamed M, Park BK, Edwards G. Glucuronidation of dihydroarte-misinin *in vivo* and by human liver microsomes and expressed UDP-glucuronosyl-transferases. Drug Metab Dispos 2002;30:1005–1012.

Johnson KA, Plumb R. Investigating the human metabolism of acetaminophen using UPLC and exact mass oa-TOF MS. J Pharm Biomed Anal 2005;39:805–810.

Kiffe M, Jehle A, Ruembeli R. Combination of high-performance liquid chromato-graphy and microplate scintillation counting for crop and animal metabolism studies: a comparison with classical on-line and thin-layer chromatography radio-activity detection. Anal Chem 2003;75:723–730.

Lautala P, Ulmanen I, Taskinen J. Radiochemical high-performance liquid chromato-graphic assay for the determination of catechol O-methyltransferase activity towards various substrates. J Chromatogr B Biomed Sci Appl 1999;736:143–151.

Lee DY. A novel dynamic flow system for sensitive radioactivity detection in LC/MS. Drug Metab Rev 2006;38 (Suppl 2):33.

Li W, Zhang D, Wang L, Zhang H, Cheng PT, Everett DW, Humphreys WG. Biotransformation of carbon-14-labeled muraglitazar in male mice: interspecies difference in metabolic pathways leading to unique metabolites. Drug Metab Dispos 2006;34:807–820.

Liberman RG, Tannenbaum SR, Hughey BJ, Shefer RE, Klinkowstein RE, Prakash C, Harriman SP, Skipper PL. An interface for direct analysis of 14-C in nonvolatile samples by accelerator mass spectrometry. Anal Chem 2004;76:328–334.

Magyar K, Lengyel J, Klebovich I, Urmos I, Grezal G. Distribution of deramciclane (EGIS-3886) in rat brain regions. Eur J Drug Metab Pharmacokinet 1998;23:125–131.

Marathe PH, Shyu WC, Humphreys WG. The use of radiolabeled compounds for ADME studies in discovery and exploratory development. Curr Pharm Des 2004;10:2991–3008.

Meneses-Lorente G, Sakatis MZ, Schulz-Utermoehl T, DeNardi C, Watt AP. A quantitative high-throughput trapping assay as a measurement of potential for bioactivation. Anal Biochem 2006;351:266–272.

Morovjan G, Dalmadi-Kiss B, Klebovich I, Mincsovics E. Metabolite analysis, isolation, and purity assessment using various liquid chromatographic techniques combined with radioactivity detection. J Chromatogr Sci 2002;40:603–608.

Mullen W, Graf BA, Caldwell ST, Hartley RC, Duthie GG, Edwards CA, Lean ME, Crozier A. Determination of flavonol metabolites in plasma and tissues of rats by HPLC-radiocounting and tandem mass spectrometry following oral ingestion of [2-(14)C]quercetin-4'-glucoside. J Agric Food Chem 2002;50:6902–6909.

Mullen W, Hartley RC, Crozier A. Detection and identification of 14C-labeled flavonol metabolites by high-performance liquid chromatography-radiocounting and tandem mass spectrometry. J Chromatogr A 2003;1007:21–29.

Nassar AE, Bjorge SM, Lee DY. On-line liquid chromatography-accurate radioisotope counting coupled with a radioactivity detector and mass spectrometer for metabolite identification in drug discovery and development. Anal Chem 2003;75:785–790.

Nassar AE, Parmentier Y, Martinet M, Lee DY. Liquid chromatography-accurate radioisotope counting and microplate scintillation counter technologies in drug metabolism-studies. J Chromatogr Sci 2004;42:348–353.

Nedderman AN, Savage ME, White KL, Walker DK. The use of 96-well Scintiplates to facilitate definitive metabolism studies for drug candidates. J Pharm Biomed Anal 2004;34:607–617.

Staack RF and Hopfgarner G. New analytical strategies in drug metabolism. Anal Bioanal Chem 2007;388:1365–1380.

Thompson DC, Perera K, London R. Quinone methide formation from para isomers of methylphenol (cresol), ethylphenol, and isopropylphenol: relationship to toxicity. Chem Res Toxicol 1995;8:55–60.

Veltkamp AC. Radiochromatography in pharmaceutical and biomedical analysis. J Chromatogr 1990;531:101–129.

Vlasakova V, Brezinova A, Holik J. Study of cytokinin metabolism using HPLC with radioisotope detection. J Pharm Biomed Anal 1998;17:39–44.

Wallace D, Hildesheim A, Pinto LA. Comparison of benchtop microplate beta counters with the traditional gamma counting method for measurement of chromium-51 release in cytotoxic assays. Clin Diagn Lab Immunol 2004;11:255–260.

Wang G, Hsieh Y, Cui X, Cheng KC, Korfmacher WA. Ultra-performance liquid chromatography/tandem mass spectrometric determination of testosterone and its metabolites in in vitro samples. Rapid Commun Mass Spectrom 2006a;20:2215–2221.

Wang J, Zhao X, Zheng Y, Kong H, Lu G, Cai Z, Xu G. Metabolite fingerprint and biomarkers identification of rat urine after dosed with ginsenoside Rg3 based on ultra high performance liquid chromatography/time-of-flight mass spectrometry (UPLC/TOF-MS)]. Se Pu 2006b;24:5–9.

White IN, Brown K. Techniques: the application of accelerator mass spectrometry to pharmacology and toxicology. Trends Pharmacol Sci 2004;25:442–447.

Zhang D, Krishna R, Wang L, Zeng J, Mitroka J, Dai R, Narasimhan N, Reeves RA, Srinivas NR, Klunk LJ. Metabolism, pharmacokinetics, and protein covalent binding of radiolabeled MaxiPost (BMS-204352) in humans. Drug Metab Dispos 2005;33:83–93.

Zhang D, Ogan M, Gedamke R, Roongta V, Dai R, Zhu M, Rinehart JK, Klunk L, Mitroka J. Protein covalent binding of maxipost through a cytochrome P450-mediated orthoquinone methide intermediate in rats. Drug Metab Dispos 2003;31:837–845.

Zhang D, Wang L, Chandrasena G, Ma L, Zhu M, Zhang H, Davis CD, Humphreys WG. Involvement of multiple cytochrome P450 and UDP-glucuronosyltransferase enzymes in the *in vitro* metabolism of muraglitazar. Drug Metab Dispos 2007a;35:139–149.

Zhang D, Wang L, Raghavan N, Zhang H, Li W, Cheng PT, Yao M, Zhang L, Zhu M, Bonacorsi S, Yeola S, Mitroka J, Hariharan N, Hosagrahara V, Chandrasena G, Shyu WC, Humphreys WG. Comparative metabolism of radiolabeled muraglitazar in animals and humans by quantitative and qualitative metabolite profiling. Drug Metab Dispos 2007b;35:150–167.

Zhang D, Zhao W, Roongta VA, Mitroka JG, Klunk LJ, Zhu M. Amide N-glucuronidation of MaxiPost catalyzed by UDP-glucuronosyltransferase 2B7 in humans. Drug Metab Dispos 2004;32:545–551.

Zhao W, Wang L, Zhang D, Zhu M. Rapid and sensitive determination of enzyme kinetics of drug metabolism using HPLC coupled with a stop-flow radioactivity flow detector. Drug Metab Rev 2004;36:257.

Zhao W, Zhang H, ZH, Zhu M, Warrack B, Ma L, Humphreys WG, Sanders M. An integrated method for quantification and identification of radiolabeled metabolites: application of chip-based nanoelectrospray and mass defect filter techniques. *ASMS Archive Abstract.*

Zhu M, Ma L, Zhang D, Ray K, Zhao W, Humphreys WG, Skiles G, Sanders M, Zhang H. Detection and characterization of metabolites in biological matrices using Mass defect filtering of liquid chromatography/High resolution mass spectrometry data. Drug Metab Dispos. 2006;34:1722–1733.

Zhu M, Wang L, Gedamke R, Zhao W, Zhang D, Gozo S. Determination of radioactivity distribution and metabolite profiles in tissues using a combination of microplate scintillation counting, capillary LC/MS and whole-body autoradiography. Drug Metab Rev 2003;35:76.

Zhu M, Zhang H, Zhao W, Mitroka J, Klunk L. A novel *in vitro* approach, using a microplate scintillation counter and LC/MS, for determining formation pathways and structures of sequential metabolites: Buspirone as an example. Drug Metab Rev 2002;34:167.

Zhu M, Zhao W, Jimenez H, Zhang D, Yeola S, Dai R, Vachharajani N, Mitroka J. Cytochrome P450 3A-mediated metabolism of buspirone in human liver microsomes. Drug Metab Dispos 2005a;33:500–507.

Zhu M, Zhao W, Vazquez N, Mitroka JG. Analysis of low level radioactive metabolites in biological fluids using high-performance liquid chromatography with microplate scintillation counting: method validation and application. J Pharm Biomed Anal 2005b;39:233–245.

Zhu M, Zhang D, Skills G. Quantification and structural elucidation of low quantities for radiolabeled metabolites using microplate scintillation counting (MSC) techniques in conjunction with LC/MS. In: Chowdhurry SK, editor. *Identification and quantification of Drugs, Metabolites and Metabolizing Enzymes.* Amsterdam: Elsevier; 2005c. p 195–223.

11

APPLICATION OF LIQUID CHROMATOGRAPHY/MASS SPECTROMETRY FOR METABOLITE IDENTIFICATION

SHUGUANG MA AND SWAPAN K. CHOWDHURY

11.1 INTRODUCTION

Metabolites are chemical entities formed as a result of biotransformation of endogenous compounds and xenobiotics. In general, metabolism leads to more hydrophilic species compared to the progenitor molecules thus facilitating the elimination from the body. Metabolic pathways are classified as nonsynthetic or Phase I (e.g., oxidation, reduction, and hydrolysis) and synthetic or Phase II (e.g., conjugation) reactions. The assessment of the metabolic fate of the drug candidates, knowledge of the routes and extent of metabolism in animals and humans, and evaluation of the biological properties of metabolites, represent important objectives in pharmaceutical research. Metabolites could be pharmacologically active, toxic, or involved in drug–drug interactions; therefore, metabolite identification plays a pivotal role in various phases of drug discovery and development. Identification of pharmacologically active metabolites leads to protection of intellectual properties as these metabolites can potentially become next generation drugs.

Metabolite characterization is complicated by the fact that metabolism often produces a multitude of relatively low concentration (nM to μM) structurally diverse compounds. In addition, biological matrices containing a large excess of

Drug Metabolism in Drug Design and Development, Edited by Donglu Zhang, Mingshe Zhu and W. Griffith Humphreys
Copyright © 2008 John Wiley & Sons, Inc.

proteins, lipids, and other endogenous material interfere with the detection of drug-derived material. Mass spectrometry (MS) has become the analytical tool of choice for the detection and identification of metabolites due to its ability to easily couple to liquid chromatography (LC) systems through atmospheric pressure ionization (API) interfaces. The high sensitivity and selectivity of liquid chromatography/mass spectrometry (LC/MS) and its ability to separate, detect, and identify many metabolites in the presence of endogenous material, makes it well-suited for drug metabolism studies. In a previous article (Ma et al., 2006), we discussed recent advances in the LC/MS technologies and current practices for profiling, characterization, and identification of drug metabolites in drug discovery and development settings. This chapter represents an extension of the previous article with the incorporation of new developments in metabolite identification and LC/MS instrumentation with particular emphasis on the detection and identification of reactive metabolites and intermediates.

11.2 LC/MS INSTRUMENTATION

11.2.1 High Performance Liquid Chromatography (HPLC)

HPLC is now one of the most widely used analytical tools to separate complex mixtures. Liquid chromatography uses differential analyte interactions between a stationary phase and a liquid mobile phase to effect separation of components in a mixture. The degree of separation depends on the extent of interaction between the solute components and the stationary phase. The interaction of the solute with mobile and stationary phases can be manipulated through choices of solvents and stationary phases. The choice of chromatography conditions for metabolite identification studies can be critical. It is important to separate the metabolites from matrix ions and from each other. A good separation can facilitate the ionization and characterization of the compounds of interest. Traditional metabolite characterization experiments are conducted using reverse phase LC with a slow gradient and long column to ensure adequate separation of drug metabolites in biological matrices.

Further advances in instrumentation and column technology are needed to achieve significant increases in resolution, speed, and sensitivity in liquid chromatography. Better resolution and reduced analysis time can be attained by reducing the particle size. In 2004, columns with smaller particle sizes ($\sim 1.7\,\mu m$) and instrumentation designed to work at 15,000 psi were developed and commercialized by Waters Corporation to achieve a new level of performance. This new technology is called ultra performance liquid chromatography (UPLC^{TM}). UPLC is rapidly gaining acceptance as an ideal separation tool for complex mixture analysis in drug metabolism studies due to its superior chromatographic resolution, sensitivity, and speed of analysis. The typical peak widths generated by the UPLC system are in the order of 1–2 s. The reduction in peak width increases the sensitivity by three to five fold. The increased efficiency

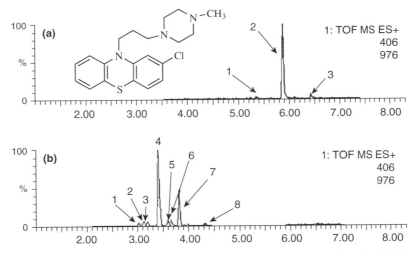

FIGURE 11.1 Comparison of the extracted ion chromatograms for doubly hydroxylated metabolites (*m/z* 406) of prochlorperazine from incubations with rat liver microsomes obtained from HPLC/MS (**a**) and UPLC/MS (**b**). Reprinted from Castro-Perez et al. (2005a) with permission of John Wiley and Sons Limited.

results in the reduction of analytical run time by a factor of 5, while still maintaining the chromatographic resolution (Johnson and Plumb, 2005; Plumb et al., 2004). The overall enhancement in the chromatographic resolution and peak capacity is translated into a reduction in the number of coeluting species. Castro-Perez et al. reported the detection of additional prochlorperazine metabolites from a rat liver microsomal incubation using UPLC coupled to a Q-TOF mass spectrometer, as shown in Fig. 11.1 (Castro-Perez et al., 2005a). In the HPLC separation, just three of the doubly hydroxylated metabolites were detected; however, with the UPLC analysis, all eight metabolites were detected within 4.5 min run time. UPLC systems with similar technology are now available from other manufacturers (Frank, 2006; Pereira et al., 2006).

11.2.2 Atmospheric Pressure Ionization Methods

A prerequisite for detection, identification, and quantification of any species by MS is that all analytes must be converted into gas-phase ions before they enter the mass analyzer. API techniques are most widely used for metabolite identification, mainly due to their ability to couple to liquid chromatography and generate intact gas-phase molecular ions at very high sensitivity (Rossi, 2002; Voyksner, 1997).

11.2.2.1 Electrospray Ionization (ESI) Much of the current importance of the electrospray technique derives from the pioneering work of Dole et al. (1968) who first recognized the possibility of generating gas-phase ions by spraying a sample solution from the tip of an electrically charged capillary in the presence of a flow of warm nitrogen to assist desolvation and then measured the ions by ion mobility techniques. The most recent resurgence of interest in ESI–MS resulted from the work of Fenn and coworkers who demonstrated, for the first time, that intact molecule ions of large peptides and proteins can be generated by ESI and their molecular weight determined by deconvolution of the mass-to-charge (m/z) envelope generated by multiple charging (Whitehouse et al., 1985).

In this technique, a solution of the analyte is passed through a capillary tip that is held at high potential (typically in the range of 2.5–5 kV) to generate a mist of highly charged droplets. Two mechanistic models, the charge residue model (CRM) and the ion evaporation model (IEM), have been proposed for the gas-phase ion formation from liquids (Kebarle, 2000; Kebarle and Ho, 1997). In the CRM, evaporation of solvents from the initially formed droplets leads to an increasing charge density on the droplet surface. Once the droplets have reached the Rayleigh limit (the point in which charge repulsion exceeds the surface tension of the droplet), Coulombic explosion occurs and the droplets are ripped apart. This process continues to the point where no further evaporation of the solvent can occur and finally individual ions or ion clusters are formed. In the IEM, when solvent evaporation and Coulomb droplet fissions reduce the size of charged droplets to a certain radius (~10–20 nm), direct emission of ions into the gas-phase begins to occur. Ion evaporation replaces Coulomb fission and evaporation of ions continues until a solid residue of ion clusters is formed. It has been postulated that small ions are more likely formed by IEM, while large protein ions are proposed to be formed by CRM.

Due to the inherent nature of electrohydrodynamic atomization and ion formation, electrospray is more controlled and yields higher sensitivity at lower flow rates. Wilm and Mann developed the "nanospray" technique in which a small volume (μL) of sample was loaded into a metal-coated capillary tip and was sprayed at the flow rate of 30–200 nL/min (Wilm and Mann, 1996). These nanospray tips can be located very close to the nozzle that samples ions into the mass spectrometer, thus increasing sensitivity. At these extremely low flow rates, small amounts of sample can undergo ESI for a prolonged period of time allowing many experiments to be performed and yielding maximum information at very high sensitivity.

ESI is often referred to as a "soft" ionization technique in that little fragmentation of intact molecular ions is produced. Despite the numerous benefits of ESI–MS, it suffers from a shortcoming in that it is susceptible to ion suppression effects from high concentrations of buffer, salt, and other endogenous material in matrix solutions.

11.2.2.2 Atmospheric Pressure Chemical Ionization (APCI)

In an APCI source, liquid effluent is nebulized into fine droplets by a high gas flow and carried into a heated chamber where the solvent droplets are evaporated.

Atmospheric gases (O_2 and N_2) and gas-phase solvent molecules initially react with electrons from a corona discharge and form secondary reactant ions (H_3O^+ and $(H_2O)_nH^+$) through ion/molecule reactions:

$$N_2 + e \rightarrow N_2^{+\cdot} + 2e$$

$$N_2^{+\cdot} + 2N_2 \rightarrow N_4^{+\cdot} + N_2$$

$$N_4^{+\cdot} + H_2O \rightarrow H_2O^{+\cdot} + 2N_2$$

$$H_2O^{+\cdot} + H_2O \rightarrow H_3O^+ + \cdot OH$$

$$H_3O^+ + H_2O^{+\cdot} \rightarrow H^+(H_2O)_2$$

$$H^+(H_2O)_{n-1} + H_2O^{+\cdot} \rightarrow H^+(H_2O)_n$$

These solvent ions then undergo ion/molecule reactions with the analyte (M), giving rise to protonated molecular ions $[M + H]^+$ through a proton transfer reaction (Horning et al., 1974).

It is interesting to note that ionization in APCI occurs mainly in the gas-phase, while ionization in ESI occurs in solution. Therefore, different compounds will show varying sensitivity and charge distributions between the two techniques. Also, due to the large excess of gas-phase reagent ions generated by the corona discharge and the lower degree of ion/molecule reaction in the gas-phase, APCI is recognized as being less susceptible to ion suppression effects and typically provides a wider dynamic detection range than ESI (Dams et al., 2003; Hsieh et al., 2001). Higher flow rates are used with APCI (1–2 mL/min) relative to conventional ESI (0.1–0.5 mL/min) in order to generate large amounts of reagent ions and maximize the collision frequency needed for ion/molecule charge transfer. Because of the need for thermal desolvation, extremely polar and large molecules (MW > 1000 Da) that may be susceptible to thermal degradation, are not well suited for APCI. ESI is more commonly used for the analysis of peptides, proteins, carbohydrates, and oligonucleotides, while both ESI and APCI can be used for small molecule pharmaceuticals (MW < 600 Da) and drug metabolism studies (Brewer and Henion, 1998).

11.2.2.3 Atmospheric Pressure Photoionization (APPI)

APPI is a relatively new API method developed by Bruins and coworkers in 2000 (Robb et al., 2000). The APPI source is similar to the APCI source in that mobile phase is vaporized using a heated nebulizer (350–500°C) to generate a dense cloud of gas-phase solvent and analyte molecules. The APPI source uses a krypton discharge lamp ($hv = 10\,\text{eV}$) to photoionize dopants added to the liquid effluent entering the ionization source. Dopants (usually toluene or acetone) are routinely added to increase ionization efficiency and to act as a charge carrier for ionizing trace levels of analyte by charge transfer (Hanold et al., 2004).

APPI is mainly used as an alternative to normal corona discharge APCI. APPI has produced equal or better detection limits compared to APCI in selected applications (Wang et al., 2005), however, the ionization process in APPI depends directly on the solvent, dopant, nebulizing gas, and impurities in the gas and surrounding atmosphere, as well as the analyte itself (Yang and Henion, 2002).

11.2.2.4 *Desorption Electrospray Ionization (DESI)*   In DESI (Fig. 11.2a), a fine spray of charged droplets is directed to the surface of interest, from which small organic molecules and large biomolecules are desorbed and ionized. Following desolvation these ions are presented to the mass spectrometer for analysis (Cooks et al., 2006; Takats et al., 2004). The mechanism of DESI is not yet clear but several processes have been suggested to account for the ionization of molecules, including, droplet pick up and chemical sputtering and evaporation followed by gas-phase ionization (Takats et al., 2005b). DESI causes minimum fragmentation, resulting in the formation of intact molecule ions. The main advantage of DESI compared to other desorption ionization methods is that it does not require sample pretreatment and it can be

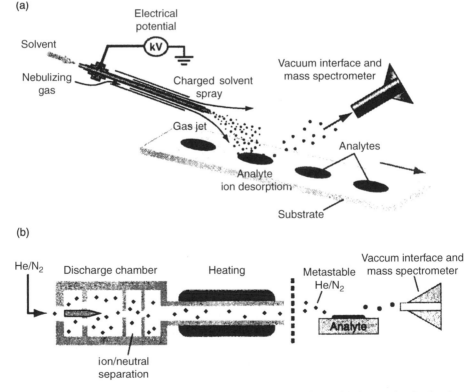

FIGURE 11.2   Schematics of a typical DESI (**a**) and a DART (**b**) desorption ionization source. Reprinted from Cooks et al. 2006 with permission of American Association for the Advancement of Science.

performed under ambient conditions. Other closely related methods have also been introduced, including desorption atmospheric pressure chemical ionization (DAPCI) (Takats et al., 2005a) and electrospray-assisted laser desorption ionization (ELDI) (Shiea et al., 2005).

11.2.2.5   Direct Analysis in Real-Time (DART)   In DART (Fig. 11.2b), a plasma of excited-state atoms/molecules, ions, and electrons are generated by an electrical discharge of several kilovolts applied to a gas (helium or nitrogen). The gas flows into a second chamber where ions and electrons are removed by a perforated electrode. The gas flow then passes through a third region that can be optically heated. The exiting gas is directed toward a liquid or solid sample surface. The electronic or vibronic excited-state species (metastable helium atoms or nitrogen molecules) are the working reagent that desorb and ionize low molecular weight molecules from the surface (Cody et al., 2005). Different ionization mechanisms occur depending upon the nature of the carrier gas, analyte, and polarity of ions (Cody et al., 2005). Although not yet extensively used for metabolite profiling and identification, this technique could be useful in identifying low molecular weight, non-polar metabolites.

11.2.3   Mass Analyzers

The function of a mass analyzer is to measure the m/z ratios of ions. There are five principal types of mass analyzers in common use today that can be divided into two groups: beam-type mass analyzers and ion-trapping mass analyzers (McLuckey and Wells, 2001). In beam-type analyzers, the ions leave the ion source in a beam and pass through the analyzing (electric or magnetic) field to the detector. Beam-type mass analyzers include time-of-flight, sector magnets, and quadrupole mass filters. In trapping-type analyzers, ions are trapped in the analyzing field after being formed in the analyzer itself or being ejected from external ion source (Glish and Vachet, 2003). Trapping analyzers includes ion traps, Fourier transform ion cyclotron resonance (FTICR) and FT-Orbitrap mass spectrometers.

11.2.3.1   Time-of-Flight (TOF)   Analysis in a TOF mass spectrometer is based on the principle that ions of different m/z values, when accelerated by the same kinetic energy, possess different velocities after acceleration out of the ion source and into a field-free drift tube (Guilhaus et al., 2003; Weickhardt et al., 1996). As a result, the time (t) required for each ion to traverse the flight tube is different and high mass ions will take longer to reach the detector than low mass ions. The equation relating the flight time of an ion with its m/z value is shown below:

$$t = \frac{L}{v} = \frac{L}{\sqrt{\frac{2zV}{m}}} = L\sqrt{\frac{m}{2zV}}$$

where L, v, and V are the ion drift length, the ion velocity, and the accelerating potential, respectively.

Despite the simplicity of the equation, an ion's initial spatial and angular kinetic energy spread complicate the precise determination of m/z. These issues are addressed to improve m/z accuracies in high performance time-of-flight mass spectrometers (Guilhaus et al., 2003).

11.2.3.2 Sector Magnets In magnetic sector instruments, ions accelerated from an ionization source are deflected by a magnetic field and adopt a constant trajectory of radius (r) around the center of the field. For a fixed magnetic field strength (B) and a fixed accelerating potential (V), only ions with a certain momentum-to-charge ratio will pass through slits placed after the magnetic field (Glish and Vachet, 2003). The equation that characterizes the radius of curvature as a function of the m/z ratio is given by the following:

$$m/z = \frac{r^2 B^2}{2V}$$

A mass-to-charge spectrum can be obtained by scanning B (at constant V) or V (at constant B), so that ions of different m/z can travel through the slits with a typical resolution of a few hundred. The low resolution is primarily due to the ions' velocity spread. To improve the resolving power, an electrostatic analyzer (ESA) is placed in the ion optic pathway before or after the magnet sector. The combination of an ESA and a magnetic sector provides both directional and energy focusing (double focusing) thus improving the m/z accuracies significantly. The resolution of a double-focusing instrument is typically in the range of 10^4–10^5.

11.2.3.3 Quadrupole Mass Filters Quadrupole mass analyzers consist of four electrodes, ideally of hyperbolic rods, that are accurately positioned in a radial array. For practical as well as economic reasons, most quadrupole mass filters have employed electrodes of circular cross section. A potential is applied to one pair of diagonally opposite rods consisting of a DC voltage and an rf voltage. To the other pair of rods, a DC voltage of opposite polarity and an rf voltage with a 180° phase shift are applied. The ion motion under the influence of this two-dimensional (2D) field can be described mathematically by the solutions to the second-order linear differential equation, known as Mathieu equation, from which the Mathieu parameters, a_u and q_u, can be derived as

$$a_u = a_x = -a_y = \frac{4zU}{mr_0^2\omega^2}$$

$$q_u = q_x = -q_y = \frac{2zV}{mr_0^2\omega^2}$$

where m/z is the mass-to-charge ratio of the ion, U is the DC voltage, V is the rf amplitude, ω is the rf frequency, and r_0 is half the distance between two diagonally opposite rods.

Separation of ions in a quadrupole mass filter is based on achieving a stable trajectory for ions of specific m/z values. Only ions that undergo stable motion in both x-and y-directions will remain within the device and be detected by the detector as they emerge from the quadrupole. When $U = 0$ ($a_u = 0$, rf only mode), a wide band of ions with m/z values below a certain cut-off corresponding to $q_u = 0.908$ are transmitted. As the value of U/V is increased, the resolution increases so that only ions within a narrow m/z window are transmitted. A mass spectrum may be generated by scanning the values of U and V at a constant U/V ratio, while keeping the rf frequency fixed (Chernushevich et al., 2001).

11.2.3.4 Quadrupole Ion Traps (QIT)

The fundamental working principle of quadrupole ion traps is the same as described for quadrupole mass filters, but in three dimensions rather than two (March, 1997). A QIT is composed of three electrodes: two end caps and one ring. One end cap electrode has a single small central aperture through which ions can be gated periodically and the other has several small apertures arranged centrally through which ions pass to a detector. The mass analysis equation for a QIT operated in the mass-selective instability mode is

$$m/z = \frac{8V}{q_z(r_0^2 + 2z_0^2)\omega^2}$$

where V is the rf potential, r_0 is the radius of the ring electrode, z_0 is the distance from the center to the end cap, and ω is its angular frequency. In contrast to quadrupole mass filters, mass spectra with ion traps are generated by making the ion trajectories unstable in a mass-selective manner in which ions are moved along the q_z axis by raising the rf voltage V until they become unstable at the boundary, where $q_z = 0.908$. Ions of progressive m/z values are ejected and detected by an electron multiplier; hence called mass-selective instability.

Due to the small footprint, low cost and ability to perform sequential fragmentation experiments (MS^n), this mass spectrometry technique gained a lot of popularity in the past decade. The main issue that limits the performance of 3D ion traps is that only a limited number of ions can be trapped in the center of the device before reaching the space-charge limit of stability. This limitation was overcome in the linear ion trap (LIT) MS, which is a geometrical variation of the 3D ion trap (Douglas et al., 2005). LIT consists of four parallel hyperbolic rods with two end caps, in which ions are confined radially by a two-dimensional radio frequency (rf) field, and axially by stopping potentials applied to end electrodes. LIT has two major advantages over 3D IT: a larger ion storage capacity and a higher trapping efficiency. LIT can hold almost 10–100 times more ions than 3D traps because ions are more radially focused in the LIT while in a 3D ion trap they are focused at the center of the trap. In addition, ions are injected into the linear trap through an end cap and then ejected from it. Doing so allows the use of two detectors on each side of the trap, therefore, doubles the ion current. Because of these advantages one can predict that in the not too distant future the LITs will replace the 3D ion traps.

11.2.3.5 Fourier Transform Mass Spectrometer (FTMS)   Fourier transformation (FT) of time-dependent image from the detector to m/z intensity is utilized for two types of mass spectrometers: ion cyclotron resonance (ICR) and Orbitrap. FTICR mass spectrometers operate based on the ion cyclotron resonance principle: ions in a magnetic field (B) move in circular orbits at frequencies (ω_c) characteristic of their m/z values as shown below (Marshall et al., 1998, 2002):

$$\omega_c = \frac{zeB}{2\pi m}$$

$$m/z = \frac{eB}{2\pi\omega_c}$$

An FTICRMS consists of three pairs of parallel plates arranged as a cube that is used for trapping, excitation, and detection of ions. As shown in the equation above, at a constant and uniform magnetic field (B) the cyclotron frequency (ω_c) of an ion is inversely proportional to its m/z value. Trapped ions within an FTICRMS are detected by applying a frequency-sweep signal. When the applied frequency becomes equal to the cyclotron frequency of ions at a given m/z, the ions absorb energy and orbit at a larger radius. These translationally excited ions move coherently between the receiver plates generating image currents. The time-dependent image currents are subjected to Fourier transformation that resolve the components of ion currents and produce the mass spectrum. The FTICRMS can easily achieve a resolution of 100,000, thus providing accurate determination of m/z values of unknown metabolites more reliably than any other mass spectrometer available today. FTICRMS systems are typically more expensive and possess a larger laboratory footprint than other mass spectrometers.

Recently, a new high performance mass analyzer, called Orbitrap™, has been developed. The Orbitrap consists of an inner and an outer electrode, which are shaped to create a quadro-logarithmic electrostatic potential. Ions rotate about the inner electrode and oscillate harmonically along its axis (the z-direction) with a frequency (ω_z) characteristic of their m/z values:

$$\omega_z = \sqrt{\frac{k}{m/z}}$$

where k is field curvature. These oscillations are observed using image current detection and are transformed into mass spectra using fast Fourier transforms (FFTs). High mass resolution (up to 150,000) and mass accuracy (2 ppm) have been demonstrated (Day et al., 2005; Makarov, 2000; Hardman and Makarov, 2003).

11.2.4   Tandem Mass Spectrometry

Tandem mass spectrometry employs two stages of mass analysis, one to preselect an ion and the second to analyze the product ions produced most commonly by collisional activation with an inert gas like argon, helium, or nitrogen. This dual analysis can be tandem-in-space or tandem-in-time (de Hoffmann, 1996). Tandem-in-space instruments are constructed with two physically separated mass analyzers. Representative tandem-in-space instruments include two-sector instruments comprised of a magnetic sector and an electric sector, triple quadrupole instruments, or hybrid tandem instruments (Q-TOF, Qq-LIT, LIT–FTICRMS, LIT–FT-OrbitrapMS, etc). Tandem-in-time MS/MS can be achieved on trapping devices such as FTICRMS and ion traps. These devices are capable of storing ions within the device. Selection of an ion of interest is achieved by ejecting all the other ions. The selected ion is then excited and fragmented during a selected time frame and the product ions can then be mass analyzed. This process can be repeated for product ions; therefore, this type of instrument is capable of performing MS^n experiments.

Four scan functions can be performed in a tandem mass spectrometry experiment (Schwartz et al., 1990). These are product ion scan, precursor ion scan (PIS), constant neutral loss scan (CNLS), and selected/multiple reaction monitoring (SRM/MRM). In product ion scan, an ion of a given m/z value is selected, activated, and fragmented. The product ions are then analyzed to provide structural information. This scan function is available for both types of tandem mass spectrometers. In precursor ion scan mode, the second mass analyzer is set to only pass the ions with a selected m/z value, while the first mass analyzer is scanned over a defined m/z range. This type of experiment is particularly useful for monitoring groups of compounds that fragment to produce common fragment ions. In CNLS experiments, both mass analyzers are scanned while the m/z difference between the mass analyzers is kept constant. Since precursor ion scan and neutral loss scan require two distinct mass analyzers, these two scan functions are only possible with tandem-in-space type instruments. In selected/multiple reaction monitoring, both of the analyzers are static as ions of interest are selected and transmitted through the first mass analyzer, while the fragment ions of interest are selected by the second mass analyzer. Liquid chromatography/tandem mass spectrometry (LC/MS/MS) is by far the most powerful technique for metabolite profiling and identification (Clarke et al., 2001; Kostiainen et al., 2003; Ma et al., 2006; Nassar and Talaat, 2004).

11.3   METABOLITE IDENTIFICATION—ROLE OF LC/MS

11.3.1   Metabolite Characterization in Drug Discovery

The utilization of LC/MS for metabolite characterization in drug discovery is vital and extensive. Different laboratories use the technology differently depending on the extent various information are sought. Early in the drug discovery process, metabolism studies are often aimed at understanding the metabolic stability of a compound or a structural series. The major goal is to locate the metabolically vulnerable spots in the structure that may increase the rate of intrinsic metabolic clearance resulting in short half-life and low oral bioavailability. Identification of the metabolic soft spots is vital in guiding synthetic chemistry efforts to design new lead compounds with optimal pharmacokinetic profiles (Ma et al., 2006).

Metabolic stability studies are performed by estimating the rate of disappearance of the parent drug by measuring the concentration before and following incubation in microsomal systems for a predetermined period (Chovan et al., 2004). Modern LC/MS instruments, such as linear ion traps (Hopfgartner et al., 2004; Xia et al., 2003) offer very high scan speed such that multiple MS/MS experiments on the parent drug, as well as putative metabolites, can be performed simultaneously not only to detect them at high sensitivity but also to extract additional information on the site of modification (Hopfgartner et al., 2003).

In vivo metabolic stability, as well as preliminary metabolite characterization, is usually performed when a compound is determined to be optimally stable in *in vitro* systems. Since radiolabeled compounds are rarely available in the early stages of drug discovery, the detection of molecular ions of drug-related components is mainly achieved from the analysis of the test and control samples in parallel followed by the comparison of the extracted ion chromatograms of potential metabolites according to predicted gains and losses in their molecular mass. The common biotransformations and corresponding m/z changes are listed in Table 11.1 (Anari et al., 2004). The structures of these putative metabolites are then determined from product ion MS/MS and/or MS^n experiments (Tozuka et al., 2003). The tandem mass spectrometry approach to revealing the molecular ions of unexpected metabolites has focused on the use of PIS and CNLS. These two experiments are particularly useful since they do not require previous knowledge of the molecular weight of metabolites. The PIS allows detection of all molecular entities that form a common product ion and signifies a structural relationship between the administered drug and its metabolites. Typically, several product ions of the parent drug are used to detect precursor ions of the metabolites. These experiments take advantage of the fact that of any one metabolite, large parts of the parent molecule do not undergo metabolic changes. The product ion representing an unchanged substructure is used as a structure-specific detection signal to detect the metabolites. As more than one substructure-specific ion is used, one can detect metabolites that have modifications on various portions of the molecule.

If a drug and its metabolites have a common structure that is lost in a MS/MS experiments as neutral species, the mass spectrometer can be set to detect the metabolite using CNLS, irrespective of its molecular weight. Several conjugated metabolites fragment to lose a distinct neutral group. This distinct

TABLE 11.1 Common biotransformations and the corresponding elemental composition and m/z changes.

Metabolic reaction	Description	Molecular formula change	m/z change
$R-CH_2C_6H_5 \rightarrow R-H$	Debenzylation	$-C_7H_6$	-90.0468
$R-^{79}Br \rightarrow R-H$	Reductive debromination	$-Br + H$	-77.9104
$R-CF_3 \rightarrow R-H$	Trifluoromethyl loss	$-CF_3 + H$	-67.9874
$2R-^{35}Cl \rightarrow 2R-H$	$2 \times$ reductive dechlorination	$-Cl_2 + H_2$	-67.9222
$R-^{79}Br \rightarrow R-OH$	Oxidative debromination	$-Br + OH$	-61.9156
$R-C(CH_3)_3 \rightarrow R-H$	*Tert*-butyl dealkylation	$-C_4H_8$	-56.0624
$R-ONO_2 \rightarrow R-OH$	Hydrolysis of nitrate ester	$-NO_2 + H$	-44.9851
$R-COOH \rightarrow R-H$	Decarboxylation	$-CO_2$	-43.9898
$R-CH(CH_3)_2 \rightarrow R-H$	Isopropyl dealkylation	$-C_3H_6$	-42.0468
$R-CH_2COCH_2CH_2CH_3 \rightarrow R-COOH$	Propyl ketone to acid	$-C_4H_8 + O$	-40.0675
$R-C(CH_3)_3 \rightarrow R-OH$	*Tert*-butyl to alcohol	$-C_4H_8 + O$	-40.0675
$2R-F \rightarrow 2R-H$	$2 \times$ reductive defluorination	$-F_2 + H_2$	-35.9811
$R-^{35}Cl \rightarrow R-H$	Reductive dechlorination	$-Cl + H$	-33.9611
$R-CH_2OH \rightarrow R-H$	Hydroxymethylene loss	$-CH_2O$	-30.0106
$R-NO_2 \rightarrow R-NH_2$	Nitro reduction	$-O_2 + H_2$	-29.9742
$R-CH_2OCH_2CH_2CH_3 \rightarrow R-COOH$	Propyl ether to acid	$-C_3H_8 + O$	-28.0675
$R-C_2H_5 \rightarrow R-H$	Deethylation	$-C_2H_4$	-28.0312
$R-CO-R' \rightarrow R-R'$	Decarbonylation	$-CO$	-27.9949
$R-CH_2COCH_2CH_3 \rightarrow R-COOH$	Ethyl ketone to acid	$-C_3H_6 + O$	-26.0519
$R-CH(CH_3)_2 \rightarrow R-OH$	Isopropyl to alcohol	$-C_3H_6 + O$	-26.0519
$R-CH_2-CH_2OH \rightarrow R-CH = CH_2$	Alcohols dehydration	$-H_2O$	-18.0105
$R-CH = N-OH \rightarrow R-CN$	Dehydration of oximes	$-H_2O$	-18.0105
$R-F \rightarrow R-H$	Reductive defluorination	$-F + H$	-17.9906
$R-^{35}Cl \rightarrow R-OH$	Oxidative dechlorination	$-Cl + OH$	-17.9662
$RR'S = O \rightarrow R-S-R'$	Sulfoxide to thioether	$-O$	-15.9949
$R-NHNHR'C = S \rightarrow R-NHNHR'C = O$	Thioureas to ureas	$-S + O$	-15.9772

(continued)

TABLE 11.1 *(Continued)*

Metabolic reaction	Description	Molecular formula change	*m/z* change
R–CH$_2$OCH$_2$CH$_3$ → R–COOH	Ethyl ether to acid	–C$_2$H$_6$ + O	–14.0519
R–CH$_3$ → R–H	Demethylation	–CH$_2$	–14.0157
R–C(CH$_3$)$_3$ → R–COOH	*Tert*-butyl to acid	–C$_3$H$_8$ + O$_2$	–12.0726
R–CH$_2$COCH$_3$ → R–COOH	Methyl ketone to acid	–C$_2$H$_4$ + O	–12.0363
R–CH$_2$CH$_3$ → R–OH	Ethyl to alcohol	–C$_2$H$_4$ + O	–12.0363
R–CH$_2$–CH$_2$–CH$_2$–CH$_2$–R′ → R–CH = CH–CH = CH–R′	Two sequential desaturation	–H$_4$	–4.0314
Hydroxylation + dehydration	Hydroxylation + dehydration	–H$_2$	–2.0157
R–CH$_2$–OH → R–CHO; R–CHOH–R′ → R–CO–R′	Primary/secondary alcohols to aldehyde/ketone	–H$_2$	–2.0157
R–CH$_2$–CH$_2$–R′ → R–CH = CH–R′	Desaturation	–H$_2$	–2.0157
C$_5$H$_7$N → C$_5$H$_5$N	1,4-Dihydropyridines to pyridines	–H$_2$	–2.0157
R–F → R–OH	Oxidative defluorination	–F + OH	–1.9957
R–CHNH$_2$–R′ → R–CO–R′	Oxidative deamination to ketone	–NH$_3$ + O	–1.0316
R–CH$_2$OCH$_3$ → R–COOH	Demethylation and oxidation to acid	–CH$_4$ + O	–0.0365
R–CH(OH)CH$_3$ → R–COOH	2-Ethoxyl to acid	–CH$_4$ + O	–0.0365
R–CH$_2$–NH$_2$ → R–CH$_2$–OH	Oxidative deamination to alcohol	–NH + O	0.9840
R–CH(CH$_3$)$_2$ → R–COOH	Isopropyl to acid	–C$_2$H$_6$ + O$_2$	1.9430
R–CH$_3$ → R–OH	Demethylation and hydroxylation	–CH$_2$ + O	1.9792

Reaction	Description	Change	Mass
R-CO-R' → R-CHOH-R'	Ketone to alcohol	$+H_2$	2.0157
R-CH$_2$-R' → R-C(O)-R'	Methylene to ketone	$-H_2 + O$	13.9792
Hydroxylation and desaturation	Hydroxylation and desaturation	$-H_2 + O$	13.9792
R-XH → R-X-CH$_3$ (X = N, O, S)	N, O, S methylation	$+CH_2$	14.0157
R-CH$_2$CH$_3$ → R-COOH	Ethyl to carboxylic acid	$-CH_4 + O_2$	15.9586
R-H → R-OH	Hydroxylation	$+O$	15.9949
R-NH-R' → R-NOH-R'; RR'RN → RR'R'NO	Second/third amine to hydroxylamine/N-oxide	$+O$	15.9949
R-S-R' → R-SO-R'; R-SO-R' → R-(O)S(O)-R'	Thioether to sulfoxide, sulfoxide to sulfone	$+O$	15.9949
R-CH = CH-R' → R-CH(O)CH-R'	Aromatic ring to arene oxide	$+O$	15.9949
R-CH$_2$CH$_3$ → R-CH-(OH)$_2$	Demethylation and two hydroxylation	$-CH_2 + O_2$	17.9741
R-CH = CH-R → R-CH$_2$-CHOH-R'	Hydration, hydrolysis (internal)	$+H_2O$	18.0106
R-CN → R-CONH$_2$	Hydrolysis of aromatic nitriles	$+H_2O$	18.0106
Hydroxylation and ketone formation	Hydroxylation and ketone formation	$-H_2 + O_2$	29.9741
C$_n$H$_m$ → C$_n$H$_{m-2}$O$_2$	Quinine formation	$-H_2 + O_2$	29.9741
R-CH$_3$ → R-COOH	Methyl to carboxylation	$-H_2 + O_2$	29.9741
R-H → R-O-CH$_3$	Hydroxylation and methylation	$+CH_2O$	30.0105
2 × hydroxylation	2 × hydroxylation	$+O_2$	31.9898
RR'S → RR'SO$_2$	Thioether to sulfone	$+O_2$	31.9898
R-CH = CH-R' → R-CH(OH)-CH(OH)-R'	Alkenes to dihydrodiols	$+H_2O_2$	34.0054
R-NH$_2$ → R-NHCOCH$_3$	Acetylation	$+C_2H_2O$	42.0106
3 × hydroxylation	3 × hydroxylation	$+O_3$	47.9847
R-SH → R-SO$_3$H	Aromatic thiols to sulfonic acids	$+O_3$	47.9847

(continued)

TABLE 11.1 *(Continued)*

Metabolic reaction	Description	Molecular formula change	m/z change
R–COOH → R–CONHCH$_2$COOH	Glycine conjugation	+ C$_2$H$_3$NO	57.0215
R–OH → R–OSO$_3$H	Sulfation	+ SO$_3$	79.9568
R–H → R–OSO$_3$H	Hydroxylation and sulfation	+ SO$_4$	95.9517
R–COOH → R–CONH–CH(CH$_2$SH)–COOH	Cysteine conjugation	+ C$_3$H$_5$NOS	103.0092
R–COOH → R–CONH–CH$_2$CH$_2$SO$_3$H	Taurine conjugation	+ C$_2$H$_5$NO$_2$S	107.0041
R–CH$_2$–R′ → RR′–CH–SCH$_2$CH(NH$_2$)–COOH	S-Cysteine conjugation	+ C$_3$H$_5$NO$_2$S	119.0041
R–COOH → R–CO–SCH$_2$CH(NHCOCH$_3$)COOH	S,N-Acetylcysteine conjugation	+ C$_5$H$_7$NO$_2$S	145.0198
–CO + C$_6$H$_8$O$_6$	Decarboxylation and glucuronidation	+ C$_5$H$_8$O$_5$	148.0372
RR′–CH$_2$ → RR′–CH–SCH$_2$CH (NHCOCH$_3$)–COOH	N-Acetylcysteine conjugation	+ C$_5$H$_7$NO$_3$S	161.0147
R–OH → R–O–C$_6$H$_{11}$O$_5$	Glucosidation	+ C$_6$H$_{10}$O$_5$	162.0528
R–OH → R–O–C$_6$H$_9$O$_6$	Glucuronide conjugation	+ C$_6$H$_8$O$_6$	176.0321
2 × hydroxylation + 2 × SO$_3$	2 × hydroxylation + 2 × sulfation	+ S$_2$O$_8$	191.9035
R–H → R–O–C$_6$H$_9$O$_6$	Hydroxylation + glucuronide	+ C$_6$H$_8$O$_7$	192.0270

R-COOH → R-CO-SG (GSH = Glutathione)	S-Acyl-glutathione conjugates	$+ C_{10}H_{15}N_3O_5S$	289.0732
$+GSH-4H$	Desaturation + GSH	$+ C_{10}H_{13}N_3O_6S$	303.0525
$+GSH-2H$	GSH conjugation	$+ C_{10}H_{15}N_3O_6S$	305.0682
$+GSH$	GSH conjugation	$+ C_{10}H_{17}N_3O_6S$	307.0839
$+GSH + O-2H$	Oxidation + GSH conjugation	$+ C_{10}H_{15}N_3O_7S$	321.0631
Epoxidation + GSH	Epoxidation + GSH conjugation	$+ C_{10}H_{17}N_3O_7S$	323.0788
$+ 2 \times C_6H_8O_6$	2 × glucuronide conjugation	$+ C_{12}H_{16}O_{12}$	352.0642

Adapted with modification from with permission of the American Chemical Society.

TABLE 11.2   Characteristic neutral losses from common Phase II conjugates under collision-induced dissociation.

Conjugation reaction/ conjugation with	Mass shift	Characteristic neutral loss	Mass of neutral loss
Acetylation	42	Ketene (CH_2=C=O)	42
Glycine	57	Glycine	75
		CO + H_2O	46
Sulfatation	80	SO_3	80
Cysteine	119	Cysteine	121
		Alanine	89
		Formic acid	46
N-Acetylcysteine	161	N-Acetylcysteine	163
		N-Acetyl-2-iminopropionic acid	129
		Acetamide	59
		Ketene	42
Glucosidation	162	Anhydroglucose	162
		Glucose	180
Cysteine–glycine	176	Cysteine–glycine	178
		Alanine–glycine	146
		Glycine	75
Glucuronidation	176	Anhydroglucuronic acid	176
		Glucuronic acid	194
N-Acetylglucosamine	203	Anhydro-N-acetylglucosamine	203
		N-Acetylglucosamine	221
Glutathione	305	Glutathione	307
		Glutathione – 2H	305
		γ-Glu-Ala-Gly	275
		γ-Glu-Ala-Gly – 2H	273
		Glutamine	146
		Anhydroglutamic acid	129
		Glycine	75

mass loss from the molecular ion can be used to specifically detect various conjugated metabolites. For examples, glucuronides, sulfates, and glutathione conjugates are often detected through CNL of 176, 80, and 129 Da, respectively. Characteristic neutral losses of common Phase II conjugates are listed in Table 11.2 (Levsen et al., 2005). The PIS and CNLS experiments only record the *m/z* values of metabolite ions showing predefined modification. To obtain detailed structural information, full scan product ion spectra are required for each metabolite. Unique fragment ions are then used to determine structural differences between the parent molecule and related metabolites. However, should a metabolite undergo a modification that alters the mass of the fragment ion or neutral fragment used in the PIS or CNLS experiments, the metabolite will not be detected. This is typically not a major concern since as a compound advances toward late stage drug discovery, a more detailed metabolite characterization with radiolabeled drugs will likely detect any missed metabolite.

During late stage drug discovery, it will be important to characterize major metabolites to establish whether there are significant metabolic differences between species and to identify potential pharmacologically active, reactive, or toxic metabolites (Nassar and Talaat, 2004). At this time, the tritium (^{3}H)-labeled parent drug is often available; therefore, radiometric detectors can be coupled in-line with mass spectrometers to facilitate metabolite identification. The radiochromatograms assist in locating the metabolites in the chromatogram, as well as to provide a semiquantitative estimate for each metabolite. Structural characterization of all major metabolites (>10% of the total chromatographic radioactivity) is attempted (Cox, 2005). Using the parent drug and its corresponding fragmentation pattern as a reference, the site(s) of biotransformation can often be narrowed down to a certain region of the molecule based on a corresponding shift in m/z value of a characteristic product ion after comparing the product ion (MS/MS or MS^n) spectra of a metabolite with that from the parent drug.

Identification of reactive intermediates/metabolites is critical in drug discovery research (Evans et al., 2004; Ma and Subramanian, 2006). Reactive intermediates are short-lived; therefore chemical trapping agents (e.g., glutathione, methoxylamine, cyanide) are often used *in vitro* to form stable adducts with reactive intermediates that can be characterized by LC/MS/MS and LC/NMR (Baillie, 2006; Evans and Baillie, 2005). These experiments provide valuable indirect information on the identity of the reactive intermediates, thereby defining a potential bioactivation mechanism and hence a rationale on which to base a synthetic intervention strategy to minimize the formation of reactive intermediates. Detecting and characterizing reactive metabolites will be discussed in more details in the later section of this chapter (Section 11.8).

11.3.2 Metabolite Identification in Preclinical and Clinical Development

In the early stages of preclinical development, metabolite profiling is performed using *in vitro* systems from animal and human, mainly to identify potential species-dependent metabolism early in the development process and to support the selection of the animal species employed in safety assessment studies. As the compound moves further into development, *in vivo* animal ADME studies are performed. The compound is usually dosed into a rodent and nonrodent species along with an efficacy model. Metabolism data from the animal studies are then used in species selection for safety assessment to insure that all expected human metabolic transformations will be represented in the animal models used in the safety study.

As the drug enters Phase I clinical trials the drug metabolism efforts focus on identifying both circulating and excreted metabolites in humans. These results are then compared to those observed in the animal species used for safety assessment (Ma et al., 2006; Nassar and Talaat, 2004). All metabolism studies in clinical drug development are conducted with great austerity to be compliant with all regulatory guidelines required for marketing registration. Mass balance studies are designed to collect most of the excreted drug-derived material where the characterization of metabolites is more complete. Typically, a single LC/MS method is used for all matrices and species so that metabolites observed in humans and animals can be directly compared using their LC elution time, single MS spectra, and MS/MS results (Ramanathan et al., 2005).

To assess human risk, qualitative and quantitative differences in metabolite profiles are examined to establish exposure in a preclinical species relative to humans. Attempts should be made as early as possible in the drug development process to identify the differences in drug metabolism between animals used in nonclinical safety assessments and humans (US Food and Drug Administration, 2005). It is important to identify unique metabolites only produced in humans or formed to a much greater extent in humans compared to that in animals as early as possible to allow for timely assessment of potential safety issues (Davis-Bruno and Atrakchi, 2006; Humphreys and Unger, 2006; Smith and Obach, 2006). Therefore, the primary objectives of metabolite profiling and identification in drug development are to compare metabolism across species, to fully characterize major metabolites, and to search for human-specific metabolites or any metabolite(s) formed to a much greater extent in humans compared to those in preclinical species used in safety testing.

The most widely used approach for metabolite profiling and identification in drug development involves administration of radiolabeled drug (i.e., ^{14}C and ^{3}H). Following collection of excreta and blood, analysis of these matrices is typically performed using LC coupled to both a mass spectrometer (MS) and a flow scintillation analyzer (FSA). An LC/MS/FSA method allows for simultaneous acquisition of mass spectra (and/or MS/MS spectra) and the radiochromatographic response for each metabolite. The former provides the identification of metabolites and the latter the relative amounts of each. Thus, circulating levels of each detected metabolite and/or unchanged drug can be estimated from the plasma metabolite profile. In addition, percentage of administered dose for each drug-derived entity excreted in urine and feces can be calculated from total excretion data and the radiochromatographic response for each metabolite in urine and feces. Recently, Ramanathan et al. reported the characterization of more than 50 metabolites of loratadine in male and female mice, rats, and monkeys using the approach described above and provided a very comprehensive interspecies comparison (Ramanathan et al., 2005). This robust approach to metabolite characterization across species not only allows for identification and semiquantification of individual metabolites, but also establishes whether all human metabolites have been adequately tested in nonclinical safety assessments.

11.4 TECHNIQUES FOR IMPROVING METABOLITE DETECTION AND IDENTIFICATION

11.4.1 Chemical Derivatization

Chemical derivatization used in metabolite identification is a process of adding ionizable groups onto molecules that are difficult to ionize to facilitate their detection by mass spectrometry (Liu and Hop, 2005). In the pre-API era, chemical derivatization was aimed at improving volatility of compounds, which were difficult to transfer to the gas-phase for analysis. Since the development of API techniques, derivatization is not very frequently required. Nonetheless, chemical derivatization of very polar, small metabolites is sometimes advantageous. It reduces their polarity and increases their molecular weight thus making them more amenable to LC separation while minimizing potential mass spectral interferences from low molecular weight endogenous material present in the sample (Dalvie and O'Donnell, 1998).

Chemical derivatization can also be very useful for metabolite structural confirmation. Two oxidative metabolites, M4 and M7, were detected in the incubation of pioglitazone with dog liver microsomes. LC/MS/MS analysis indicated the introduction of a hydroxyl group on the 2-(5-ethyl-2-pyridyl) ethyl moiety (Shen et al., 2003). In order to assign the exact site of hydroxylation, chemical oxidation by Jones reagent was applied to distinguish the terminal hydroxyl (M7) from ω-1 hydroxyl (M4, Scheme 11.1). Oxidation of M4 would give rise to a ketone derivative that is two mass units lower than M4. On the contrary, M7 would give rise to a carboxylic acid derivative following oxidation, which is 14 mass units higher than M7 itself. This approach was applied to unequivocally distinguish the two isomeric metabolites (Liu and Hop, 2005).

Since N-oxide metabolites have the same elemental composition as those metabolites resulting from hydroxylation, differentiation by mass spectrometry

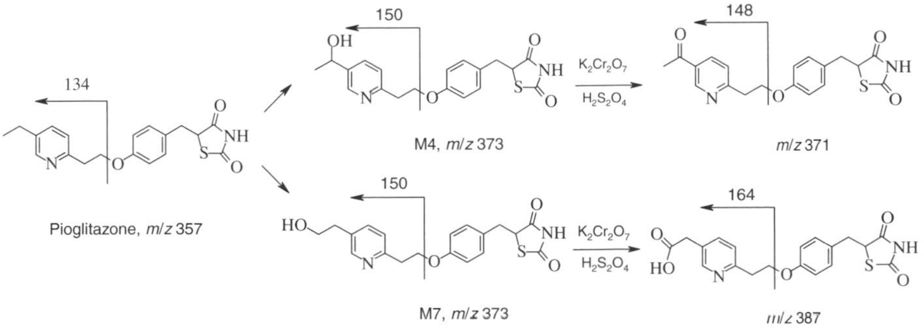

SCHEME 11.1 Oxidation of M4 and M7 of pioglitazone using Jones reagent for structural characterization.

(also see later in Section 11.7.2) alone is a challenging task because these analytes exhibit the same m/z for their protonated or deprotonated molecular ions, and their product ion mass spectra are usually very similar. Chemical reduction of N-oxides by $TiCl_3$ provides an efficient and selective way to differentiate N-oxides from hydroxylated metabolites (Kulanthaivel et al., 2004; Seaton et al., 1984). The reduction can be carried out in the presence of biological matrices, sulfoxides, and other labile groups (Kulanthaivel et al., 2004).

11.4.2 Stable Isotope Labeling

Stable-isotope-labeled (^2H, ^{13}C, ^{15}N, ^{18}O, ^{34}S, etc.) xenobiotics can facilitate metabolite detection and identification by mass spectrometry, especially when radiolabeled parent drug is not available. Custom-designed isotopic clusters resulting from the mixture of natural and synthetically enriched isotopes can greatly facilitate the detection and identification of metabolites (Mutlib and Shockcor, 2003; Wienkers et al., 1995, 1996).

Chowdhury et al. utilized stable isotope-labeled drug to aid in the detection and identification of ribavirin metabolites in rats (Chowdhury et al., 2005). A mixture containing equal amounts of ribavirin ($RTCONH_2$) and $^{13}C_3$-$RTCONH_2$ and a relatively small mass of ^{14}C-$RTCONH_2$ was dosed to male Sprague-Dawley rats (60 mg/kg) to facilitate metabolite identification. Two major metabolites, RTCOOH and $TCONH_2$ (Scheme 11.2), were detected in the pooled urine. The detection of a doublet peak with an m/z difference of 3 Th at 246 and 249 Th in the mass spectrum suggested the presence of RTCOOH (M1). Its identity was further confirmed by MS/MS of 246 and 249 Th. The respective major product ions at 114 and 116 Th were consistent with the unlabeled and $^{13}C_2$-labeled carbons of carboxylic acid and triazole

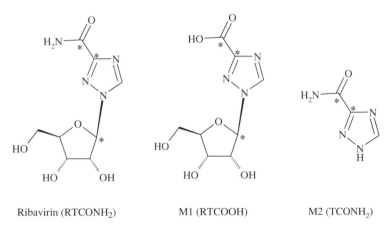

Ribavirin ($RTCONH_2$)   M1 (RTCOOH)   M2 ($TCONH_2$)

SCHEME 11.2 Chemical structures of ribavirin, metabolite M1, and metabolite M2
*denotes the position of ^{13}C label.

ring in RTCOOH. Similarly, the detection of doublet peaks with an m/z difference of 2 Th at 113 and 115 Th in the mass spectrum suggested the presence of TCONH$_2$ (M2). In the absence of a diagnostic isotopic pattern, the presence of RTCOOH (M1) could not be confirmed and its MH$^+$ ion could not be distinguished from interfering ions in the mass spectrum. This customized stable isotope-labeling technique provided unambiguous identification and characterization of ribavirin metabolites by the detection of diagnostic doublet ions in the LC/MS spectra.

Stable isotope labeling is usually not necessary when the compound contains Cl or Br, each of which has a unique natural isotopic abundance that can be easily recognized in the mass spectra. Assuming that these halogens remain intact during biotransformation, their unique isotope pattern can be used to search for metabolites in a complex biological matrix. Several software packages have been developed to analyze the MS data based on the isotopic pattern recognition, which significantly improve throughput in metabolite identification.

11.4.3   Hydrogen/deuterium (H/D) Exchange MS

The exchange of labile hydrogen with deuterium in small organic molecules has been widely used in structural characterization by mass spectrometry. This technique now finds widespread use because it can be utilized under LC/MS conditions. Biotransformation of xenobiotics usually involves introduction of polar functional groups (such as −OH, −SH, −N(R)H, −NH$_2$, −COOH) and leads to changes in the number of exchangeable hydrogens. Determination of the number of exchangeable hydrogens in metabolites can provide additional information to facilitate structural elucidation (Olsen et al., 2000). This approach was applied to differentiate sulfoxide and sulfone metabolites from the isomeric mono and dihydroxylated metabolites, respectively (Liu et al., 2001). Ohashi et al. used the H/D exchange method for drug metabolism studies of denopamine and promethazine, in which they were able to successfully differentiate N- or S-oxide from the hydroxylated metabolites (Ohashi et al., 1998). Miao et al. demonstrated the use of H/D exchange LC/MS and LC/MS/MS in characterization of a novel metabolite of ziprasidone (Miao et al., 2005). A major metabolite, designated as M11, was detected from the incubation of ziprasidone and hepatic cytosolic fractions of rat, dog, and human liver. Full scan MS of M11 revealed a protonated molecular ion at m/z 415; 2 Th higher than the parent drug. MS/MS analysis suggested that the addition of two mass units had occurred in the benzisothiazole moiety (Scheme 11.3). H/D exchange LC/MS analysis of M11 showed a shift for the protonated molecular ion from m/z 415 to 419, which suggested that M11 contains three exchangeable hydrogen atoms. Comparison of the product ion mass spectra of M11 at m/z 415 to that of m/z 419 in the D$_2$O mobile phase (Fig. 11.3) indicated that the addition of two exchangeable hydrogen atoms had occurred at the benzisothiazole moiety, potentially due to reductive cleavage

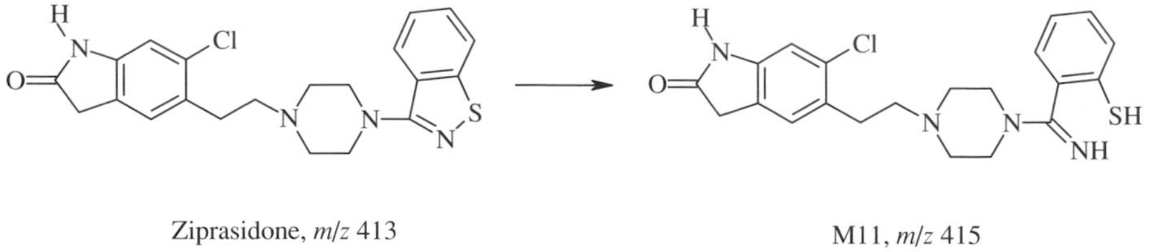

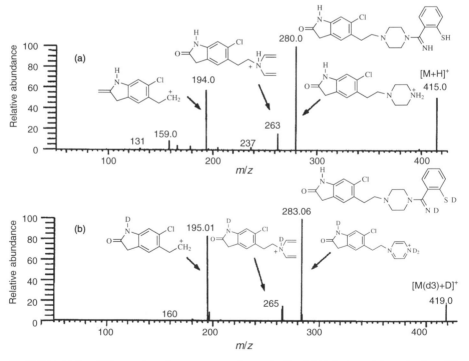

Ziprasidone, *m/z* 413   M11, *m/z* 415

SCHEME 11.3 Structures of ziprasidone and its major metabolite M11 in hepatic cytosolic fractions.

of N–S bond. Based on these data, the major metabolite M11 was proposed as dihydroziprasidone (Scheme 11.3).

11.4.4 Accurate Mass Measurement

Accurate mass measurements are often used for the identification or confirmation of the molecular formulas of drug metabolites. The analytical utility of high resolution mass spectrometry began with the introduction of double-focusing magnetic sector instruments, which could routinely achieve

FIGURE 11.3 CID product ion spectra of metabolite M11 before (**a**) and after (**b**) H/D exchange. Reprinted from Miao et al. (2005) with permission of the American Society for Pharmacology and Experimental Therapeutics.

mass resolution of 10^4–10^5 with a mass accuracy better than 2 ppm. During the past decade, the application of the sector instruments has declined owing to the rapid advancement of TOF and Fourier transform (FT) mass spectrometry, which are capable of high resolution but more importantly, are more compatible with atmospheric pressure ionization sources.

Liquid chromatography coupled with hybrid quadrupole time-of-flight mass spectrometers has emerged as a powerful tool in accurate mass measurement for metabolite identification, owing to its ease of use, fast acquisition speed, high sensitivity, and high mass accuracy in both TOF–MS and MS/MS mode (Chernushevich et al., 2001). Eckers and co-workers reported a mass accuracy of within a few mTh of the theoretical value for protonated and deprotonated molecule ions as well as product ions using a Q-TOF instrument (Eckers et al., 2000; Wolff et al., 2001). The hybrid LTQ–FTICR and LTQ-FT-Orbitrap mass spectrometers integrate ease of operation, standard HPLC flow rate compatibility and automatically triggered data-dependent MS^n experiments with high mass resolving power and mass accuracy. These high resolution instruments have demonstrated great value for rapid confirmation of expected metabolites or structural elucidation of unusual metabolites (Gratz et al., 2006; Peterman et al., 2006; Sanders et al., 2006).

11.4.5 Nanospray Ionization (NSI) MS for Metabolite Identification

Nanospray ionization is a variation of regular ESI in that the typical flow rate is reduced to between 30 and 200 nL/min. Because of the very low flow rate at which samples are consumed, this technique is rapidly being integrated in many analytical applications including proteomics, metabolite characterization, and pharmaceutical analysis. An approach of combining fraction collection with automated chip-based NSI MS was recently introduced for metabolite identification (Hop, 2006; Staack et al., 2005). LC effluent was collected into a 96-well plate and the fractions of interest were infused using an automated chip-based nanospray system for structure elucidation.

Compared to conventional LC/MS analysis, the NSI technique has some distinct advantages. For example, a complete mass spectral analysis of a sample using online LC/MS usually requires multiple injections to acquire the necessary data, while NSI typically requires only a single infusion of selected chromatographic fractions. Due to the low flow rates involved, mass spectral acquisition on any given peak can be extended for prolonged periods. This allows many different types of mass spectrometric experiments to be performed including, constant neutral loss scan, precursor ion scan, product ion scan, MS^n, and even polarity switching. In addition, online LC/MS typically requires nanogram quantities of drug metabolites to acquire high quality spectra suitable for structural characterization, while NSI, due to the ability to signal average for a long period of time, consumes picogram quantities to generate the same high quality MS/MS spectra (Borts et al., 2004). NSI also exhibits more uniform ionization efficiencies between metabolites and parent drug. This results in mass spectral peak intensities that more accurately reflect relative amounts of material as compared to traditional online ESI–LC/MS. Hop et al. used a chip-based nanospray ionization source (NanoMate™) to compare the ionization efficiencies of compounds with different physical–chemical characteristics and found that the ionization efficiencies varied to a much lesser extent in NSI MS compared to that in LC/MS (Hop et al., 2005). A set of 25 compounds from six distinct structural series was examined using conventional LC/MS and the NanoMate. The LC/MS data (Fig. 11.4a) showed that the

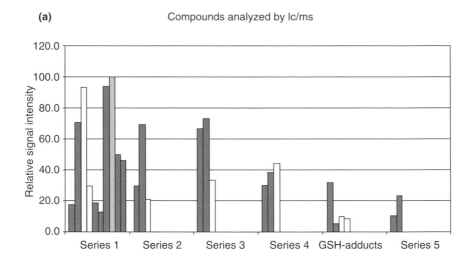

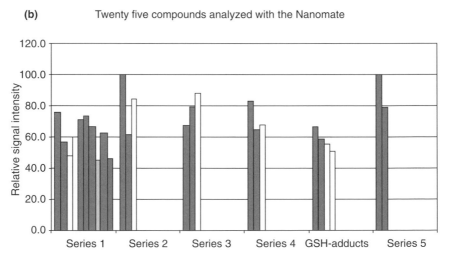

FIGURE 11.4 Relative ion intensities obtained for 25 compounds from 6 distinct structural series using conventional LC/MS (**a**) and NanoMate™ (**b**). Reprinted from Hop et al. (2005) with permission of John Wiley and Sons Limited.

responses varied by a 21-fold range, while in the NSI MS (Fig. 11.4b) the responses varied only by 2.2-fold across all compounds. Thus, the metabolite identification studies with NSI systems can potentially be semiquantitative in nature (Hop et al., 2005; Valaskovic et al., 2006).

11.5 SOFTWARE-ASSISTED METABOLITE IDENTIFICATION

11.5.1 Data-Dependent Acquisition (DDA)

Detection, characterization, and identification of metabolites require multiple mass spectrometry experiments such as full scan LC/MS and LC/MS/MS. The process becomes very time consuming, particularly when a large number of metabolites are formed. The interpretation of LC/MS/MS data is very laborious and inefficient, and can be a rate-limiting step in the metabolite identification process. Therefore, new approaches to data acquisition that would minimize the need for multiple experiments, and data processing tools that would simplify mass spectral interpretation, are highly desired.

2w? > Intelligent data-dependent acquisition (DDA) processes have been developed to maximize the qualitative mass spectral information that may be obtained from a single run. In general, a survey scan (full scan MS, precursor ion scan, or constant neutral loss scan) is acquired as each analyte elutes from the LC column in real-time. The data system then analyzes the mass spectra to determine if the precursor ion should be isolated for subsequent MS/MS experiments based on the predetermined selection criteria. If the precursor ion meets predetermine user-defined criteria, then it automatically switches to MS/MS mode for a predefined time. Once the product ion mass spectra are acquired, the system returns back to the survey scan and acquires data until the next MS/MS experiment is triggered. This approach has been widely used for the rapid identification of drug metabolites in complex biological matrices (Fernandez-Metzler et al., 1999; Gu and Lim, 2001; Lopez et al., 1998; Ramanathan et al., 2002; Tiller et al., 1998; Yu et al., 1999). DDA significantly reduces both the data acquisition and the data interpretation process and, therefore, dramatically increases the throughput of metabolite identification.

The data-dependent metabolite identification strategies can be configured to incorporate the masses of expected or predicted metabolites into the mass inclusion list. The use of an inclusion list increases the probability of acquiring product ion MS/MS spectra for low level drug-derived materials in complex biological matrices by forcing the trigger to occur even at low signal-to-noise ratios. Two types of software packages, database systems (MDLI metabolite database and Accelrys' biotransformation database) and expert systems (META, MetabolExpert, and METEOR) have been developed and are commercially available for the prediction of xenobiotic metabolism (Anari and Baillie, 2005; Langowski and Long, 2002). Anari and co-workers reported the

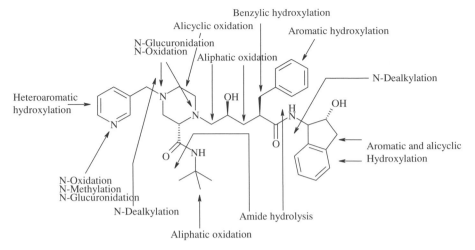

SCHEME 11.4   Knowledge-based metabolic prediction of indinavir. Adapted from Anari et al. (2004) with permission of the American Chemical Society.

integration of knowledge-based metabolic prediction with data-dependent LC/MS^n to study the biotransformation of indinavir (Anari et al., 2004). Potential metabolic pathways of indinavir were predicted from MDLI metabolite database, which included two hydrolytic, two N-dealkylation, three N-glucuronidation, one N-methylation, and several aromatic and aliphatic oxidation pathways (Scheme 11.4). The m/z values of potential metabolites resulting from a single or multiple common metabolic pathways were calculated. This m/z list was then used to set up an inclusion list-dependent LC/MS^n instrument method. This approach led to the identification of 18 metabolites of indinavir following incubation of the drug with human hepatic S9 fractions. The inclusion list-specific data-dependent analysis for potential metabolites was found to be effective in triggering MS^n data acquisition for the corresponding metabolites even though the signal intensities of these metabolites were very low.

11.5.2   Mass Defect Filter (MDF)

Mass defect is defined as the difference between the exact molecular weight and nominal molecular weight of an element. The atomic mass scale defines carbon-12 with a mass of exactly 12.0000 Da, therefore all other elements will have a uniquely different mass defect. For example, the mass defects of hydrogen and oxygen are 0.007825 and −0.005085 Da, respectively. Therefore, oxidation will introduce a mass defect of −5.1 mDa, while glucuronidation will introduce a mass defect of +32 mDa. The mass defects of common nonsynthetic (Phase I) and synthetic (Phase II) metabolites typically fall within 50 mDa. Therefore with LC/MS instruments capable of high mass accuracy, it is possible to filter out matrix-related interference ions whose mass defects lie outside of the window specified for drug-related ions. With a mass defect window set approximately ± 50 mDa from that of the parent drug, the metabolite profile of dog bile was obtained with the majority of interference ions removed (Zhang et al., 2003). The filtered ion chromatogram dramatically simplified data interpretation and therefore facilitated the identification of both common and uncommon metabolites. A similar approach was applied to reduce the false positive entries in analyzing metabolites in liver microsomal incubations (Mortishire-Smith et al., 2005).

Recently, Zhu and co-workers reported an improved MDF method employing both drug and core structure filter templates to the processing of high resolution LC/MS data for the detection and structural characterization of metabolites with mass defects similar to, or significantly different from, those of the parent drugs (Zhu et al., 2006). The effectiveness of MDF for detecting metabolites in complex biological samples was further demonstrated from the analysis of omeprazole metabolites in human plasma. The unprocessed total ion chromatogram showed no distinct metabolite peaks (Fig. 11.5a). After MDF processing, however, the metabolite peaks were easily

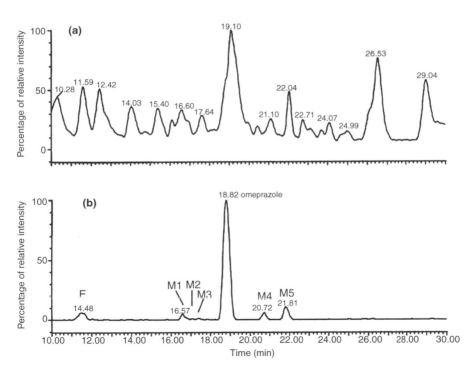

FIGURE 11.5 Chromatographic profiles of omeprazole samples analyzed by a Q-TOF LC/MS system and after examination by the MDF approach. (**a**) and (**b**) are TIC profiles of plasma spiked with omeprazole metabolites obtained without and with MDF processing, respectively. Reprinted from Zhu et al. (2006) with permission of the American Society for Pharmacology and Experimental Therapeutics.

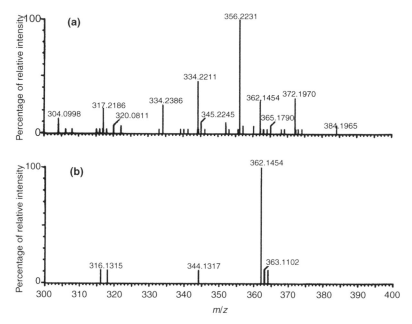

FIGURE 11.6   Mass spectra of omeprazole metabolite M3 in plasma obtained by Q-TOF LC/MS without (**a**) and with (**b**) MDF processing (chromatograms shown in Fig. 11.5). Reprinted from Zhu et al. (2006) with permission of the American Society for Pharmacology and Experimental Therapeutics.

identified in the ion chromatogram (Fig. 11.5b). In addition, mass spectra of the metabolite peaks were greatly simplified after MDF processing. The protonated molecule of M3 was embedded amongst many other predominant endogenous ions in the unprocessed spectrum (Fig. 11.6a). After applying MDF, the majority of the interference ions were removed and the protonated molecule of M3 became predominant in the mass spectrum (Fig. 11.6b).

Mass defect filter is solely dependent on the availability of accurate mass data. Therefore, it can detect unusual metabolites resulting from novel biotransformation pathways and is especially useful in metabolite profiling studies when radiolabeled drug is not available.

11.6   ADDITIONAL MS-RELATED TECHNIQUES FOR METABOLITE IDENTIFICATION

11.6.1   LC/NMR/MS

Although mass spectrometry (LC/MS and LC/MS/MS) is by far the most widely used technique for metabolite profiling and identification, it does not always provide sufficient information for unequivocal structural elucidation of metabolites. NMR spectroscopy is an excellent complementary tool for deriving detailed structural information once preliminary data on the nature of the metabolic change is available from LC/MS. Static NMR analysis is often performed on a metabolite at high purity and thus requires time-consuming isolation and purification steps to generate a suitable sample from *in vivo* or *in vitro* experiments. The direct coupling of LC with NMR eliminates the need for extensive sample purification steps and, therefore, increases its capability of solving structural problems in complex mixtures. The simultaneous coupling of LC with NMR and MS, enables the acquisition of both NMR and MS data from a single chromatographic analysis and provides comprehensive data for unambiguous structural determination (Dear et al., 2000; Sidelmann et al., 2001). The application of this technique in pharmaceutical research and development was recently reviewed in the literature (Lindon et al., 2000; Yang, 2006).

11.6.2 LC/ICPMS

In the absence of radiolabeled drug or reference standards, the detection and quantification of drug-related material in a biological sample is problematic. Liquid chromatography combined with inductively coupled plasma MS (ICPMS) could, in principle, offer an analytical solution (Marshall et al., 2004; Montes-Bayon et al., 2003). The temperature in the ICP source is approximately 5000 K, resulting in the production of singly charged elemental ions with high efficiency. LC/ICPMS provides a sensitive and highly selective platform for the detection and quantitation of drug metabolites containing atoms such as bromine, chlorine, sulfur, and selenium, which are not commonly found in endogenous material (Corcoran et al., 2000; Jensen et al., 2005; Nicholson et al., 2001). Unfortunately, ICPMS does not provide structural information and if one of the isotopes mentioned is not present in the drug, the technique has limited value for metabolite profiling.

11.7 CHARACTERIZATION OF UNSTABLE METABOLITES

Identification and quantification of unstable metabolites pose a serious challenge for mass spectrometric investigations. The unstable metabolites may undergo decomposition, rearrangement, and/or adduct formation during sample processing and mass spectrometric analysis. This may thereby lead to incorrect determination of metabolite quantification and structural characterization. In these cases, extra precautions are needed during sample preparation, while conducting the LC/MS experiment, and when interpreting LC/MS results.

11.7.1 Glucuronides

Glucuronides are often thermally unstable in the ion source of mass spectrometers and can break down to the aglycone via thermally induced in-source fragmentation. This may lead to an overestimation of the aglycone concentration if the chromatographic separation between the glucuronide and aglycone is not achieved (Yan et al., 2003). APCI and APPI typically result in in-source fragmentation of glucuronide to a much greater extent than ESI.

Acyl-glucuronide metabolites, formed by conjugation of glucuronic acid with a carboxylic acid moiety on a drug or metabolite, contain an ester group that is susceptible to both hydrolysis and intramolecular acyl migration (Stachulski et al., 2006). The unstable nature of acyl glucuronides makes careful sample handling crucial. The common practice is to immediately cool the sample on ice after collection and acidified to pH 2–4 to prevent hydrolysis and minimize acyl migration.

11.7.2 *N*-Oxides

Amine *N*-oxides are thermally labile species that undergo decompositions in many ways (Albini, 1993). Deoxygenation is often observed for *N*-oxides during atmospheric pressure chemical ionization, and occurs due to thermal energy activation in the vaporizer of the APCI source. Deoxygenation does not occur during the softer ESI process and thereby may represent a potential way to differentiate *N*-oxides from hydroxylated metabolites since the latter usually do not undergo thermal deoxygenation (Ramanathan et al., 2000).

In addition to deoxygenation, *tert N*-oxides containing an alkyl or benzyl group on the *N*-oxide nitrogen also undergo an N- to O- rearrangement (Meisenheimer arrangement), followed by elimination of an aldehyde (or a ketone) through an internal hydrogen transfer (Scheme 11.5). This has been observed under both APCI and APPI conditions (Ma et al., 2005). The elimination of an aldehyde or a ketone results from thermal energy activation at the vaporizer and is not induced by collisional activation. Significant in-source fragmentation was observed in the APCI and APPI mass spectra of clozapine *N*-oxide (Fig. 11.7). Notably, two major fragment ions at m/z 313 and 327, observed in the APCI and APPI mass spectra, were not detected in the ESI mass spectrum (Fig. 11.7). The fragment ion at m/z 327 ($MH^+ - O$) arose

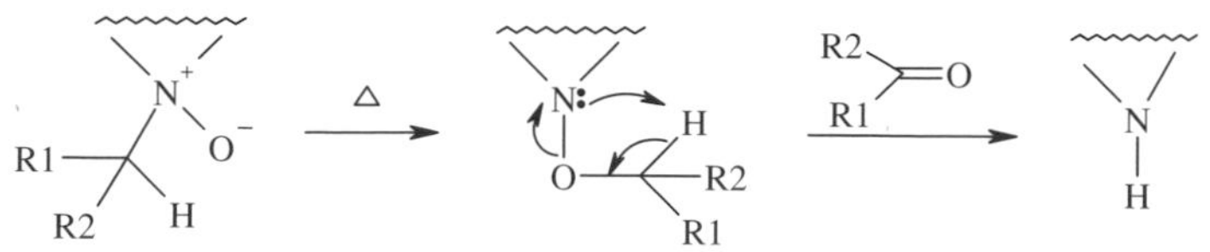

SCHEME 11.5 Proposed mechanism of aldehyde or ketone elimination following Meisenheimer rearrangement of *tert*-amine *N*-oxides.

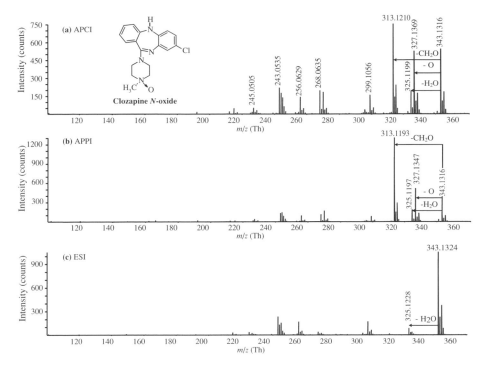

FIGURE 11.7 (a) APCI, (b) APPI, and (c) ESI mass spectra of clozapine *N*-oxide. Reprinted from Ma et al. (2005) with permission of the American Chemical Society.

from thermal energy activation in the vaporizer of the APCI and APPI sources. The most prominent fragment ion (m/z 313) resulted from the loss of formaldehyde. This fragment ion resulted from a thermally induced Meisenheimer rearrangement with migration of the *N*-methyl group to the corresponding *N*-alkoxylamine, followed by elimination of formaldehyde through an internal hydrogen transfer. This fragmentation would not, however, be expected from hydroxylated metabolites. Applying the Meisenheimer rearrangement reaction scheme, this fragment can be used to distinguish not only *N*-oxides from isomeric hydroxylated metabolites, but potentially the exact site of *N*-oxidation in a setting where multiple sites of differing chemical environment are present. The fragment ion at m/z 313 in APCI and APPI indicates that the position of *N*-oxygenation is at the nitrogen atom of the piperazine ring where the methyl group is attached, and not in the diazepine ring.

The tertiary amine *N*-oxides with a β-hydrogen in the alkyl substituted group may also undergo Cope elimination to give a hydroxylamine and an olefin. The *N*-oxide metabolite of SC-57461 was found to be unstable in organic

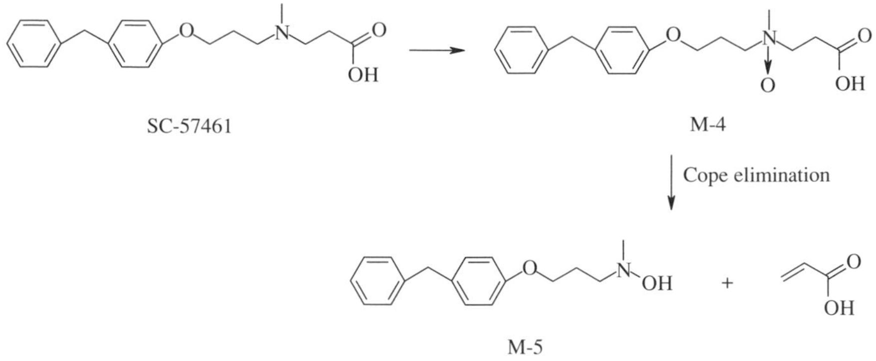

SC-57461   M-4

Cope elimination

M-5

SCHEME 11.6   Cope elimination of SC-57461-*N*-oxide.

solvents and rapidly underwent Cope elimination leading to the corresponding hydroxylamine metabolite M-5 and acrylic acid (Scheme 11.6) (Yuan et al., 1996).

11.7.3   Differentiation of Molecular Ions from In-Source Fragment Ions by the Presence of Alkali Adducts

Recently, Tong and co-workers reported the identification of unstable metabolites of Lonafarnib using ion source temperature alteration (Tong et al., 2006). Three major drug-derived components (A, B, and C) were characterized following *in vitro* incubation of Lonafarnib with recombinant human CYP3A4. The mass spectra for these three metabolites, obtained on an ion trap mass spectrometer, exhibited a prominent ion at m/z 635, which was 2 Th lower than that of the molecular ion for Lonafanib (m/z = 637). The LC/MS spectra obtained on a QSTAR™ mass spectrometer at relatively low TurboSpray™ probe temperature (250 and 350°C) provided evidence that the protonated molecular ion for metabolite A was at m/z 653 with confirmatory Na$^+$ and K$^+$ adduct ions observed at m/z 675 and 691, respectively. At higher probe temperatures, metabolite A showed no corresponding Na$^+$ and K$^+$ adducts for m/z 635 and therefore the ion at m/z 635 was most likely the in-source fragment ion formed due to facile elimination of H$_2$O. Similarly, the true molecular ion for metabolite B was determined to be at m/z 667, while the ion at m/z 635 was formed in the ion source owing to the elimination of a molecule of methanol. The true [M + H]$^+$ for metabolite C was confirmed as m/z 635 from the existence of the Na$^+$ and K$^+$ adducts at m/z 657 and 673, respectively. The detection of the alkali metal adduct ions at low ion source temperature facilitated the identification of the true molecular ion for the unstable metabolites. The definitive structures (Scheme 11.7) were further characterized from accurate mass measurement, stable isotope incorporation, and MS/MS fragmentation studies.

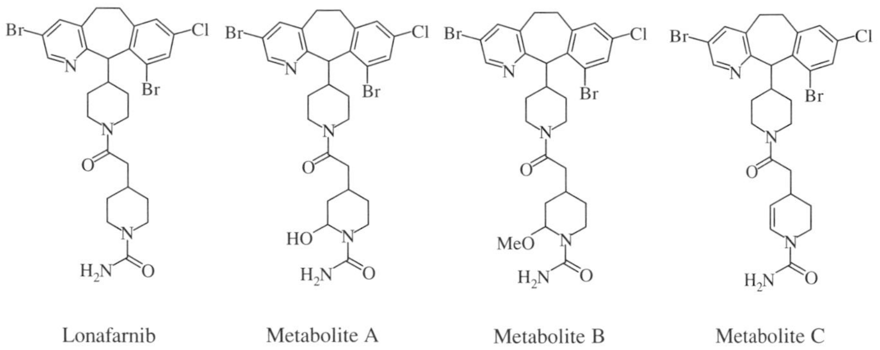

Lonafarnib   Metabolite A   Metabolite B   Metabolite C

SCHEME 11.7   Chemical structures of Lonafarnib, metabolite A, B, and C.

11.8   DETECTION AND CHARACTERIZATION OF REACTIVE METABOLITES AND INTERMEDIATES

Bioactivation is a long standing issue for drug development because of potential risks associated with drug–protein adduct and organ toxicity (Williams et al., 2002). There has been an increasing emphasis on screening drug candidates for their tendency to generate reactive metabolites and characterizing the nature of the reactive metabolites. Potential bioactivation mechanism(s) can then be defined and rationalized, which may provide a synthetic intervention strategy at an early stage to modify or screen out problematic compounds.

11.8.1   Trapping Reactive Metabolites

Most reactive metabolites are electrophiles that can react with nucleophiles (such as proteins). *In vitro* chemical trapping approaches are generally employed to examine the bioactivation potential of drug candidates. These experiments are often conducted in liver microsomes with NADPH and appropriate nucleophilic trapping agents, such as thiols (glutathione, its ethyl ester derivative, cysteine, or *N*-acetylcysteine), amines (semicarbazide and methoxylamine), or cyanide anion (Evans et al., 2004; Kalgutkar and Soglia, 2005). Glutathione (GSH) contains a free sulfhydryl group, a soft nucleophile capable of reacting with a broad range of reactive electrophiles, including Michael acceptors, epoxides, arene oxides, nitrenium ions, and alkyl halides. The use of the corresponding ethyl ester analogue of GSH has been shown to increase the MS sensitivity of the detection of reactive metabolites (Soglia et al., 2004). Semicarbazide and methoxylamine are hard nucleophiles, which will preferentially react with hard electrophiles such as aldehydes (Chauret et al., 1995; Zhang et al., 1996). The cyanide anion is a hard nucleophile that can be used to effectively trap iminium species (Argoti et al., 2005).

11.8.2 Screening for Glutathione Conjugates

Glutathione (GSH, γ-glutamylcysteinylglycine) is present virtually in all mammalian tissues and therefore serves as a natural scavenger for chemically reactive metabolites. A screen for the formation of glutathione conjugates could potentially identify a significant portion of reactive metabolites formed from a drug. Fragmentation of glutathione conjugates following collision-induced dissociation (CID) is illustrated in Scheme 11.8 (Haroldsen et al., 1988; Levsen et al., 2005; Murphy et al., 1992). Similar to CID of peptides, the fragmentations of GSH conjugates mainly result from the cleavage of the peptide backbone of the glutathione moiety. Even though the relative abundances of different types of fragment ions are sometimes dependent on the nature of the conjugated species, glutathione conjugates generally undergo a neutral loss of 129 Da (pyroglutamic acid) to produce e-type fragment ions (Scheme 11.8). Therefore, glutathione conjugates can be easily detected by a constant neutral loss (CNL) scan of 129 Da. One of the main disadvantages of this technique is its poor selectivity, as endogenous compounds presented in biological matrices may also give rise to a neutral loss of 129 Da. Therefore, false positives are routinely detected. To overcome this drawback, Yan and Caldwell developed a novel approach that demonstrated great selectivity and reliability for high-throughput screening for reactive metabolites using a stable-isotope-labeled trapping agent (Yan and Caldwell, 2004; Yan et al.,

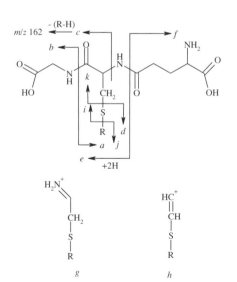

Fragment ion	m/z
a	$MH^+ - 75$
b	76
c	$MH^+ - 146$
$c - (R-H)$	162
$c - H_2O$	$MH^+ - 164$
d	$MH^+ - 273/275$
e	$MH^+ - 129$
$e - H_2O$	$MH^+ - 147$
f	130
g	$MH^+ - 232$
$g - H_2O$	$MH^+ - 250$
h	$MH^+ - 249$
i	308
j	$MH^+ - 305/307$
k	274

SCHEME 11.8   Characteristic fragment ions of glutathione conjugates under collision-induced dissociation. Adapted from Ma and Subramanian (2006) with permission of John Wiley and Sons Limited.

2005). An equimolar ratio of glutathione and $^{13}C_2/^{15}N$-labeled glutathione (stable-isotope label in glycine residue) was applied to trap reactive metabolites from microsomal incubations. Unambiguous identification of GSH-conjugates was realized by the presence of a unique doublet isotopic peak with m/z difference of 3 Th in the mass spectra. Additionally, in subsequent MS/MS experiments, the loss of 75 Da (glycine) and 129 Da (pyroglutamic acid) from GSH-adducts, and 78 Da and 129 Da from stable-isotope-labeled GSH-adducts, substantiated the presence of GSH-conjugates.

This approach was applied to characterize reactive metabolites of acetaminophen. Acetaminophen is well known to form a reactive metabolite, N-acetyl-p-benzoquinone imine (NAPQI, Scheme 11.9). When a 1:1 mixture of natural GSH and stable-isotope-enriched $^{13}C_2/^{15}N$-GSH was added to the human liver microsmal incubations, the reactive metabolites/intermediates were trapped by GSH. The sample was then analyzed using a CNL of 129 Da to detect the glutathione conjugates in an LC/MS/MS experiment. The total ion chromatogram included three major peaks and several other minor peaks; but only one neutral loss scan MS/MS spectrum showed doublet peaks at m/z 457 and 460 Th with approximately equal intensity. This unique ion cluster pattern suggested the formation of GSH-conjugates. In contrast, the CNL mass spectra of other peaks did not show these unique doublet peaks, indicating they were false positives. The formation of GSH-conjugates was further confirmed by full scan MS/MS spectra (Fig. 11.8). The product ion spectrum of m/z 457 showed major fragment ions at m/z 382 and 328, resulting from neutral losses of glycine (75 Da) and pyroglutamic acid (129 Da), while the product ion spectrum of m/z 460 showed fragment ions at m/z 382 and 331 resulting from losses of $^{13}C_2/^{15}N$-labeled glycine (78 Da) and pyroglutamic acid. This approach can be fully automated by employing isotopic pattern recognition and data-dependent acquisition methods, which would greatly

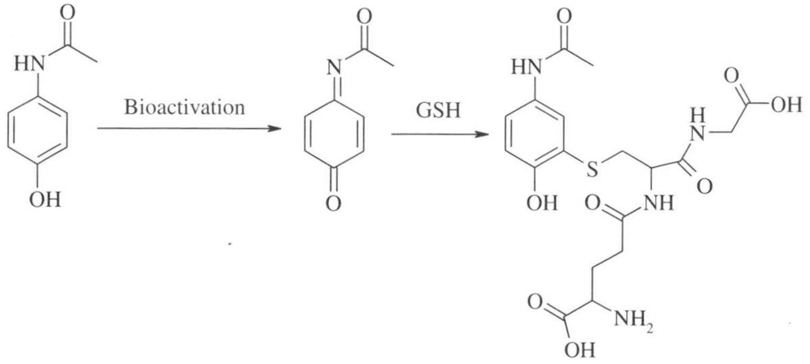

SCHEME 11.9 Bioactivation of acetaminophen to form a reactive N-acetyl-p-benzoquinone imine (NAPQI) and trapping of the electrophilic intermediate by GSH.

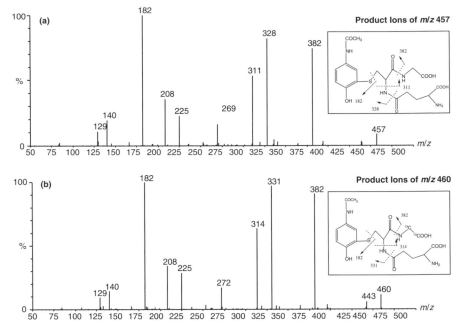

FIGURE 11.8 Collision induced product ion spectra of natural (**a**, m/z 457) and isotope-labeled GSH adduct (**b**, m/z 460). Reprinted from Yan et al. (2004) with permission of the American Chemical Society.

facilitate high throughput screening for reactive metabolites of molecules in the discovery stage programs.

Recently, Castro-Perez et al. (2005b) reported an LC/MS/MS method for screening glutathione conjugates using exact mass neutral loss of 129.0426 Da (corresponds to the exact mass of pyroglutamic acid) in a hybrid quadrupole time-of-flight mass spectrometer to eliminate false positives. The neutral losses of endogenous compounds in biological matrices may have the same nominal mass of 129 Da but less likely to have the same exact mass; therefore, exact mass measurements can play a pivotal role in excluding false positives. The instrument acquired survey mass spectra sequentially at low and high energy by switching the collision energy from 5 to 20 eV. The data system was set to examine these mass spectra in real-time to look for ions with a mass difference of 129.0426 Th within a narrow mass tolerance window in the high energy MS survey scan. Whenever an exact neutral loss was detected, the instrument was automatically switched to MS/MS mode to acquire a full scan MS/MS spectrum that was then used to elucidate the structures of detected GSH adducts. This approach allowed a selective detection and identification of GSH-adducts. In addition, a single injection provided accurate mass MS and MS/MS data on each detected GSH-conjugate.

Different classes of GSH-conjugates appear to behave differently upon collision-activated dissociation; not all afford a neutral loss of 129 Da as the primary fragmentation pathway. For example, aliphatic thioether conjugates may eliminate a glutathione molecule (307 Da) as a neutral fragment and/or yield protonated glutathione (m/z 308) as the product ion; thioester conjugates, on the other hand, typically fragment by a neutral loss of 147 Da (loss of pyroglutamic acid followed by a loss of H_2O) (Baillie and Davis, 1993; Dieckhaus et al., 2005). Hence, there is a need for a more broadly applicable MS/MS survey scan for the detection of GSH-conjugates of different structural classes. Dieckhaus et al. demonstrated that negative ion MS/MS showed promise in overcoming this limitation since MS/MS spectra of the deprotonated molecular ions $[M-H]^-$ of glutathione and major classes of GSH-conjugates afforded a common fragment anion at m/z 272 (deprotonated γ-glutamyl-dehydroalanyl-glycine), derived by loss of H_2S from the glutathionyl moiety. Therefore, precursor ion scan of m/z 272 in negative ion mode could provide a generally applicable technique for the detection of GSH-adducts (Dieckhaus et al., 2005). The utility of this approach was demonstrated in a variety of compounds that are known to form reactive metabolites. It should be noted that the MS/MS spectra of deprotonated molecular ions of GSH-conjugates are dominated by fragment ions from the tri-peptide moiety of glutathione, and few structurally informative ions from the xenobiotics. Therefore, an approach that could provide more information would be a combination of precursor ion scan of m/z 272 in negative ion mode as a survey scan for unambiguous detection of GSH-adducts, with polarity switching to positive ion mode to acquire full scan product ion spectra of MH^+ of the conjugates for structural elucidation (Dieckhaus et al., 2005).

11.9　CONCLUSIONS AND FUTURE DIRECTIONS

Metabolism studies play a pivotal role in drug discovery and development and LC/MS has become an indispensable tool to elucidate metabolite structures and metabolism pathways. Characterization of metabolic "hot spots" as well as reactive and pharmacologically active metabolites is critical to designing new drug candidates with improved metabolic stability, toxicological profile, and efficacy. Metabolite identification in the preclinical species used for safety evaluation is required in order to determine whether human metabolites have been adequately tested during nonclinical safety assessment. High performance liquid chromatography coupled with tandem mass spectrometry has emerged as a cardinal apparatus for the identification of metabolites in biological matrices. Additional techniques, such as chemical derivatization, H/D exchange, stable isotope labeling, accurate mass measurements, and software-assisted data acquisition and processing methods, have proved to be useful for improving metabolite detection and identification.

Emergence of new technologies will further accelerate drug metabolism studies in both drug discovery and development. In drug discovery settings, where rapid metabolite characterization is required to speed up lead optimization and candidate selection for development, software-assisted data acquisition and processing are extremely valuable and will continue to play a pivotal role. FTICRMS and FT-OrbitrapMS are rapidly gaining acceptance as tools of choice for drug metabolism studies, due to their ultrahigh mass accuracy, which make them highly attractive for identification of unknown metabolites. The ability of DESI–MS technique to desorb intact molecule ions directly from the surface is an exciting new feature and can be used to detect, identify, and quantitate drugs and metabolites directly from tissue slices without the need for extraction of drug-derived material.

In drug development, the availability of human metabolism data early in the program has significant merit from a business as well as regulatory perspective (Baillie et al., 2002). Understanding of metabolism in humans early in the clinical development program allows for defining the enzymology responsible for major *in vivo* human biotransformation pathway(s). Traditionally, these findings have been obtained from a single-dose metabolism study with radiolabeled drug that provides both mass balance (excretion of radioactivity in urine and feces) and metabolism information from plasma and excreta. Unfortunately, due to the significant cost and effort involved, radiolabeled studies are not conducted prior to establishing a therapeutic "proof of principle" in humans. Although the detection of metabolites from nonradiolabeled first-in-human (FIH) studies can be achieved with the current LC/MS technologies, it is far more difficult to obtain relative amounts of metabolites present in any matrix. Metabolism often results in structural changes that can dramatically change LC/MS response of a metabolite. Chip-based NSI systems (e.g., NanoMate™) may overcome this issue. Recent studies (Hop et al., 2005) indicated that the degree of variability associated with the ion intensities of a variety of compounds is much smaller with NSI compared to that with ESI, which makes metabolite identification studies with NSI potentially semiquantitative in nature.

There is also a demand for a universal detector that will allow quantification of drug metabolites without the need for radioisotopes or authentic reference standards. Online coupling of LC with inductively coupled plasma mass spectrometry (LC/ICPMS) offers the ability to quantify metabolites. ICPMS is an element-specific detector with almost uniform response independent of molecular structure. The response is only related to the molar content of the detected element (Axelsson et al., 2001). Unfortunately, the use of ICPMS for metabolite profiling and identification is limited to compounds containing specific elements such as P, I, F, Br, S, and selected metals. We predict that in not too distant future technologies will be available to reliably detect, identify, and quantitate "major" human metabolites routinely from first-in-human studies.

ACKNOWLEDGMENTS

The authors would like to thank Dr. Matt McLean for reviewing the manuscript and providing valuable suggestions.

REFERENCES

Albini A. Synthetic utility of amine *N*-oxides. Synthesis 1993;3:263–277.

Anari MR, Baillie TA. Bridging cheminformatic metabolite prediction and tandem mass spectrometry. Drug Discov Today 2005;10:711–717.

Anari MR, Sanchez RI, Bakhtiar R, Franklin RB, Baillie TA. Integration of knowledge-based metabolic predictions with liquid chromatography data-dependent tandem mass spectrometry for drug metabolism studies: application to studies on the biotransformation of indinavir. Anal Chem 2004;76:823–832.

Argoti D, Liang L, Conteh A, Chen L, Bershas D, Yu C-P.Vouros P, Yang E. Cyanide trapping of iminium reactive intermediates followed by detection and structure identification using liquid chromatography–tandem mass spectrometry (LC–MS/MS). Chem Res Toxicol 2005;18:1537–1544.

Axelsson BO, Jornten-Karlsson M, Michelsen P, Abou-Shakra F. The potential of inductively coupled plasma mass spectrometry detection for high-performance liquid chromatography combined with accurate mass measurement of organic pharmaceutical compounds. Rapid Commun Mass Spectrom 2001;15:375–385.

Baillie TA. Future of toxicology-metabolic activation and drug design: challenges and opportunities in chemical toxicology. Chem Res Toxicol 2006;19:889–893.

Baillie TA, Cayen MN, Fouda H, Gerson RJ, Green JD, Grossman SJ, Klunk LJ, LeBlanc B, Perkins DG, Shipley LA. Drug metabolites in safety testing. Toxicol Appl Pharmacol 2002;182:188–196.

Baillic TA, Davis MR. Mass spectrometry in the analysis of glutathione conjugates. Biol Mass Spectrom 1993;22:319–325.

Borts DJ, Cook ST, Bowers GD, O'Mara MJ, Quinn KE. Use of automated nanoelectrospray for drug metabolite structure elucidation.Proceedings of the 52nd ASMS Conference on Mass Spectrometry and Allied Tpoics;2004. TN: Nashville.

Brewer E, Henion J. Atmospheric pressure ionization LC/MS/MS techniques for drug disposition studies. J Pharm Sci 1998;87:395–402.

Castro-Perez J, Plumb R, Granger JH, Beattie I, Joncour K, Wright A. Increasing throughput and information content for *in vitro* drug metabolism experiments using ultra-performance liquid chromatography coupled to a quadrupole time-of-flight mass spectrometer. Rapid Commun Mass Spectrom 2005a;19:843–848.

Castro-Perez J, Plumb R, Liang L, Yang E. A high-throughput liquid chromatography/tandem mass spectrometry method for screening glutathione conjugates using exact mass neutral loss acquisition. Rapid Commun Mass Spectrom 2005b;19:798–804.

Chauret N, Nicoll-Griffith D, Friesen R, Li C, Trimble L, Dube D, Fortin R, Girard Y, Yergey J. Microsomal metabolism of the 5-lipoxygenase inhibitors L-746,530 and L-739,010 to reactive intermediates that covalently bind to protein: the role of the 6,8-dioxabicyclo[3.2.1]octanyl moiety. Drug Metab Dispos 1995;23: 1325–1334.

Chernushevich IV, Loboda AV, Thomson BA. An introduction to quadrupole-time-of-flight mass spectrometry. J Mass Spectrom 2001;36:849–865.

Chovan LE, Black-Schaefer C, Dandliker PJ, Lau YY. Automatic mass spectrometry method development for drug discovery: application in metabolic stability assays. Rapid Commun Mass Spectrom 2004;18:3105–3112.

Chowdhury SK, Blumenkrantz N, Zhong R, Kulmatycki K, Wirth M, McNamara P, Gopaul VS, Alton KB, Patrick JE. Detection and characterization of polar metabolites by LC-MS: proper selection of LC column and use of stable isotope-labeled drug to study metabolism of ribavirin in rats. In: Chowdhury S K, editor. Identification and Quantification of Drugs, Metabolites and Metabolizing Enzymes by LC-MS. Oxford: Elsevier; 2005. p 277–293.

Clarke NJ, Rindgen D, Korfmacher WA, Cox KA, Systematic LC/MS metabolite identification in drug discovery. Anal Chem 2001;73:430A–439A.

Cody RB, Laramee JA, Durst HD. Versatile new ion source for the analysis of materials in open air under ambient Conditions. Anal Chem 2005;77:2297–2302.

Cooks RG, Ouyang Z, Takats Z, Wiseman JM. Detection technologies. Ambient mass spectrometry. Science 2006;311:1566–1570.

Corcoran O, Nicholson JK, Lenz EM, Abou-Shakra F, Castro-Perez J, Sage AB, Wilson ID. Directly coupled liquid chromatography with inductively coupled plasma mass spectrometry and orthogonal acceleration time-of-flight mass spectrometry for the identification of drug metabolites in urine: application to diclofenac using chlorine and sulfur detection. Rapid Commun Mass Spectrom 2000;14: 2377–2384.

Cox KA. Special requirements for metabolite characterization. In: Korfmacher WA, editor. Using Mass Spectrometry for Drug Metabolism Studies. CRC Press; Boca Raton: 2005. p 229–252.

Dalvie DK, O'Donnell JP. Characterization of polar urinary metabolites by ionspray tandem mass spectrometry following dansylation. Rapid Commun Mass Spectrom 1998;12:419–422.

Dams R, Huestis MA, Lambert WE, Murphy CM. Matrix effect in bio-analysis of illicit drugs with LC-MS/MS: influence of ionization type, sample preparation, and biofluid. J Am Soc Mass Spectrom 2003;14:1290–1294.

Davis-Bruno KL, Atrakchi A. A regulatory perspective on issues and approaches in characterizing human metabolites. Chem Res Toxicol 2006;19:1561–1563.

Day SH, Mao A, White R, Schulz-Utermoehl T, Miller R, Beconi MG. A semi-automated method for measuring the potential for protein covalent binding in drug discovery. J Pharmacol Toxicol Methods 2005;52:278–285.

deHoffmann E. Tandem mass spectrometry: a primer. J Mass Spectrom 1996;31:129–215.

Dear GJ, Plumb RS, Sweatman BC, Ayrton J, Lindon JC, Nicholson JK, Ismail IM. Mass directed peak selection, an efficient method of drug metabolite identification using directly coupled liquid chromatography-mass spectrometry-nuclear magnetic resonance spectroscopy. J Chromatogr B 2000;748:281–293.

Dieckhaus CM, Fernandez-Metzler CL, King R, Krolikowski PH, Baillie TA. Negative ion tandem mass spectrometry for the detection of glutathione conjugates. Chem Res Toxicol 2005;18:630–638.

Dole M, Mach LL, Hines RL, Mobley RC, Ferguson LD, Alice MB. Molecular beams of macroions. J Chem Phys 1968;49:2240–2249.

Douglas DJ, Frank AJ, Mao D. Linear ion traps in mass spectrometry. Mass Spectrom Rev 2005;24:1–29.

Eckers C, Wolff JC, Haskins NJ, Sage AB, Giles K, Bateman R. Accurate mass liquid chromatography/mass spectrometry on orthogonal acceleration time-of-flight mass analyzers using switching between separate sample and reference sprays. 1. Proof of concept. Anal Chem 2000;72:3683–3688.

Evans DC, Baillie TA. Minimizing the potential for metabolic activation as an integral part of drug design. Curr Opin Drug Discov Devel 2005;8:44–50.

Evans DC, Watt AP, Nicoll-Griffith DA, Baillie TA. Drug-protein adducts: an industry perspective on minimizing the potential for drug bioactivation in drug discovery and development. Chem Res Toxicol 2004;17:3–16.

Fernandez-Metzler CL, Owens KG, Baillie TA, King RC. Rapid liquid chromatography with tandem mass spectrometry-based screening procedures for studies on the biotransformation of drug candidates. Drug Metab Dispos 1999;27:32–40.

Frank M. Achieving fastest analyses with the Agilent 1200 series rapid resolution LC system and 2.1-mm id columns. Agilent Technologies Application Note 5989-4502EN,2006.

Glish GL, Vachet RW. The basics of mass spectrometry in the twenty-first century. Nat Rev Drug Discov 2003;2:140–150.

Gratz SR, Gamble BM, Flurer RA. Accurate mass measurement using Fourier transform ion cyclotron resonance mass spectrometry for structure elucidation of designer drug analogs of tadalafil, vardenafil and sildenafil in herbal and pharmaceutical matrices. Rapid Commun Mass Spectrom 2006;20:2317–2327.

Gu M, Lim HK. An intelligent data acquisition system for simultaneous screening of microsomal stability and metabolite profiling by liquid chromatography/mass spectrometry. J Mass Spectrom 2001;36:1053–1061.

Guilhaus M, Selby D, Mlynski V. Orthogonal acceleration time-of-flight mass spectrometry. Mass Spectrom Rev 2003;19:65–107.

Hanold KA, Fischer SM, Cormia PH, Miller CE, Syage JA. Atmospheric pressure photoionization. 1. General properties for LC/MS. Anal Chem 2004;76:2842–2851.

Hardman M, Makarov AA. Interfacing the orbitrap mass analyzer to an electrospray ion source. Anal Chem. 2003;75:1699–1705.

Haroldsen PE, Reilly MH, Hughes H, Gaskell SJ, Porter CJ. Characterization of glutathione conjugates by fast atom bombardment/tandem mass spectrometry. Biomed Environ Mass Spectrom 1988;15:615–621.

Hop CE. Use of nano-electrospray for metabolite identification and quantitative absorption, distribution, metabolism and excretion studies. Curr Drug Metab 2006;7:557–563.

Hop CE, Chen Y, Yu LJ. Uniformity of ionization response of structurally diverse analytes using a chip-based nanoelectrospray ionization source. Rapid Commun Mass Spectrom 2005;19:3139–3142.

Hopfgartner G, Husser C, Zell M. Rapid screening and characterization of drug metabolites using a new quadrupole-linear ion trap mass spectrometer. J Mass Spectrom 2003;38:138–150.

Hopfgartner G, Varesio E, Tschappat V, Grivet C, Bourgogne E, Leuthold LA. Triple quadrupole linear ion trap mass spectrometer for the analysis of small molecules and macromolecules. J Mass Spectrom 2004;39:845–855.

Horning EC, Carroll DI, Dzidic I, Haegele KD, Horning MG, Stillwell RN. Atmospheric pressure ionization (API) mass spectrometry. Solvent-mediated ionization of samples introduced in solution and in a liquid chromatograph effluent stream. J Chromatogr Sci 1974;12:725–729.

Hsieh Y, Chintala M, Mei H, Agans J, Brisson JM, Ng K, Korfmacher WA. Quantitative screening and matrix effect studies of drug discovery compounds in monkey plasma using fast-gradient liquid chromatography/tandem mass spectrometry. Rapid Commun Mass Spectrom 2001;15:2481–2487.

Humphreys WG, Unger SE. Safety assessment of drug metabolites: characterization of chemically stable metabolites. Chem Res Toxicol 2006;19:1564–1569.

Jensen BP, Smith CJ, Bailey CJ, Rodgers C, Wilson ID, Nicholson JK. Investigation of the metabolic fate of 2-, 3- and 4-bromobenzoic acids in bile-duct-cannulated rats by inductively coupled plasma mass spectrometry and high-performance liquid chromatography/inductively coupled plasma mass spectrometry/electrospray mass spectrometry. Rapid Commun Mass Spectrom 2005;19:519–524.

Johnson KA, Plumb R. Investigating the human metabolism of acetaminophen using UPLC and exact mass oa-TOF MS. J Pharm Biomed Anal 2005;39:805–810.

Kalgutkar AS, Soglia JR. Minimising the potential for metabolic activation in drug discovery. Expert Opin Drug Metab Toxicol 2005;1:91–142.

Kebarle P. A brief overview of the present status of the mechanisms involved in electrospray mass spectrometry. J Mass Spectrom 2000;35:804–817.

Kebarle P, Ho Y. On the mechanism of electrospray mass spectrometry. In: Cole RB, editor. Electrospray Ionization Mass Spectrometry: Fundamentals, Instrumentation, and Applications. New York: Wiley-Interscience; 1997. p 1–63.

Kostiainen R, Kotiaho T, Kuuranne T, Auriola S. Liquid chromatography/atmospheric pressure ionization-mass spectrometry in drug metabolism studies. J Mass Spectrom 2003;38:357–372.

Kulanthaivel P, Barbuch RJ, Davidson RS, Yi P, Rener GA, Mattiuz EL, Hadden CE, Goodwin LA, Ehlhardt WJ. Selective reduction of N-oxides to amines: application to drug metabolism. Drug Metab Dispos 2004;32:966–972.

Langowski J, Long A. Computer systems for the prediction of xenobiotic metabolism. Adv Drug Deliv Rev 2002;54:407–415.

Levsen K, Schiebel HM, Behnke B, Dotzer R, Dreher W, Elend M, Thiele H. Structure elucidation of phase II metabolites by tandem mass spectrometry: an overview. J Chromatogr A 2005;1067:55–72.

Lindon JC, Nicholson JK, Wilson ID. Directly coupled HPLC-NMR and HPLC-NMR-MS in pharmaceutical research and development. J Chromatogr B 2000;748:233–258.

Liu DQ, Hop CE. Strategies for characterization of drug metabolites using liquid chromatography-tandem mass spectrometry in conjunction with chemical derivatization and on-line H/D exchange approaches. J Pharm Biomed Anal 2005;37:1–18.

Liu DQ, Hop CE, Beconi MG, Mao A, Chiu SH. Use of on-line hydrogen/deuterium exchange to facilitate metabolite identification. Rapid Commun Mass Spectrom 2001;15:1832–1839.

Lopez LL, Yu X, Cui D, Davis MR. Identification of drug metabolites in biological matrices by intelligent automated liquid chromatography/tandem mass spectrometry. Rapid Commun Mass Spectrom 1998;12:1756–1760.

Ma S, Chowdhury SK, Alton KB. Thermally induced N- to-O rearrangement of *tert-N*-oxides in atmospheric pressure chemical ionization and atmospheric pressure photoionization mass spectrometry: differentiation of N-oxidation from hydroxylation and potential determination of N-oxidation site. Anal Chem 2005;77: 3676–3682.

Ma S, Chowdhury SK, Alton KB. Application of mass spectrometry for metabolite identification. Curr Drug Metab 2006;7:503–523.

Ma S, Subramanian R. Detecting and characterizing reactive metabolites by liquid chromatography/tandem mass spectrometry. J Mass Spectrom 2006;41:1121–1139.

Makarov A. Electrostatic axially harmonic orbital trapping: a high-performance technique of mass analysis. Anal Chem 2000;72:1156–1162.

March RE. An introduction to quadrupole ion trap mass spectrometry. J Mass Spectrom 1997;32:351–369.

Marshall AG, Hendrickson CL, Jackson GS. Fourier transform ion cyclotron resonance mass spectrometry: a primer. Mass Spectrom Rev 1998;17:1–35.

Marshall AG, Hendrickson CL, Shi SD. Scaling MS plateaus with high-resolution FT-ICRMS. Anal Chem 2002;74:252A–259A.

Marshall PS, Leavens B, Heudi O, Ramirez-Molina C. Liquid chromatography coupled with inductively coupled plasma mass spectrometry in the pharmaceutical industry: selected examples. J Chromatogr A 2004;1056:3–12.

McLuckey SA, Wells JM. Mass analysis at the advent of the 21st century. Chem Rev 2001;101:571–606.

Miao Z, Kamel A, Prakash C. Characterization of a novel metabolite intermediate of ziprasidone in hepatic cytosolic fractions of rat, dog, and human by ESI-MS/MS, hydrogen/deuterium exchange, and chemical derivatization. Drug Metab Dispos 2005;33:879–883.

Montes-Bayon M, DeNicola K, Caruso JA. Liquid chromatography-inductively coupled plasma mass spectrometry. J Chromatogr A 2003;1000:457–476.

Mortishire-Smith RJ, O'Connor D, Castro-Perez JM, Kirby J. Accelerated throughput metabolic route screening in early drug discovery using high-resolution liquid chromatography/quadrupole time-of-flight mass spectrometry and automated data analysis. Rapid Commun Mass Spectrom 2005;19:2659–2670.

Murphy CM, Fenselau C, Gutierrez PL. Fragmentation characteristic of glutathione conjugates activated by high-energy collisions. J Am Soc Mass Spectrom 1992;3: 815–822.

Mutlib AE, Shockcor JP. Application of LC/MS, LC/NMR, NMR and stable isotopes in identifying and characterizing metabolites. In: Lee JS, Obach RS and Fischer MB editors. Drug Metabolizing Enzymes. Boca Raton: CRC Press; 2003. p 33–86.

Nassar AE, Talaat RE. Strategies for dealing with metabolite elucidation in drug discovery and development. Drug Discov Today 2004;9:317–327.

Nicholson JK, Lindon JC, Scarfe GB, Wilson ID, Abou-Shakra F, Sage AB, Castro-Perez J. High-performance liquid chromatography linked to inductively coupled plasma mass spectrometry and orthogonal acceleration time-of-flight mass spectrometry for the simultaneous detection and identification of metabolites of 2-bromo-4-trifluoromethyl. Anal Chem 2001;73:1491–1494.

Ohashi N, Furuuchi S, Yoshikawa M. Usefulness of the hydrogen–deuterium exchange method in the study of drug metabolism using liquid chromatography-tandem mass spectrometry. J Pharm Biomed Anal 1998;18:325–334.

Olsen MA, Cummings PG, Kennedy-Gabb S, Wagner BM, Nicol GR, Munson B. The use of deuterium oxide as a mobile phase for structural elucidation by HPLC/UV/ESI/MS. Anal Chem 2000;72:5070–5078.

Pereira L, Blythe C, Sherant R, Ritchie H. 2006. Use of small particles in ultra high pressure liquid chromatography. Available at http://www.thermo.com/eThermo/CMA/PDFs/Various/File_30599.pdf.

Peterman SM, Duczak N, Jr.Kalgutkar AS, Lame ME, Soglia JR. Application of a linear ion trap/orbitrap mass spectrometer in metabolite characterization studies: examination of the human liver microsomal metabolism of the non-tricyclic antidepressant nefazodone using data-dependent accurate mass measurements. J Am Soc Mass Spectrom 2006;17:363–375.

Plumb R, Castro-Perez J, Granger J, Beattie I, Joncour K, Wright A. Ultra-performance liquid chromatography coupled to quadrupole-orthogonal time-of-flight mass spectrometry. Rapid Commun Mass Spectrom 2004;18:2331–2337.

Ramanathan R, Alvarez N, Su AD, Chowdhury S, Alton K, Stauber K, Patrick J. Metabolism and excretion of loratadine in male and female mice, rats and monkeys. Xenobiotica 2005;35:155–189.

Ramanathan R, McKenzie DL, Tugnait M, Siebenaler K. Application of semi-automated metabolite identification software in the drug discovery process for rapid identification of metabolites and the cytochrome P450 enzymes responsible for their formation. J Pharm Biomed Anal 2002;28:945–951.

Ramanathan R, Su AD, Alvarez N, Blumenkrantz N, Chowdhury SK, Alton K, Patrick J. Liquid chromatography/mass spectrometry methods for distinguishing N-oxides from hydroxylated compounds. Anal Chem 2000;72:1352–1359.

Robb DB, Covey TR, Bruins AP. Atmospheric pressure photoionization: an ionization method for liquid chromatography-mass spectrometry. Anal Chem 2000;72:3653–3659.

Rossi DT. The Impact of atmospheric pressure ionization. In: Rossi DT, Sinz MW, editors. Mass Spectrometry in Drug Discovery. New York: Marcel Dekker; 2002. p 15–24.

Sanders M, Shipkova PA, Zhang H, Warrack BM. Utility of the hybrid LTQ-FTMS for drug metabolism applications. Curr Drug Metab 2006;7:547–555.

Schwartz JC, Wade AP, Enke CG, Cooks RG. Systematic delineation of scan modes in multidimensional mass spectrometry. Anal Chem 1990;62:1809–1818.

Seaton QF, Lawley CW, Akers HA. The reduction of aliphatic and aromatic N-oxides to the corresponding amines with titanium(III) chloride. Anal Biochem 1984;138:238–241.

Shen Z, Reed JR, Creighton M, Liu DQ, Tang YS, Hora DF, Feeney W, Szewczyk J, Bakhtiar R, Franklin RB, Vincent SH. Identification of novel metabolites of pioglitazone in rat and dog. Xenobiotica 2003;33:499–509.

Shiea J, Huang MZ, Hsu HJ, Lee CY, Yuan CH, Beech I, Sunner J. Electrospray-assisted laser desorption/ionization mass spectrometry for direct ambient analysis of solids. Rapid Commun Mass Spectrom 2005;19:3701–3704.

Sidelmann UG, Bjornsdottir I, Shockcor JP, Hansen SH, Lindon JC, Nicholson JK. Directly coupled HPLC-NMR and HPLC-MS approaches for the rapid character-isation of drug metabolites in urine: application to the human metabolism of naproxen. J Pharm Biomed Anal 2001;24:569–579.

Smith DA, Obach RS. Metabolites and safety: what are the concerns, and how should we address them? Chem Res Toxicol 2006;19:1570–1579.

Soglia JR, Harriman SP, Zhao S, Barberia J, Cole MJ, Boyd JG, Contillo LG. The development of a higher throughput reactive intermediate screening assay incorporating micro-bore liquid chromatography-micro-electrospray ionization-tandem mass spectrometry and glutathione ethyl ester as an *in vitro* conjugating agent. J Pharm Biomed Anal 2004;36:105–116.

Staack RF, Varesio E, Hopfgartner G. The combination of liquid chromatography/tandem mass spectrometry and chip-based infusion for improved screening and characterization of drug metabolites. Rapid Commun Mass Spectrom 2005;19:618–626.

Stachulski AV, Harding JR, Lindon JC, Maggs JL, Park BK, Wilson ID. Acyl glucuronides: biological activity, chemical reactivity, and chemical synthesis. J Med Chem 2006;49:6931–6945.

Takats Z, Cotte-Rodriguez I, Talaty N, Chen H, Cooks RG. Direct, trace level detection of explosive on ambient surfaces by desorption electrospray ionization mass spectrometry. Chem Commun 2005a;1950–1952

Takats Z, Wiseman JM, Cooks RG. Ambient mass spectrometry using desorption electrospray ionization (DESI): instrumentation, mechanisms and applications in forensics, chemistry, and biology. J Mass Spectrom 2005b;40:1261–1275.

Takats Z, Wiseman JM, Gologan B, Cooks RG. Mass spectrometry sampling under ambient conditions with desorption electrospray ionization. Science 2004;306:471–473.

Tiller PR, Land AP, Jardine I, Murphy DM, Sozio R, Ayrton A, Schaefer WH. Application of liquid chromatography-mass spectrometry(n) analyses to the characterization of novel glyburide metabolites formed in vitro. J Chromatogr A 1998;794:15–25.

Tong W, Chowdhury SK, Su AD, Feng W, Ghosal A, Alton KB. Identification of unstable metabolites of Lonafarnib using liquid chromatography-quadrupole time-of-flight mass spectrometry, stable isotope incorporation and ion source temperature alteration. J Mass Spectrom 2006;41:1430–1441.

Tozuka Z, Kaneko H, Shiraga T, Mitani Y, Beppu M, Terashita S, Kawamura A, Kagayama A. Strategy for structural elucidation of drugs and drug metabolites using (MS)n fragmentation in an electrospray ion trap. J Mass Spectrom 2003;38:793–808.

US Food and Drug Administration, Draft guidance for industry on safety testing of drug metabolites. Available at www.fda.gov/cder/guidance.

Valaskovic GA, Utley L, Lee MS, Wu JT. Ultra-low flow nanospray for the normalization of conventional liquid chromatography/mass spectrometry through equimolar response: standard-free quantitative estimation of metabolite levels in drug discovery. Rapid Commun Mass Spectrom 2006;20:1087–1096.

Voyksner RD. Combining liquid chromatography with electrospray mass spectrometry. In: Cole RB, editor. Electrospray Ionization Mass Spectrometry: Fundamentals, Instrumentation, and Applications. New York: Wiley-Interscience; 1997. p 323–341.

Wang G, Hsieh Y, Korfmacher WA. Comparison of atmospheric pressure chemical ionization, electrospray ionization, and atmospheric pressure photoionization for the determination of cyclosporin a in rat plasma. Anal Chem 2005;77:541–548.

Weickhardt C, Moritz F, Grotemeyer J. Time-of-flight mass spectrometry: state-of-the-art in chemical analysis and molecular science. Mass Spectrom Rev 1996;15:139–162.

Whitehouse CM, Dreyer RN, Yamashita M, Fenn JB. Electrospray interface for liquid chromatographs and mass spectrometers. Anal Chem 1985;57:675–679.

Wienkers LC, Steenwyk RC, Mizsak SA, Pearson PG. *In vitro* metabolism of tirilazad mesylate in male and female rats. Contribution of cytochrome P4502C11 and delta 4–5 alpha-reductase. Drug Metab Dispos 1995;23:383–392.

Wienkers LC, Steenwyk RC, Sanders PE, Pearson PG. Biotransformation of tirilazad in human: 1. Cytochrome P450 3A-mediated hydroxylation of tirilazad mesylate in human liver microsomes. J Pharmacol Exp Ther 1996;277:982–990.

Williams DP, Kitteringham NR, Naisbitt DJ, Pirmohamed M, Smith DA, Park BK. Are chemically reactive metabolites responsible for adverse reactions to drugs? Curr Drug Metab 2002;3:351–366.

Wilm M, Mann M. Analytical properties of the nanoelectrospray ion source. Anal Chem 1996;68:1–8.

Wolff JC, Eckers C, Sage AB, Giles K, Bateman R. Accurate mass liquid chromatography/mass spectrometry on quadrupole orthogonal acceleration time-of-flight mass analyzers using switching between separate sample and reference sprays. 2. Applications using the dual-electrospray ion source. Anal Chem 2001;73:2605–2612.

Xia YQ, Miller JD, Bakhtiar R, Franklin RB, Liu DQ. Use of a quadrupole linear ion trap mass spectrometer in metabolite identification and bioanalysis. Rapid Commun Mass Spectrom 2003;17:1137–1145.

Yan Z, Caldwell GW. Stable-isotope trapping and high-throughput screenings of reactive metabolites using the isotope MS signature. Anal Chem 2004;76:6835–6847.

Yan Z, Caldwell GW, Jones WJ, Masucci JA. Cone voltage induced in-source dissociation of glucuronides in electrospray and implications in biological analyses. Rapid Commun Mass Spectrom 2003;17:1433–1442.

Yan Z, Maher N, Torres R, Caldwell GW, Huebert N. Rapid detection and characterization of minor reactive metabolites using stable-isotope trapping in combination with tandem mass spectrometry. Rapid Commun Mass Spectrom 2005;19:3322–3330.

Yang C, Henion J. Atmospheric pressure photoionization liquid chromatographic-mass spectrometric determination of idoxifene and its metabolites in human plasma. J Chromatogr A 2002;970:155–165.

Yang Z. Online hyphenated liquid chromatography-nuclear magnetic resonance spectroscopy-mass spectrometry for drug metabolite and nature product analysis. J Pharm Biomed Anal 2006;40:516–527.

Yu X, Cui D, Davis MR. Identification of *in vitro* metabolites of Indinavir by "intelligent automated LC-MS/MS" (INTAMS) utilizing triple quadrupole tandem mass spectrometry. J Am Soc Mass Spectrom 1999;10:175–183.

Yuan JH, Birkmeier J, Yang DC, Hribar JD, Liu N, Bible R, Hajdu E, Rock M, Schoenhard G. Isolation and identification of metabolites of leukotriene A4 hydrolase inhibitor SC-57461 in rats. Drug Metab Dispos 1996;24:1124–1133.

Zhang H, Zhang D, Ray K. A software filter to remove interference ions from drug metabolites in accurate mass liquid chromatography/mass spectrometric analyses. J Mass Spectrom 2003;38:1110–1112.

Zhang KE, Naue JA, Arison B, Vyas KP. Microsomal metabolism of the 5-lipoxygenase inhibitor L-739,010: evidence for furan bioactivation. Chem Res Toxicol 1996;9:547–554.

Zhu M, Ma L, Zhang D, Ray K, Zhao W, Humphreys WG, Skiles G, Sanders M, Zhang H. Detection and characterization of metabolites in biological matrices using mass defect filtering of liquid chromatography/high resolution mass spectrometry data. Drug Metab Dispos 2006;34:1722–1733.

12

INTRODUCTION TO NMR AND ITS APPLICATION IN METABOLITE STRUCTURE DETERMINATION

Xiaohua Huang, Robert Powers, Adrienne Tymiak, Robert Espina, and Vikram Roongta

12.1 INTRODUCTION

The broad utility of nuclear magnetic resonance (NMR) was first recognized in 1951 from a series of experiments that observed a relationship between chemical structure and corresponding shifts in the NMR resonances. Since that time, NMR has evolved from continuous wave (CW) spectroscopy to Fourier transform (FT) spectroscopy, from permanent magnets to super conducting magnets and from one dimensional to multidimensional NMR spectra. The range of NMR applications has also grown significantly and includes (i) chemical structure elucidation (Breitmaier, 2002; Martin and Zektzer, 1988) (ii) three-dimensional conformational studies of biomolecules (Betz, 2006; Pellecchia, 2005), (iii) analysis of enzyme kinetics (Schutz et al., 2005), (iv) determination of reaction mechanisms (Lyčka et al., 2007; Moreno et al., 1996; Munro and Craik, 1994; Schaefer 1982), and (v) ligand binding screening for drug discovery (Hajduk, 2006; Lepre et al., 2004; Orry et al., 2006; Zartler et al. 2006). In the field of drug metabolism (Corcoran and Spraul, 2003), NMR is an extremely valuable tool in understanding biotransformation pathways and the potential involvement of metabolites in observed drug toxicities. Metabolite structure elucidation is particularly useful when the site of metabolism cannot be readily assigned by other techniques.

Drug Metabolism in Drug Design and Development, Edited by Donglu Zhang, Mingshe Zhu and W. Griffith Humphreys

369

In this chapter we provide a brief introduction to NMR theory and describe the most common experiments used for metabolite structure determination by NMR. Selected examples of metabolite structure elucidation are presented to illustrate relevant hardware, key parameters, and NMR methodology. Detailed explanations of NMR fundamentals have been previously reviewed in a series of books that are listed at the end of this chapter (Berger and Brunn, 2004; Breitmaier, 2002; Freeman, 1997; Friebolin, 2005; Homans, 2005; Keeler, 2005; Martin and Zektzer, 1988).

12.2 THEORY

An NMR signal arises from an interaction between the nuclear spin (I) of an atom with an external magnetic field (B_0). Elements with even atomic mass and number have a zero nuclear spin, where only nuclei with $I \neq 0$ produce an observable NMR signal. Nuclei with $I > \frac{1}{2}$ generate broad resonance lines in an NMR spectrum and are not generally useful for structure elucidation. Fortunately, most of the elements that comprise metabolites correspond to nuclei that are observable by NMR. They are listed in Table 12.1 along with their relevant NMR properties.

The intensity of the NMR signal depends on both the strength of the magnetic field (B_0) and the magnitude of the magnetogyric ratio (γ), an intrinsic physical property for each nucleus. Of the elements listed in Table 12.1, ^3H is the most sensitive nucleus but has limited application due to its low natural abundance and radioactivity. ^1H has the second highest sensitivity, has 100% natural abundance, is prevalent in the majority of organic and bioorganic molecules, and correspondingly, is the most frequently detected nucleus for structure characterization of organic molecules, ^{13}C, ^{15}N, ^{31}P, and ^{19}F NMR spectra are also commonly collected as part of a structure determination project.

TABLE 12.1 NMR Properties for common nuclei.

Nuclide	Spin I	Natural abundance (%)	Gyromagnetic ratio gamma (10^7 rad T^{-1} s^{-1})	NMR frequency (MHz) ($B_0 = 14.09$ T)
^3H	1/2	—	28.535	639.978
^1H	1/2	99.98	26.7519	600.000
^2H	1	0.016	4.1066	92.106
^{13}C	1/2	1.108	6.7283	150.864
^{15}N	1/2	0.37	−2.712	60.798
^{19}F	1/2	100	25.181	564.462
^{31}P	1/2	100	10.841	242.886
^{10}B	3	19.58	2.8746	64.476
^{11}B	3/2	80.42	8.5843	192.504

The intensity of the NMR signal is extremely small when compared to all other types of spectroscopy due to the small energy gap between nuclear spin states (α, β). For a given nucleus, the increase in sensitivity is directly proportional to the magnetic field produced by the magnet; therefore, greater sensitivity is achieved as the magnetic field strength is increased.

When a chemical sample is placed within this magnetic field, the nuclear spins either align with the external magnetic field (α) or align against it (β), The α spin state is at a lower energy and is more populated than the β spin state under equilibrium conditions. In a classical description of spinning particles in a magnetic field, a net magnetization created by an ensemble of spins is depicted as a vector along the Z-axis (Fig. 12.1). The NMR signal results from perturbing this equilibrium and inducing a spin transition from the α to β state by applying a radio frequency pulse (rf) perpendicular to the Z- axis. The frequency of the rf pulse must be proportional to the energy gap separating the α and β spin states. Since an rf pulse also contains a magnetic vector, the process can also be viewed as the net magnetization precessing about the new B_1 field created by the rf pulse (Fig. 12.2). The duration and direction of the rf pulse determines the orientation of the net magnetization vector when the rf pulse is terminated. A 90° pulse is typically applied and results in the net magnetization along the Z-axis being "flipped" into the X,Y plane.

After the rf pulse, the net magnetization will continue to precess in the X,Y plane about the external magnetic field (B_0). The system will also slowly relax back to the Z-axis equilibrium position. The frequency of the X,Y precession, also known as the Larmor frequency, is related to B_0, γ, and more importantly, to the local chemical environment of the nucleus. Small differences in the chemical environment of a nucleus will result in parts per million (ppm) differences in the Larmor frequency, which is known as the NMR Chemical

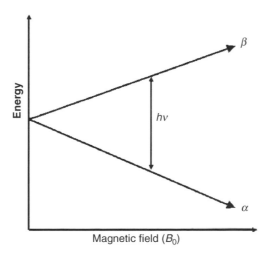

FIGURE 12.1 Relationship between the magnetic field strength and NMR sensitivity.

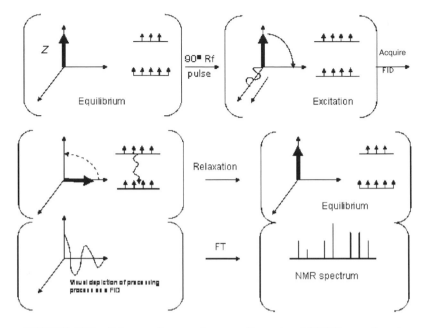

FIGURE 12.2   Scheme of one pulse, one-dimensional NMR experiment.

shift. This precessing process is captured as a free induction decay (FID) by monitoring the variation in an induced current in receiver coils positioned perpendicular to the X-,Y- axis. The FID is typically monitored for duration of milliseconds to seconds. This basic NMR process is called a one pulse, one-dimensional (1D) experiment.

A high power, short duration (hard) rf pulse will simultaneously perturb all the ^1H, ^{13}C, or ^{15}N nuclei in a molecule creating a FID that includes the NMR resonances of all these nuclei. The observed FID is a complex combination of sine and cosine wave oscillations at multiple frequencies in the time domain. The FID is transformed to a conventional frequency-domain NMR spectrum by applying a FT as depicted in Fig. 12.2.

12.3   NMR HARDWARE

The major components of an NMR spectrometer include a with a magnet, probe, rf electronics, and a desktop computer as depicted in Fig. 12.3 and further described in Table 12.2. Significant advancements in NMR hardware over the years have yielded dramatic increases in experimental sensitivity and resolution. One such advancement has been the recent introduction of the shielded 900 MHz magnet. A 900 MHz magnet results in an 84% improvement in sensitivity and a 150% improvement in resolution relative to a standard 600 MHz magnet (Kupce, 2001). The shielded aspect of the magnet reduces the

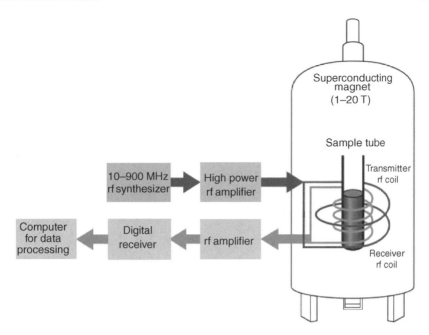

FIGURE 12.3 Scheme of a NMR system and its components.

5-gauss line by one-fourth, which needs to be shielded from personnel and other equipment, resulting in a significant reduction in laboratory space and cost. Another major advancement has been the introduction of cryoprobes (Bruker Biospin, 2002; Varian Inc, 2002), (Bieri et al., 2006; Keun et al., 2002) that has resulted in a three to five fold increase in sensitivity by reducing electronic noise by cooling the receiver electronics to near liquid helium temperature.

12.4 NMR OBSERVABLES

An NMR spectrum contains four important experimental observables, chemical shifts, coupling constants, peak intensity (integration), and peak width (relaxation time), that provide quantitative measurements of the structure, dynamics, and nuclei count of the chemical entity being studied. Chemical shifts and coupling constants provide direct evidence for identifying what functional groups are present in the structure and for assigning bond connectivity and spatial relationships between atoms (stereochemistry). Intensities of the NMR peak provide a relative measure of the quantity or count of the specific nuclei in the sample. The peak width is proportional to the T_2 relaxation time, which is related to dynamic processes such as chemical exchange or reaction kinetics. A second relaxation time (T_1) is related to the net magnetization relaxing back to its equilibrium position after an rf pulse. A summary of these basic NMR characteristics is tabulated in Table 12.3.

TABLE 12.2 Common NMR hardware and its new developments.

	Conventional design	New developments	Advantages
Magnet	Super conducting magnet.	Shielded and ultra shielded magnet. The extra outer magnet coil cancels the stray field from the inner coil of the magnet to reduce the 5-gauss line significantly	Field strength could go up to 900 MHz. High stability and homogeneity. Shielded magnets need much smaller space to site a magnet It is easy to couple the NMR spectrometer with mass spectrometer or HPLC systems.
Probe	5 mm or 3 mm CH dual probe, QNP probe, TXI probe, Broadband probe.	Cryocooled probe in which the probe rf coil and the preamplifier are close to liquid helium temperature	The sensitivity of ^1H and ^{13}C is three to four times higher than that of a conventional probe. Significant reduction in time for 2D NMR experiments1D ^1H can be obtained with as little as 10 ng of compound
		Capillarycoil probe that with a receiver coil that is 10–100 times smaller than that of the conventional probe.	High mass sensitivity Less sample and solvent consumption
		Flow (tubeless) probe that use direct injection (manual or automatic) to deliver sample to the flow cell.	Can be directly coupled with HPLC and MS spectrometer (NMR–HPLC–MS). Is used for 96- or 384-plate analysis in a high throughput mode.
Console and computer	It consists of robust components: frequency generator, radio frequency and gradient amplifiers, interface board, shimming board and signal generator and detector.	Digital-signal processor.Faster computer.Noise free and linear amplifiers. Gradient shimming.	Increases linearity of excitation over wider range of excitation Improvements in baselineReduced processing time for large data sets

TABLE 12.3 Properties of common NMR spectral characteristics.

Chemical shifts	Coupling constants	Integration	Relaxation times
Chemical shift of a nucleus depends on its electronic environment and the magnitude of the local magnetic field	Interaction of nonequivalent atoms through bonds, generally 1-, 2- and 3-bonds distances.	Integration of a ^1H spectrum provides quantitative information	Nuclei return to the equilibrium state after excitation by various mechanisms. Two most common ones are described below
$B_{eff} = B_0 - \sigma B_0$ B_0 = main field B_{eff} = reduced field $nu = \gamma/2\pi\,(1 - \sigma)B_0$	Coupling patterns or multiplicity: $M = 2nI + 1$ M = Multiplicity n = number of equivalent nuclei $I = \frac{1}{2}$ for most of the nuclei Common homo-nuclear coupling $^1H-^1H$ and $^{19}F-^{19}F$	In ^1H spectra, integral of each peak determines the number of hydrogens present in the structure The equivalent Hydrogens on a methyl, CH_3 residue integrate to three times the area compared to a methine CH residue	T_1 relaxation Spin–Spin relaxation Determines the interaction of spins that are close to each other T_1 relaxation time depends on the other spins in the molecule
Chemical shift scale is a parts per million (ppm) scale. The ^1H and ^{13}C scale is referenced to 4-trimethyl silane (TMS) at 0 ppm		External compound of known amount is used to quantitate the amount of compound in the solution	T_1 is extremely critical for optimizing all the NMR experiments

(*continued*)

TABLE 12.3 (*Continued*)

Chemical shifts	Coupling constants	Integration	Relaxation times
$> \delta = (v_{sample} - v_{ref})/v_{ref} \times 10^6$ For a spectrometer $\delta = \Delta v/$spectrometer frequency $\times 10^6$	Heter-nuclear coupling $^1H-^{13}C$, $^1H-^{15}N$, $^1H-^{19}F$, $^{13}C^{19}F$ Magnitude of coupling constants	Integration is often used to quantitate impurities or estimate the percentage of residual solvents Integral is not commonly measured for carbons	T_2 relaxation Spin–lattice relaxation T_2 relaxation times depend on the media of the molecule.
The scale goes from 0 ppm on left to higher ppm on the right ^1H range 0 ppm to 20 ppm ^{13}C range 0 ppm to 220 ppm	$^1H-^1H$, 1, 2, and 3 bonds ~1 Hz to 20 Hz $^1H-^{13}C$ bond ~140 Hz $^1H-^{15}N$ bond ~90 Hz		
The chemical shift value of a residue in a particular molecule does not change at different field strengths	The magnitude of a coupling-constant between two residues in a particular molecule does not change with field strength	Ratio of integral values between residues in a particular molecule does not change with field strength	Solvent, viscosity, temperature affects the line width. Short T_2 value results in broader peaks and Long T_2 results in narrower peaks

12.4.1 Chemical Shifts

The electronic environment of a nucleus is determined by the spatial arrangement of adjacent atoms and electrons. This local environment influences the effective magnetic field experienced by the nucleus. Typically, a nucleus is shielded, to a certain extent, from the external magnetic field (B_0) by its local environment and experiences only a fraction of the strength of B_0 and correspondingly exhibits a lower Larmor frequency. The result is a distinct chemical shift for each nucleus present in the molecule that exists in a unique local environment. The observed chemical shifts are characteristics of these environments and are very sensitive to subtle changes, where a ^1H in an aromatic ring differs from a ^1H attached to an electrophile (O, N, etc.). Among the NMR active nuclei, chemical shifts of the ^1H nuclei in a molecule are the most sensitive to the changes in solvent, pH, ionization state, temperature, and aggregation state of the molecule. Examples of ^1H and ^{13}C chemical shift ranges for different functional groups are listed in Table 12.4.

12.4.2 Coupling Constants

Chemically nonequivalent nuclei in a molecule can couple with each other through bonds, where the magnitude of this interaction (Hz) is smaller than a chemical shift (ppm) and is described by a coupling constant (J). The coupling of spins causes a mixing of the α and β spin states associated with each nucleus in the coupled system. The result is a multiplet ($n + 1$) peak pattern where the number of peaks for each coupled spin is determined by the number of nuclei (n) coupled to it. The separation between the peaks in the multiplet corresponds to J and the relative intensities of the peaks follows Pascal's triangle. The strength of this through bond coupling interaction decreases proportionally with the number of intervening bonds, and is generally observable for coupling between one, two, and three bonds, but has been observed for nuclei separated by as many as nine bonds. NMR coupling can be

TABLE 12.4 Typical ^1H and ^{13}C chemical shift ranges for organic functional groups.

Residues	^1H (ppm)	^{13}C (ppm)
R−CH$_3$, R−CH$_2$, RCH, R=C, S,	0.5–3.5	10–30
R−CH$_3$, R−CH$_2$, RCH, R=N	2.5–4.5	35–55
R−CH$_3$, R−CH$_2$, RCH, R=O	3.5–5.5	60–75
Vinyl HRC=CHR	5–7	100–150
Aromatic	6–9	110–160
Amide NH hydrogens	7–15	
Carbonyls carbons		155–220
Aldehydes	9–10	180–200

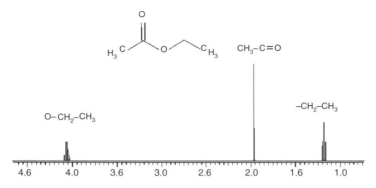

FIGURE 12.4   Simulated ^1H NMR spectrum of ethyl acetate.

illustrated with a simulated stick spectrum as demonstrated for ethyl acetate in Fig. 12.4.

The methyl signal (lowest chemical shift) is split into a triplet because of coupling to the two adjacent hydrogens of the methylene group, and the methylene is split into a quartet because of coupling to the three hydrogens of the adjacent methyl group. The methyl group next to the carbonyl group appears as a singlet since it is isolated from through bond interactions with the other hydrogens.

$^1H-^1H$ coupling constants are commonly used for structure elucidation to help identify which NMR resonances (functional groups) are chemically bonded as evidenced by observation of a shared coupling constant. Using the ethyl acetate example in Fig. 12.4, the observation that the peak separation between the methyl triplet and methylene quartet are equivalent (same J) indicates that these two functional groups must be chemical bonded in the ethyl acetate structure. The magnitude of the ^1H coupling constant also reflects the electronic environment, bond angle between the hydrogens involved, bond distance, and the hybridization state of the carbon to which the hydrogen atoms are attached. Some of the common $^1H-^1H$ coupling constants are listed in Table 12.5.

TABLE 12.5   Common $^1H-^1H$ coupling constants.

Alkane vicinal	Geminal	Double bonds
$(^1H)R2C-CR2(^1H)$	$RC(^1H)(^1H)$	Geometry
Gauche: $^3J \sim 5\,Hz$	$^2J \sim 0-30\,Hz$	$^3J_{trans} = 12-18\,Hz$
Trans: $^3J \sim 10\,Hz$		$^3J_{cis} = 6-12\,Hz$
Aromatic protons	Cyclo hexane type rings	Small rings cyclo propane, butane and pentane
$^3J_{ortho} = 7-8\,Hz$	a,a $\sim 8-14\,Hz$	$^3J_{cis\ or\ trans} = 4-5\,Hz$, $n=5$
$^4J_{meta} = 1-3\,Hz$	a,e $\sim 2-3\,Hz$	$^3J_{cis\ or\ trans} = 6-10\,Hz$, $n=4$
	e,e $\sim 2-3\,Hz$	$^3J_{cis\ or\ trans} = 3-5\,Hz$, $n=3$

Although, heteronuclear coupling constants can complicate the 1D ^{1}H NMR spectrum, they are not typically observed due to the relatively low natural abundance of ^{13}C (1.1%) and ^{15}N (0.37%). Also, ^{19}F and ^{31}P are not predominant in any given organic compound and do not usually interfere with the interpretation of ^{1}H or ^{13}C NMR spectra.

However, one and multiple bond heteronuclear coupling constants form the basis of heteronuclear two-dimensional (2D) experiments that are extremely useful for structure determination (Keeler, 2005). Specifically, a significant increase in resolution is achieved by dispersing a 1D NMR spectrum into 2D, especially as the complexity of the molecule increases. Secondly, accurately measuring coupling constants can be challenging in complicated 1D NMR spectra that contain overlapping peaks and complex coupling patterns. A typical 2D heteronuclear NMR experiment simply correlates ^{1}H and ^{13}C resonances that are coupled by the observation of a peak that has X,Y coordinates in the 2D spectra equal to the chemical shifts of the coupled resonances in the 1D NMR spectra. For example, a ^{1}H nucleus at 8.31 ppm that is bonded to a ^{13}C nucleus at 132.5 ppm would exhibit a peak in a 2D experiment with X,Y coordinates of 8.31 and 132.5 ppm.

12.4.3 Integration

NMR is a quantitative technique and the integral of each NMR resonance can provide an indication of the number of nuclei associated with each functional group within a compound. The ratios of integrals comparing different signals arising from the same molecule provide an indication of the number of nuclei contributing to each peak. An example of the use of integrals is shown in Fig. 12.5 and some characteristics of integration are summarized in Table 12.3.

If the compound contains impurities or residual solvents, integration of the impurity peak(s) will not yield an integer multiple of the other peaks arising from the major component. This provides a straightforward and easy protocol to identify the peaks associated with the compound of interest and eliminates

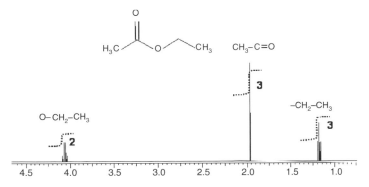

FIGURE 12.5 ^{1}H NMR spectrum with indication of integral for ethyl acetate.

spurious peaks from impurities in the structural analysis. It also provides an approach to quantitate the purity of the compound and the percentage of each impurity present by relative peak integration. A similar approach can be applied to follow dynamic processes, such as the rate of degradation of a compound or chemical exchange between different conformers or isoforms (Sridharan et al., 2005).

12.5 SAMPLE REQUIREMENTS FOR NMR

Traditionally, NMR analysis is performed in high quality NMR tubes using pure compounds and pure deuterated solvents. While this requirement has not changed over the years, the minimum concentration of a compound needed for NMR spectroscopy has been significantly reduced due to the sensitivity enhancements described above. The requirements for metabolite structure determination are the same, but it is often a challenge to obtain sufficient quantities of pure metabolites to conduct traditional NMR measurements. The minimum amount of a metabolite needed for structure determination depends on various factors that include: compound purity, complexity of the structure, the type of biotransformation, 1D versus 2D NMR experiments, and the utility of NMR sample tubes compared to hyphenated NMR approaches such as LC-NMR. With newly emerging cryoprobe technology and probes with lower sample volume (in some cases as low as 30–60 μL), it has become possible to analyze the structure of a metabolite using 1D NMR with a minimum of 500 ng of pure compound. For any detailed follow-up characterization requiring 2D NMR data collection, the amount of pure material needed ranges from 1 μg to several mg, depending on the specifics of the NMR experiments.

In the case of hyphenated NMR methods (Klaus, 2003; Exarchou et al., 2005), compound purification is done online with the structure determination process. This increases throughput and efficiency for analyzing numerous samples prepared in a comparable manner. LC–NMR incorporates a flow-probe that connects the chromatography with the NMR data collection. Simply, as a peak of interest eludes from the HPLC column, the peak continues to travel through additional "plumbing" until it is properly positioned in the fixed sample chamber (30–120 μL) in the NMR probe. Generally, the HPLC process is halted (stop-flow) while the NMR data is collected. Alternatively, the NMR data can be collected continuously (on-flow) during the HPLC process, but this is more technically challenging, limits the NMR methodology to quick 1D experiments and requires larger compound concentrations. LC–NMR requires an HPLC method that can separate the compound of interest in high enough concentration, where the solvent system also needs to be compatible with NMR data collection and analysis. In addition to good separation, the quality of deuterated or protonated solvents used for HPLC, the stability of HPLC columns and prior knowledge of retention time and the molecular weight are all crucial for metabolite identification using LC–NMR.

12.6   MOST COMMONLY USED NMR EXPERIMENTS AND TECHNIQUES

Four general classes of NMR experiments are routinely used to analyze metabolites: (1) 1D NMR experiments; (2) 2D NMR experiments; (3) Solvent suppression methods; and (4) Hyphenated NMR experiments. The 1D and 2D NMR experiments are commonly used for metabolite structure determination. The various solvent suppression techniques (Gaggelli and Valensin, 1993; Hwang and Shaka, 1995; Smallcombe and Patt, 1995) are crucial for dilute metabolite samples where the solvent peak is the most intense peak in the NMR spectrum. These solvent suppression techniques can be incorporated as needed in both 1D and 2D NMR experiments. Since their introduction in the 1990s, hyphenated NMR methods have become common tools in the identification of metabolites. These methods include LC–NMR (Albert, 1995; Spraul et al., 1993, 1994), LC–NMR–MS (Mass Spectrometry) (Shockcor et al., 1996) and LC/SPE (solid phase extraction)/NMR (Alexander et al., 2006; Bieri et al., 2006; Xu et al., 2005; Wilson et al., 2006).

12.6.1   1D NMR Experiments

A simple 1D NMR spectrum can offer rich structural information for metabolites. For example, a ^1H NMR spectrum identifies the functional groups present in the metabolite structure from ^1H chemical shifts, determines the structural connectivity of these functional groups from coupling patterns and coupling constants, provides a relative atom count from peak integrals, and suggests the number and type of exchangeable hydrogens from broadened peak widths. The main disadvantage of 1D ^1H NMR spectroscopy is signal overlap due to the narrow dispersion of ^1H chemical shifts.

Correspondingly, a 1D ^{13}C NMR spectrum overcomes the ^1H NMR resolution problem and is complimentary to ^1H NMR. The typical chemical shift range of a ^{13}C NMR spectrum is $\sim$220 ppm compared to 15 ppm for ^1H NMR spectrum. Also, a number of carbon types (carbonyl, carboxylic acid, aromatic, etc.) are not observable in a ^1H NMR spectrum because of the lack of attached hydrogen(s). The major disadvantage of ^{13}C NMR spectroscopy is its extremely low sensitivity compared to ^1H NMR experiments. This occurs because of the low natural abundance of ^{13}C nuclei (1.1%) and the low magnetogyric ratio ($\gamma^1H/\gamma^{13}C$). As a result, a ^1H NMR spectrum is $\sim$64,000 times more sensitive than a ^{13}C NMR spectrum. Also, because of the low ^{13}C abundance, ^{13}C NMR spectra are generally collected in a "decoupled" mode, which removes the strong (and generally uniform) one bond $^1H-^{13}C$ coupling. This increases the sensitivity of a ^{13}C NMR spectrum and reduces its complexity removing peak splitting, but it also eliminates the important bond connectivity information that is valuable for metabolite structure determinations.

In general, both 1D ^1H and ^{13}C NMR spectra are collected to resolve a metabolites structure. 1D NMR spectra of other heteronuclei (^{15}N, ^{19}F, ^{31}P)

are only collected to verify their presence in the metabolite or to greatly simplify the interpretation of the NMR spectra since these 1D heteronuclear NMR spectra typically only contain a few peaks.

A 1D nuclear overhauser experiment (NOE) (Gaggelli and Valensin, 1993) is an important and valuable variation on the simple 1D NMR experiment that provides spatial relationship between each nucleus in the structure. Coupling constants observed in a 1D ^{1}H NMR spectra provide connectivity for directly bonded nuclei; whereas, a 1D NOE experiment identifies nuclei that are close in space (≤ 6 Å). Briefly, a 1D NOE experiment requires the addition of a second lowpowered rf pulse that selectively "saturates" a specific peak in the NMR spectrum. The saturated peak becomes a null in the spectrum and any other nuclei that are coupled through space via a dipole–dipole interaction to the saturated peak will experience a change in peak intensity. A 1D NOE experiment requires collecting two NMR spectra, with and without saturation, to monitor changes in peak intensity. A summary of common 1D NMR experiments and their applications are listed in Table 12.6.

12.6.2 2D NMR Experiments

A fundamental component of the interpretation of NMR data is deciphering the NMR assignments, which correlates an observable NMR resonance with a specific atom in the molecular structure of the metabolite. This process is illustrated using the structure and ^{1}H NMR spectrum of 1,3-dimethylnaphthalene as an example (Fig. 12.6). The two methyl groups have distinct ^{1}H NMR chemical shifts because of their unique local environments. The NMR assignment process results in attributing the NMR peak at 2.57 ppm to methyl (a) and NMR peak 2.39 ppm to methyl (b).

The 1D NMR experiment provides the basic information, chemical shifts, coupling constants, and peak integration required for assigning NMR spectra, 2D experiments have become common for complete structure determination of organic compounds, natural products, and metabolites. 2D NMR experiments are generally used to confirm assignments derived from 1D NMR experiments, to resolve spectral ambiguities and provide new assignments that were not apparent in the 1D NMR experiments because of peak overlap or complex coupling patterns, and to provide ^{1}H–^{1}H and ^{1}H–^{13}C correlations that help confirm the structure of a compound.

2D NMR experiments have two important advantages over 1D NMR experiments. First, 2D NMR experiments provide a significant increase in resolution from the added dimensionality, which helps in resolving overlapped resonances in 1D NMR spectra. Second, 2D NMR experiments contain additional information that directly correlates NMR resonances that are coupled either through bonds or through space. Generally, a 2D NMR experiment contains a diagonal that corresponds to the standard 1D NMR spectrum. Diagonal peaks are correlated by off-diagonal "crosspeaks" that arise from a coupling constant or an NOE. A major disadvantage of 2D NMR

TABLE 12.6 Common 1D NMR experiments and their characteristics.

1D NMR experiment	Application	Advantages	Limitations	Comments
1D ^{1}H	Verify or eliminate proposed structures by comparing the changes in chemical shifts, coupling constants, and integrations with that of the parent molecule.	Simple, fast, and highly sensitive. It can be observed on a sample as little as 0.5 µg.	Signal overlaps cannot be resolved. Can not differentiate HC−O from HC−N or HC−O from CH=CR	Should be used as the first experiment for any meta bolite ID work when compared to its parent molecule
1D ^{1}H with solvent suppression	Same as above.	Improve the sensitivity of the proton spectrum for which the solvent signals is extremely strong	In addition to the above some resonances may not be observed due to overlap with solvent resonances	Should be used when the solvent signal is strong and when the sample concentration is very low
1D NOE	It provides stereochemistry information Some times, it is used to verify connectivity of the molecule.	Extremely powerful for stereochemistry determination and connectivity verification	Low sensitivity, stable spectrometer required for long data collection, >20 µg of compound required	Use only where required or appropriate

(continued)

TABLE 12.6 *(Continued)*

1D NMR experiment	Application	Advantages	Limitations	Comments
1D ^{19}F	Verify the modification site is near or at the fluorine atom. Verify the relative ratio of total metabolites.	High sensitivity. It could provide additional structure information when the sample amount is limited for other NMR experiment. Can be used for mass balance issues	Limited applications	Use for compounds containing ^{19}F.
1D ^{13}C	Verify carbon skeleton and functional group modifications in the molecule.	Chemical shifts provide characteristic range for different functional groups. It is a powerful tool to distinguish C–N versus C–O and C–O versus C=C	Insensitive, >100 µg sample required for reasonable spectra	Rarely performed on isolated metabolites due to sample limitations

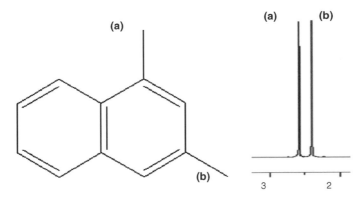

FIGURE 12.6   1,3-Dimethylnaphthalene structure with NMR assignments of methyl resonances.

experiments is the increase in experimental time (hours–days) compared to minutes for a typical 1D NMR spectrum. Also, 2D NMR experiments have a higher sample requirement (≥ 1 μg) because of the inherently lower sensitivity. Some common 2D NMR experiments that are routinely used for structure characterization of metabolites are described in Table 12.7.

12.6.3   Solvent Suppression Techniques

A significant concentration difference between the compound and the solvent (deuterated and protonated) may create a large dynamic range problem for an NMR experiment since the NMR signal intensity is proportional to concentration. Thus, dilute metabolite samples require NMR experiments that can suppress the relatively intense solvent peaks. This issue is further aggravated in cases of metabolite structure determination due to challenges in isolating critical metabolites from biological systems in sufficient quantities and purity for NMR analysis. In some cases, the dynamic range difference between the solvent and the metabolite is close to 1000 : 1 ratio. This large difference in signal intensity results in the saturation of the receiver, where only the solvent signal is observed. The signals from the metabolite are lost in the baseline because of the digital limitations of the receiver (Fig. 12.7).

To address the solvent dynamic range problem, one or multiple signals from the solvent are *selectively* suppressed in the NMR spectrum. Solvent suppression is not a perfect solution. Compound peaks that are proximal to the solvent are also completely or partially suppressed. Similarly, hydrogens that readily exchange with water are also equally suppressed with the water solvent peak. Additionally, solvent suppression causes artifacts and streaking in 2D NMR spectra. This streaking may obscure cross peaks that fall near the solvent chemical shift in either spectral dimension. The commonly used solvent suppression NMR techniques, such as, PRESAT, WET (Smallcombe and Patt,

TABLE 12.7 Common 2D NMR experiments and their applications.

2D NMR experiment	Application	Advantages	Limitations	Comments
COSY (correlated spectroscopy)	Provides 2 and 3 bond $^1H–^1H$ connectivity	Simple to run, resolves overlapped proton resonances. Minimum amount required 1–2 µg	Greater than three bond correlations cannot be observed	Any ambiguity in the 1D ^1H assignment should be followed by a COSY
TOCSY (total correlated spectroscopy)	Provides 2, 3, 4 and 5 bond $^1H–^1H$ correlations	Simple to run, provides long through-bond connectivites. Minimum amount required 2–5 µg	Sometimes, some long bond correlation may not be observed. High power may be required for the spin lock pulse	Confirms the COSY as signments or can be performed instead of a COSY. Used for identifying contiguous proton spin systems
NOESY (NOE spectroscopy)	Reveals through space interactions	Stereochemical questions can be answered for appropriate molecules	Not a simple experiment to perform. Data can be complicated due to exchange peaks Insensitive	Only used where appropriate
ROESY (rotating frame noe spectroscopy)	Same as NOESY, through-space interactions	Same as NOESY plus can distinguish between exchange and ROE peaks	Same as NOESY and also insensitive	Same as NOESY
HMQC (heteronuclear multiple quantum correlation) or HSQC (heternuclear single quantum correlation)	Provides 1 bond $^1H–^{13}C$ or $^1H–^{15}N$ correlation	Can identify all the protonated carbons and nitrogens in a molecule	Less sensitive experiment due to low sensitivity of ^{13}C and ^{15}N Compound requirement is high Minimum 20 µg or more	Can assist in identifying complete unknowns, unexpected products or metabolites
HMBC (heteronuclear multiple bond correlation)	Provides 2 to 3 bonds $^1H–^{13}C$ and $^1H–^{15}N$ correlation.	Crucial for identifying nonprotonated carbons and nitrogens	Most insensitive experiment. 100 µg and higher amount of material required	Useful for complete unknowns, unexpected products and metabolites

386

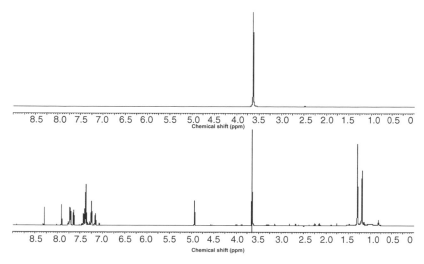

FIGURE 12.7   1D ^{1}H spectrum of a metabolite sample in DMSO-d_6, with and without solvent suppression. Regular single pulse sequence (top). WET sequence used to suppress residual H_2O and DMSO signals (bottom). The NMR spectra are plotted on the same scale.

1995) and excitation sculpting (Hwang, 1995) and their advantages and disadvantages are listed in Table 12.8.

12.6.4   Hyphenated NMR Methods

High performance liquid chromatography has played a major role in the separation and purification of compounds, especially in the pharmaceutical industry. With the introduction of NMR flowprobes in late 1990s (Albert, 1995), it became possible to link HPLC directly to a NMR flowprobe and introduce HPLC separated fractions directly into an NMR spectrometer for analysis. The LC–NMR combination provides a unique advantage for unstable metabolites by permitting real time monitoring of the purification and structure analysis process. Since the introduction of LC–NMR (Lindon et al., 1996), several additional hyphenated techniques LC–MS–NMR (Shockcor et al., 1996; Yang, 2006), and LC–MS–SPE–NMR (Alexander et al., 2006; Bieri et al., 2006; Xu and Alexander, 2005; Seger et al., 2006; Wilson et al., 2006), have been introduced that are now routinely used in the pharmaceutical industry to support the characterization of metabolites and impurities.

MS provides the added capability of identifying a molecularweight to an unknown metabolite chromatographic peak, which significantly simplifies the NMR structural analysis. Similarly, solid-phase extraction addresses a common limitation of hyphenated techniques. NMR is an inherently insensitive technique requiring a large sample size (≥ 500 ng) to observe a simple 1D NMR spectrum, which may not be achieved for biological

TABLE 12.8 Common solvent suppression methods.

Suppression sequence	Setup	Advantages	Limitations
Presaturation	Traditionally, the solvent signal is irradiated for a period of time with a continuous wave rf field	Can be easily set up Useful for eliminating single solvent signal.	Extremely sensitive to spectrometer stability and shimming. Not easy to suppress multiple solvent peaks simultaneously. Also, suppresses exchange peaks Suppresses NMR signals from the compound that overlaps with the solvent.
Excitation sculpting	The solvent resonances are extracted from the spectra with selective pulses	Involved setup, could be automated, exchange resonances are not suppressed, multiple solvent peaks can be suppressed. Extremely flat baselines.	Setup is involved. Several elements in the pulse sequence need to be optimized Suppresses NMR signals from the compound that overlap with the solvent
Watergate exitation technique (WET)	The solvent resonances are selectively suppressed at the beginning of the NMR experiment. This pulse sequence is the most widely used solvent suppression technique for LC–NMR experiments	Involved setup, could be automated, exchange resonances are not suppressed, multiple solvent peaks can be suppressed. ^{13}C decoupling	Suppresses NMR signals from the compound that overlaps with the solvent Baseline distortions

metabolites in the LC–NMR mode. SPE provides a simple approach to collect and concentrate individual HPLC peaks prior to NMR analysis. The purified compound is extracted from an SPE column in a small volume (~30 μL) and can be further concentrated by collecting peaks from multiple HPLC injections. This also reduces the amount of deuterated solvents used, since only a small volume (~300 μL) of solvent is required to elute the compound from the SPE column.

The most common HPLC solvents used for LC–NMR are acetonitrile and water. Both protonated and deuterated acetonitrile and water are commercially available and routinely used, where cost is a major consideration in minimizing the utility of deuterated solvents in HPLC experiments. Commonly, a low percentage of trifloroacetic acid (TFA) is also added to the HPLC solvent system to improve the LC peak lineshape. In some cases, a methanol–water solvent mixture replaces the water–acetonitrile system if the compound of interest does not behave well (low solubility, aggregation, broad peak shape). If an MS system is attached to the LC–NMR to monitor the molecular weight than formic acid is preferred as a modifier. Table 12.9 provides a list of commonly used hyphenated techniques and their advantages and disadvantages.

The LC–NMR component of hyphenated systems can be performed in three different modes; on-flow, stop-flow and loop storage/transfer. In the on-flow mode, the NMR spectra are collected continuously at predefined intervals during the chromatographic run. The NMR experiments are typically limited to simple 1D ^{1}H NMR experiments that can be rapidly collected during the limited time. Also, NMR spectra may contain multiple compounds since there is no correlation between peak elution and NMR data collection. An NMR spectrum could be collected during the time period when one peak is finishing eluting from the column and a second peak has started eluting from the column. So, both compounds are present during some fraction of the NMR data collection time.

Conversely, stop-flow stops the chromatography after a specific peak has eluded and the compound has been transferred to the NMR flowprobe. This enables longer, more complicated NMR experiments to be collected on a pure metabolite. The major disadvantage of this mode occurs if multiple metabolites are analyzed. The compounds and the chromatographic peaks still remaining on the HPLC column may deteriorate during the NMR data collection time. The loop storage/transfer mode is similar to stop-flow except for the inclusion of a SPE column to concentrate the sample prior to NMR analysis. The loop storage/transfer mode has similar issues regarding compound stability. The three modes and their advantages and limitations are highlighted in the Table 12.10.

12.7 GENERAL PROTOCOL FOR NMR ANALYSIS OF UNKNOWN COMPOUNDS OR METABOLITES

The application of NMR for the analysis of metabolite structures is a multi-step process that is fundamentally based on spectral comparisons. The first step

TABLE 12.9 Advantages and limitations of the most common hyphenated NMR methods.

Technique	Setup	Advantages	Limitations
LC–NMR	HPLC system connected to the NMR flow probe (30–120 µl) via a UV detector. Experiments can be performed in on-flow, stop-flow, and loop collection modes DAD UV detector used for peak detectionRegular reversed-phase C18 (4.6 × 150 mm) columns CH_3CN or $CH_3OH:H_2O$ with trifloroacetic acid or formic acid as a modifier. At least one of the solvents is deuterated.	Eliminates the need for sample isolation and purification associated with regular tube NMR Useful for analysing compounds that may degrade during isolation Isocratic and gradient HPLC methods can be used	Use of deuterated solvents for LCSample limited to column size, NMR flow cell and chromatographic resolution Solvent suppression techniques are necessary even when both solvents are deuterated Need for NMR friendly solvent modifiers Disconnect between UV and MS retention time due to the use of two separate instruments
LC–NMR–MS	20 : 1 flow splitter added after the LC column resulting in 5% of the flow going into the MS source and 95% going into the NMR flow cell All three modes can be performed	Disconnect between the UV and MS retention times are addressed MS and MS/MS data can be collected during analysis Can be used to determine number of exchangeable protons in the metabolite	Need for MS friendly solvent modifier (i.e., formic acid) Complex setup. Maintenance of the hardware is time consuming

| LC-SPE-MS-NMR | HPLC system connected to a solid phase extraction (SPE) system Compound of interest is "trapped" in a SPE cartridge and later transferred into the NMR flow probe for analysis MS can be used during the LC analysis and during the analysis of the cartridge | Fraction of interest is concentrated in the cartridge prior to the NMR analysis Chromatography can be performed using protonated solvents with any modifier Sample is loaded into the NMR flow cell using deuterated solvents minimizing the need for solvent suppression Various solid phase options available in SPE cartridge to optimize for different compounds | Not all compounds can be trapped into the SPE cartridge, limiting its use with certain type of compounds Complex setup. Maintenance of the hardware is time consuming |

TABLE 12.10 Advantages and limitations of the common LC–NMR–MS methods.

Mode	Setup	Advantages	Limitations
On-flow	In this mode the HPLC column is connected directly to the NMR probe via a DAD detector. A series of 1D spectra are collected with a predefined number of scans without stopping the chromatographic experiment	Provides a metabolite profile of the sample. Can be used with ^{1}H and ^{19}F to monitor metabolite degradation. With ^{19}F monitoring, as most of the peaks observed are from the compound solvent suppression is not needed	Limited to compounds with on column loadings >10 μg. Limited to 1D NMR experiments. Broadening of the NMR signal due to on-flow conditions. Solvent suppression is a challenge due to solvent gradients used for LC
Stop-flow	In this mode the LC chromatography is temporarily stopped at the peak of interest using the UV or MS (LC–NMR–MS) signal. The timing between the UV or MS detector and the NMR flow cell is critical. Either the UV or the MS signal can be used to pause the LC run.	Detailed structure determination can be performed on the peak of interest by performing 1D and 2D NMR. >100 ng for ^{1}H experiments and >1 μg for 2D ^{1}H–^{1}H experiments is required.	The LC peak shapes deteriorate for the latter peaks of interest. LC peaks of interest must be well resolved. >2 min retention time difference. Cross contamination from various LC fractions

TABLE 12.10 (*Continued*)

Mode	Setup	Advantages	Limitations
Loop-storage or SPE/transfer	LC chromatography is performed first with collection of peaks in a loop cassette or SPE cartridge using the UV or MS signals The stored fractions in the loops or SPE are then transferred into the NMR flow probe for analysis Computer control of the instrument is essential in this mode of operation	LC chromatography could be performed independently from the NMR experiments Since the chromatography is not stopped, issues with peak broadening in the stop-flow mode are eliminated. Eliminates cross contamination problems Multiple experiments on each sample are possible including Heteronuclear experiments such as $^1H-^{15}N$	Sample could decompose during the storage period Requires the use of special equipment to direct the flow through different pathways

of metabolite structure identification is to assign all the ^{1}H and ^{13}C NMR resonances of the *parent compound* to the corresponding atoms in the molecular structure. This is accomplished using the complete repertoire of 1D- and 2D-NMR experiments discussed in detail in Sections 12.6.1 and 12.6.2. The next step of the process is to compare the ^{1}H NMR spectrum of the metabolite to the corresponding NMR spectrum for the parent compound.

From this spectral comparison, it is relatively straightforward to identify key changes in chemical shift (e.g., upfield versus downfield), coupling patterns (e.g., doublet versus triplet), coupling constants (e.g., 2 Hz versus 8 Hz), peak integration (e.g., 1 H versus 2 H) and the disappearance or appearance of peaks in the metabolite NMR spectrum. The third step is to link these NMR spectral changes to specific site(s) on the parent molecule's structure. (Are these changes on an aromatic ring or near an alkene double bond?) In most cases, the comparison of 1D spectrum, combined with MS spectral analysis, provides preliminary answers or inferences to the location and type of modifications that occur between metabolites and the parent compound. These preliminary results can be confirmed by performing 2D NMR experiments that specifically address issues identified from the 1D NMR analyses.

The application of NMR to analyze metabolite structures is illustrated using a 6-hydroxy buspirone as an example. Figure 12.8 shows the ^{1}H spectrum of 6-hydroxy buspirone. It is clear from the complexity of the 1D ^{1}H NMR spectrum, that the primary structure of 6-hydroxy buspirone could not be easily determined by only using 1D NMR data. The 2D homonuclear experiments, TOCSY (Figs. 12.9 and 12.10) and DQF–COSY (Fig. 12.11), were collected on 6-hydroxy buspirone to identify ^{1}H NMR resonances that are coupled and correspondingly chemically bonded. The spin systems for protons 7–10 were identified by the combination of coupling patterns observed in both the TOCSY and DQF–COSY spectra. The spin system and coupling pattern are highlighted in both NMR spectra. The ^{13}C chemical shifts were determined from 2D heteronuclear data, HSQC and HMBC, as shown in Figs. 12.11 and 12.12, respectively. The 2D ^{1}H–^{13}C HSQC spectrum shows cross peaks for all one bonded ^{1}H–^{13}C pairs. The ^{13}C NMR assignments for all protonated carbons are obtained by simply correlating the assigned ^{1}H NMR resonances with a corresponding ^{13}C NMR resonance by the cross peaks observed in the 2D ^{1}H–^{13}C HSQC spectrum (Fig. 12.10).

The HMBC experiment correlates long-range (two to three bond) ^{1}H–^{13}C pairs (Fig. 12.11) three is used to determine the ^{13}C chemical shifts and structural connectivity of quaternary and carbonyl carbons. Quaternary and carbonyl carbons do not have directly bonded hydrogens and as a result do not have a cross peak in the 2D ^{1}H–^{13}C HSQC spectrum. Fig. 12.11 shows cross peaks for the correlation between hydrogens 10 and 13 with carbonyl carbon 12, and hydrogen 20 with carbonyl carbon 19. Also shown in the HMBC spectrum is the correlation between hydrogens 23 and 25 with quaternary carbon 21. Despite a lack of correlations to spiro-carbon 14, the overwhelming body of evidence from interpretations of multiple NMR

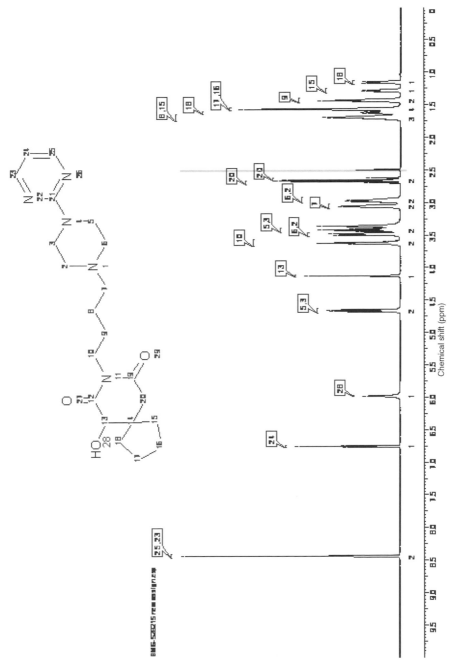

FIGURE 12.8 ^{1}H spectrum and its assignment of 6-hydroxy Buspirone in DMSO.

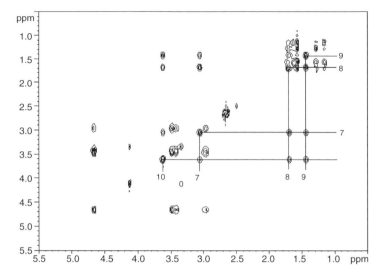

FIGURE 12.9   2D ^{1}H$-^1$H TOCSY spectrum of 6-hydroxy buspirone.

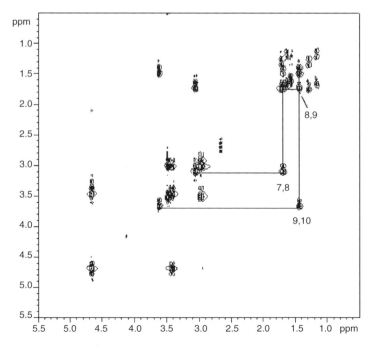

FIGURE 12.10   2D ^{1}H$-^1$H DQF–COSY spectrum of 6-hydroxy buspirone.

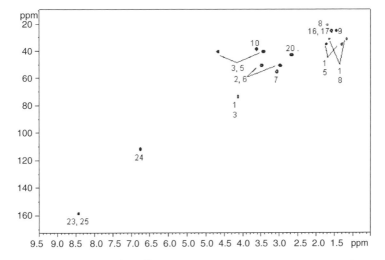

FIGURE 12.11 2D ^{1}H–^{13}C HSQC spectrum of 6-hydroxy buspirone.

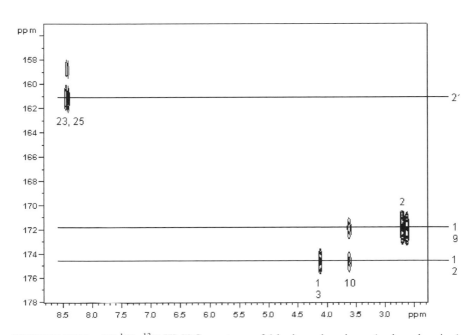

FIGURE 12.12 2D ^{1}H–^{13}C HMBC spectrum of 6-hydroxy buspirone (carbonyl region).

TABLE 12.11 Metabolite identification by using NMR spectroscopy.

Metabolic reaction	Metabolite identification examples	NMR indications	Confirmation method and comments
1. Dealkylation	Fluoxetine → Nor fluxetine	The methyl group at $\delta2.98$ (s, 3H) disappeared The α-CH$_2$ group shifted from $\delta2.82$ to $\delta2.98$ ppm	There is no need for further confirmation
2. Aliphatic hydroxylation (Gerhard et al., 2003)		The methyl group at $\delta2.45$(s, 3H) disappeared A new peak appeared at $\delta4.71$ ppm (s, 2H), which is characteristic for a methylene alcohol	There is no need for further confirmation
(Chando et al., 1998)	Irbesartan → hydroxyl-irbesartan	The methyl proton shifted from $\delta0.80$ to $\delta1.00$ ppm and its coupling pattern changed from a triplet to a doublet The α-CH group shifted from $\delta1.23$ (m, 2H) to $\delta3.58$ (m, 1H), which is consistent with a secondary alcohol	The methyl and methine groups indeed belonged to the same spin system could be confirmed by the 2D COSY experiment

3. Aromatic hydroxylation (Gerhard et al., 2003)		The hydroxylation position at C2 or at C4 was not obvious from the 1D proton spectrum due to the presence of the F–H couplings 2D COSY spectrum indicated the larger doublet at δ7.25 (8.1 Hz) was due to vicinal coupling of F–H while the smaller coupling (2.3 Hz) was the result of H2–H6 meta coupling	No further experiment is needed to confirm the regiochemistry
		Based on the coupling constant, the hydroxylation occurred at the C4 position	
4. Epoxidation (Garcia et al., 2004)	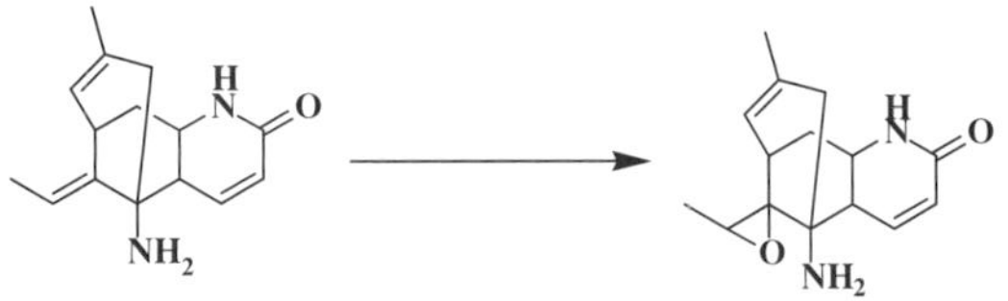	The olefinic protons shifted from δ6.12 to δ2.79 ppm. This observation is characteristic for epoxide in which the proton normally appears at δ2.2–2.9 ppm	The formation of epoxide can be confirmed by determining the shifts of C13 and C14 carbons involved in epoxide formation either by performing ^{1}H–^{13}C HMQC or direct ^{13}C 1D experiments. Both carbons should shift from ~130 to ~50 ppm.

Huperzine A

13,14-Epoxy Huperzine A

(continued)

TABLE 12.11 (*Continued*)

Metabolic reaction	Metabolite identification examples	NMR indications	Confirmation method and comments
5. *N*-oxidation (Jairaj et al., 2002)	Pholcodine → Pholcodine-N-Oxide	The large downfield carbon ($\Delta\delta > 12$ ppm) and proton ($\Delta\delta > 0.6$ ppm) shifts for C16, C17, and C9 were indicative of *N*-oxide formation at the *N*15 nitrogen (chemical shifts changes are listed in the Table 12.12)	The *N*-oxidation position could be confirmed by ^{15}N HMBC experiment. The downfield 10–30 ppm nitrogen shift is indicative of the *N*-oxide. However, this experiment requires >100 µg of sample
6. *S*-Oxidation and *S*-reduction	Nizatidine → Nizatidin Sulfoxide	The methylene carbons that connected to the sulfur group shifted from ~35 to ~52 ppm The α-methylene protons shifted from ~2.3 to ~3.0 ppm	There is no direct NMR method to confirm the existence of the sulfoxide group as sulfur is not an NMR active nucleus
7. Alkene reduction	Cinnamic acid → Phenylpropionic acid	The olefinic protons shifted from 6.36 and 7.73 to 2.59 and 2.89 ppm, respectively The coupling pattern changed from large doublet (14.8 Hz) to the smaller multiplets The integration of the corresponded signals increased from one proton to two protons	The alkene reduction can be further confirmed by observing the changes in the carbon shifts from 110–140 ppm to about 30 ppm by performing an $^{1}H-^{13}C$ HMQC experiment

8. Glutathione conjugation (Dagnino-Subiabre et al., 2000)	The C-5 aromatic proton was missing and the coupling pattern of the remaining two aromatic protons changed to a 1.8 Hz doublet The β-methylene of the cysteine shifted from 2.87 to 3.10 ppm	The position of the GSH attachment could be confirmed by an NOE experiment. NOE may be observed between the β-methylene of the cysteine and the aromatic C4 proton. The *H1*−*C13* HMBC experiment may also help. This experiment requires require >100 ug of the metabolite
9. Methylation (Basker et al., 1990) Analog of Catecholic Cephalosporin	A new sharp methoxyl signal at δ3.86 (s, 3H) ppm was observed	The structure was confirmed by HMBC and NOE experiments
10. Sulfation (Daykin et al., 2005) Pyrogallol → dihydroxyphenyl-2-O-sulfate	All three aromatic protons shifted slightly downfield. H4 and H6 protons shifted by 0.10 ppm and the H5 shifted by 0.21 ppm The aromatic protons of the metabolite showed two groups of signals as did the parent compound, which is consistent with the symmetrical metabolite formed by sulfation at the C2 hydroxyl	No further experiments can be performed due to lack of NMR active nuclei in the −SO3H group.

(continued)

TABLE 12.11 (Continued)

Metabolic reaction	Metabolite identification examples	NMR indications	Confirmation method and comments
11. Hydrolysis	Enalapril → Enalaprilate	The ethylene protons (δ4.5, q, 2H; δ1.2, t, 3H) disappeared from the proton spectrum of the Enalaprilate and the rest of the protons and the spin system were intact	There is no need for confirmation
12. Glucuronide conjugation (Froehlich et al., 2005)	N-hydroxydebrisoquenine	The O versus N-glucuronidation was not determined from the proton spectrum as the anomeric protons for both of the regio isomers were in the 4.5–6 ppm range.	However, measurement of anomeric carbon chemical shifts provided the answer due to significant chemical shift differences between the O-glucoronide (100–110 ppm) and the N-glucoronide (80–90 ppm).
(Zhang et al., 2004)	BMS-204352 → or → Confirmed	The carbon chemical shift (δ82.9 ppm) for the anomeric carbon C1′ was indicated from the HMQC experiment, which is consistent with the N-conjugation	The N-glucuronide structure was confirmed by the HMBC experiment. Observation of the long range H–C correlation from the anomeric proton to the indolinone carbonyl supported the N-glucuronide conjugation

402

13. Amino acid conjugation

Benzoic Acid — Glycine → Hippuric Acid

The proton signal for the methylene proton was at $\delta4.3$ ppm, compared with the same proton for glycine at $\delta3.8$ ppm
The amide proton ($\delta9.4$, t, 1H) was also observed
All the aromatic protons were present
Integration of all the peaks was consistent

No further experiments are needed

spectral data sets led to the establishment of the metabolite structure as 6-hydroxy buspirone.

12.8 EXAMPLES OF METABOLITE STRUCTURE DETERMINATION FROM KNOWN BIOTRANSFORMATIONS

The availability of the chemical structure and NMR assignment s for the parent compound, as illustrated in Section 12.7 for Buspirone, significantly simplifies the identification of related metabolites. A significant amount of information has already been accumulated and presented in the literature that describes numerous metabolic reactions associated with a variety of metabolic pathways. Examples of the most common metabolic reactions are tabulated in Table 12.11. These metabolic reactions identify a range of possible chemical modifications that may be applied to the parent compound and generate a variety of related metabolites.

The metabolites that are generated from the parent compound can be viewed as simply incurring an addition and/or subtraction of functional groups while maintaining most of the intact parent structure. Similarly, the addition and/or subtraction of functional groups will result in corresponding changes in the NMR spectra, while the majority of the NMR spectrum is unperturbed relative to the parent compound. Again, the comparison of the NMR spectra between the parent compound and the metabolite will easily highlight these structural changes while confirming the parts of the structure that are unaffected. Representative NMR methods to determine the structure of metabolites resulting from various metabolic reactions have been described in the literature and are also listed in Table 12.11.

Occasionally, xenobiotics undergo metabolic activation to produce reactive intermediates, where these intermediates rearrange to form an unpredictable metabolite. For example, the metabolic activation of DPC 963 in rat formed a highly reactive oxirene intermediate (Chen et al., 2002). This intermediate rapidly rearranged to form an unstable cyclobutenyl ketone through possible intermediates a or b (Fig. 12.13). This reactive intermediate was prone to nucleophilic attack and in this case reacted with glutathione via a 1,4 Michael addition that resulted in two isomeric GSH adducts M3 and M4. The ^{1}H NMR spectrum provided evidence in support of the M3 and M4 structures. First, both the aromatic hydrogens were still present in the metabolite. Second, the

TABLE 12.12 Proton and carbon chemical shifts of pholcodine and its *N*-oxide metabolite.

Protons	Parent	*N*-oxide	$\Delta\delta$	Carbons	Parent	*N*-oxide	$\Delta\delta$
δH-16	2.35, t 2.72 dt	3.25, t, 3.38, dt	0.66–0.90	δC-16	47.5	60.4	12.9
δH-17	2.58 s	3.44, s	0.86	δC-17	43.2	58.5	15.3
δH-9	3.31, dd	3.92, dd	0.61	δC-9	60.2	75.9	15.7

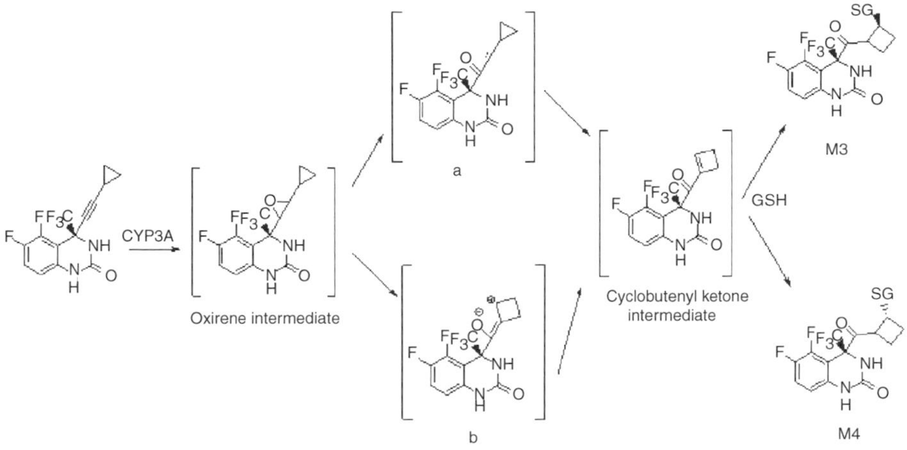

FIGURE 12.13 Proposed metabolic pathways leading to the formation of glutathione adducts in rat.

two methylene proton signals for the cyclopropyl ring (two triplets in 0.2–0.4 ppm range) were missing. Third, the TOCSY experiment indicated a new spin system that involved eight protons and their chemical shift were consistent with a disubstituted cyclobutyl group. These observations pointed toward the cyclopropyl group as the point of modification. Finally, the characteristic ^{1}H NMR signals for the GSH group were also clearly observed and the integration of the GSH hydrogens matched with the rest of the molecule. Based on a combination of NMR evidence, the structures of these two unusual GSH adducts were determined.

In conclusion, NMR is an essential tool for the successful determination of crucial metabolite structures and is routinely used in the pharmaceutical industry. As discussed, metabolite structure problems could be as simple as hydroxylation on an aromatic ring or as complex as a rearrangement depicted in the formation of glutathione adducts. NMR provides a vast and continually expanding combination of techniques applicable to the analysis of metabolite structures. The judicious choice of NMR experiments based on the particulars of the system and the nature of the metabolites can be combined with mass spectrometry and liquid chromatography to successfully analyze a variety of biological metabolites to benefit drug discovery.

REFERENCES

Albert K. Online use of NMR detection in separation chemistry. J Chromatogr A 1995;703:123–147.

Alexander AJ, Xu F, Bernard C. The design of a multidimensional LC–SPE–NMR system (LC2–SPE–NMR) for complex mixture analysis. Magn Reson Chem 2006;44(1):1–6.

Basker MJ, Finch SC, Tyler JW. Use of ^{1}H NMR in the identification of a metabolite of a catecholic cephalosporin excreted in rat bile. J Pharm Biomed Anal 1990;8: 573–576.

Berger S, Braun S. 200 and More NMR Experiments. second ed. John Wiley & Sons; 2004.

Betz M, Saxena K, Schwalbe H. Biomolecular NMR: a chaperone to drug discovery. Curr Opin Chem Biol 2006;10(3):219–225.

Bieri S, Varesio E, Veuthey JL, Muñoz O, Tseng LH, Braumann U, Spraul M, Christen P. Identification of isomeric tropane alkaloids from *Schizanthus grahamii* by HPLC–NMR with loop storage and HPLC–UV–MS/SPE–NMR using a cryogenic flowprobe. Phytochem Anal 2006;17(2):78–86.

Borlak J, Walles M, Elend M, Thum T, Preiss A, Levsen K. Verapamil: identification of novel metabolites in cultures of primary human hepatocytes and human urine by LC–MS(n) and LC–NMR. Xenobiotica 2003;33(6):655–676.

Breitmaier E. Structure Elucidation by NMR in Organic Chemistry: A Practical Guide. Third ed. John Wiley & Sons; 2002.

Bruker Biospin GmbH. Bruker BioSpin announces novel capillary LC–NMR system 2002. Forthcoming.

Buker Biospin GmbH. Bruker BioSpin introduces the CryoFlowProbe, the world's first cryogenic flow-injection NMR probe 2002. Forthcoming, Available at http://www.bruker-biospin.com/nmr/news/pdf_files/cryoflowprb.pdf.

Chando TJ, Everett DW, Kahle AD, Starrett AM, Vachharajani N, Shyu WC, Kripalani KJ, Barbhaiya RH. Biotransformation of irbesartan in man. Drug Metab Dispos 1998;26:408–417.

Chen H, Shockcor J, Chen W, Espina R, Gan LS, Mutlib AE. Delineating novel metabolic pathways of DPC 963, a nonnucleoside reverse transcriptase inhibitor, in rats. Characterization of glutathione conjugates of postulated oxirene and benzoquinone imine intermediates by LC–MS and LC–NMR. Chem Res Toxicol 2002;15:388–399.

Corcoran O, Spraul M. LC–NMR–MS in drug discovery. Drug Discov Today 2003; 8(14):624–631.

Dagnino-Subiabre A, Cassels BK, Baez S, Johansson AS, Mannervik B, Segura-Aguilar J. Glutathione transferase M2-2 catalyzes conjugation of dopamine and dopa *O*-quinones. Biochem Biophys Res Commun 2000;274:32–36.

Daykin CA, VanDuynhoven JPM, Groenewegen A, Dachtler M, Van Amelsvoort JMM, Mulder TPJ. Nuclear magnetic resonance spectroscopic based studies of the metabolism of black tea polyphenols in humans. J Agric Food Chem 2005;53: 1428–1434.

Exarchou V, Godejohann M, van Beek TA, Gerothanassis IP, Vervoort J. LC–UV–solid-phase extraction-NMR–MS combined with a cryogenic flow probe and its application to the identification of compounds present in Greek oregano. Anal Chem 2003;75(22):6288–6294.

Exarchou V, Krucker M, van Beek TA, Vervoort J, Gerothanassis IP, Albert K. LC–NMR coupling technology: recent advancements and applications in natural products analysis. Magn Reson Chem 2005;43(9):681–687.

Freeman R, A Handbook of Nuclear Magnetic Resonance. second ed. Longman Publishing Group; 1997.

Friebolin H, Basic One- and Two-Dimensional NMR Spectroscopy. fourth ed. John Wiley & Sons; 2005.

Froehlich AK, Girreser D, Clement B. Metabolism of *N*-hydroxyguanidines (*N*-hydroxydebrisoquine) in human and porcine hepatocytes: reduction and formation of glucuronides. Drug Metabo Dispos 2005;33:1532–1537.

Gaggelli E, Valensin G. Methods of single- and double-selective excitation: theory and applications. Part II. Conc Magn Reson 1993;5(1):19–42.

Garcia GE, Hicks RP, Skanchy D, Moorad-Doctor DR, Doctor BP, Ved HS. Identification and characterization of the major huperzine a metabolite in rat blood. J Anal Toxicol 2004;28:379–383.

Gerhard U, Thomas S, Mortishire-Smith R. Accelerated metabolite identification by "extraction-NMR". J Pharm Biomed Anal 2003;32:531–538.

Godejohann M, Tseng LH, Braumann U, Fuchser J, Spraul M. Characterization of a paracetamol metabolite using on-line LC–SPE–NMR–MS and a cryogenic NMR probe. J Chromatogr A 2004;1058(1–2):191–196.

Hajduk PJ. SAR by NMR: putting the pieces together. Mol Interv 2006;6(5):266–272.

Homans SW. A Dictionary of Concepts in NMR. Second ed. Oxford University Press; 2005.

Hwang TL, Shaka AJ. Water suppression that works. Excitation sculpting using arbitrary wave-forms and pulsed-field gradients. J Magn Reson Series A 1995; 112(2):275–279.

Iwasa K, Kuribayashi A, Sugiura M, Moriyasu M, Lee DU, Wiegrebe W. LC–NMR and LC–MS analysis of 2,3,10,11-oxygenated protoberberine metabolites in Corydalis cell cultures. Phytochemistry 2003;64(7):1229–1238.

Jairaj M, Watson DG, Grant MH, Gray AI, Skellern GG. Comparative biotransformation of morphine, codeine and pholcodine in rat hepatocytes: identification of a novel metabolite of pholcodine. Xenobiotica 2002;32:1093–1107.

Keeler J. Understanding NMR Spectroscopy. first ed. UK: John Wiley and Sons; 2005.

Keun HC, Beckonert O, Griffin JL, Richter C, Moskau D, Lindon JC, Nicholson JK. Cryogenic probe ^{13}C NMR spectroscopy of urine for metabonomic studies. Anal Chem 2002;74:4588–4593.

Klaus A. On-line LC–NMR and Related Techniques, first ed. E-Book;2003.

Kupee E. NMR at 900 MHz. Chem Heter Comp 2001;37(11):1429–1438.

Lepre CA, Moore JM, Peng JW. Theory and applications of NMR-based screening in pharmaceutical research. Chem Rev 2004;104(8):3641–76

Lindon JC, Nicholson JK, Wilson ID. The development and application of coupled HPLC–NMR spectroscopy. Adv Chromatogr 1996;36:315–382.

Lyčka A, Fryšová I, Slouka J. An anomalous course of the reduction of 2-(3-oxo-3,4-dihydroquinoxalin-2-yl)benzene diazonium salt: a reinvestigation. Magn Reson Chem 2007;45(1):46–50.

Martin G, Zektzer A, Two-Dimensional NMR Methods for Establishing Molecular Connectivity. First ed. VCH; New York: 1998.

Moreno A, Daz-Ortiz A, Dez-Barra E, de la Hoz A, Langa F, Prieto P, Claridge TDW. Determination of the stereochemistry of four spirodiastereoisomers by one- and two-dimensional NOE studies. Magn Reson Chem 1996;34(1):52–58.

Munro SLA, Craik DJ. NMR conformational studies of fenamate non-steroidal antiinflammatory drugs. Magn Reson Chem 1994;32(6):335–342.

NMR probes. Varian Inc, Palo Alto, CA, USA;2002. Available at http://www.varianinc.com/cgi-bin/nav?varinc/docs/nmr/products/probes.

Orry AJ, Abagyan RA, Cavasotto CN. Structure-based development of target-specific compound libraries. Drug Discov Today 2006;11(5–6):261–266.

Pellecchia M. Solution nuclear magnetic resonance spectroscopy techniques for probing intermolecular interactions. Chem Biol 2005;12(9):961–971.

Reilly CA, Ehlhardt WJ, Jackson DA, Kulanthaivel P, Mutlib AE, Espina RJ, Moody DE, Crouch DJ, Yost GS. Metabolism of capsaicin by cytochrome P450 produces novel dehydrogenated metabolites and decreases cytotoxicity to lung and liver cells. Chem Res Toxicol 2003;16(3):336–349.

Schaefer M, Faller P, Nicole D. Regio- and stereoselective alkylation of a pyrrolidinic system: structural and conformational studies by high field NMR techniques. Org Magn Reson 1982;19(2):108–111.

Schlotterbeck G, Ross A, Hochstrasser R, Senn H, Kuhn T, Marek D, Schett O. High-resolution capillary tube NMR. A miniaturized 5 mL high-sensitivity TXI probe for mass-limited samples, off-line LC NMR, and HT NMR. Anal Chem 2002;74: 4464–4471.

Schutz A, Golbik R, Konig S, Hubner G, Tittmann K. Intermediates and transition states in thiamin diphosphate-dependent decarboxylases. A kinetic and NMR study on wild-type indolepyruvate decarboxylase and variants using indolepyruvate, benzoylformate, and pyruvate as substrates. Biochemistry 2005;26:44916):6164–79.

Seger C, Godejohann M, Spraul M, Stuppner H, Hadacek F. Reaction product analysis by high-performance liquid chromatography-solid-phase extraction-nuclear magnetic resonance Application to the absolute configuration determination of naturally occurring polyyne alcohols. J Chromatogr A 2006;1136(1):82–88.

Sem DS, Pellecchia M. NMR in the acceleration of drug discovery. Curr Opin Drug Discov Devel 2001;4(4):479–492.

Shockcor PJ, Unger SH, Wilson ID, Foxall PJD, Nicholson JK, Lindon JC. Combined hyphenation of HPLC, NMR spectroscopy and ion trap mass spectrometry (LC–NMR–MS) with application to the detection and characterization of xenobiotic and endogenous metabolites in human urine. Anal Chem 1996;68:4431–4435.

Smallcombe SH, Patt SL. WET solvent suppression and its applications to LC–NMR and high-resolution NMR spectroscopy. J Magn Reson Series A 1995;117(2):295–303.

Spraul M, Hofmann M, Lindon JC, Farrant RD, Seddon MJ, Nicholson JK, Wilson ID. Evaluation of liquid-chromatography coupled with high-field ^{1}H NMR spectroscopy for drug metabolite detection and characterization: the identification of paracetamol metabolites in urine and bile. NMR Biomed 1994;7:295–303.

Spraul M, Hofmann M, Dvortsak P, Nicholson JK, Wilson ID. High-performance liquid chromatography coupled to high-field proton nuclear magnetic resonance spectroscopy: application to the urinary metabolites of ibuprofen. Anal Chem 1993;65:327–330.

Spraul M, Freund AS, Nast RE, Withers RS, Maas WE, Corcoran O. Advancing NMR sensitivity for LC–NMR–MS using a cryoflow probe: application to the analysis of acetaminophen metabolites in urine. Anal Chem 2003;75(6):1536–1541.

Sridharan V, Saravanan S, Muthusubramanian S, Sivasubramanian S. NMR investigation of hydrogen bonding and 1,3-tautomerism in 2-(2-hydroxy-5-substituted-aryl) benzimidazoles. Magn Reson Chem 2005;43(7):551–556.

Valente AP, Miyamoto CA, Almeida FC. Implications of protein conformational diversity for binding and development of new biological active compounds. Curr Med Chem 2006;13(30):3697–3703.

Wilson SR, Malerd H, Petersen D, Simic N, Bobu MM, Rise F, Lundanes E, Greibrokk T. Controlling LC–SPE–NMR systems. J Sep Sci 2006;29(4):582–589.

Xu F, Alexander AJ. The design of an on-line semipreparative LC–SPE–NMR system for trace analysis. Magn Reson Chem 2005;43(9):776–782.

Yang Z. Online hyphenated liquid chromatography–nuclear magnetic resonance spectroscopy–mass spectrometry for drug metabolite and nature product analysis. J Pharm Biomed Anal 2006;40(3):516–527.

Zartler ER, Shapiro MJ. Protein NMR-based screening in drug discovery. Curr Pharm Des 2006;12(31):3963–3972.

Zhang D, Zhao W, Roongta VR, Mitroka JG, Klunk LJ, Zhu M. Amide N-glucuronidation of MaxiPost catalyzed by UDP-glucuronosyltransferase 2B7 in humans. Drug Metab Dispos 2004;32:545–551.

Zhou CC, Hill DR. The keto–enol tautomerization of ethyl butylryl acetate studied by LC–NMR. Magn Reson Chem 2006;45(2):128–132.

PART IV

COMMON EXPERIMENTAL APPROACHES AND PROTOCOLS

13

DETERMINATION OF METABOLIC RATES AND ENZYME KINETICS

ZHI-YI ZHANG AND LAURENCE S. KAMINSKY

13.1 INTRODUCTION

In vitro biochemical studies of drug metabolism, particularly studies associated with pharmacokinetic (PK) properties and potential toxicological consequences, are essential in drug discovery and development. For over a decade, the major pharmaceutical companies have been promoting such efforts for incorporation into early stages in discovery (Obach, 2001; Smith and van de Waterbeemd, 1999). Consequently, the failure rate of new chemical entities (NCEs) due to absorption, distribution, metabolism, and excretion (ADME) shortcomings in clinical trials, in contrast to some of the other major hurdles including poor efficacy and intolerable toxicity, has continuously decreased (Apic et al., 2005).

Studies of metabolic stability were integrated into critical paradigms in all discovery phases, particularly during early lead identification and optimization. The nature of discovery demands that such metabolism studies at these early stages be conducted with short turnaround times, and the studies should be simple rather than comprehensive and thus less time consuming. Metabolism is usually only expressed as the intrinsic clearance (CL_{int}), or as disappearance or half-life ($t_{1/2}$). Characterization of biotransformation, enzyme-kinetically, is the main theme for metabolic studies during preclinical development, as opposed to discovery. The *in vitro* kinetic parameters, including those directly used for the selections of incubation condition in enzyme reaction phenotyping and the

Drug Metabolism in Drug Design and Development,   Edited by Donglu Zhang, Mingshe Zhu and W. Griffith Humphreys

413

prediction of human metabolic clearance, such as K_m and V_{max}, need to be accurately determined. The features of metabolic stability studies in both discovery and development will be described in this chapter.

High throughput screening (HTS) through automation is generally desired in pharmaceutical research, and is feasible for most steps of stability studies, including reaction preparation, incubation, sample cleanup, and even sample detection utilizing radiometric and fluorometric means (Crespi and Stresser, 2000; Jenkins et al., 2004). However, the desired reliable chromatography-based detection may not be applicable to HTS, as the data analyses for quantification are unlikely to be completely automated without human intervention. This is particularly true for metabolic investigations during development. Additionally, computational approaches are being increasingly applied to delineate the complex kinetic behaviors of metabolic enzymes (Mei et al., 2002; Zhang et al., 2002). Besides classic biochemical plots, traditionally applied to characterize enzyme kinetics, nonlinear regression analyses will be emphasized largely due to their unbiased nature and user-friendly execution.

Finally, predictions of human pharmacokinetics using *in vitro* metabolic data, one of the ultimate goals for *in vitro* metabolic studies, are described, both conceptually and in case studies.

13.2 DETERMINATION OF METABOLIC STABILITY

13.2.1 Aims

In early drug discovery, one of the properties required for lead candidates is reasonable metabolic stability. The need for stability is based on the presumption that metabolism or metabolic clearance tends to be the major, if not the only determinant, for total body clearance when the pharmacokinetic and pharmaceutical properties for the discovery compound are unknown. To obtain metabolism data, liver subcellular fractions, microsomal preparations in particular, are frequently chosen as standard enzyme preparations for metabolic studies *in vitro*, because of the abundance and the broad spectrum of drug-metabolizing enzymes expressed in hepatocytes, principally localized in the endoplasmic reticulum. On the contrary, lead candidate stability in blood may also need to be studied, even prior to the microsomal assays, particularly for compounds with potential hydrolytic susceptibilities, such as those containing ester or amide bonds. Therapeutic agents, regardless of the route of dosing, must be sufficiently metabolically stable in the blood to be able to reach the intended biological targets. Determination of stability in blood (or plasma), which is at risk from the hydrolytic activities, is straightforward, with no requirements for supplements of any of the constituents required for metabolism mediated by the majority of enzymes in liver subcellular fractions (Liederer and Borchardt, 2005; Skopp et al., 2001). Moreover, hydrolytic degradations can take place in both blood and liver, since the latter also expresses several classes of hydrolyases, such as carboxylesterases (Satoh and Hosokawa, 1998; Satoh et al., 2002). The metabolic instability due to hydrolysis could be potentially detected in the liver subcellular fractions, particularly S9 fractions. Therefore, the potential instability in the blood is usually minor for typical drug-like compounds to the overall metabolism.

13.2.2 Experimental Procedures

13.2.2.1 Assay Conditions

Buffers Tris–HCl and potassium phosphate buffer (or phosphate-buffered saline, PBS) are used for the microsomal stability assays, while PBS is often the choice for the metabolism in hepatocyte suspensions.

i. Tris–HCl buffer (10×, or 500 mM) and potassium phosphate buffer (10×, or 1 M): Both buffers can be purchased commercially or prepared (Ackley et al., 2004), and may require a slight pH adjustment prior to use. The stock solutions are usually stored at 4°C after being filtered, if necessary.

ii. NADPH-regencrating system: The reagents are usually obtained commercially. The system in the final reaction mixtures contains $NADP^+$ (1.3 mM), glucose-6-phosphate (Glc-6-PO$_4$) (3.3 mM), $MgCl_2$ (3.3 mM), and glucose-6-phosphate dehydrogenase (G6PDH; 0.4 U/mL).

Enzymes

i. Liver subcellular fractions: The microsomal preparations, S9 fractions, and cytosolic fractions are prepared using differential ultracentrifugation procedures (Raucy and Lasker, 1991). Liver microsomal preparations contain the majority of the drug-metabolizing enzymes, particularly cytochromes P450 (CYPs), and are the most common enzyme sources for *in vitro* metabolic studies. CYPs play a preeminent role in drug metabolism. For example, the metabolic clearance of nearly half of the drugs on the market is contributed by CYP3A4, the most abundant hepatic CYP member in humans (Shimada et al., 1994). This suggests that use of microsomal preparations as an *in vitro* surrogate for liver metabolism is a reasonable approximation. Liver microsomal preparations also provide several practical benefits: a simple system that provides clear-cut results; ease of use with high throughput adaptability; a concentrated enzyme system, particularly for monooxidases (CYPs and flavin-containing monooxygenases or FMOs), thereby permitting marked turnover for unambiguous detection (if response enzymes are microsomal); and a commercially steady high quality supply. Human liver microsomes (HLM) are the most relevant and are often viewed as the standard enzyme source used in metabolic stability studies. However, the *in vitro* system that uses liver S9 fractions, which comprises both endoplasmic reticulum-associated and cytosolic enzymes, is also of value, particularly for chemicals with readily conjugated functional groups and/or susceptibility to hydrolysis. S9 fractions, although likely to exhibit slower turnover capacities for CYP/FMO/UDP-glucuronosyltransferase (UGT)-mediated biotransformation relative to microsomal fractions due to lower enzyme contents, are useful, particularly for cofactor-independent metabolism such as that mediated by certain hydrolyases. Therefore, S9-mediated metabolic stability is also routinely evaluated for compounds at early discovery stages (Hewawasam et al., 2002; Lai and Khojasteh-Bakht, 2005; Shen et al., 2003). Both CYPs and FMOs require electron donation by the cofactor, NADPH, to elicit their catalytic capabilities. Such a NADPH-dependency is generally considered an indicator of CYP/FMO-catalyzed reactions. In contrast to the broad spectrum of substrates for CYPs, potential FMO substrates are usually limited to nitrogen- and sulfur-containing nucleophiles, for example, tertiary amines (Krueger and Williams, 2005; Ziegler, 2002). Nonetheless, it does not appear essential to differentiate between CYP- and FMO-mediated biotransformations in early metabolic stability studies. FMOs tend to be thermolabile, and thus the reactions mediated by FMOs are usually initiated by the addition of the enzymes instead of NADPH, as is the case for CYP reactions. CYPs and FMOs both catalyze their reactions at low to moderate rates, from an enzyme kinetic perspective (Kruger and Williams, 2005); a 15- or 30-min incubation is often required for these enzymes to produce quantifiable metabolites.

ii. Other *in vitro* enzyme sources: cDNA-expressed recombinant proteins, including both Phase I and Phase II enzymes, are commercially available. While helpful for the identification of the enzymes responsible for catalysis of specific metabolic reactions and for enzyme kinetic characterization, recombinant enzymes should not be the primary source for the metabolic stability studies, because their matrices are inappropriate surrogates for those of naturally occurring enzymes; this mismatch could lead to skewed enzyme kinetics. On the contrary, primary human hepatocytes, either freshly isolated or cryopreserved, are being increasingly applied to studies of drug metabolism, including metabolic stability. It has been shown that the assays employing these hepatocytes are straightforward to perform, and often lead to high-quality quantitative and qualitative results (Soars et al., 2002). Human hepatocytes have intrinsic limitations, such as inconsistent supply, batch-to-batch or vendor-to-vendor variations, the inability to be used for automation/HTS, and interindividual donor variability.

These various *in vitro* systems each have value for certain applications and provide information that might be used to predict metabolic stability *in vivo*.

Factors of Concern   Pharmacological or physiological relevance is always a concern for *in vitro* experiments, including microsomal stability assays. The primary issue is the substrate concentration, but other issues, such as reaction mixture constituents and the availability of these constituents (such as NADPH) should also be considered. Microsomal incubations are carried out either in PBS or in Tris–HCl buffer, while NADPH can be added directly in excess or through regeneration from NADP *in situ*. The NADPH-regenerating system is comprised of $NADP^+$, G6PDH, and the substrate Glc-6-PO_4. Cofactor regeneration can be more physiologically relevant when it is carried out in a PBS-based reaction matrix. However, the influence of NADPH source differences, the effect of different buffers, and the stability of G6PDH on the rates of CYP-mediated biotransformation are not known (Adediran, 1991). The selection of reaction conditions will depend on the circumstances. Direct NADPH addition is a reasonable alternative to the regeneration at early discovery stages, when the simplicity, reproducibility, and rigidity of the assay system are usually predominant considerations. Relevant substrate concentrations and the use of the relevant enzymes, in contrast, are essential for the prediction of pharmacokinetics (Zhang and Wong, 2005). The anticipated drug levels in the circulation of a patient on therapy frequently dictate the use of low substrate concentrations *in vitro*. The blood concentration, however, may not always be consistent with that in the tissues, particularly in the liver, *in vivo*. It is not uncommon for the hepatic concentrations of some of the therapeutic agents to be much higher than those in the circulation (Levine et al., 2001; Sugie et al., 2004). Apparently, pharmacologically relevant concentrations are compound-dependent, and thus no general concentration standard is applicable.

Nevertheless, 1 and 10 μM are often arbitrarily selected as the initial concentrations for metabolic stability assays (MacKenzie et al., 2002; Obach and Reed-Hagen, 2002). The effects of organic solvents on enzyme activities have been carefully investigated (Busby et al., 1999; Chauret et al., 1998; Easterbrook et al., 2001). For the purpose of data comparability for enzyme activities, the levels of organic vehicles should, in principle, be kept as low as possible, and consistent in all incubations in a given experiment. Dimethyl sulfoxide (DMSO) and methanol (or acetonitrile), the most commonly used vehicle solvents, should be kept at levels equal to, or preferably less than, 0.2 and 2% (v/v), respectively (Easterbrook et al., 2001; Hickman et al., 1998).

Studies on metabolic stability using hepatocyte suspensions are not feasible for automation/HTS, but these studies do provide rather complete profiles of hepatic biotransformation without the supplements of cofactors and cosubstrates. The use of S9 in metabolic stability studies can be evaluated in a manner similar to that used for the microsomal assays, but with the possible addition of a broader panel of cofactors or cosubstrates. These include NADPH for CYP/FMO-mediated reactions, NADH for xanthine oxidoreductase and quinone oxidoreductase 2, NADPH-dependent reductions by carbonyl reductases, and NADPH/NADH-dependent reductions catalyzed by aldo–keto reductases, uridine 5'-diphosphate glucuronic acid (UDPGA) for glucuronidation, adenosine $3'$-phosphate $5'$-phosphosulfate (APPS) for sulfation, glutathione (GSH) for glutathione conjugation, and acetyl CoA for N-acetylation.

Some of these potential factors will be further discussed when appropriate.

Typical Assay Procedures   The experimental procedures for determining CYP- and UGT-mediated metabolic stability are provided. These procedures, mainly for microsomal assays, should be amended according to the aims of the specific study and particularly the *in vitro* system in use.

i. NADPH-dependent CYP-mediated metabolism: In a test tube (12 × 75 mm), 25 μL of the reaction buffer stock (10× Tris–HCl buffer) is mixed into 170 μL of the distilled and deionized (d.d.) H_2O, followed by the addition of 25 μL of prethawed pooled HLM (20 mg/mL) obtained from a commercial source. After 5 μL of the substrate stock solution is introduced, the test tube is pre-incubated at 37°C for 1 min in a gently shaking water bath. The reaction is initiated by the addition of 25 μL of freshly prepared NADPH solution (20 mg/mL). The protein content can be adjusted by varying the quantity of HLM. For example, if the final protein concentration required is 0.4 mg/mL instead of 2 mg/mL, 5 μL of HLM should be added, along with 190 μL of water to make up 195 μL total volume. The final concentration of substrate is 50 times diluted from the stock. Test tubes during the incubation should be kept open to the air, the source of oxygen required for the reactions.

ii. UGT-mediated (or UDPGA-dependent) metabolism: In a test tube (12 × 75 mm), 25 μL of the reaction buffer stock (10× Tris–HCl buffer) is mixed into 140 μL of d.d. H_2O, followed by the addition of 25 μL of prethawed pooled HLM or HLS9 (20 mg/mL) obtained commercially. After the addition of 5 μL of the substrate stock, 30 μL of the mixture of alamethicin (5 μL of a 10 mg/mL solution) and saccharic acid-1, 4-lactone (25 μL of a 500 mM solution) is added into the reaction mixture. After the test tube is preincubated at 37°C for 1 min in a gently shaking waterbath, the reaction is initiated by the introduction of 25 μL of freshly prepared UDPGA (5 mM). The concentration of the proteins can be adjusted by varying the volume of the subcellular liver preparations. Glucuronidation by the subcellular fractions tends to be alleviated, and thus the reaction might often require to be optimized with the adjustments of the incubation condition such as pH.

iii. Sample clean up: At each of a set of appropriate incubation time points (e.g., 0, 5, 15, 30, 45, 60, and 90 min), the reaction is terminated by quickly cooling the test tube on ice, followed by the immediate addition of an equal volume (250 μL) of ice-cold methanol containing an internal standard. The mixture is transferred into an Eppendorf tube (1.5 mL) after being vortexed. The tube is centrifuged for 5 min in a desktop microcentrifuge at maximum speed. The supernatant is filtered through a spin filter (0.22 μm), and transferred into an HPLC injection vial. The filtrate is analyzed using either a high performance liquid chromatography/ultraviolet/fluorescence (HPLC/UV/FL) or preferably a high performance liquid chromatography/triple quadrupole mass spectrometry (LC/MS/MS).

Zero-time samples, containing the intact compounds (100%), are prepared by mixing all reaction components on ice, followed by the immediate addition of 250 μL methanol containing the appropriate internal standard. The samples are prepared for the analyses in the same manner as described above.

iv. Metabolism in hepatocyte suspensions: The hepatocytes (either freshly isolated or cryopreserved) are thawed and resuspended in PBS (1×) following one of the protocols provided by the manufacturers; 1 μL of the compound stock (e.g., 250× DMSO stock) is mixed into 249 μL PBS containing $0.25–1 \times 10^6$ cells (or $1–4 \times 10^6$ cells/mL) in a test tube (Soars et al., 2002). The suspension is incubated at 37°C in a gently shaking waterbath for up to 4 h, followed by reaction quenching and sample preparation as described for HLM assays.

13.2.2.2 Determination of Metabolic Stability

Substrate depletion and *in vitro* half-life ($t_{1/2}$) are common descriptors for metabolic stabilities (Obach and Reed-Hagen et al., 2002). These descriptive terms are useful, particularly for the ranking of stabilities among a series of synthetic analogs (MacKenzie et al., 2002).

Substrate Depletion and In vitro Half-Life   Studies to detect substrate disappearance are readily undertaken. The results obtained are directly used for estimating $t_{1/2}$ (Dordal et al., 2005).

As mentioned, such experiments are frequently and increasingly carried out in hepatocyte suspensions and in reaction mixtures containing liver subcellular fractions, especially microsomal preparations. In a standard metabolic study, the extent of linear increases of metabolic rates with enzyme concentrations (e.g., 0.2–4 mg/mL), and with reaction periods (e.g., 0–90 min), should be predetermined. Such preliminary data, while essential for kinetic characterizations, may not be feasible in discovery, or may not be crucial when substrate disappearance or half-life is determined. Therefore, it may be possible to perform stability assays under a condition of default setting, that is, a 30-min incubation at the protein concentration of 1 mg/mL (or any concentration between 0.4 and 2.0 mg/mL) (Mandagere et al., 2002). To minimize the effects of nonspecific protein or lipid binding, a potential variable (Kalvass et al., 2001), lower protein concentrations are generally preferable. However, such nonspecific binding is likely *in vitro*, and even more so *in vivo*. Thus, the effects of protein or lipid binding of substrates have not been completely resolved (Andersson et al., 2004; Obach, 1999). The selection of the quantity of the protein or enzyme should be largely based on the requirement to achieve a measurable overall rate of the metabolism at low micromolar substrate concentrations, that is, 1 or 10 μM, together with the need to minimize nonspecific protein binding (Li et al., 2003; Naritomi et al., 2001).

For a fixed-time study, samples are usually run in triplicate, to enable standard deviations to be calculated. In studies with five or six time points, for example, 0, 10, 30, 45, 60, (90) min, duplicate samples are adequate (Li et al., 2003). The time-course studies are preferred for the determination of *in vitro* $t_{1/2}$, although estimates are possible using two time points (Zhao et al., 2005). The last time point, and preferably the last two time points selected ideally should permit 50% or more of compound disappearance.

In a single exponential decay model,

$$C_t/C_0 = \exp(-k_e t) \qquad (13.1)$$

where C_t, is the concentration of the parent compound at time t; C_0, the initial incubation concentration; and k_e, the elimination rate constant.

$C_t/C_0 = R$, the relative concentration at time t

$$\ln R = -k_e t, \quad \text{thus} \quad k_e = -\ln R/t.$$

Therefore, k_e is the slope of a linear regression curve in a semilog plot (Fig. 13.1a), while the $t_{1/2}$ is the time when $R = 0.5$, or

$$\begin{aligned} t &= -\ln R/k_e \\ t_{1/2} &= -(\ln 0.5)/k_e, \quad \text{or} \\ t_{1/2} &= 0.693/k_e. \end{aligned} \qquad (13.2)$$

As shown in Table 13.1, the reaction period t and the corresponding amount (or concentration) of remaining compound R are tabulated for the construction of plots for $t_{1/2}$ determination (Fig. 13.1). Although k_e is readily estimated in the semilog plots, it is preferably calculated through a fitting of the data with the model of a single exponential decay, using computer software. Many application software packages are commercially available, including Scientist, SigmaPlot, WinNonlin, ADAPT II, and KaleidaGraph (Holleran et al., 2003; Slaughter et al., 2003). SigmaPlot (Version 9.0, Syst Inc.) is selected as the demonstration software throughout this text due to its applicability, flexibility and user-friendly operation.

After the raw data, that is, the time points and the corresponding determined quantities, are entered into a SigmaPlot data sheet, Regression Wizard under Statistics is selected. Exponential decay and single with two parameters are chosen for Equation Category and Equation Name, respectively, followed by Next. *XY* Pair is selected following the assignments of the independent variable *x* as time *t* and the dependent variable *y* as *R* (the remaining amount relative to that at time 0), to the equation ($y = a \exp(-bx)$).

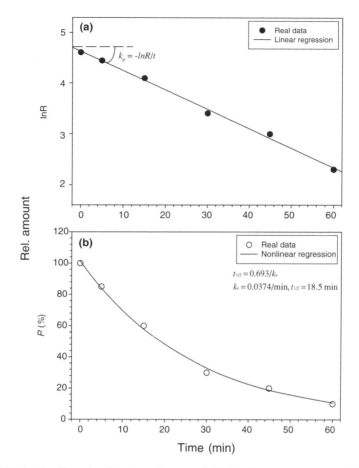

FIGURE 13.1 Semilog plot for the estimates of elimination rate constants (**a**), and the direct time-concentration plot for nonlinear regression analysis (**b**). The semilog plots are conventionally used to convert the exponential decay to the linear first order process facilitating the k_e determination (**a**). The semilog format is no longer necessary if nonlinear regression analysis is applied for the quantification, since k_e is direct calculated following the exponential decay model, as is $t_{1/2}$ (**b**).

The nonlinear regression analyses with default options (automatic initial parameter estimations with the constraints of $a > 0$ and $b > 0$; no weight ht; and iterations of 200 with the step size of 1 and the tolerance of $1e^{-10}$) are initiated following selection of Finish.

TABLE 13.1 Data required for k_e and $t_{1/2}$ determination.

t (min)	0	5	15	30	45	60
R (%)	100	85	60	30	20	10
$\ln R$	4.61	4.44	4.09	3.40	3.00	2.30

TABLE 13.2   Results of nonlinear regression analysis for k_e and $t_{1/2}$ determination.

Equation: Single exponential decay ($y = a\exp(-bx)$)				
r	r^2	a		b
0.999	0.997	101.3		0.037
		($P < 0.0001$)		($P < 0.0001$)
Regression				
	DF	SS	MS	F	P
	1	6702.5	6702.5	1464.5	<0.0001
Statistical test				
	Normality			Constant variance
	Pass			Pass
Kinetic parameter				
	k_e			$t_{1/2}$ (min)
	0.037			18.7

The typical results of such an analysis, in which the data in Table 13.1 were used, are shown in Table 13.2.

Intrinsic Clearance

  i. Intrinsic clearance (CL$_{int}$), the key parameter for *in vitro–in vivo* correlation, can be directly derived from *in vitro* $t_{1/2}$ (Obach et al., 1997; Reddy et al., 2005; Slaughter et al., 2003; Zhao et al., 2005):

$$\ast \quad \mathrm{CL_{int}} = k_e \mathrm{SF, or}$$

$$\mathrm{CL_{int}} = 0.693/t_{1/2}\mathrm{SF}. \tag{13.3}$$

If liver microsomal preparations are used as the enzyme source,

$$\mathrm{SF} = (G_{\mathrm{micrs}}/W_{\mathrm{liver}})(W_{\mathrm{liver}}/W_{\mathrm{body}})/C_{\mathrm{protein}}, \quad \text{with}$$

SF, scaling factor (mL/kg); G_{micrs}, the average quantity of microsomal protein in the liver (mg), W_{liver}, liver weight (g); W_{body}, body weight (kg); and C_{protein}, protein concentration in the reaction mixture (mg/mL).

$G_{\mathrm{micrs}}/W_{\mathrm{liver}}$ is usually considered to be around 50 mg/g (Iwatsubo et al., 1997), while $W_{\mathrm{liver}}/W_{\mathrm{body}}$ is approximately 20 g/kg in humans; thus,

$$\mathrm{CL_{int}} = 1000 \times 0.693/(t_{1/2}\,C_{\mathrm{protein}}) = \underline{693/(t_{1/2}\,C_{\mathrm{protein}})}, \text{ in } \underline{\text{units}} \text{ of } \underline{\text{mL}}/\mathrm{min/kg},$$

or,

$$\mathrm{CL_{int}} = 0.693/(t_{1/2}\,C_{\mathrm{protein}}) \text{ in units of } \underline{\mathrm{L}}/\mathrm{min/kg}. \tag{13.4}$$

If metabolic stability is determined in a suspension of hepatocytes,

$$SF = (G_{cell}/W_{liver})(W_{liver}/W_{body})/C_{cell}$$

where G_{cell} is the average number of cells in the liver, C_{cell}, the cell density (or concentration) in the reaction mixtures (e.g., 10^6 cells/mL).

G_{cell}/W_{liver} is approximately 120×10^6 cells/g for humans (Bayliss et al., 1990); thus,

$$CL_{int} = 2400 \times 0.693/(t_{1/2} C_{cell}) = 1663/(t_{1/2} C_{cell}) \text{ (mL/min/kg), or}$$

$$CL_{int} = 1.663/(t_{1/2} C_{cell}) \text{ with units of L/min/kg.} \qquad (13.5)$$

It should be noted that the SFs used here are physiologically based. However, despite their general acceptance, these SFs may not always lead to reasonable *in vivo* predictions, and SFs other than the physiologically-based should be also considered (Ito and Houston, 2005; Naritomi et al., 2001).

The previously described assays for detecting substrate disappearance or half-life, however, are not suitable for the determination of metabolic rates, which are required for the kinetic delineation of individual metabolic pathways.

ii. CL_{int} can be estimated from rates of metabolism, usually gauged from the formation of the metabolites, if authentic metabolite standards or radioactively labeled parent compounds are available. However, the metabolite standards or radioactively labeled compounds are often unavailable, particularly in early discovery. The rates of product formation should be determined as the initial step towards an understanding of the enzymatic basis of metabolism.

Rate calculation is straightforward when the amount of enzyme in the reaction system is known:

$$V = M/(tE) \qquad (13.6)$$

where V is reaction rate, M is the quantity of metabolite formed during the incubation period t, and E is the quantity of the enzymes or proteins in the incubation mixture. V is often in units of pmol metabolites/min/mg protein or pmol metabolites/min/pmol enzyme.

The incubation time t, although theoretically as short as possible, is typically between 15 and 45 min. Such a reaction period is based largely on the following factors: the metabolite formation rates, the linearity of metabolite formation rates, the responsible enzyme thermostabilities, and substrate and enzyme concentrations. For CYP-mediated reactions the rates of metabolite formation are often moderate, and the CYP enzymes are often thermostable, resulting in linear metabolite production of up to 45 min of incubation. Nevertheless, the concentrations of the compounds and microsomal proteins can largely govern the suitable incubation time, as well as the overall metabolic rates.

Michaelis–Menten (M–M) hyperbolic kinetics (Eq. 13.7) is often assumed and directly applied in rate determination, particularly in early discovery when the enzyme kinetic behavior is usually unknown:

$$V = V_{max}S/(S + K_m) \qquad (13.7)$$

$$\text{If } S \ll K_m, \quad V = V_{max}S/K_m = CL'_{int}S, \quad \text{or}$$

$$CL'_{int} = V/S. \qquad (13.8)$$

Equation 13.8 provides one of several means for estimating CL'_{int}. Again, this approach is valid only when the initial substrate concentration is much lower than the anticipated K_m, that is, $S < K_m/5$ (Soars et al., 2002). Here V is reaction rate, V_{max} is the maximal rate, and K_m is the substrate concentration at half maximum rate, and S is substrate concentration.

Alternatively, reaction rates can be expressed as follows (Zhang and Wong, 2005):

$$V = (k_{cat}/K_m)ES \qquad (13.9)$$

where k_{cat} is the maximal catalytic rate constant, and k_{cat}/K_m is defined as catalytic efficiency.

The rate is thus a first-order function of the substrate and enzyme concentrations (Eq. 13.9). Although governed by two variables, enzyme concentrations in the steady state are pseudoconstant during a normal incubation period, that is, 15 or 30 min, while the substrate concentrations are continuously reduced as a consequence of the formation of metabolites.

If the substrate concentration (C_t) is reduced to 60% of the initial concentration (C_0) during a given reaction period (t), the rate at the end of that period (V_t) will theoretically be 60% of the initial rate (V_0) (Eq. 13.9). Therefore, the initial rates are apparently underestimated when Equation 13.6 is used; instead, this equation determines the average rates during the reaction period. To minimize such a potential bias, the substrate concentration during the reaction periods should be pseudoconstant or, at most, reduced by 20%; such reduction can often be achieved by reducing enzyme quantities and/or incubation time.

Besides the possible inaccuracy of the calculation, the determination of rate can be further complicated by potential atypical enzyme kinetics or "allosteric" effect, which is not uncommon for several important drug-metabolizing CYP members, especially CYP3A4 (Shou et al., 2001). While atypical kinetics is poorly understood mechanistically, awareness of some characteristics of atypical kinetics might be beneficial in the recognition of such kinetic behaviors. For instance, certain forms of atypical kinetics, substrate inhibition and activation in particular, tend to manifest at relatively high substrate concentrations *in vitro* (Lin et al., 2001; Tracy, 2003), and thus have uncertain pharmacological relevance (Zhang and Wong, 2005).

13.3 CHARACTERIZATION OF ENZYME KINETICS

One apparent difference between drug discovery and development is the level of comprehension. Speed dictates the early discovery process, while in-depth understanding, which requires the synthetic standards of metabolites and/or the radiolabeled drug candidates, is emphasized in later discovery and particularly development. Therefore, metabolic stability is studied in the context of enzyme kinetics, that is, the determination of K_m and V_{max}, in later discovery and development. These parameters are derived from the Michaelis–Menten kinetics, and thus may not be directly applicable to atypical enzyme kinetics.

13.3.1 Basic Theory

As described earlier, V_{max} is the maximal rate, and K_m is the Michaelis constant, which can in practice be viewed as the substrate concentration at half maximal rate. A brief review of the relevant background follows.

In this scheme, here E, S, [ES] and P are the concentrations of the enzymes, substrates, substrate-bound enzymes, and enzyme products (metabolites), respectively, k_1 is the association rate constant, and k_2 and k_3 are the dissociation rate constants from [ES] to here E, S, [ES] and P, respectively.

As shown in Scheme 13.1,

$$V = k_3[ES] \tag{13.10}$$

At steady state,

$$k_1 ES = k_2[ES] + k_3[ES],$$

$$E + S \underset{k_2}{\overset{k_1}{\rightleftharpoons}} ES \xrightarrow{k_3} E + P$$

SCHEME 13.1   The classic enzyme catalytic process.

thus,

$$[ES] = ESk_1/(k_2 + k_3). \tag{13.11}$$

$(k_2 + k_3)/k_1$ is defined as the Michaelis constant (K_m).

Equation 13.11 becomes $[ES] = ES/K_m$, and Equation 13.10 is converted to $V = (k_3/K_m)ES$, or $V = (k_{cat}/K_m)ES$, as well as $V = (V_{max}/K_m)ES$ as an alternative.

k_{cat}, the maximal catalytic rate constant, is usually expressed as the number of metabolite molecules formed by an enzyme molecule in 1 s (or simply in units of s^{-1}) and is one of the intrinsic properties of a given enzymatic reaction. However, it is not always readily determined *in vitro*, for example, in microsomal assays, since the quantities of the individual enzymes involved are usually unknown. In contrast, the protein concentrations are easily determined. Thus, V_{max}, an alternative to k_{cat}, expressed in nmol/min/mg and, is more readily determined. However, there is a subtle difference between k_{cat} and V_{max} due to the differences between defined and undefined enzyme quantities. V_{max}, in contrast to the constant k_{cat}, will likely vary for a given enzymatic reaction, due to the potentially variable enzyme content in the enzyme preparations.

13.3.2 Experimental Design

Prior to enzyme kinetic investigation, the linearity of the relationship between the metabolite formation and the enzyme quantity and reaction time should be determined. The time courses and enzyme concentration dependencies of CYP activities are often linear when determined at concentration ranges for microsomal proteins of 0.2–2 mg/mL, although this should be confirmed for each metabolic pathway being studied.

As the rates for CYP-mediated reactions are usually either low or moderate from a standard enzymology perspective, the reaction time for the linearity assay could be extended to 60 min, or longer. Six time points for such assays are common, for example, 0, 5, 15, 30, 45, and 60 min.

After the testing for linearity (at typically 0.2 or 1 mg/mL of microsomal proteins and a 15- or 30-min period of incubation), experiments to determine other parameters can be designed. The rationale for a suitable assay condition is that such a condition will permit the formation of quantifiable amounts of the metabolites without markedly depleting (i.e., by less than 20%) the substrate.

Estimates of K_m and V_{max} with reasonable accuracy probably require detection of the rates of metabolite formation with substrate concentrations spanning the range from 0.3 to $3 \times K_m^{app}$. K_m^{app} is the apparent value of K_m. Unfortunately, the K_m^{app} must be determined, since it is not readily predicted. The alternative, therefore, is the use of a default set of substrate concentrations, based largely on the general understanding of the kinetic characteristics of metabolic enzymes, particularly CYP-mediated reactions (Guengerich,

2001). For instance, six, or preferably eight, concentrations, covering a range of 0–4 mM, could be considered for the initiation of the study. These concentrations could be 0, 1, 10, 50, 200, 500, 1000, and 4000 μM. The range of substrate concentrations is adjustable, and can be narrowed if K_m^{app} can be approximated. However, for accurate calculations, the metabolic rates generated at the lower substrate concentration (i.e., 1 μM), or the higher substrate concentrations (i.e., 4 mM), may have to be disregarded if, during the incubation, the lower concentrations are markedly changed, or if substrate inhibition, a common form of atypical kinetics, is observed at the higher concentrations. Therefore, if K_m^{app} is presumed to fall within the range of 20–400 μM, the above substrate concentration selection could be revised to 0, 10, 20, 50, 100, 200, 500, and 1200 μM.

13.3.3 Determination of Kinetic Parameters

13.3.3.1 Biochemical Plots Several methods are readily applied to the determination of kinetic parameters (K_m and V_{max}). Traditionally, these terms are determined using the classic biochemical plots, particularly those transformed from the well-known Michaelis–Menten plot, for example, Lineweaver–Burk and Eadie–Hofstee plots (Li et al., 1995; Nakajima et al., 2002; Nnane et al., 2003; Yamamoto et al., 2003).

Lineweaver–Burk plots have been widely used in biochemical studies, although their intrinsic limitations are occasionally overlooked. For instance, as shown in Fig. 13.2, the data points in Lineweaver–Burk plots tend to be unevenly distributed, thus potentially leading to unreliable reciprocals at lower metabolic rates ($1/V$). These lower rates dictate the linear regression curves and, therefore, the apparent values of K_m and V_{max}. In contrast, the data points in Eadie–Hofstee plots are usually homogeneously distributed, and thus tend to be more accurate. Eadie–Hofstee plots are diagnostic for biphasic kinetics that arises from the involvement of multiple enzymes having differing kinetic properties, or from atypical kinetics, such as homotropic cooperation (Fig. 13.3) (Zhang and Kaminsky, 1995; Zhang and Wong, 2005). One example is theophylline 8-hydroxylation, the major primary metabolic pathway of theophylline in humans (Campbell et al., 1987; Zhang and Kaminsky, 1995). The kinetics of theophylline 8-hydroxylation in a HLM system appeared biphasic (Campbell et al., 1987). Multiple CYPs were implicated in the metabolism, including CYP1A2, CYP2D6, CYP2E1, and CYP3A4, characterized by markedly differing substrate binding affinities, as indicated by the K_m^{app} values: 0.6 mM for CYP1A2, but more than 10 mM for the others (Zhang and Kaminsky, 1995). The intrinsic clearance (CL_{int}) estimated by V_{max}/K_m for the individual CYP enzymes, and the correlation analyses of the CYP form-specific activities with theophylline 8-hydroxylation activity at 5 and 40 mM of theophylline using a panel of HLM preparations, confirmed the primary role of CYP1A2 in this major biotransformation of theophylline in humans. There are, however, potential contributions from the other species, CYP2E1, and

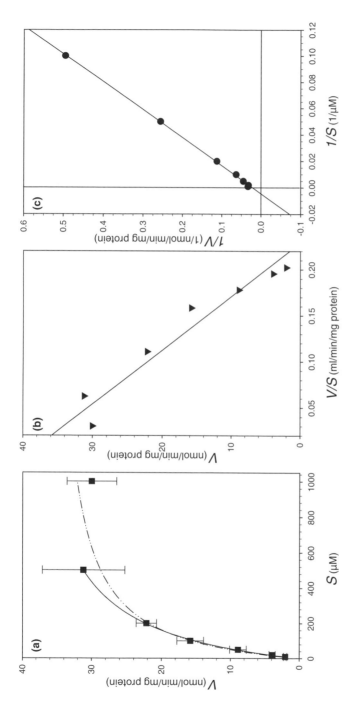

FIGURE 13.2 Biochemical plots for the enzyme kinetic characterizations of biotransformation. (a) Direct concentration-rate or Michaelis–Menten plot; (b), Eadie–Hofstee plot; (c), double-reciprocal or Lineweaver–Burk plot. The Michaelis–Menten plot (a), typically exhibiting hyperbolic saturation, is fundamental to the demonstration of the effects of substrate concentration on the rates of metabolism, or metabolite formation. Here, the rates at 1 mM were excluded for the parameter estimation because of the potential for substrate inhibition. Eadie–Hofstee (b) and Lineweaver–Burk (c) plots are frequently used to analyze kinetic data. Eadie–Hofstee plots are preferred for determining the apparent values of K_m and V_{max}. The data points in Lineweaver–Burk plots tend to be unevenly distributed and thus potentially lead to unreliable reciprocals of lower metabolic rates ($1/V$); these lower rates, however, dictate the linear regression curves. In contrast, the data points in Eadie–Hofstee plot are usually homogeneously distributed, and thus tend to be more accurate.

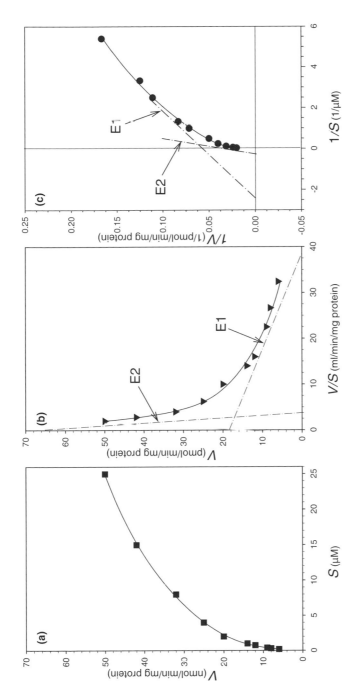

FIGURE 13.3 Determination of the potential involvements of multiple enzymes in a biotransformation pathway using the common biochemical plots. As shown by the plots, (a) Michaelis–Menten plot; (b) Eadie–Hofstee plot; and (c) Lineweaver–Burk plot, at least two enzymatic components (E1 and E2) are responsible for the substrate's biotransformation: one high affinity and low capacity, and the other low affinity and high capacity. Of the three plots shown, the Eadie–Hofstee plot most apparently demonstrates the biphasic kinetics due to either multiple enzymes or possibly the deviations from Michaelis–Menten kinetics, that is, homotropic cooperation.

possibly CYP3A4, at very high theophylline concentrations. The observed biphasic kinetics, as indicated, could be either due to the contribution of multiple enzymes or to atypical kinetics. The key to differentiating between these possibilities is to resolve whether single or multiple enzymes are involved. Although biphasic in the HLM preparations (Campbell et al., 1987), the kinetics of theophylline 8-hydroxylation mediated by each of the responsible CYP enzymes including cDNA-expressed CYP1A2, CYP2D6, CYP2E1, and CYP3A4, was indeed monophasic, suggestive of a multiple enzyme involvement rather than atypical enzyme kinetics.

13.3.3.2 Computational Approach   Computational nonlinear regression analysis, in which the data points in the Michaelis–Menten plots are directly fitted, is the preferred approach for analysis of metabolic kinetic data. Such an approach should be utilized as often as possible, given its unbiased nature. Regression analyses for the determination of K_m and V_{max}, or CL_{int} (or K_m/V_{max}) are described below, for the SigmaPlot (Version 9.0, Syst Inc.) software.

The data, including substrate concentrations, metabolite concentrations, and the calculated rates of metabolite formation, are entered, and Regression Wizard is selected in the pop-up menu of Statistics. Hyperbola and single rectangular with two parameters are chosen for Equation Category and Equation Name, respectively, followed by Next. *XY* Pair is selected and the variables for the equation, $y = ax/(b + x)$, are assigned. The independent variable x is designated as the substrate concentrations (μM) and the dependent variable y as the rates (or the means of rates) of metabolite formation, respectively. The constants of the equation, a and b, thus represent V_{max} and K_m in Eq. 13.7. Following the selection of Finish, the nonlinear regression analysis is performed with a default option setting (automatic initial parameter estimations with the constraints of $a > 0$ and $b > 0$, no weight fit, and iterations of 200 with the step size of 1 and the tolerance of $1e^{-10}$).

Representative results of such analysis, in which the data in Table 13.3 were analyzed, are provided in Table 13.4.

TABLE 13.3   Data transformations for enzyme kinetic characterization.

Primary data		Transformed data		
S (μM)	V^a (nmol/min/mg)	V/S (mL/min/mg)	$1/S$ (L/μM)	$1/V$ (L/nmol/min/mg)
10.0	2.0	0.202	0.10	0.50
20.0	3.9	0.196	0.05	0.26
50.0	8.9	0.178	0.020	0.11
100.0	15.8	0.158	0.010	0.063
200.0	22.2	0.111	0.005	0.045
500.0	31.2	0.0624	0.002	0.032
(1000.0)	(30.0)	(0.0300)	(0.001)	(0.033)

^{a}Mean ($N = 3$); the data in parentheses, due to potential substrate inhibition, were detected but not used for parameter determination.

TABLE 13.4   Results of nonlinear regression analysis for K_m and V_{max} determination.

Equation: Hyperbola ($y = ax/(b + x)$)

r	r^2	a	b
0.999	0.999	42.5 ($P < 0.0001$)	180.0 ($P < 0.0001$)

Regression

	DF	SS	MS	F	P
	1	637.1	637.1	3635.5	<0.0001

Statistical test

Normality	Durbin–Watson	Constant variance
Pass	Pass	Pass

Kinetic parameter

K_m^{app} (μM)	V_{max}^{app} (nmol/min/mg)
180.0	42.5

The apparent V_{max} and K_m were determined to be 42.5 nmol/min/mg protein and 180.0 μM, respectively, indicating a high capacity with a moderate affinity reaction. To ensure the suitability and accuracy of the parameter determination provided by the nonlinear regression analyses, the traditional biochemical plots, such as Michaelis–Menten or Eadie–Hofstee plots, can be used (Fig. 13.2).

Biphasic kinetics should preferably be analyzed using such a computational approach. The above mathematic model can be revised to comprise two independent hyperbolic components: $y = ax/(b + x) + cx/(d + x)$. Here a and b are, respectively, V_{max} and K_m for one kinetic component, and c and d are those for the other, respectively. After the raw data (S and V), used for the construction of the Michaelis–Menten plot shown in Fig. 13.3, were processed using SigmaPlot, following the procedure as described, the results for the nonlinear regression analyses were generated, they are summarized in Table 13.5. Of the

TABLE 13.5   Results of nonlinear regression analyses for K_m and V_{max} determination with two enzymes involved in the catalysis.

Equation: Hyperbola with two components ($y = ax/(b + x) + cx/(d + x)$)

r	r^2	a	b	c	d
0.999	0.999	17.5	0.46	65.1	24.5
		($P < 0.0001$)	($P < 0.0001$)	($P < 0.0001$)	($P < 0.0001$)

Regression

	DF	SS	MS	F	P
	3	2079.2	693.1	1697.0	<0.0001

Statistical test

Normality	Durbin–Watson	Constant variance
Pass	Pass	Pass

Kinetic parameter

K_m^{app} (μM)		V_{max}^{app} (nmol/min/mg)	
0.46	24.5	17.5	65.1

CYPs involved in the metabolism, one has a high affinity and low capacity ($K_{m1}^{app} = 0.46\ \mu M$; $V_{max1}^{app} = 17.5\ nmol/min/mg$ protein), and the other has a low affinity and high capacity ($K_{m2}^{app} = 24.5\ \mu M$; $V_{max2}^{app} = 65.1\ nmol/min/mg$ protein). The basis for these differences has not been resolved.

13.4 QUANTITATIVE ANALYTICAL METHODS

Accurate data acquisition is essential for successful metabolic studies. The current trend from HPLC-based to LC/MS/MS-based analytical tools in ADME research, particularly in the pharmaceutical industry, demonstrates this requirement for accurate data acquisition. General aspects of the operation of LC/MS/MS are described below, together with a brief overview of traditional HPLC systems.

13.4.1 HPLC/UV/FL

These systems, traditionally used for the quantitative determination of metabolites and/or parent compounds, require the compounds to exhibit UV absorbance and/or FL emission, as well as to be absolutely resolved chromatographically. The use of the extended run times necessary to achieve requisite resolution is counterproductive for rapid-throughput analysis. Aromatic compounds are readily analyzed using a HPLC/UV system because of their chromophores with high molar absorptivity, typically eliciting UV absorbance between 240 and 260 nm (single rings), 270 and 290 nm (two fused rings particularly for the heterocyclic systems) and at more than 300 nm (multiple conjugated ring systems). Moreover, compounds that exhibit strong UV absorbance can potentially be monitored by fluorescence. The fluorescent emission wavelengths are longer than the excitation wavelengths, due to the lower emission energy, often by at least 25 nm. For example, if a compound absorbs UV at 320 nm, it is expected to emit fluorescence at wavelengths between 350 and 400 nm, as occurs for the fluorescent warfarin metabolites, 6- and 7-hydroxywarfrin (Takahashi et al., 1997).

Conversely, the HPLC/UV systems have limited applicability for monitoring aliphatic compounds without conjugated π systems due to the rather poor UV absorbance, and an alternative tool, LC/MS/MS in particular, is recommended if available.

13.4.2 LC/MS/MS

13.4.2.1 System and Principle   LC/MS/MS provides a standard analytic system for DMPK studies in the pharmaceutical industry. This system is versatile, and does not have the prerequisites of LC/UV/FL systems. Instead, MS ionization and ion fragmentation of the molecules of interest become essential. The typical instrumental settings and operational conditions for LC/MS/MS systems are described below.

An instrument comprises a HPLC, often consisting of an autosampler, a column compartment, and a binary pumping unit, and a triple quadrupole mass spectrometer (MS/MS). The column effluents are directly diverted into the MS/MS, serving as the detector for the HPLC. The quasi-molecular ions (e.g., a protonated molecules), formed after the molecules in the effluents are ionized in the ion source (Q0), are selected based on the mass-to-charge (m/z) ratios at the first quadrupole (Q1), and are fragmented upon the collisions with molecules of the inert gases, for example, nitrogen, helium, or argon in the second quadrupole (Q2; also called the collision cell). The fragmented product ions are detected at the third quadrupole (Q3). Although a tandem MS can serve as a single quadrupole MS when the function of either Q1 or Q3 is used, sensitive and specific quantitative methods require the use of MS/MS functions.

13.4.2.2 General Operations

HPLC Conditions Columns: Reverse phase, and particularly C18 columns such as Waters Symmetry Shield RP18, are most frequently used (Li et al., 2003; Soars et al., 2002; Zhang et al., 2002).

Mobile phases: There are several frequently used combinations of aqueous and organic solvents. The options for the organic mobile phases are rather limited, usually either acetonitrile (AcCN) or methanol (MeOH). The aqueous mobile phases are often lightly acidified and/or buffered (Li et al., 2003; Zhang et al., 2002). To limit the potential for ion suppression, the salt concentrations in aqueous mobile phases are usually in the low mM range, much lower than what typically applied in the traditional HPLC/UV methods (Fasco et al., 1977).

Flow rate: 250 ± 100 μL/min for a narrow bore column (with 2.0–3.0 mm diameter).

Gradient (*B%*): Depending largely on the physicochemical properties of the analytes, the following gradients, assuming a 10-min run, can be used to initiate the method development.

General: 10% (0–1 min), 90% (4–9.5 min), and 10% (10 min).
Hydrophobic compounds: 20% (0–0.5 min), 95% (2–9.5), and 20% (10 min).
Hydrophilic compounds: 5% (0–2 min), 90% (6–9.5 min), and 5% (10 min).

The HPLC run time dictates the gradient program. If the run time is required to be very short, a mobile phase in the isocratic mode should be considered, with 30–50% organic solvents.

Internal standard (IS): Commercially available small molecules (MW: 200–800), exhibiting the appropriate chromatographic properties and the potential to be ionized in the MS, besides the structural analogs of the analytes, are in principle applicable as internal standards.

MS Condition   The ion source is heated (e.g., $400 \pm 50°C$) with gas flows. Ion spray voltage is usually high (typically 5000 eV in $+$ ESI or -4500 eV in $-$ESI).

The parameters of multiple reaction monitor (MRM) for quantification are varied upon the analytes, with normally the collision energy (CE) of 20–50 eV.

In brief, the short column (10–50 mm) and sharp mobile phase gradients or isocratic elution with high organic solvent content ($\geq 30\%$), facilitating a quick HPLC run (2–10 min), are often used for the detection of single or a few compounds, as is the case for microsomal stability studies. On the contrary, the conventional column (50–150 mm) and the smooth gradients, resulting in a slow HPLC run (15–30 min), are preferably applied for the detection of several analytes such as both parent and metabolites. Moreover, because of the specificity of LC/MS/MS (MRM) methods, it is possible to quantify several analytes in the injection samples pooled from the different reaction mixtures during a single LC run (Hakala et al., 2005).

13.5 PREDICTION OF HUMAN HEPATIC CLEARANCE

13.5.1 Aims

For *in vitro* data, PK properties, particularly for the liver, are potentially predictable (Lin and Lu, 1997). Compared to other possible predictive approaches, such as allometric scaling using animal models, predictions based on *in vitro* human metabolic data have indeed been shown to be reasonably accurate and cost effective (Zuegge et al., 2001). However, it must be borne in mind that the intrinsic differences between *in vitro* and *in vivo* systems inevitably challenge the consistency between pharmacokinetics and *in vitro* results (Masimirembwa et al., 2003; Zhang and Wong, 2005). Human physiology is complex and dynamic, and can by no means be completely simulated *in vitro*. The following should, therefore, be viewed as a simplification.

While several fundamental PK terms including hepatic clearance (CL_h), extraction (E_h) and availability (F_h), could be estimated from *in vitro* intrinsic clearance (CL'_{int}), a pharmacokinetic model describing the concentration of the drugs/xenobiotics in the liver, serving as a link, should be employed. Such potential predictive models include the well-stirred (Rowland et al., 1973; Wilkinson and Shand, 1975), the parallel-tube (Pang and Rowland, 1977a, 1977b; Winkler et al., 1973), the distributed (Roberts and Rowland 1986a; Bass et al., 1978), and the dispersion models (Roberts and Rowland, 1986b, 1986c). Generally speaking, the predictive powers of these models appear to be drug clearance-dependent, with few differences when applied to low clearance drugs; however, they show potential disparities for high clearance drugs (Iwatsubo et al., 1997; Ito and Houston, 2004). The traditional well-stirred model, although not always the most accurate, has been shown to be generally suitable for *in vitro–in vivo* correlations (Ito and Houston, 2004); thus, it is chosen as the one for demonstration in the following content.

The *in vitro–in vivo* correlations have been focused on the use of hepatocytes and liver microsomal preparations to predict those parameters describing metabolic stability, namely, intrinsic clearance (CL_{int}) and hepatic clearance (CL_h) (Houston and Galetin, 2003), supported by the successful *in vivo* predictions reported for the clinical drugs such as verapamil, diazepam, lidocaine, and phenacetin (Bargetzi et al., 1989; Iwatsubo et al., 1997; Kroemer et al., 1992; Seddon et al., 1989; Smith and Timbrell 1974). Basically, two stepwise processes are involved in the prediction of PK properties using these *in vitro* data: scaling-up from CL_{int} determined *in vitro* (CL'_{int}), and estimates of the PK parameters using one of the previously-mentioned mathematical models.

CL'_{int} can be estimated via either the depletion of substrate or the formation of metabolites as described. Substrate disappearance, based on presumed exponential decay is often the choice in early discovery. Mechanisms of metabolite formation, founded primarily on the understanding of enzyme kinetic behavior, can be complicated, and is typically the preference in later discovery and development. In fact, CL'_{int} is appropriately determined using either approach, given the consistency of their results (Obach and Reed-Hagen, 2002).

13.5.2 Procedure of *In vitro–In vivo* Correlation

13.5.2.1 Scaling-Up

Weighing   As previously described, CL'_{int} is equivalent to V_{max}/K_m. However, compounds frequently undergo metabolism via more than one metabolic pathway, possibly mediated by several enzymes. The relationship between CL_{int} and V_{max}/K_m has accordingly been modified:

$$CL'_{int} = \Sigma CL'_{int,i} = \Sigma(E_i/E_T)V_{max,i}/K_{m,i}. \tag{13.12}$$

The values of $V_{max,i}$ and $K_{m,i}$ are assigned to one of the relevant enzymes i involved in one of the primary biotransformation pathways (Tang et al., 2001). E_i is the quantity (or level of protein expression) of the enzyme i, while E_T is the total quantity (or the total protein expression level) of all relevant enzymes. Therefore, $E_i/E_T = 1$ when only one enzyme is significantly contributing to the overall metabolism. As discussed previously, the kinetic parameters (K_m and V_{max}) for individual contributing enzymes can be determined using either the classic biochemical plots or, preferably, nonlinear regression analyses. Alternatively, CL'_{int} can be estimated by measuring the rates at a single, relatively low, substrate concentration, for example, 1 μM. The substrate concentration relevant to the anticipated therapeutic levels will presumably be much lower than the K_m.

As shown earlier,

$$\text{If } S \ll K_m, V/S = V_{max}/K_m = CL'_{int}, \quad \text{or}$$

$$CL'_{int} = \Sigma(E_i/E_T)V_i/S, \text{ if multiple enzymes are involved.} \quad (13.13)$$

V/S is in essence the slope of the pseudolinear portion of the Michaelis–Menten plot.

Again, V_i would preferably be detected, for each metabolic pathway, in HLM or hepatocyte suspensions, if appropriate.

Scaling Factor CL_{int} determined *in vitro* (CL'_{int}) must be multiplied by a SF to be converted to CL_{int} in units of L/min/kg from units of ml/min/mg protein (determined using liver microsomal preparations), ml/min/million cells (determined using hepatocyte suspensions), or ml/min/nmol enzymes (determined using enzyme preparations with known enzyme quantities). SFs based on human physiologic parameters have been described previously, and are listed in Table 13.6.

SFs could be derived using several different approaches (Naritomi et al., 2003; Ito and Houston, 2005). Physiologically based SFs, although widely used as indicated, do not necessarily lead to appropriate *in vivo* predictions, the empirically derived SFs, sometime, might be considered as the alternative (Ito and Houston, 2005). SFs, unlike typical constants, are potentially compound dependent, due in part to their physicochemical properties (Ito and Houston, 2005; Naritomi et al., 2001). Moreover, CL'_{int}, when calculated with nonspecific protein binding taken into account, might correlate better with liver clearance in some cases (Andersson et al., 2001; Ito and Houston, 2005; Obach, 2000; Soars et al., 2002). Therefore, it has been suggested for serum or albumin to be included in the reaction mixtures, to alleviate the discrepancy in protein binding between *in vitro* assay systems and human liver, particularly for acidic

TABLE 13.6   Basic conversion factors for the scaling-up of CL_{int} determined *in vitro*.

Content			References
CYPs[a]	HLM[b]	0.32 nmol/mg protein	Broly et al. (1990), Richard et al. (1991)
	Hepatocytes	0.14 nmol/mill cells	Grant et al. (1987)
	Liver	14.4 nmol/g liver	Obach et al. (1997)
		16.8 nmol/g liver	Iwatsubo et al. (1997)
Microsomal proteins	Liver	45 mg/g liver	Obach et al. (1997)
		52.5 mg/g liver	Iwatsubo et al. (1997)
Hepatocytes	Liver	120 mill cells/g liver	Bayliss et al. (1990)

[a]CYP3A4, the abundant human hepatic CYP form, at the level of approximately 5 nmol/g liver, in average. (Iwatsubo et al., 1997).
[b]Human liver microsomal preparations.

compounds. However, the outcomes of such remedies have been mixed, providing one of the current challenges for *in vitro–in vivo* correlations (Shibata et al., 2000; Soars et al., 2002).

13.5.2.2 In vivo *Prediction* Hepatic clearance (CL_h) in the well-stirred model can be expressed as

$$CL_h = Q_h CL_{int} f_u / (Q_h + CL_{int} f_u). \tag{13.14}$$

Thus,

$$F_h = 1 - CL_h/Q_h = 1/(1 + CL_{int} f_u/Q_h) \tag{13.15}$$

$$E_h = 1 - F_h = 1/[1 + Q_h/(CL_{int} f_u)]. \tag{13.16}$$

If CL_{int} is determined *in vitro* and if f_u is known, the important PK parameters (CL_h, F_h, and E_h) can be directly estimated using Eq. 13.14–13.16, assuming that Q_h is constant, that is, 1.5 L/min, or approximately 21 mL/min/kg for humans (Davies and Morris, 1993). Moreover, if nonspecific protein binding in the *in vitro* assay systems is factored into the equations, f_u should be replaced by $f_u/f_u{}'$, where $f_u{}'$ is the unbound fraction in the *in vitro* metabolic systems, which can be determined using equilibrium dialysis (Soars et al., 2002). Therefore, Eq. 13.14 can be expressed as

$$CL_h = Q_h CL_{int} f_u / f_u{}' / (Q_h + CL_{int} f_u / f_u). \tag{13.17}$$

For many drugs, clearance from the body is due mainly to liver metabolism and renal excretion:

$$CL_T = CL_h + CL_r$$

where CL_T and CL_r are total body and renal clearances, respectively. CL_r is usually determined *in vivo* using urinary excretion and blood concentration data for the parent compound.

If the compound is being developed for oral administration,

$$CL_{po} = CL_T/(F_h f_a) = (CL_h + CL_r)/(F_h f_a)$$

where f_a is the fraction absorbed into the portal vein from the small intestine, and is equal to 1 if the compound is completely absorbed. In the case of the high hepatic clearance with negligible urinary excretion of the parent compound,

$$CL_{po} = CL_h/(F_h f_a) = CL_{int} f_u/f_a \quad \text{or} \tag{13.18}$$

$$CL_{po} = CL_{int} f_u/f_u{}'/f_a \tag{13.19}$$

$$AUC_{po} = Dose/(CL_{int} f_u/f_u{}'/f_a). \tag{13.20}$$

Therefore, the oral clearance or the plasma-concentration-time curve for an oral administration (AUC$_{po}$) can be directly estimated from CL$_{int}$ for those compounds that are exclusively eliminated in the liver.

Nevertheless, the reliability of prediction of *in vivo* hepatic clearance using *in vitro* metabolic data, due largely to its multifactorial nature, is uncertain. The variables, which may potentially affect the accuracy of such an estimate, include, but are not limited to, extrahepatic metabolism, biliary excretion, active transporters, compound solubility, and permeability, gastrointestinal milieu conditions, applicability of the predictive PK models, and interindividual variability.

13.5.3 Examples

Development of *in vivo* metabolic prediction using *in vitro* data, one of the major goals for metabolic stability studies, is accorded high priority, yet remains challenging. Current approaches are still being refined (Andersson et al., 2004). However, successful cases using both human liver microsomal and hepatocyte metabolism data have been reported (Ito and Houston, 2004; Shibata et al., 2000). Recent studies suggest that while human liver microsomal results often yield qualitative predictions (Li et al., 2003), quantitative correlations are likely to be obtained if hepatocytes are used (Ito and Houston, 2004; Iwatsubo et al., 1997; Soars et al., 2002). As indicators of current understanding, several recent studies are presented to illustrate anticipated and unexpected outcomes.

In vitro–in vivo correlation studies have mainly been undertaken for CYP-mediated metabolism. One example is the study reported by Li et al. (2003). Using Eqs. 13.4, 13.8 and 13.14 and physiological SFs, Li et al. predicted the CL$_h$ values of 15 antiparasitic drugs from studies using HLM. With the exception of a single drug with a high CL$_h$, they were able to classify each drug into a category of low, intermediate, or high hepatic clearances based on the HLM CL$_{int}$ values (Li et al., 2003). A similar correlation study that, employed the HLMs from individual donors ($N = 12$) and recombinant enzymes (CYP3A4, CYP2C9 and their allelic variants), was reported as an effort for the assessment of the genotypic impacts on CYP2C9-mediated celecoxib metabolism in man (Tang et al., 2001). Celecoxib is predominantly metabolized by CYP2C9 and CYP3A4 in humans; the potential changes in kinetic properties (K_m and V_{max}) due to CYP2C9 polymorphism were thereby negated by the involvement of CYP3A4 (Eq. 13.12). Thus, the plasma AUC of celecoxib was increased by up to five fold in those individuals carrying the allelic variants, namely CYP2C9*2 (Cys^{144}Ile^{359}) and CYP2C9*3 (Arg^{144}Leu^{359}), consistent with predictions based on the *in vitro* results. In contrast, the average 2.2-fold increase for the homozygous and heterozygous variant CYP2C9*3 (Arg^{144}Leu^{359}) and no apparent changes for the heterozygous variant CYP2C9*2 (Cys^{144}Ile^{359}) were determined in a subsequent clinical study. The authors demonstrated that the CL$'_{int}$ (V_{max}/K_m) for

CYP2C9-mediated celecoxib hydroxylation was reduced by the amino acid substitutions of Ile[359] to Leu[359] and Arg[144] to Cys[144] in the recombinant CYP2C9 proteins, while an *in vivo* prediction of the impact of CYP2C9 genotype was attempted using the *in vitro* results (Tang et al., 2001).

Correlation studies have been also undertaken for metabolism primarily mediated by Phase II metabolic enzymes, UGTs in particular (Andersson et al., 2001; Soars et al., 2002). Soars and coworkers claimed to find significant correlations between human hepatic clearance, predicted using compound depletion in HLM ($t_{1/2}$ and k_e), and clearance determined *in vivo* for 11 UGT substrates, when protein binding and blood partitioning were factored into the correlation scheme. However, the *in vivo* clearance was consistently under-predicted by an order of magnitude. Similar prediction was obtained using the results from either freshly isolated ($N = 4$) or cryopreserved human hepatocytes ($N = 3$). Interestingly, relative to the marked under-prediction when they used the HLM data, the predictions based on the hepatocyte results were more accurate (Soars et al., 2002).

In contrast, studies reported by Andersson et al. (2001, 2004) raise some concerns as to the reliability and utility of *in vivo* prediction using *in vitro* results for both Phase I and Phase II enzyme-mediated metabolism. Using precision-cut liver slices, primary hepatocytes, and liver microsomal preparations, they demonstrated that glucuronidation is the major pathway of biotransformation of almokalant in humans. The metabolic clearance *in vivo*, however, was constantly underpredicted when the results from these *in vitro* systems were used (Andersson et al., 2001). In another study, using four well-characterized CYP2C9 substrates, they showed generally poor predictability of *in vivo* CL_h using CL'_{int} determined in HLMs based on the well-stirred predictive model; these predictions were not improved by inclusion of protein binding. Therefore, the authors questioned *in vitro* metabolic screening as a compound selection tool, in the absence of a proven *in vitro–in vivo* correlation (Andersson et al., 2004).

The studies described in this section demonstrate both success and challenges, and reflect the current understanding of prediction of hepatic clearance using *in vitro* human metabolic data. In contrast to the well-defined concepts, readily applicable biochemical methods, and sensitive analytical techniques available to studies of metabolic stability and enzyme kinetics, efforts to accurately predict *in vivo*, using *in vitro* data, require further refinement.

ABBREVIATIONS

AcCN	acetonitrile
ADME	absorption, distribution, metabolism, and elimination.
APPS	adenosine 3′-phosphate 5′-phosphosulfate

AUC_{po}	the plasma-concentration-time curve of following an oral administration
CL_h	hepatic clearance
CL_{int}	intrinsic clearance
CL'_{int}	intrinsic clearance determined *in vitro*
CL_r	renal clearance
CL_T	total body clearance
CYP	cytochrome P450
DMSO	dimethyl sulfoxide
E_h	hepatic extraction
F_h	hepatic bioavailability
ESI	electrospray ionization
FL	fluorescence
FMO	flavin-containing monooxygenases
HTS	high-throughput screening
f_a	the fraction absorbed in the intestinal tract
f_u	unbound fraction
f_u'	the unbound fraction in the *in vitro* metabolic systems
$Glc\text{-}6\text{-}PO_4$	glucose-6-phosphate
G6PDH	glucose-6-phosphate dehydrogenase
GSH	glutathione (reduced form)
HLM	human liver microsomal preparations
HTS	high-throughput screening
IS	internal standard
k_{cat}	the maximal catalytic rate constant
k_e	the elimination rate constant
K_m	Michaelis constant, or the substrate concentration at half maximum rate
LC/MS/MS	liquid chromatography-triple quadrupole mass spectrometry
MeOH	methanol
MRM	multiple reaction monitoring
NADH	β-nicotinamide adenine dinucleotide (reduced form)
NADPH	β-nicotinamide adenine dinucleotide phosphate (reduced form)
NCE	new chemical entity
PBS	phosphate-buffered saline
PK	pharmacokinetics
Q_h	hepatic flow rate
SF	scaling factor
$t_{1/2}$	half-life

UDPGA	uridine 5′-diphosphate glucuronic acid
UGT	UDP-glucuronosyltransferase
UV	ultraviolet
V_{max}	the maximal rate

REFERENCES

Ackley DC, Rockich KT, Baker TR. Metabolic stability assessed by liver microsomes and hepatocytes. In: Yan Z, Caldwell GW editors. Methods in Pharmacology and Toxicology: Optimization in Drug Discovery: *In Vitro Method.* Totowa, NJ: Human Press Inc; 2004; p151–162.

Adediran SA. Kinetic properties of normal human erythrocyte glucose-6-phosphate dehydrogenase dimmers. Biochimie 1991;73:1211–1218.

Andersson TB, Bredberg E, Ericsson H, Sjoberg H.An evaluation of the *in vitro* metabolism data for predicting the clearance and drug-drug interaction potential of CYP2C9 substrates. Drug Metab Dispos 2004;32:715–721.

Andersson TB, Sjoberg H, Hoffmann KJ, Boobis AR, Watts P, Edwards RJ, Lake BG, Price RJ, Renwick AB, Gomez-Lechon MJ, Castell JV, Ingelman-Sundberg M, Hidestrand M, Goldfarb PS, Lewis DF, Corcos L, Guillouzo A, Taavitsainen P, Pelkonen O. An assessment of human liver-derived *in vitro* systems to predict the *in vivo* metabolism and clearance of almokalant. Drug Metab Dispos 2001;29:712–720.

Apic G, Ignjatovic T, Boyer S, Russell RB. Illuminating drug discovery with biological pathways. FEBS Lett 2005;579:1872–1877.

Bargetzi MJ, Aoyama T, Gonzalez FJ, Meyer UA. Lidocaine metabolism in human liver microsomes by cytochrome P450IIIA4. Clin Pharmacol Ther 1989;46:521–527.

Bass L, Robinson P, Bracken AJ. Hepatic elimination of flowing substrates: the distributed model. J Theor Biol 1978;72:161 184.

Bayliss MK, Bell JA, Jenner WN, Wilson K.Prediction of intrinsic clearance of loxtidine from kinetic studies in rat, dog and human hepatocytes. Biochem Soc Trans 1990;18:1198–1199.

Broly F, Libersa C, Lhermitte M. Mexiletine metabolism *in vitro* by human liver. Drug Metab Dispos 1990;18:362–368.

Busby Jr WF, Ackermann JR, Crespi CL. Effect of methanol, ethanol, dimethyl sulfoxide, and acetonitrile on *in vitro* activities of cDNA-expressed human cytochromes P-450. Drug Metab Dispos 1999;27:246–249.

Campbell ME, Grant DM, Inaba T, Kalow W.Biotransformation of caffeine, paraxanthine, theophylline, and theobromine by polycyclic aromatic hydrocarbon-inducible cytochrome(s) P-450 in human liver microsomes. Drug Metab Dispos 1987;15:237–249.

Chauret N, Gauthier A, Nicoll-Griffith DA. Effect of common organic solvents on *in vitro* cytochrome P450-mediated metabolic activities in human liver microsomes. Drug Metab Dispos 1998;1:1–4.

Crespi CL, Stresser DM. Fluorometric screening for metabolism-based drug–drug interactions. J Pharmacol Toxicol Methods 2000;44:325–331.

Davies B, Morris T. Physiological parameters in laboratory animals and humans. Pharm Res 1993;10:1093–1095.

Dordal A, Lipkin M, Macritchie J, Mas J, Port A, Rose S, Salgado L, Savic V, Schmidt W, Serafini MT, Spearing W, Torrens A, Yeste S. A preliminary study of the metabolic stability of a series of benzoxazinone derivatives as potent neuropeptide Y5 antagonists. Bioorg Med Chem Lett 2005;15:3679–3684.

Easterbrook J, Lu C, Sakai Y, Li AP. Effects of organic solvents on the activities of cytochrome P450 isoforms, UDP-dependent glucuronyl transferase, and phenol sulfotransferase in human hepatocytes. Drug Metab Dispos 2001;29:141–144.

Fasco MJ, Piper LJ, Kaminsky LS. Biochemical applications of a quantitative high-pressure liquid chromatographic assay of warfarin and its metabolites. J Chromatogr 1977;131:365–373.

Grant MH, Burke MD, Hawksworth GM, Duthie SJ, Engeset J, Petrie JC. Human adult hepatocytes in primary monolayer culture. Maintenance of mixed function oxidase and conjugation pathways of drug metabolism. Biochem Pharmacol 1987; 36:2311–2316.

Guengerich FP. Common and uncommon cytochrome P450 reactions related to metabolism and chemical toxicity. Chem Res Toxicol 2001;14:611–650.

Hakala KS, Suchanova B, Luukkanen L, Ketola RA, Finel M, Kostiainen R. Rapid simultaneous determination of metabolic clearance of multiple compounds catalyzed *in vitro* by recombinant human UDP-glucuronosyltransferases. Anal Biochem 2005;341:105–112.

Hewawasam P, Erway M, Moon SL, Knipe J, Weiner H, Boissard CG, Post-Munson DJ, Gao Q, Huang S, Gribkoff VK, Meanwell NA. Synthesis and structure–activity relationships of 3-aryloxindoles: a new class of calcium-dependent, large conductance potassium (maxi-K) channel openers with neuroprotective properties. J Med Chem 2002;45:1487–1499.

Hickman D, Wang JP, Wang Y, Unadkat JD. Evaluation of the selectivity of *in vitro* probes and suitability of organic solvents for the measurement of human cytochrome P450 monooxygenase activities. Drug Metab Dispos 1998;3:207–215.

Holleran JL, Egorin MJ, Zuhowski EG, Parise RA, Musser SM, Pan SS. Use of high-performance liquid chromatography to characterize the rapid decomposition of wortmannin in tissue culture media. Anal Biochem 2003;323:19–25.

Houston JB, Galetin A. Progress towards prediction of human pharmacokinetic parameters from *in vitro* technologies. Drug Metab Rev 2003;35:393–415.

Ito K, Houston JB. Comparison of the use of liver models for predicting drug clearance using *in vitro* kinetic data from hepatic microsomes and isolated hepatocytes. Pharm Res 2004;21:785–792.

Ito K, Houston JB. Prediction of human drug clearance from *in vitro* and preclinical data using physiologically based and empirical approaches. Pharm Res 2005;22:103–112.

Iwatsubo T, Hirota N, Ooie T, Suzuki H, Shimada N, Chiba K, Ishizaki T, Green CE, Tyson CA, Sugiyama Y.Prediction of *in vivo* drug metabolism in the human liver from *in vitro* metabolism data. Pharmacol Ther 1997;73:147–171.

Jenkins KM, Angeles R, Quintos MT, Xu R, Kassel DB, Rourick RA. Automated high throughput ADME assays for metabolic stability and cytochrome P450 inhibition profiling of combinatorial libraries. J Pharm Biomed Anal 2004;34:989–1004.

Kalvass JC, Tess DA, Giragossian C, Linhares MC, Maurer TS. Influence of microsomal concentration on apparent intrinsic clearance: implications for scaling *in vitro* data. Drug Metab Dispos 2001;29:1332–1336.

Kroemer HK, Echizen H, Heidemann H, Eichelbaum M. Predictability of the *in vivo* metabolism of verapamil from *in vitro* data: contribution of individual metabolic pathways and stereoselective aspects. J Pharmacol Exp Ther 1992;260: 1052–1057.

Krueger SK, Williams DE. Mammalian flavin-containing monooxygenases: structure/ function, genetic polymorphisms and role in drug metabolism. Pharmacol Ther 2005;106:357–387.

Lai F, Khojasteh-Bakht SC.Automated online liquid chromatographic/mass spectrometric metabolic study for prodrug stability. J Chromatogr B Analyt Technol Biomed Life Sci 2005;814:225–232.

Levine B, Zhang X, Smialek JE, Kunsman GW, Frontz ME. Citalopram distribution in postmortem cases. J Anal Toxicol 2001;25:641–644.

Li AP, Rasmussen A, Xu L, Kaminski DL.Rifampicin induction of lidocaine metabolism in cultured human hepatocytes. J Pharmacol Exp Ther 274:1995;673–677.

Li XQ, Bjorkman A, Andersson TB, Gustafsson LL, Masimirembwa CM. Identification of human cytochrome P(450)s that metabolise anti-parasitic drugs and predictions of *in vivo* drug hepatic clearance from *in vitro* data. Eur J Clin Pharmacol 2003;59:429–442.

Liederer BM, Borchardt RT. Stability of oxymethyl-modified coumarinic acid cyclic prodrugs of diastereomeric opioid peptides in biological media from various animal species including human. J Pharm Sci 2005;94:2198–2206.

Lin JH, Lu AY. Role of pharmacokinetics and metabolism in drug discovery and development. Pharmacol Rev 1997;49:403–449.

Lin Y, Lu P, Tang C, Mei Q, Sandig G, Rodrigues AD, Rushmore TH, Shou M.Substrate inhibition kinetics for cytochrome P450-catalyzed reactions. Drug Metab Dispos 2001;29:368–374.

MacKenzie AR, Marchington AP, Middleton DS, Newman SD, Jones BC.Structure-activity relationships of 1-alkyl-5-(3,4-dichlorophenyl)- 5-[2-[(3-substituted)-1-azetidinyl]ethyl]-2-piperidones. 1. Selective antagonists of the neurokinin-2 receptor. J Med Chem 2002;45:5365–5377.

Mandagere AK, Thompson TN, Hwang KK. Graphical model for estimating oral bioavailability of drugs in humans and other species from their Caco-2 permeability and *in vitro* liver enzyme metabolic stability rates. J Med Chem 2002;45:304–311.

Masimirembwa CM, Bredberg U, Andersson TB.Metabolic stability for drug discovery and development: pharmacokinetic and biochemical challenges. Clin Pharmacokinet 2003;42:515–528.

Mei Q, Tang C, Lin Y, Rushmore TH. Shou M. Inhibition kinetics of monoclonal antibodies against cytochromes P450. Drug Metab Dispos 2002;30:701–708.

Nakajima M, Tanaka E, Kwon JT, Yokoi T.Characterization of nicotine and cotinine N-glucuronidations in human liver microsomes. Drug Metab Dispos 2002;30:1484–1490.

Naritomi Y, Terashita S, Kimura S, Suzuki A, Kagayama A, Sugiyama Y. Prediction of human hepatic clearance from *in vivo* animal experiments and *in vitro* metabolic studies with liver microsomes from animals and humans. Drug Metab Dispos 2001;29:1316–1324.

Naritomi Y, Terashita S, Kagayama A, Sugiyama Y. Utility of hepatocytes in predicting drug metabolism: comparison of hepatic intrinsic clearance in rats and humans *in vivo* and *in vitro*. Drug Metab Dispos 2003;31:580–588.

Nnane IP, Damani LA. Involvement of cytochrome P450 and the flavin-containing monooxygenase(s) in the sulphoxidation of simple sulphides in human liver microsomes. Life Sci 2003;73:359–369.

Obach RS. Prediction of human clearance of twenty-nine drugs from hepatic microsomal intrinsic clearance data: An examination of *in vitro* half-life approach and nonspecific binding to microsomes. Drug Metab Dispos 1999;11:1350–1359.

Obach RS. Metabolism of ezlopitant, a nonpeptidic substance P receptor antagonist, in liver microsomes: enzyme kinetics, cytochrome P450 isoform identity, and *in vitro–in vivo* correlation. Drug Metab Dispos 2000;28:1069–1076.

Obach RS. The prediction of human clearance from hepatic microsomal metabolism data. Curr Opin Drug Discov Devel 2001;4:36–44.

Obach RS, Baxter JG, Liston TE, Silber BM, Jones BC, MacIntyre F, Rance DJ, Wastall P. The prediction of human pharmacokinetic parameters from preclinical and *in vitro* metabolism data. J Pharmacol Exp Ther 1997;283:46–58.

Obach RS, Reed-Hagen AE. Measurement of Michaelis constants for cytochrome P450-mediated biotransformation reactions using a substrate depletion approach. Drug Metab Dispos 2002;30:831–837.

Pang KS, Rowland M. Hepatic clearance of drugs. I. Theoretical considerations of a "well-stirred" model and a "parallel tube" model. Influence of hepatic blood flow, plasma and blood cell binding, and the hepatocellular enzymatic activity on hepatic drug clearance. J Pharmacokinet Biopharm 1977a;5:625–653.

Pang KS, Rowland M. Hepatic clearance of drugs. II. Experimental evidence for acceptance of the "well-stirred" model over the "parallel tube" model using lidocaine in the perfused rat liver *in situ* preparation. J Pharmacokinet Biopharm 1977b;5:655–680.

Raucy JL, Lasker JM. Isolation of P450 enzymes from human liver. Methods Enzymol 1991;206:577–587.

Reddy A, Heimbach T, Freiwald S, Smith D, Winters R, Michael S, Surendran N, Cai H. Validation of a semi-automated human hepatocyte assay for the determination and prediction of intrinsic clearance in discovery. J Pharm Biomed Anal 2005;37:319–326.

Richard B, Fabre G, De Sousa G, Fabre I, Rahmani R, Cano JP. Interspecies variability in mitoxantrone metabolism using primary cultures of hepatocytes isolated from rat, rabbit and humans. Biochem Pharmacol 1991;41:255–262.

Roberts MS, Rowland M. Correlation between *in vitro* microsomal enzyme activity and whole organ hepatic elimination kinetics: analysis with a dispersion model. J Pharm Pharmacol 1986a;38:177–181.

Roberts MS, Rowland M.A dispersion model of hepatic elimination: 2. Steady-state considerationsinfluence of hepatic blood flow, binding within blood, and hepatocellular enzyme activity. J Pharmacokinet Biopharm 1986b;14:261–288.

Roberts MS, Rowland M.A dispersion model of hepatic elimination: 3. Application to metabolite formation and elimination kinetics. J Pharmacokinet Biopharm 1986c;14: 289–308.

Rowland M, Benet LZ, Graham GG. Clearance concepts in pharmacokinetics. J Pharmacokinet Biopharm 1:1973;123–136.

Satoh T, Hosokawa M. The mammalian carboxylesterases: from molecules to functions. Annu Rev Pharmacol Toxicol 1998;38:257–288.

Satoh T, Taylor P, Bosron WF, Sanghani SP, Hosokawa M, La Du BN. Current progress on esterases: from molecular structure to function. Drug Metab Dispos 30:2002;488–493.

Seddon T, Michelle I, Chenery RJ. Comparative drug metabolism of diazepam in hepatocytes isolated from man, rat, monkey and dog. Biochem Pharmacol 1989;38:1657–1665.

Shen M, Xiao Y, Golbraikh A, Gombar VK, Tropsha A. Development and validation of k-nearest-neighbor QSPR models of metabolic stability of drug candidates. J Med Chem 2003;46:3013–3020.

Shibata Y, Takahashi H, Ishii Y.A convenient *in vitro* screening method for predicting *in vivo* drug metabolic clearance using isolated hepatocytes suspended in serum. Drug Metab Dispos 2000;28:1518–1523.

Shimada T, Yamazaki H, Mimura M, Inui Y, Guengerich FP. Interindividual variations in human liver cytochrome P-450 enzymes involved in the oxidation of drugs, carcinogens and toxic chemicals: studies with liver microsomes of 30 Japanese and 30 Caucasians. J Pharmacol Exp Ther 1994;270:414–423.

Shou M, Lin Y, Lu P, Tang C, Mei Q, Cui D, Tang W, Ngui JS, Lin CC, Singh R, Wong BK, Yergey JA, Lin JH, Pearson PG, Baillie TA, Rodrigues AD, Rushmore TH.Enzyme kinetics of cytochrome P450-mediated reactions. Curr Drug Metab 2001;2:17–36.

Skopp G, Klingmann A, Potsch L, Mattern R.*In vitro* stability of cocaine in whole blood and plasma including ecgonine as a target analyte. Ther Drug Monit 2001;23:174–181.

Slaughter D, Takenaga N, Lu P, Assang C, Walsh DJ, Arison BH, Cui D, Halpin RA, Geer LA, Vyas KP, Baillie TA.Metabolism of rofecoxib *in vitro* using human liver subcellular fractions. Drug Metab Dispos 2003;31:1398–1408.

Smith DA, van de Waterbeemd H.Pharmacokinetics and metabolism in early drug discovery. Curr Opin Chem Biol 1999;3:373–378.

Smith RL, Timbrell JA.Factors affecting the metabolism of phenacetin. I. Influence of dose, chronic dosage, route of administration and species on the metabolism of (1-14C-acetyl)phenacetin. Xenobiotica 1974;4:489–501.

Soars MG, Burchell B, Riley RJ. *In vitro* analysis of human drug glucuronidation and prediction of *in vivo* metabolic clearance. J Pharmacol Exp Ther 2002;301:382–390.

Sugie M, Asakura E, Zhao YL, Torita S, Nadai M, Baba K, Kitaichi K, Takagi K, Takagi K, Hasegawa T.Possible involvement of the drug transporters P-glycoprotein and multidrug resistance-associated protein Mrp2 in disposition of azithromycin. Antimicrob Agents Chemother 2004;48:809–814.

Takahashi H, Kashima T, Kimura S, Muramoto N, Nakahata H, Kubo S, Shimoyama Y, Kajiwara M, Echizen H.Determination of unbound warfarin enantiomers in human plasma and 7-hydroxywarfarin in human urine by chiral stationary-phase liquid chromatography with ultraviolet or fluorescence and on-line circular dichroism detection. J Chromatogr B Biomed Sci Appl 1997;701:71–80.

Tang C, Shou M, Rushmore TH, Mei Q, Sandhu P, Woolf EJ, Rose MJ, Gelmann A, Greenberg HE, De Lepeleire I, Van Hecken A, De Schepper PJ, Ebel DL, Schwartz JI, Rodrigues AD.*In vitro* metabolism of celecoxib, a cyclooxygenase-2 inhibitor, by allelic variant forms of human liver microsomal cytochrome P450 2C9: correlation with CYP2C9 genotype and *in vivo* pharmacokinetics. Pharmacogenetics 2001;11: 223–235.

Tracy TS.Atypical enzyme kinetics: their effect on *in vitro–in vivo* pharmacokinetic predictions and drug interactions. Curr Drug Metab 2003;4:341–346.

Wilkinson GR, Shand DG.Commentary: a physiological approach to hepatic drug clearance. Clin Pharmacol Ther 1975;18:377–390.

Winkler K, Keiding S, Tygstrup N.The liver. In: Paumgartner G, Preisig R. editors. Quantitative Aspects of Structure and Function. Basel: S Karger; 1973. p 144–155.

Yamamoto T, Hagima N, Nakamura M, Kohno Y, Nagata K, Yamazoe Y. Differences in cytochrome P450 forms involved in the metabolism of *N,N*-dipropyl-2-[4-methoxy-3-(2-phenylethoxy)phenyl]ethylamine monohydrochloride (NE-100), a novel sigma ligand, in human liver and intestine. Drug Metab Dispos 2003;31:60–66.

Zhang ZY, Kaminsky LS. Characterization of human cytochromes P450 involved in theophylline 8-hydroxylation. Biochem Pharmacol 1995;50:205–211.

Zhang ZY, King BM, Mollova NN, Wong YN. *In vitro* interactions between a potential muscle relaxant E2101 and human cytochromes P450. Drug Metab Dispos 2002;30:805–813.

Zhang ZY, Wong YN. Enzyme kinetics for clinically relevant CYP inhibition. Curr Drug Metab 2005;6:241–257.

Zhao SX, Forman D, Wallace N, Smith BJ, Meyer D, Kazolias D, Gao F, Soglia J, Cole M, Nettleton D.Simple strategies for reducing sample loads in *in vitro* metabolic stability high-throughput screening experiments: a comparison between traditional, two-time-point and pooled sample analyses. J Pharm Sci 2005;94:38–45.

Ziegler DM. An overview of the mechanism, substrate specificities, and structure of FMOs. Drug Metab Rev 2002;34:503–511.

Zuegge J, Schneider G, Coassolo P, Lave T. Prediction of hepatic metabolic clearance: comparison and assessment of prediction models. Clin Pharmacokinet 2001;40:553–563.

14

PROTOCOLS FOR ASSESSMENT OF *IN VITRO* AND *IN VIVO* BIOACTIVATION POTENTIAL OF DRUG CANDIDATES

ZHOUPENG ZHANG AND JINPING GAN

Xenobiotics including drugs can be biotransformed *in vivo* to various metabolites. Most of the metabolic pathways are considered as detoxification processes, in which polar metabolites are formed and then eliminated from the body. However, xenobiotics can also be metabolized to form reactive intermediates that may react with macromolecules, including protein or DNA. Thus, some reactive intermediates formed from bioactivation of xenobiotics may be one of the causes of the observed adverse drug reactions, including drug-induced liver toxicities (Liebler and Guengerich, 2005; Park et al., 2005; Walgren et al., 2005). For example, Cytochrome P450-mediated bioactivation of acetaminophen results in acute liver damage in both experimental animals and humans (James et al., 2003). Bioactivation of some drugs can result in the observed inhibition of P450 activity (Correia and Ortiz de Montellano, 2005). Although formation of reactive metabolites and covalent protein binding do not necessarily lead to toxicity, they are suspected to play a role in the idiosyncratic reactions of drugs such as troglitazone, clozapine, and diclofenac (Park et al., 2005). Thus, in order to minimize attrition due to bioactivation-related toxicity and associated high development cost, a low propensity to form reactive intermediates may be considered as a desirable property of drug candidates.

Drug Metabolism in Drug Design and Development, Edited by Donglu Zhang, Mingshe Zhu and W. Griffith Humphreys

Bioactivation, or metabolic activation, is a process where xenobiotics are metabolized to proximate and ultimate reactive intermediates (Kalgutkar et al., 2002, 2005). In general, these reactive intermediates are electrophiles and ready to react with nucleophilic macromolecules, including protein and DNA. From chemistry point of view, electrophiles include polarized double bonds, epoxides, carbonium, and iminium ions (Coles, 1984–1985). The nucleophilic portions of macromolecules in biological systems include the thiol groups of cysteinyl residues in proteins and glutathione, amino groups in proteins, DNA and RNA, and phosphate oxygen atoms in DNA and RNA. The chemical hardness and softness of the electrophiles and nucleophiles are mainly dependent on their polarizability. The polarized double bonds and epoxides are considered as soft electrophiles, and iminium ions hard electrophiles. On the contrary, the thiol group is the soft nucleophile in biological systems. The amino and the phosphate groups in DNA and RNA are considered as hard nucleophiles. Reactions occur more readily between electrophiles and nucleophiles of similar hardness (Coles, 1984–1985).

Bioactivation can also lead to the formation of free radicals and reactive oxygen (ROS)/nitrogen/chlorine species, and these species may cause oxidative stress that is deemed as one of the factors in the development of drug-induced toxicity (Comporti, 1989; Aust et al., 1993; Guengerich, 2001). Exposure to ROS has been implicated in diseases such as cancer, Alzheimer's, and atherosclerosis (Butterfield, 1997; Harrison et al., 2003; Klaunig and Kamendulis, 2004). Classic examples of free radical formation/oxidative stress are cytochrome P450-mediated semiquinone radical formation from acetaminophen and reductive dehalogenation of CCl_4 and halothane, DT-diaphorase-mediated semiquinone radical formation from menadione, and peroxidase-mediated radical formation from phenylbutazone and butylated hydroxytoluene (Parkinson, 2001 and references therein). At present time, it is not a common practice to screen new chemical entities (NCEs) for free radical/ROS formation in the pharmaceutical industry. However, there have been excellent reviews in the recent years on the methodologies of detection and measurement of free radical/ROS (Degli Esposti, 2002; Halliwell and Whiteman, 2004). In this section, measurement of intracellular GSH/GSSG levels, one of the surrogate biomarkers of oxidative stress, will be discussed.

Common approaches to evaluate bioactivation potential include screening for glutathione (GSH, Fig. 14.1), N-acetylcysteine (NAc, Fig. 14.1), and cyanide adducts of unlabeled compounds in incubations with liver microsomal preparations or recombinant cytochrome P450 enzymes in the presence of corresponding trapping agents and evaluating covalent binding of radiolabeled compounds to proteins of liver microsomal preparations or hepatocytes. Evaluation of covalent binding of radiolabeled compounds gives a quantitative measure of the extent of bioactivation, hence provides valuable guidance in compound selection for further development. However, covalent protein binding studies are resource intensive and do not provide mechanistic insight of the bioactivation pathways. Trapping studies using trapping agents such as

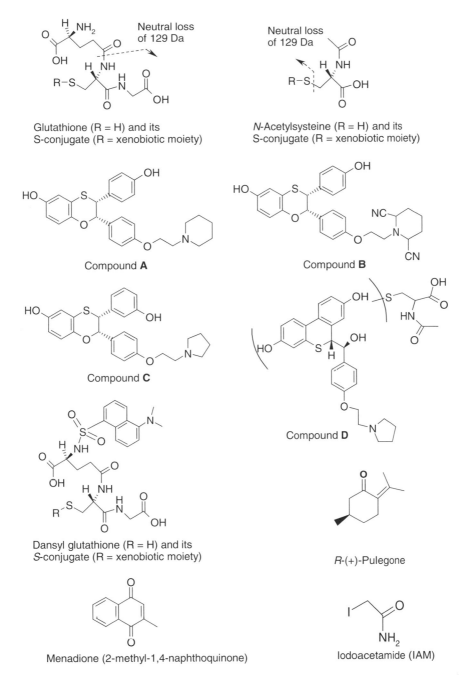

FIGURE 14.1 The structures of glutathione and its conjugate, *N*-acetylcysteine and its conjugate, (2*S*,3*R*)-3-(4-hydroxyphenyl)-2-[4-(2-piperidin-1-ylethoxy)phenyl]-2,3-dihydro-1,4-benzoxathiin-6-ol (**A**) and its bis–cyano adduct (**B**), (2*S*,3*R*)-(+)-3-(3-hydroxyphenyl)-2-[4-(2-pyrrolidin-1-ylethoxy)-phenyl]-2,3-dihydro-1,4-benzoxathiin-6-ol (**C**) and its *N*-acetylcysteine adduct (**D**), dansyl glutathione, *R*-(+)-pulegone, menadione, and iodoacetamide.

GSH and NAc are complementary to covalent protein binding studies because trapping studies are easily implemented and amenable to high throughput screening. Mass spectra and/or NMR spectra of the trapped adducts often provide valuable information of bioactivation pathways to guide further structural modifications. In this section, the use of these approaches will be reviewed, and generic protocols will be described along with typical examples. Future advances in this field will be discussed briefly at the end of this section.

14.1 GLUTATHIONE, *N*-ACETYLCYSTEINE, AND POTASSIUM CYANIDE AS TRAPPING AGENTS

14.1.1 Introduction

It is known that most chemically reactive intermediates are electrophiles that can react with cellular proteins to form covalently bound drug–protein adducts. In order to provide guidance for chemical modification of lead compounds, it is desirable to characterize structures of reactive intermediates for understanding of bioactivation mechanism. However, reactive intermediates are generally unstable in biological systems, and readily react with proteins. On the contrary, reactive intermediates can also react with some trapping agents *in vitro* to form corresponding adducts. The most commonly used agents for trapping reactive intermediates are glutathione, *N*-acetylcysteine, and cyanide ion. These trapping agents can have nucleophilic addition to reactive intermediates to form corresponding adducts. Glutathione is an endogenous tripeptide consisting of L-γ-glutamyl-L-cysteinyl-glycine (Glu-Cys-Gly) with a molecular weight of 307 Da. It is the most commonly used trapping agent in qualitative analysis of GSH-reactive intermediate adducts, because the resulting adducts may be conveniently detected via constant neutral loss scanning of m/z 129 at positive ion mode with triple quadrupole mass spectrometry (Tang and Miller, 2005). The sensitivity and selectivity of analysis of GSH adducts may be increased using an exact neutral loss detection of m/z 129.0426 at positive ion mode via sequential low and high energy MS acquisition in a hybrid quadrupole time-of-flight mass spectrometer (Castro-Oerez et al., 2005) or precursor scanning of m/z 272 at negative ion mode in triple quadrupole mass spectrometer (Dieckhaus et al., 2005). Glutathione ethyl ester (GSH–EE) may also be used as a trapping agent. The use of GSH–EE improved the detection capacities for reactive intermediates (Soglia et al., 2004), because an approximately 10-fold increase in sensitivity in mass spectrometry analysis was observed for adducts containing the GSH–EE moiety compared to GSH. In most cases, mass spectra of GSH adducts offer less structural information than anticipated because most fragment ions of GSH adducts are derived from the GSH moiety. Alternatively, NAc, a thiol molecule with a molecular weight of 163 Da, may be used as a trapping agent. Because the NAc moiety has less fragments than the GSH moiety, the MS–MS spectra of NAc adducts may provide more structural information of modification site on the drug moiety. Even though fragmentations of the NAc adducts also result in a neutral loss of 129 Da (Fig. 14.1), the sensitivity of the constant neutral loss scanning of 129 Da for NAc adducts may not be as high as that for the GSH adducts. Also, NAc may be less efficient than GSH in trapping reactive intermediates if glutathione-*S*-transferase enzymes are involved in the formation of adducts (Evans et al., 2004).

Both GSH and NAc, are thiol-containing soft nucleophiles, which are commonly used to trap soft electrophiles (e.g., quinones and epoxides). On the contrary, cyanide ion is a hard nucleophile, and can readily react with certain hard electrophiles, such as iminium ions. It is applicable to use a mixture of $KCN/K^{13}C^{15}N$ (1:1) in trapping studies, and the isotopic patterns of the corresponding cyanide adducts might be observed in LC/MS/MS analysis (Evans et al., 2004). Inhibitory effects of trapping agents on the metabolism of compounds may need to be evaluated before conducting trapping studies. It was reported that higher concentration of sodium cyanide inhibited metabolism and adduct formation (Ho and Castagnoli, 1980; Ward et al., 1982). However, sodium cyanide at 1 mM or less did not significantly inhibit cytochrome P450 activity in liver microsomes. In addition, investigators may need to pay attentions to some cases, where the formation of trapped adducts was due to artifacts in trapping studies (Gorrod and Sai, 1997).

Even though radiolabeled compounds are not required for trapping studies, because investigators may use LC/MS/MS and NMR techniques to qualitatively characterize the trapped adducts, from which the structures of reactive intermediates may be inferred, the use of the ^{3}H- or ^{14}C-labeled compounds provides an easy means for identifying, characterizing, and quantifying potential adducts in trapping studies. In addition, dansyl glutathione (dGSH, Fig. 14.1), a fluorescence labeled GSH, can also be used as a trapping agent for both qualitative and quantitative analysis of thiol adducts of unlabeled compounds (Gan et al., 2005). Fluorescence detection after HPLC separation allows universal quantitation of adducts without a need of synthetic standards or radiolabels. In addition, online HPLC/MS/UV detection provides another means of semiquantitative analysis of trapped adducts of unlabeled compounds.

It was reported that some GSH adducts formed in trapping studies are highly unstable, and readily undergo degradation or rearrangement (Alt and Eyer, 1998; Doss et al., 2005). Thus, the constant neutral loss scanning of m/z 129 with triple quadrupole mass spectrometry to screen GSH adducts may yield false negative results. Investigators may consider other alternative thiol-containing trapping agents (e.g., NAc, GSH–EE). It was reported that a mixture of GSH and GSH–EE was effective in trapping studies for detecting reactive intermediates (Ge et al., 2006). Two separate GSH and GSH–EE adducts at a mass difference of m/z 28 were detected in LC–MS analysis. This approach may increase the probability for detection of trapped adducts. On the contrary, compounds which are known to form reactive intermediates resulting in covalent protein binding *in vivo* may not be detected in incubations with common trapping agents *in vitro* (Zhang et al., 2003). In such cases, false negative results might be obtained from *in vitro* trapping experiments.

14.1.2 Detection of Glutathione/N-Acetylcysteine or Cyano Adducts Using Ion-Trap Mass Spectrometer

14.1.2.1 Incubation   Human or rat liver microsomes (1 mg protein/mL) are suspended in phosphate buffer (100 mM, pH 7.4) containing EDTA (1 mM), $MgCl_2$ (0.1 mM), and GSH (5 mM, Note 1) or NAc (5 mM) or $KCN/K^{13}C^{15}N$ (1 mM) in a total volume of 1 mL. A test compound in methanol is added to a final concentration of 50 μM, such that the concentration of methanol in the incubation mixture does not exceed 0.2% (v/v). Incubations are performed in the presence of NADPH (1.2 mM) at 37 °C for 60 min. Control experiments contain liver microsomes and a test compound in the absence of either NADPH or trapping agents (Note 2). The reaction is quenched by adding 2 mL of acetonitrile. The suspension is then sonicated for 5 min and centrifuged at $20,800 \times g$ for 10 min. The supernatants are removed and the pellets are extracted twice with 1 mL of methanol–water (3:1, v/v). The extracts are combined with the above supernatants and are evaporated to dryness under nitrogen at room temperature. The residues are dissolved in 300 μL of methanol—water (3:1, v/v), and an aliquot (75 μL) is loaded onto an HPLC column for LC/MS/MS analysis (Zhang et al., 2005).

14.1.2.2 Instrumentation   LC/MS/MS is carried out on a Finnigan LCQ Deca XP Plus ion-trap mass spectrometer (San Jose, CA) interfaced to a HPLC system consisting of two Shimadzu LC–10AD pumps (Kyoto, Japan), a Shimadzu SIL–10AD autoinjector and a Finnigan UV6000LP photodiode array detector (San Jose, CA). The electrospray ionization is employed in the positive ion mode. The heated capillary temperature is set at 275 °C, the normalized collision energy is 42%, the sheath gas flow rate is 60 units, and the auxiliary gas flow rate is 20 units. The ion spray voltage, the capillary voltage, and the tube lens offset are adjusted to achieve maximum sensitivity using the test compound. The collision gas is helium.

14.1.2.3 LC/MS/MS Analysis   The samples (75 μL) are loaded onto an Agilent Zorbax RX-C8 column (4.6 mm × 250 mm, 5 μm, Wilmington, DE). The flow rate is set at 1 mL/min with a 1:5 split to the mass spectrometer ion source and a waste container, respectively. The mobile phase consists of solvent A (5 mM ammonium acetate in water–acetonitrile–acetic acid, 95:5:0.05, v/v/v) and solvent B (5 mM ammonium acetate in acetonitrile–water–acetic acid, 95:5:0.05, v/v/v). The HPLC runs are programmed by a linear increase from 0% to 80% of solvent B during a 30 min period. The mass spectra are recorded in full scan and data-dependent scans of MS^n (n = 2, 3, or 4). The MS^n spectra are recorded by collision-induced dissociation (CID) of molecular ion of the test compound (MH$^+$) species. Several approaches might be used to detect molecular ions of potential adducts. First, by comparing UV spectra of the samples in the presence and absence of trapping agents, the additional UV peaks shown in the samples in the presence of trapping agents might be corresponding to the trapped adducts. Subsequent MS^n analysis of the peaks of interest should be performed to further confirm and characterize the structures of the potential adducts. Secondly, different molecular ions of potential adducts can be generally screened in MS and MS^n analysis. The m/z values commonly used in screening are the molecular weights of the trapping agents plus MH$^+$, or plus molecular weights of modified test compounds (MH$^+$ + 14, MH$^+$ + 16, MH$^+$ + 18, MH$^+$ + 32) (Note 3). Any potential adducts should be further characterized in MS^n analysis. Thirdly, postrun processing of mass spectra generated from data-dependent MS^n acquisition for the neutral fragment (loss) of m/z 129 or 27/29 can be performed in MS^2 analysis for the detection of potential GSH/NAc adducts and cyano adducts, respectively. Subsequent MS^n (n = 3 and 4) analysis of the molecular ions of potential adducts should be performed for further characterization. Two examples of LC/MS/MS analyses of trapped adducts are shown below.

Figure 14.2 shows LC/MS/MS analysis of the bis-cyano adduct **B** from incubation of (2*S*,3*R*)-3-(4-hydroxyphenyl)-2-[4-(2-piperidin-1-ylethoxy)phenyl]-2,3-dihydro-1,4-benzoxathiin-6-ol (**A**, Fig. 14.1) in human liver microsomes in the presence of KCN/K^{13}C^{15}N (Zhang et al., 2005). The MS spectrum extracted in chromatography at m/z 531 (MH$^+ \cdot$NH$_3$) in total ion chromatogram revealed a significant peak at 26.1 min (Fig. 14.2a). The MS spectrum of this adduct showed isotopic peaks at m/z 531, 533, and 535 at an approximate ratio of 1:2:1 (Fig. 14.2b), suggesting the formation of a bis-cyano adduct. The MS^2 spectrum of this adduct, obtained by CID of the m/z 531 species, exhibited product ions at m/z 514 (loss of NH$_3$) and m/z 487 (loss of NH$_3$ and HCN) (Fig. 14.2c). Two minor fragments at m/z 347 and 391 associated with the right-hand side of the dihydrobenzoxathiin moiety were also detected. The MS^3 spectrum, obtained by CID of m/z 487, showed loss of a second HCN molecule (27 Da) to give an ion at m/z 460 (Fig. 14.2d). The other product ions are also shown in Fig. 14.2d. The MS^2 CID spectrum of the m/z 533 species exhibited product ions at m/z 516 (loss of NH$_3$), m/z 489 (loss of NH$_3$ and HCN), and m/z 487 (loss of NH$_3$ and H^{13}C^{15}N) (Fig. 14.2e). A minor fragment at m/z 259 is associated with the loss of the 4-(2-piperidin-1-ylethoxy)phenyl portion. Furthermore, the MS^3 CID spectrum of m/z 489 showed loss of the H^{13}C^{15}N moiety (29 Da) to give an ion at m/z 460 (Fig. 14.2f). Collectively, these LC/MS/MS data suggest the presence of two cyano groups in the piperidine ring, the most logical positions being α- to the ring nitrogen in the adduct **B** (Fig. 14.1). This structure was further confirmed by NMR analysis (Zhang et al., 2005). These results suggest that compound **A** is metabolized by cytochrome P450 in human liver microsomes to form reactive iminium ions, and the proposed mechanism for the formation of the bis–cyano adduct **B** is shown in Fig. 14.3a.

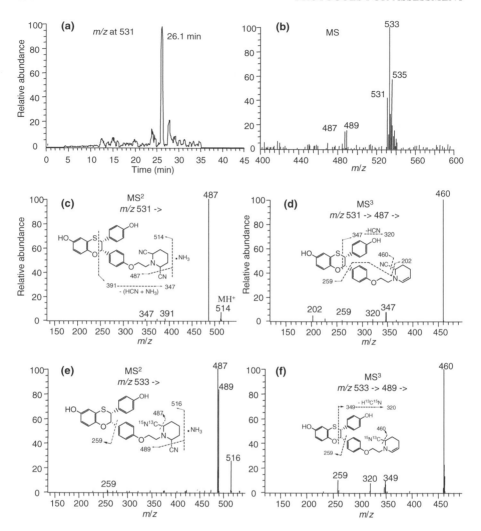

FIGURE 14.2 LC/MS/MS analysis of the bis–cyano adduct **B** of (2S,3R)-3-(4-hydroxyphenyl)-2-[4-(2-piperidin-1-ylethoxy)phenyl]-2,3-dihydro-1,4-benzoxathiin-6-ol (**A**). (**a**) Total ion chromatogram of the mass scanning at m/z 531. (**b**) ESI–MS spectrum of the adduct **B** at 26.1 min; (**c** and **e**), ESI–MS^2 spectra of the adduct **B** obtained by CID of the MH$^+$ · NH$_3$ ion at m/z 531 and 533, respectively. (**d** and **f**), ESI–MS^3 spectra of the adduct **B** obtained by CID of the MH$^+$ · NH$_3$ ion at m/z 487 and 489, respectively. The origin of characteristic fragment ions is as noted.

In trapping studies of (2S,3R)-(+)-3-(3-hydroxyphenyl)-2-[4-(2-pyrrolidin-1-ylethoxy)-phenyl]-2,3-dihydro-1,4-benzoxathiin-6-ol (**C**) in human liver microsomes in the presence of N-acetylcysteine (Zhang et al., 2005), a molecular ion at m/z 611 (MH$^+$ + 161) was detected at 16.7 min (Fig. 14.4a

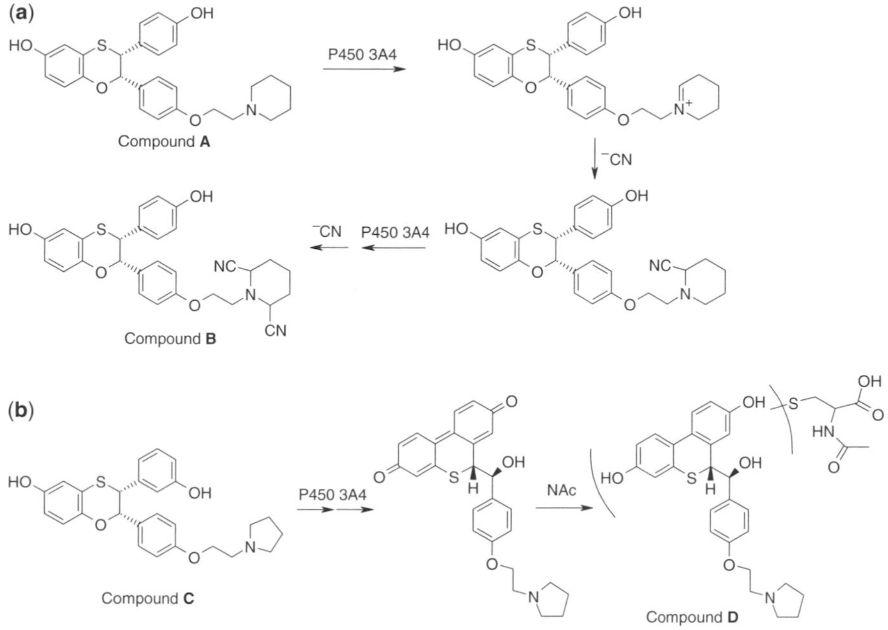

FIGURE 14.3 Proposed mechanism for the cytochrome P450 3A4-mediated bioactivation of dihydrobenzoxathiins **A** (**a**) and **C** (**b**).

and 14.4b), consistent with the addition of the element of *N*-acetylcysteine to the parent molecule. The MS^2 spectrum of *m/z* 611 reveals a loss of the element of water (18 Da) to give a product ion at *m/z* 593 (Fig. 14.4c). The other fragment ion at *m/z* 482 was ascribed to cleavage adjacent to the thioether moiety. The product ion at *m/z* 464 resulted from elimination of the element of water from the fragment ion at *m/z* 482. The MS^3 spectrum of *m/z* 464 generated product ions at *m/z* 393 (loss of the pyrrolidinyl group) and 273 (loss of the 4-(2-pyrrolidin-1-ylethoxyl) group) (Fig. 14.4d). Further NMR analysis of the *N*-acetylcysteine adduct **D** revealed that the element of *N*-acetylcysteine was added to a biphenyl hydroquinone metabolite with the same molecular weight as the parent molecule **C** through structural rearrangements (Zhang et al., 2005). These results suggest that compound **C** is metabolized by cytochrome P450 in human liver microsomes to form a reactive extended quinone intermediate, and the proposed bioactivation mechanism of **C** is shown in Fig. 14.3b.

14.1.3 Protocol for Detection of Glutathione or Cyano Adducts Using Constant Neutral Loss Scanning of Triple Quadrupole Mass Spectrometer

14.1.3.1 Incubation Incubation procedures of test compound in human or rat liver microsomes in the presence of GSH or NAc or KCN and NADPH

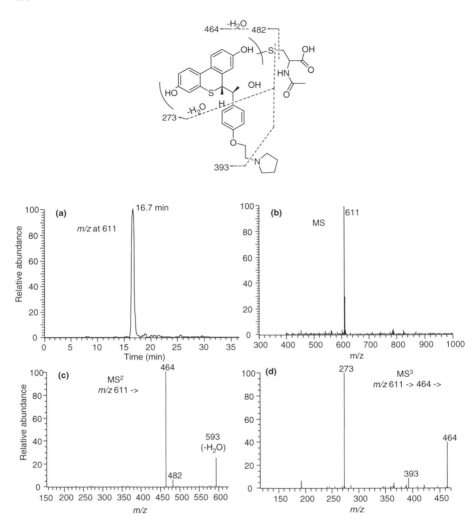

FIGURE 14.4 LC/MS/MS analysis of the *N*-acetylcysteine adduct **D** of (2*S*,3*R*)-(+)-3-(3-hydroxyphenyl)-2-[4-(2-pyrrolidin-1-ylethoxy)-phenyl]-2,3-dihydro-1,4-benzoxathiin-6-ol (**C**). (**a**) Total ion chromatogram of the mass scanning at *m/z* 611. (**b**) ESI–MS spectrum of the adduct D at 16.7 min. (**c** and **d**) ESI–MS–MS spectra of the adduct **D** obtained by CID of the MH⁺ ion at *m/z* 611 (MS²) and 464 (MS³), respectively. The origin of characteristic fragment ions is as noted.

are the same as that described in Section 14.1.2 (the protocol for detection of glutathione/*N*-acetylcysteine or cyano adducts using ion-trap mass spectrometer).

14.1.3.2 Instrumentation LC/MS/MS is carried out on a Perkin-Elmer SCIEX API 3000 tandem mass spectrometer (Toronto, Canada) interfaced to an HPLC system consisting of a Perkin-Elmer Series 200 quaternary pump and a Series 200 autosampler (Norwark, CT). A Turbo IonSpray probe is employed for constant neutral loss scanning of *m/z* 129 in positive ion mode. The instrument parameters are set as follows: declustering potential, 21; focusing potential, 160; entrance potential, 10; collision energy, 40; collision cell exit potential, 12; nebulizer gas, 6; curtain gas, 10; collision gas, 4; ionspray voltage, 4500; temperature, 300 °C; step size, 0.1 Da; pause time, 5 ms; scan rate, 2 s. The collision gas is nitrogen.

Samples (75 μL) are loaded onto an Agilent Zorbax Rx-C8 column (4.6 mm × 250 mm, 5 μm, Wilmington, DE). The flow rate is set at 1 mL/min with a 1:5 split to the mass spectrometer ion source and a waste container, respectively. The mobile phase consists of solvent A (5 mM ammonium acetate in water–acetonitrile–acetic acid, 95:5:0.05, v/v/v) and solvent B (5 mM ammonium acetate in acetonitrile–water–acetic acid, 95:5:0.05, v/v/v). The HPLC runs are programmed by a linear increase from 0% to 80% of solvent B during a 30 min period. The constant neutral loss scanning of *m/z* 129 or 27/29 in LC/MS/MS analysis may afford initial detection of GSH/NAc or cyano adducts. Subsequent MS–MS analysis by CID of MH$^+$ species is required for further structural elucidation of the adducts.

14.1.4 Protocol for Qualitative and Quantitative Analysis of Thiol Adducts Using Dansyl Glutathione (dGSH)

14.1.4.1 Preparation of dGSH   Dansyl glutathione (Fig. 14.1) is prepared by derivatization of glutathione disulfide (GSSG) with dansyl chloride and subsequent cleavage of the disulfide bond by dithiothreitol (Gan et al., 2005).

14.1.4.2 Incubation

(1) Prepare NADPH stock solution (20 mM) in water, test compound stock solution (20 mM) in appropriate solvent, *R*-(+)-pulegone (Fig. 14.1) stock solution (20 mM) in water:acetonitrile (1:1), dGSH stock solution (20 mM, Note 4), potassium phosphate buffer (0.1 M, pH 7.4), and chilled quenching solution (5 mM dithiothreitol in methanol). Human liver microsomes (20 mg/mL) are thawed on ice.

(2) Dilute test compound stock solution to 1 mM with water (or ≤20% organic solvent if necessary) and *R*-(+)-pulegone stock solution to 1 mM with water.

(3) Prepare incubation mixture by adding 10 μL each of stock solutions of human liver microsomes, dGSH, and a test compound to 160 μL of 0.1 M phosphate buffer in 1 mL deep well 96-well plate and preincubate at 37 °C for 5 min. A blank incubation without addition of a test compound is used to prepare a baseline chromatogram for identification of adducts. A control incubation without dGSH is used for identification of potential fluorescence interference from the test

compound and its metabolites. R-($+$)-Pulegone is used as a positive control.

(4) Add 10 μL of the NADPH stock solution to start the reactions. After 30 min of incubation, 400 μL of 5 mM dithiothreitol in methanol is added into each well to stop the reactions (Note 5).

(5) After vortexing and centrifugation, 60 μL of supernatants is injected into a HPLC column for analyses.

14.1.4.3 Instrumentation LC/MS/MS is carried out on a Finnigan LCQ ion-trap mass spectrometer interfaced to a Shimadzu LC-10Avp HPLC system consisting of two pumps, a refrigerated autoinjector and a fluorescence detector. The electrospray ionization is employed in negative ion mode. Full scan mass monitoring is employed at mass range of 500–1100 Da. The heated capillary temperature is set at 270 °C, the sheath gas flow rate is 70 units, and the auxiliary gas flow rate is 20 units. The ion spray voltage, the capillary voltage, and the tube lens offset are adjusted to achieve maximum sensitivity using dGSH. When MS^n (n = 2, 3, or 4) spectra of dGSH adducts are collected, the normalized collision energy is 20–30%. The collision gas is helium.

14.1.4.4 LC/MS/MS Analysis The samples (60 μL) are loaded onto a Phenomenex Prodigy ODS-2 column (4.6 mm × 150 mm). The flow rate is set at 1 mL/min. Fluorescence signal is recorded at excitation wavelength of 340 nm and emission wavelength of 525 nm. The HPLC eluent after fluorescence detector is split to allow a flow of 200 μL/mL into the electrospray source of the ion-trap mass spectrometer. The mobile phase consists of solvent A (0.1% formic acid in water) and solvent B (0.1% formic acid in acetonitrile). The HPLC runs are programmed at 20% of B for 3 min, followed by a linear gradient to 50% B at 23 min, a linear gradient to 90% B at 40 min, 90% B at 43 min, 20% B at 43 min, and 20% B at 50 min. Total run time is 50 min.

14.1.4.5 Data Analysis A calibration curve of the fluorescence detector should be obtained upon installation and once every month using a concentration range of dGSH in water:5 mM dithiothreitol in methanol (1:2). QC standards (preferable with a concentration of 1 μM) should be included in each batch of analysis to ensure accuracy of calibration. The fluorescence response factor is usually very consistent. Thus, daily calibration is not recommended.

Fluorescence chromatograms from the blank, test compound control, and sample incubations are superimposed. Additional peaks present only in the sample chromatogram are considered adducts. Mass spectra (full scan and MS^n) of the peak provide additional confirmation. Quantitation is based on comparison of fluorescence peak area of adduct versus that of dGSH standard (external calibration). Quantitative analysis should not be performed if the test compound or its metabolites have fluorescence interference at the wavelength pair of 340 and 525 nm. The final quantitative result of dGSH adduct formation is expressed as percentage of initial compound concentration. The positive control, *R*-(+)-pulegone, usually produces 8–10% of dGSH adduct under the above conditions (Fig. 14.5). The quantitative analysis of dGSH adducts of some model compounds are shown in Table 14.1.

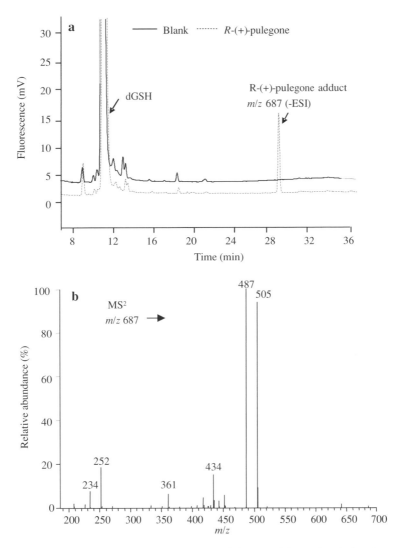

FIGURE 14.5   LC/Fluorescence/MS analysis of dGSH adduct of *R*-(+)-pulegone. (**a**) Fluorescence chromatograms of samples from HLM incubation of *R*-(+)-pulegone in the presence of NADPH and dGSH. Dashed profile was from blank sample without *R*-(+)-pulegone. (**b**), ESI–MS^2 spectrum of *R*-(+)-pulegone adduct obtained by CID of the [M-H]$^-$ ion at *m/z* 687 (negative ionization mode). The fragment ions were all originated from cleavage of the dGSH moiety [(Gan et al, 2005).]

TABLE 14.1 Quantitative analysis of dGSH adducts of model compounds.[a]
(Gan et al, 2005)

Compound	Retention time (min)	Mass (Da)	Postulated adduct composition	Adduct concentration (μM)	% of substrate concentration
	19.5			0.16	
	22.5			0.05	
Troglitazone	23.0			0.04	
	29.3	979	M + dGSH-2H	5.98	
	Total			6.23	12.5
Acetaminophen	11.9	689	M + dGSH-2H	0.26	0.5
Bromobenzene	18.5	726	M + dGSH + 2O-2H	0.10	0.2
	15.1	864	M + dGSH-2H	4.39	
	16.2	880	M + dGSH + O-2H	0.44	
Clozapine	19.5			0.09	
	Total			4.92	9.8
	22.7	815	M + dGSH + O-HCl	0.64	
Diclofenac	23.8	849	M + dGSH + O-2H	0.11	
	Total			0.75	1.5
	14.7			0.16	
	15.1	732	M + dGSH + O-CH$_2$	0.95	
Precocene I	16.4	732	M + dGSH + O-CH$_2$	1.44	
	17.7	762	M + dGSH + 2O	0.29	
	22.4	746	M + dGSH + O	0.38	
	Total			3.22	6.4
R-(+)-Pulegone	28.9	688	M + dGSH-4H	4.90	9.8

[a]Adduct concentrations are taken as average of two determinations. Adduct masses were determined by full scan mass spectrometry in the negative ion mode.

14.1.5 Notes

(1) Investigators may use a mixture of GSH and its stable labeled analog (L-γ-glutamyl-L-cysteinyl-glycine-^{13}C$_2$-^{15}N) at a ratio of 1:1 (Yan and Caldwell, 2004).

(2) Control experiments are conducted to evaluate the NADPH dependency for adduct formations.

(3) In some cases, bis-adducts may be formed in trapping studies. It is recommended to extract MS spectra of these potential adducts in chromatography at a molecular weight of two molecules of the trapping agents plus the molecular weight of the test molecule in MS spectrum analysis.

(4) Because of the natural tendency of oxidation of free thiols, dGSH should be stored at $-80°C$ prior to use. Aliquots of dGSH in water are stable at $-80°C$ for at least 1 year, but not very stable at $-20°C$.

(5) Use of dithiothreitol in MeOH reduces the oxidized dGSH to obtain cleaner chromatograms. Dithiothreitol should be omitted when the potential adduct of interest has a disulfide bond.

14.2　PROTOCOLS FOR *IN VITRO* AND *IN VIVO* COVALENT PROTEIN BINDING STUDIES

14.2.1　Introduction

Bioactivation of xenobiotics leads to the formation of reactive intermediates, which may covalently bind to cellular proteins. General approaches to quantitatively assess the bioactivation potential of compounds include *in vitro* covalent protein binding studies in liver microsomal preparations, and hepatocytes from human and preclinical species, and *in vivo* studies in selected preclinical species. In cases wherein the compounds of interest are mainly metabolized by Phase I metabolism, *in vitro* binding studies in liver microsomal preparations should provide relatively relevant assessment for covalent protein binding potentials *in vivo*. On the contrary, if compounds are mainly cleared through Phase II metabolism or mixed Phase I/II metabolism, *in vitro* binding studies in hepatocytes would be relatively more valuable. *In vivo* covalent protein binding studies have been often performed in rats for practical reasons. At Merck, a high concentration of $10\,\mu M$ and a dose of $20\,mg/kg$ of compounds are commonly used for *in vitro* and *in vivo* covalent protein binding studies, respectively (Evans et al., 2004). This reflects the desire to balance maximizing analytical sensitivity with standardizing protocols. Investigators may use lower doses depending on the intended clinical use of testing compounds.

Covalent protein binding studies require the use of radiolabeled compounds. ^{3}H-tracers are often used because they are relatively easy to synthesize and are of high specific radioactivity. However, possibilities exist that the ^{3}H-label may be lost if oxygenation of the test compound occurs directly on the carbons where the ^{3}H-label is attached. As a consequence, observed binding values could be underestimated. [14]C- tracers, on the contrary, are less likely to lose the [14]C- label, and therefore are preferred. However, possibilities of losing a radiolabel should be considered for any given radiolabeled tracer. Therefore, evaluation of the loss of radiolabels needs to be performed before covalent protein binding studies, particularly when such a possibility is obvious theoretically. Alternative radiolabeled tracers should be made for covalent protein binding, drug metabolism, and disposition studies, if the radiolabel loss is significant. Instead of using radiolabeled compounds of interest, [14C]potassium cyanide is also used in a quantitative high throughput trapping assay for assessing bioactivation potential of some unlabeled compounds (Meneses-Lorente et al., 2006).

At Merck, a conservative "threshold" value of 50 pmol drug equiv./mg protein is currently used as a target upper limit for covalent protein binding evaluations (Evans et al., 2004). It is recommended that this figure should not be considered as an absolute threshold for selecting a compound for development. Other factors, regarded as "qualifying considerations," should be considered on a case-by-case basis. Readers may refer to an industry perspective on minimizing the potential for drug bioactivation in drug discovery and development (Evans et al., 2004).

14.2.2 Protocol for *In vitro* Covalent Protein Binding in Human or Rat Liver Microsomes—A Test-Tube Method

14.2.2.1 In vitro *Covalent Protein Binding in Human or Rat Liver Microsomes*

Rat or human liver microsomes (1 mg protein/mL) are suspended in phosphate buffer (100 mM, pH 7.4) containing EDTA (1 mM) and $MgCl_2$ (0.1 mM) in a total volume of 1 mL. The stock solution of 5 mM of a ^{3}H-labeled compound in methanol is prepared by mixing unlabeled material with radiolabeled material with a final specific radioactivity of 100 Ci/mol, and is added to the above suspension with a final concentration of 10 μM, such that the concentration of methanol in the incubation mixture does not exceed 0.2% (v/v). Incubations are performed in duplicate in the presence of NADPH (1.2 mM) at 37 °C, and quenched with 5 mL of acetonitrile at 0, 30, and 60 min. The duration of the incubations conducted in the presence of sodium cyanide (1 mM) and glutathione (5 mM) is 60 min. Samples are centrifuged at $2500 \times g$ to afford protein pellets, which then are suspended in 1 mL of water and sonicated for 10 min. Four milliliters of ethanol is added to the above suspension and the mixture is vortexed and sonicated for 10 min. The samples are placed at −20 °C for 30 min, and are centrifuged at 4 °C for 10 min. Supernatants are aspirated and the residues are resuspended in 1 mL of water. The above washing procedures are repeated until radioactivity in the supernatant is less than two-fold background. The protein pellets are then dissolved in 0.1 M sodium hydroxide (1 mL), 50% of which is neutralized with 0.1 M hydrochloric acid (0.5 mL) and analyzed by a Beckman Counter liquid scintillation counter (LS6500, Fullerton, CA). The protein concentration in the remaining aliquot is determined using a Pierce bicinchoninic (BCA) protein assay kit (Rockford, IL). Covalent protein binding values in pmol-equiv./mg protein are estimated based on the residual radioactivity in the protein pellets. Control experiments are performed in the absence of NADPH for 60 min. It

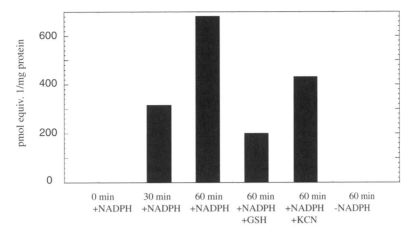

FIGURE 14.6 Covalent protein binding of [^{3}H]**E** in human liver microsomes in the presence or absence of NADPH, glutathione and sodium cyanide. Incubations were performed, as described in Section 14.1.2.

was reported that the covalent protein binding values of reference model compounds in human and rat liver microsomes were ~136 and 574 pmol-equiv./mg protein for [^{3}H]imipramine, 68 and 413 pmol-equiv./mg protein for [^{14}C]diclofenac, 1405 and 1578 pmol-equiv./mg protein for [^{14}C]naphthalene, respectively (Day et al., 2005). An example of covalent protein binding studies of the tritium-labeled dihydrobenzoxazhiin analog **E** in human liver microsomes is shown in Fig. 14.6. The covalent protein binding of **E** was both NADPH and time dependent. In the presence of NADPH, the binding was 685 pmol-equiv./mg protein and was attenuated significantly by glutathione or potassium cyanide (Fig. 14.6).

14.2.2.2   Evaluation of the Loss of the ^{3}H-Label in Incubations of ^{3}H-Tracers with Human or Rat Liver Microsomes   Incubations of ^{3}H-tracers with human or rat liver microsomes are performed for 60 min under the same conditions as described above. Final incubation volume is 1 mL. Reactions are terminated by adding 100 μL of 10% trifluoroacetic acid (TFA) to the final TFA concentration of 1%. Water (400 μL) is then added to the above mixture, and the resulting sample is sonicated for 10 min and vortexed for 10 min. After centrifugation at 2500 × *g* for 20 min, a portion of supernatant (10 μL) is counted for radioactivity. The remaining sample is loaded on a preconditioned 3 cmcc Oasis HLB column (prewashed with water, MeOH, and water), and is slowly passed through the column by gravity (Note 1). The eluate is vortexed and loaded onto a second preconditioned Oasis HLB column. The collected eluate is then vortexed and loaded onto a third preconditioned Oasis HLB column. One portion (0.5 mL) of the last eluate is counted for radioactivity. The second portion (0.5 mL) of the last eluate is evaporated to dryness under nitrogen overnight. The residue is reconstituted in 0.5 mL of MeOH–H$_2$O (2:1) and is counted for radioactivity. The difference in radioactivity between prior and after evaporations is considered as a loss of the amount of the tritium label. By dividing the amount of the tritium loss by the radioactivity in the original sample, an extent of the tritium loss can be estimated.

14.2.3 Protocol for *In vitro* Covalent Protein Binding in Human or Rat Hepatocytes

Cryopreserved human hepatocytes from three male and two female donors or freshly isolated male rat hepatocytes are analyzed for viabilities (75–85%) using the trypan blue exclusion methods. Incubations are performed by suspending the hepatocytes in Krebs-biocarbonate buffer followed by addition of a ^{3}H-labeled compound in methanol. The specific radioactivity of the compounds is 100 Ci/mol. The final concentration of test compound in the suspension is 10 μM in a final volume of 1 mL (1 × 10^6 cells/mL), and the final concentration of methanol does not exceed 0.2% (v/v). Incubations are allowed to proceed at 37 °C for 1 h, and are quenched with acetonitrile (5 mL). The remaining procedures are the same as described in Section 14.2.2 (the protocol for *in vitro* covalent protein binding in human or rat liver microsomes – a test-tube method). Covalent protein binding values in pmol-equiv./mg protein are estimated based on the residual radioactivity in the protein pellets.

14.2.4 Protocol for *In vitro* Covalent Protein Binding in Human or Rat Liver Microsomes—A Semiautomated Method

A grand pool of suspension is prepared by mixing human or rat liver microsomes, a ^{3}H-tracer of a test compound or a reference ^{3}H-tracer (Evans et al., 2004) in phosphate buffer (100 mM, pH 7.4) containing EDTA (1 mM) and MgCl$_2$ (0.1 mM). The specific radioactivity of the tracers is 100 Ci/mol. Aliquots (160 μL) of the suspension are transferred to individual wells of a 96-well plate (Note 2). Reactions are initiated by adding 40 μL of NADPH solution (5 mM). Incubations in duplicate are performed at 37 °C with a final protein concentration of 1 mg/mL, a test compound concentration of 10 μM, a NADPH concentration of 1 mM in a total volume of 200 μL. Reactions are quenched at 0 and 60 min by adding 800 μL of acetone into individual wells and are vortexed for 30 s. Proteins are collected by aspirating samples from the 96-well plate onto a prewet filter mat by a Brandel cell harvester (Brandel, Gaithersburg, MD), and are washed with MeOH–H$_2$O (8:2) for 4 times. Individual filter disks are punched from the filter mat, and are placed in plastic scintillation vials. One milliliter of 7.5% SDS is added to each vial. Caped vials are placed in a water bath shaker at 55 °C and 200 rpm overnight. After the vials are cooled to room temperature, an aliquot (20 μL) of each sample in duplicate is taken for protein assays using the BCA method. The remaining sample in each vial is then mixed with 5 mL of scintillation cocktail Pico-Fluor, and radioactivity is counted by a Beckman Counter liquid scintillation counter. Covalent protein binding in pmol-equiv./mg protein is estimated based on the residual radioactivity in the proteins (Note 3). It was reported that the covalent protein binding values of reference model compounds in human and rat liver microsomes were ~127 and 463 pmol-equiv./mg protein for [^{3}H]imipramine, 57 and 290 pmol-equiv./mg protein for [^{14}C]diclofenac, 1234 and 916 pmol-equiv./mg protein for [^{14}C]naphthalene, respectively (Day et al., 2005).

14.2.5 Protocol for *In vivo* Covalent Protein Binding in Rats

14.2.5.1 **In vivo *Covalent Protein Binding in Plasma and Liver of Rats***   A dose solution of 4 mg/mL is prepared by dissolving/suspending a test compound in ethanol/PEG400/water (1:4:5, v/v/v) with a final specific radioactivity of 3–10 Ci/mol (^{3}H-tracer) or 1.5–3 Ci/mol (^{14}C-tracer). Nine male Sprague–Dawley rats are orally dosed with a test compound at 20 mg/kg. Blood and liver samples of rats are taken at 2, 6, and 24 h postdosing and urine samples are collected at 24 h. Plasma samples are prepared by centrifugation of blood at 4 °C for 30 min. Liver is suspended in PBS buffer (3 mL/g tissue) and the mixture is homogenized to give liver homogenate. The washing procedures are as follows:

(1) Aliquots of samples (0.5 mL of liver homogenates or 0.15 mL of plasma) in duplicate are placed in test tubes, and 1.5 mL of acetonitrile is added to each tube.

(2) The mixtures are sonicated, vortexed for 10 min, and then are placed at −20°C for 30 min.

(3) Samples are centrifuged at 2500 × g at 4 °C for 10 min, and 50 μL of supernatant in each tube is taken for measurement of radioactivity.

(4) The remaining supernatant is carefully aspirated under vacuum, and the remaining pellets are resuspended in 1 mL of water, sonicated, and vortexed.

(5) Four milliliters of ethanol is added to each tube, and the resulting suspensions are sonicated, vortexed, and placed at −20 °C for 30 min.

(6) Steps 3–5 are repeated for four to six times until the radioactivity of the supernatant (0.5 mL) in the last washing is below two times of background readout.

The protein pellets are then dissolved in 4 mL of 0.1 M sodium hydroxide. The remaining procedures for measurement of radioactivity and protein concentrations of the resulting samples are the same as described in Section 14.2.2 (the protocol for *in vitro* covalent protein binding in human or rat liver microsomes—a test-tube method). Covalent protein binding in pmol-equiv./mg

TABLE 14.2   Covalent protein binding of [³H]E in liver and plasma of rats.[a]

	Covalent protein binding (pmol equiv./mg protein)		
Tissue	2 h	6 h	24 h
Liver	7 ± 1	16 ± 2	2 ± 2
Plasma	1 ± 0.2	2 ± 0.2	2 ± 0.2

[a]Male rats were orally dosed with [³H]E at 20 mg/kg. Three rats per time point.

protein is estimated based on the residual radioactivity in the proteins. An example of covalent protein binding studies in liver and plasma of male rats orally dosed with the tritium-labeled dihydrobenzoxazhiin analog **E** at 20 mg/ kg is shown in Table 14.2. The covalent protein binding of **AE** was below 20 pmol-equiv./mg protein in rat liver and plasma samples. All procedures on these animals were in accordance to established guidelines and were reviewed and approved by the Institutional Animal Care and Use Committee (IACUC).

14.2.5.2   *Evaluation of the Loss of the ³H-Label in Rat Urine*   Urine samples of rats ($n = 3$) at 24 h time point are pooled and centrifuged for 10 min. A portion of urine sample (3 mL) is taken, and 10 μL of the sample is counted for radioactivity. The remaining sample is loaded onto a preconditioned 3 cm³ᶜᶜ Oasis HLB column (prewashed with water, MeOH, and water), and is slowly passed through the column by gravity (Note 1). The eluate is vortexed and loaded onto a second preconditioned Oasis HLB column. The collected eluate is then vortexed and loaded onto a third preconditioned Oasis HLB column. One portion (0.5 mL) of the last eluate is counted for radioactivity. The second portion (0.5 mL) of the last eluate is evaporated to dryness under nitrogen overnight. The residue is reconstituted in 0.5 mL of MeOH–H₂O (2:1) and is counted for radioactivity. The difference in radioactivity between prior- and after evaporation samples is the loss of the amount of the tritium label. By dividing the amount of the lost tritium by the radioactivity in the original urine sample, an extent of tritium loss for the urine sample is estimated (Note 4).

14.2.6   Notes

(1) For the optimal performance of retaining drug-related components on the solid phase extraction columns, samples should be passed through the columns by gravity. However, if needed, very low vacuum may be applied.

(2) Additional incubations may also be conducted in the presence of sodium cyanide (1 mM) or glutathione (5 mM) for 60 min.

(3) Readers may also refer to the published procedures for measuring the covalent protein binding using this semiautomated method (Day et al., 2005).

(4) If ^{14}C-tracers are used in *in vivo* covalent protein binding studies in rats, evaluation of the loss of the ^{14}C-label in rat urine may also be warranted if the portion of the molecule with the ^{14}C-label attached is likely to be cleaved to a significantly smaller molecular weight molecule in the metabolism processes.

14.3   PROTOCOL FOR MEASUREMENT OF INTRACELLULAR GSH AND GSSG CONCENTRATIONS IN HEPATOCYTES

14.3.1   Introduction

Bioactivation of drugs and xenobiotics leads to the formation of reactive intermediates. Most of reactive intermediates are electrophiles (e.g, *p*-quinones) and readily react with nucleophiles to form corresponding adducts. Figure 14.7 is a schematic presentation of redox cycling and GSH adduct formation of reactive *p*-quinones. Mechanistically, reactive *p*-quinones formed from bioactivation of drugs and xenobiotics can be reduced by cytochrome P450 and P450 reductase to semiquinone free radicals through one-electron reduction mechanism or by quinone oxidoreductases to hydroquinones (Iskander and Jaiswal, 2005). Semiquinone free radicals can be reoxidized to the quinone form by molecular oxygen with the concomitant formation of superoxide anion radicals. Superoxide anion radicals can be converted to hydrogen peroxide by either enzymatic (superoxide dismutase) or spontaneous dismutation (Bolton et al., 2000). In the presence of trace amount of transition metals, hydrogen peroxide can react with superoxide anion radical to form hydroxyl radical (O'Brien, 1991). Both hydrogen peroxide and hydroxyl

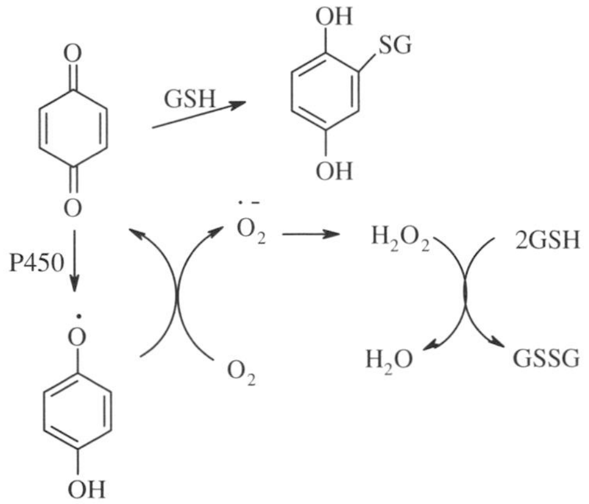

FIGURE 14.7   Redox cycling and formation of GSH adducts of quinones.

radical generated from above redox cycling of quinones can oxidize cysteine residues of proteins, causing oxidative stress in cells.

Glutathione, an endogenous tripeptide (Glu-Cys-Gly), is the most abundant nonprotein sulfhydryl compound in the human body. It exists mainly as the reduced form (GSH) with a little portion as the oxidized glutathione disulfide form (GSSG) *in vivo*. In addition to its role as a nucleophile to react with reactive electrophile intermediates (e.g., *p*-quinones) to form corresponding GSH-hydroquinone adducts (Fig. 14.7), GSH also plays a critical role in the detoxication of reactive oxygen species (e.g., hydrogen peroxide) by the reduction of these species. As a consequence, GSH is oxidized by reactive oxygen species to GSSG.

Collectively, the above protective mechanisms result in a decrease in cellular GSH concentrations and an increase in GSSG concentrations. Therefore, changes of GSH and GSSG concentrations in cells may be used as biomarkers for oxidative stress (Afzal et al., 2002). Freshly isolated rat hepatocytes are commonly used for *in vitro* evaluation of oxidative stress potential of a test compound. Menadione (Fig. 14.1), a quinone compound that can induce oxidative stress in cells (Smith et al., 1985), might be used as a positive control. A negative control that does not use any test compound is also warranted, due to the fact that GSH also can be oxidized to GSSG by molecular oxygen. By comparing the concentrations of GSH and GSSG or the ratios of GSH/GSSG of the test compound with those of menadione as well as the negative control, a relative potential of a compound to induce oxidative stress in cells might be assessed.

Several chemical, enzymatic, and chromatographic methods have been reported for the measurement of GSH and GSSH contents in biological samples. The enzymatic method for total GSH (GSH + GSSG) involves both oxidation of GSH by 5, 5′-dithiobis(2-nitrobenzoic acid) (DTNB) to form GSSG with stoichiometric formation of 5-thio-2-nitrobenzoic acid (TNB) and reduction of GSSG by GSSG reductase to GSH (Anderson, 1985). The total GSH concentration is calculated based on the rate of TNB formation monitored at 412 nm. Chromatographic methods for total GSH (GSH + GSSG) involves initial derivatization of GSH with 2-vinylpyridine, followed by comparison with GSH standard samples in ion-exchange chromatography or high performance liquid chromatography (Anderson, 1985). Total GSH concentration can also be measured by converting GSSG to GSH with $NaBH_4$, followed by derivatization with monobromobimane and fluorescence detection (Svardal et al., 1990). The major disadvantage of these assays is the inability to conveniently measure both GSH and GSSG. Afzal et al. 2002 developed a rapid HPLC–UV method for the quantification of GSH and GSSG in biological samples. It has been reported that both GSH and GSSG concentrations in hepatocytes were measured through LC/MS/MS analysis (Loughlin et al., 2001). In this assay, GSH was derivatized by iodoacetic acid to form S-carboxymethyl derivative (GS–CM). The GS–CM conjugate, together with GSSG, were subjected to direct LC/MS/MS analysis.

Chemically, GSH can also be derivatized by iodoacetamide (IAM, Fig. 14.1) to form the corresponding GS–AM conjugate. The following is a modified protocol using IAM for the analysis of individual GSH and GSSG concentrations in freshly isolated rat hepatocytes.

14.3.2 Measurement of Intracellular GSH/GSSG in Hepatocytes

14.3.2.1 Rat Hepatocytes Rat hepatocytes are freshly isolated by perfusion and are diluted with Krebs-bicarbonate buffer (pH 7.4) to 1×10^6 cells/mL. Cell viability is analyzed by the trypan blue exclusion method.

14.3.2.2 The GS–AM Standard Curve IAM stock solution (10 mM, 100 μL) in 10 mM ammonium bicarbonate buffer (pH 10) is added to amber centrifuge tubes, to which 100 μL of various GSH stock solutions in Krebs-bicarbonate buffer (pH 7.4) is added. The final GSH concentrations are 1, 5, 10, 25, 50, and 100 μM. All samples are vortexed and are placed in dark at room temperature for 60 min. Derivatization reaction is quenched by the addition of 400 μL of acetonitrile. The resulting mixture is mixed with 25 μL of γ-Glu-Glu stock solution (0.25 mM, internal standard) in water/acetonitrile (1:1). After centrifugation, the supernatant (300 μL) is transferred to a 96-well plate for LC/MS/MS analysis.

14.3.2.3 The GSSG Standard Curve The procedures for preparation of the GSSG standard curve are the same as described in the Section 14.3.2.2 (*the GS–AM standard curve*), except that various GSSG stock solutions are used. The final GSSG concentrations are 1, 5, 10, 25, 50, and 100 μM.

14.3.2.4 GSH/GSSG Measurement in Rat Hepatocytes in the Presence of a Test Compound IAM stock solution (10 mM, 100 μL) in 10 mM ammonium bicarbonate buffer (pH 10) is added to amber centrifuge tubes. Menadione stock solutions (6 and 60 mM) and a test compound (2, 10, 20, and 60 mM) in methanol are added to rat hepatocytes. The final concentrations are 30 and 300 μM for menadione and 10, 50, 100, and 300 μM for a test compound. Aliquots (100 μL) of incubation samples are added to the above amber tubes at various time points (0, 5, 10, 20, 30, and 60 min). All samples are vortexed and are placed in the dark at room temperature for 60 min. Derivatization reaction is quenched by addition of 400 μL of acetonitrile. The resulting mixture is then mixed with 25 μL of γ-Glu-Glu stock solution (0.25 mM, internal standard) in water/acetonitrile (1:1). After centrifugation, the supernatant (300 μL) is transferred to a 96-well plate for LC/MS/MS analysis.

14.3.2.5 Instrumentation LC/MS/MS analysis is carried out on a Perkin-Elmer SCIEX API 3000 triple quadrupole mass spectrometer (Toronto, Canada) interfaced to an HPLC system consisting of a Perkin-Elmer Series 200 quaternary pump and a Series 200 autosampler (Norwark, CT). The mass spectrometer employs an electrospray ionization probe with positive ion detection. A Turbo IonSpray probe is employed in positive ion mode. The instrument parameters are optimized and set as follows: declustering potential, 36 (GS–AM), 46 (GSSG), 26 (γ-Glu-Glu); focusing potential, 250 (GS–AM), 340 (GSSG), 230 (γ-Glu-Glu); entrance potential, 10 (GS–AM, GSSG, γ-Glu-Glu); collision energy, 27; collision cell exit potential, 16; nebulizer gas, 6; curtain gas, 10; collision gas, 4; ionspray voltage, 3000; temperature, 150 °C. The collision gas is nitrogen. Samples (100 μL) are loaded onto an Agilent Zorbax SB-phenyl column (4.6 × 250 mm, 5 μm, Wilmington, DE) at the flow rate of 1 mL/min with a 1:5 split to the mass spectrometer ion source and a waste container. The mobile phase consists of 5 mM ammonium acetate in H_2O–acetonitrile–trifluoroacetic acid (95:5:0.1, v/v/v). The HPLC run time is 7 min. The quantitation is based on multiple reaction monitoring (MRM) of the transitions of m/z 365.2 –>236.1 (GS–AM), 613.1 –>355.1 (GSSG), and 277.0 –>148.0 (γ-Glu-Glu).

14.4 PERSPECTIVES

In addition to their propensity to covalent modification of proteins, some reactive metabolites also can react with DNA molecules to form corresponding adducts (Guengerich, 1992; Miller and Miller, 1981; Pereg et al., 2002; Wogan et al., 2004). Therefore, screening of DNA adduct formation could provide some guidance in lead optimization when certain structural modifications can be made to eliminate the potential risk of genotoxicity of a suspected mutagen based on structural information of DNA adducts. Some reactive intermediates (e.g., epoxide, quinone, quinone methide, and carbocation) readily react with nucleosides. Other reactive metabolites (e.g., N-hydroxylamines of aromatic amines), however, usually do not initially react with nucleosides, but require additional bioactivation steps (e.g., O-acetylation, glucuronidation, or sulfation). Simple incubation systems containing liver fractions in the presence of cofactors and nucleosides could be useful for generation of adducts, and LC/MS/MS is the method of choice for characterization of adduct structures. Because of the availability of several well-established platforms for the prediction of mutagenicity (Muller et al., 1999), the widespread use of the DNA adduct screening is not expected.

With significant advances of the LC/MS/MS and other bioanalytical technologies, it becomes possible to qualitatively characterize and monitor covalent drug–protein adducts (Yang et al., 2006). Elucidation of the protein target would not only yield information leading to possible mechanism of toxicity but also provide lead proteins to search for antibodies as a potential marker of toxicity (hypersensitivity) in subjects (Boelsterli et al., 1995; Cohen et al., 1997; Welch et al., 2005). In addition, the covalent modification of cytochrome P450 enzymes is often mechanistically intriguing for mechanism-based cytochrome

P450 inhibitors (Bateman et al., 2004; Blobaum et al., 2002; Correia and Ortiz de Montellano, 2005; Koenigs et al., 1999). A potentially higher success rate is expected in protein adduct characterization by applying the modern proteomic approaches such as tryptic digestion followed by multidimensional separation and high resolution mass spectrometry (Zhou, 2003). However, a few problems intrinsic to protein adduct identification cannot be solved by just the advancement of analytical techniques. These problems include the specific difficulty in proteomic research of membrane proteins such as cytochrome P450 enzymes, instability of some drug–protein adducts, and the lack of specificity of modification sites. Direct mass measurement by MALDI–TOF may circumvent the adduct instability problem, but no detailed information on modification site of proteins can be obtained. Because of the mechanistic nature of these studies and the resources and expertise requirement, experimental protocols for proteomic research will need some time to be optimized.

In addition to reactive intermediates formed in Phase I metabolism, some metabolites (e.g., acyl glucuronides) resulted from Phase II metabolism might also react with proteins to form corresponding adducts, which might result in observed drug-related adverse effects. Acyl glucuronides and xenobiotic acyl thioesters are two important classes of electrophilic intermediates (Boelsterli, 2002). Two major mechanisms have been proposed for the reactions of acyl glucuronides with proteins. The first mechanism is a nucleophilic acyl substitution, resulting in acylation of proteins. The second is involved in a glycation via chain open acyl migration processes (Boelsterli, 2002; Ritter, 2000; Shipkova et al., 2003). Recently, a few groups have shown interesting approaches for the screening of adduct formation from acyl glucuronides (Bolze et al., 2002; Wang et al., 2004). One of these approaches is to use a synthetic lysine-containing peptide to react with acyl glucuronide to form a Schiff's base that can be assayed by LC/MS/MS, and the relative amount of peptide adduct formation may be correlated to the relative amount of protein adduct formation by a variety of acyl glucuronides. This approach requires use of acyl glucuronides, which may be obtained either via purification from incubations with *in vitro* metabolism systems or from chemical synthesis. Alternatively, investigators may consider performing *in vitro* instability studies of acyl glucuronides to assess their potential liability as reactive intermediates (Wen et al., 2006). In addition, acyl glucuronide metabolites may also react with DNA to cause potential genotoxicity. It was reported that the acyl glucuronides of two widely used fibrate hypolipidemics, clofibric acid, and gemfibrozil, caused a concentration-dependent decrease in the transfection efficiency of the DNA in *Escherichia Coli*, with a greater than 80% decrease in phage survival in the presence of 5 mM glucuronides (Sallustio et al., 1997). It was also demonstrated that the acyl glucuronides of probenecid and clofibric may have induced DNA damage in isolated hepatocytes (Sallustio et al., 2006). The mechanism for the acyl CoA thioesters involves in the reaction with proteins through nucleophilic acyl substitution. It was reported that the acyl CoA-thioester of zomepirac, together with the acyl glucuronide of zomepirac, may contribute to formation of the potentially toxic covalent zomepirac–protein adducts *in vitro* in rat hepatocytes and *in vivo* in rats (Olsen et al., 2005).

The techniques described in this chapter can be of use to the optimization of lead compounds during drug discovery processes. However, the link between drug–protein adduct formation and toxicity has not been well defined. Additional considerations should be taken, such as therapeutic area and clinical doses. It has been suggested that a 10 mg daily dose rarely results in drug-related adverse effects (Uetrecht, 1999). Thus, the propensity of bioactivation should only be considered as one of the factors in the selection of developmental drug candidates.

ACKNOWLEDGMENTS

ZZ thanks Ying Li, Jason Ngui, and Qing Chen for their contributions to the protocol development and to example presentations. JG thanks Qinling Qu for his contribution to example presentations. The authors also thank Dr. Wei Tang for critical review of the manuscript.

REFERENCES

Afzal M, Afzal A, Jones A, Armstrong D. A rapid method for the quantification of GSH and GSSG in biological samples. In:Armstrong D, editor. Oxidative Stress Biomarkers and Antioxidant Protocols. Totowa, NJ: Human Press Inc; 2002. p 117–122.

Alt C, Eyer P. Ring addition of the α-amino group of glutathione increases the reactivity of benzoquinone thioethers. Chem Res Toxicol 1998;11:1223–1233.

Anderson ME. Determination of glutathione and glutathione disulfide in biological samples. Meth Enzymol 1985;113:548–555.

Aust SD, Chignell CF, Bray TM, Kalyanaraman B, Mason RP. Free radicals in toxicology. Toxicol Appl Pharmacol 1993;120:168–178.

Bateman KP, Baker J, Wilke M, Lee J, LeRiche T, Seto C, Day S, Chauret N, Oqullet M, Nicoll-Griffith DA. Detection of covalent adducts to cytochrome P450 3A4 using liquid chromatography mass spectrometry. Chem Res Toxicol 2004;17:1356–1361.

Blobaum AL, Kent UM, Alworth WL, Hollenberg PF. Mechanism-based inactivation of cytochrome P450 2E1 and 2E1 T303A by tert-butyl acetylenes: characterization of reactive intermediate adducts to the heme and apoprotein. Chem Res Toxicol 2002;15:1561–1571.

Boelsterli UA, Zimmerman HJ, Kretz-Rommel A. Idiosyncratic liver toxicity of nonsteroidal antiinflammatory drugs: molecular mechanisms and pathology. Crit Rev Toxicol 1995;25:207–235.

Boelsterli UA. Xenobiotic acyl glucuronide and acyl CoA thioesters as protein-reactive metabolites with the potential to cause idiosyncratic drug reactions. Curr Drug Metab 2002;3:439–450.

Bolton JL, Trush MA, Penning TM, Dryhurst G, Monks TJ. Role of quinones in toxicity. Chem Res Toxicol 2000;13:135–160.

Bolze S, Bromet N, Gay-Feutry C, Massiere F, Boulieu R, Hulot T. Development of an *in vitro* screening model for the biosynthesis of acyl glucuronide metabolites and the assessment of their reactivity toward human serum albumin. Drug Metab Dispos 2002;30:404–413.

Butterfield DA. β-Amyloid-associated free radical oxidative stress and neurotoxicity: implications for Alzheimer's disease. Chem Res Toxicol 1997;10:495–506.

Castro-Oerez J, Plumb R, Liang L, Yang E. A high-throughput liquid chromatography/ tandem mass spectrometry method for screening glutathione conjugates using exact mass neutral loss acquisition. Rapid Commun Mass Spectrom 2005;19:798–804.

Cohen SD, Pumford NR, Khairallah EA, Boekelheide K, Pohl LR, Amouzadeh HR, Hinson JA. Selective protein covalent binding and target organ toxicity. Toxicol Appl Pharmacol 1997;143:1–12.

Coles B. Effects of modifying structure on electrophilic reactions with biological nucleophiles. Drug Metab Rev 1984–1985;15:1307–1334.

Comporti M. Three models of free radical-induced cell injury. Chem Biol Interact 1989;72:1–56.

Correia MA, Ortiz de Montellano, PR. Inhibition of cytochrome P450 enzymes. In: Ortiz de Montellano PR, editor. *Cytochrome P450: structure, mechanism, and biochemistry* third. New York: Kluwer Academic/Plenum Publishers; 2005; p 247–322.

Day SH, Mao A, White R, Schulz-Utermoehl T, Miller R, Beconi MG. A semiautomated method for measuring the potential for protein covalent binding in drug discovery. J Pharmacol Toxicol Methods 2005;52:278–285.

Degli Esposti M. Measuring mitochondrial reactive oxygen species. Methods 2002;26:335–340.

Dieckhaus CM, Fernandez-Metzler CL, King R, Krolikowski PH, Baillie TA. Negative ion tandom mass spectrometry for the detection of glutathione conjugates. Chem Res Toxicol 2005;18:630–638.

Doss GA, Miller RR, Zhang Z, Teffera Y, Nargund RP, Palucki B, Park MK, Tang YS, Evans DC, Baillie TA, Stearns RA. Metabolic activation of a 1,3-disubstituted piperazine derivative: evidence for a novel ring contraction to an imidazole. Chem Res Toxicol 2005;18:271–276.

Evans DC, Watt AP, Nicoll-Griffith DA, Baillie TA. Drug–protein adducts: an industry perspective on minimizing the potential for drug bioactivation in drug discovery and development. Chem Res Toxicol 2004;17:3–16.

Gan J, Hsueh MM, Qu Q, Harper TW, Humphreys WG, Dansyl glutathione as a trapping agent for the quantitative estimation and identification of reactive metabolites. Chem Res Toxicol 2005;18:896–903.

Ge XS, Shen J, Subramanyam B, Tseng JL, editors. The development of a strategy for unambiguous identification of reactive intermediate/metabolite. 54th ASMS Conference on Mass Spectrometry and Allied Topics; 2006 May 28–June 1; Seattle. Washington: Poster # ThP 166;2006.

Gorrod JW, Sai Y. Recognition of novel artifacts produced during the microsomal incubation of secondary alicyclic amines in the presence of cyanide. Xenobiotica 1997;27:389–399.

Guengerich FP. Metabolic activation of carcinogens. Pharmacol Ther 1992;54:17–61.

Guengerich FP. Common and uncommon cytochrome P450 reactions related to metabolism and chemical toxicity. Chem Res Toxicol 2001;14:611–650.

Halliwell B, Whiteman M. Measuring reactive species and oxidative damage *in vivo* and in cell culture: how should you do it and what do the results mean? Br J Pharmacol 2004;142:231–255.

Harrison D, Griendling KK, Landmesser U, Hornig B, Drexler H. Role of oxidative stress in atherosclerosis. Am J Cardiol 2003;91(3A):7A–11A.

Ho B, Castagnoli N. Trapping of metabolically generated electrophilic species with cyanide ion: metabolism of 1-benzylpyrrolidine. J Med Chem 1980;23:133–139.

Iskander K, Jaiswal AK. Quinone oxidoreductases in protection against myelogenous hyperplasia and benzene toxicity. Chem Biol Interact 2005;153–154:147–157.

James LP, Mayeux PR, Hinson JA. Acetaminophen-induced hepatotoxicity. Drug Metab Dispos 2003;31:1499–1506.

Kalgutkar AS, Dalvie DK, O'Donnell JP, Taylor TJ, Sahakian C. On the diversity of oxidative bioactivation reactions on nitrogen-containing xenobiotics. Curr Drug Metab 2002;3:379–424.

Kalgutkar AS, Gardner I, Obach RS, Shaffer CL, Callegari E, Henne KR, Mutlib AE, Dalvie DK, Lee JS, Nakai Y, O'Donnell JP, Boer J, Harriman SP. A comprehensive listing of bioactivation pathways of organic functional groups. Curr Drug Metab 2005;6:161–225.

Klaunig JE, Kamendulis LM. The role of oxidative stress in carcinogenesis. Ann Rev Pharmacol Toxicol 2004;44:239–267.

Koenigs LL, Peter RM, Hunter AP, Haining RL, Rettie AE, Friedberg T, Pritchard MP, Shou M, Rushmore TH, Trager WF. Electrospray ionization mass spectrometric analysis of intact cytochrome P450:identification of tienilic acid adducts to P450 2C9. Biochemistry 1999;38:2312–2319.

Liebler DC, Guengerich FP. Elucidating mechanisms of drug-induced toxicity. Nat Rev Drug Discov 2005;4:410–420.

Loughlin AF, Skiles GL, Alberts DW, Schaefer WH. An ion exchange liquid chromatography/mass spectrometry method for the determination of reduced and oxidized glutathione and glutathione conjugates in hepatocytes. J Pharm Biomed Anal 2001;26:131–142.

Meneses-Lorente G, Sakatis MZ, Schulz-Utermoehl T, De Nardi C, Watt A. A quantitative high-throughput trapping assay as a measurement of potentialfor bioactivation. Anal Biochem 2006;351:266–272.

Miller EC, Miller JA. Mechanisms of chemical carcinogenesis. Cancer 1981;47(5 S): 1055–1064.

Muller L, Kikuchi Y, Probst G, Schechtman, L, Shimada, H, Sofuni, T, Tweats, D. ICH-harmonised guidances on genotoxicity testing of pharmaceuticals: evolution, reasoning and impact. Mutat Res 1999;436:195–225.

O'Brien PJ. Molecular mechanisms of quinone cytotoxicity. Chem Biol Interact 1991;80:1–41.

Olsen J, Li C, Bjornsdottir I, Sidenius U, Hansen SH, Benet LZ. *In vitro* and *in vivo* studies on acyl-coenzyme A-dependent bioactivation of zomepirac in rats. Chem Res Toxicol 2005;18:1729–1736.

Park BK, Kitteringham NR, Maggs JL, Pirmohamed M, Williams DP. The role of metabolic activation in drug-induced hepatotoxicity. Annu Rev Pharmacol Toxicol 2005;45:177–202.

Parkinson A. Biotransformation of xenobiotics. In: Klaassen, CD, editor. *Casarett & Doull's Toxicology: The Basic Sciences of Poisons.* New York: McGraw-Hill Companies; 2001. p 133–224.

Pereg D, Robertson LW, Gupta RC. DNA adduction by polychlorinated biphenyls: adducts derived from hepatic microsomal activation and from synthetic metabolites. Chem Biol Interact 2002;139:129–144.

Ritter JK. Role of glucuronidation and UDP-glucuronosyltransferase in xenobiotic bioactivation reactions. Chem Biol Interact 2000;129:171–193.

Sallustio BC, Harkin LA, Mann MC, Krivickas SJ, Burcham PC. Genotoxicity of acyl glucuronide metabolites formed from clofibric acid and gemfibrozil: a novel role for phase II-mediated bioactivation in the hepatocarcinogenicity of the parent aglycones? Toxicol Appl Pharmacol 1997;147:459–464.

Sallustio BC, DeGraaf YC, Weekley JS, Burcham PC. Bioactivation of carboxylic acid compounds by UGT-glucuronosyltransferase to DNA-damaging intermediates: role of glycoxidation and oxidative stress in genotoxicity. Chem Res Toxicol 2006;19:683–691.

Shipkova M, Armstrong VW, Oellerich M, Wieland E. Acyl glucuronide drug metabolites: toxicological and analytical implication. *Therap Drug Metab* 2003; 25:1–6.

Smith MT, Evans CG, Thor H, Orrenius S. Quinone-induced oxidative injury to cells and tissues. In: Sies H, editor. *Oxidative Stress.* Orlando, FL: Academic Press Inc; 1985. p 91–113.

Soglia JR, Harriman SP, Zhao S, Barberia J, Cole MJ, Boyd JG, Contillo LG. The development of a higher throughput reactive intermediate screening assay incorporating micro-bore liquid chromatography–micro-electrospray ionization–tandem mass spectrometry and glutathione ethyl ester as an *in vitro* conjugating agent. J Pharm Biomed Anal 2004;36:105–116.

Svardal AM, Mansoor MA, Ueland PM. Determination of reduced, oxidized, and protein-bound glutathione in human plasma with precolumn derivatization with monobromobimane and liquid chromatography. Anal Biochem 1990;184: 338–346.

Tang W, Miller RR. In vitro drug metabolism: thiol conjugation. In:Yan Z, Caldwell GW, editors. *Methods in Pharmacology and Toxicology Optimization in Drug Discovery: In Vitro* Methods. Totowa, NJ: Human Press Inc; 2005. p 369–383.

Uetrecht J. New concepts in immunology relevant to idiosyncratic drug reactions: the "danger hypothesis" and innate immune system. Chem Res Toxicol 1999;12:387–395.

Walgren JL, Mitchell MD, Thompson. Role of metabolism in drug-induced idiosyncratic hepatotoxicity. Crit Rev Toxicol 2005;35:325–361.

Wang J, Davis M, Li F, Azam F, Scatina J, Talaat R. A novel approach for predicting acyl glucuronide reactivity via Schiff base formation: development of rapidly formed peptide adducts for LC/MS/MS measurements. Chem Res Toxicol. 2004;17:1206–1216.

Ward D, Kalir A, Trevor A, Adams J, Baillie T, Castagnoli N. Metabolic formation of iminium species: the metabolism of phencyclidine. J Med Chem 1982;25:491–492.

Welch KD, Wen B, Goodlett BR, Yi EC, Lee H, Reilly TP, Nelson SD, Pohl LR. Proteomic identification of potential susceptibility factors in drug-induced liver diseases. Chem Res Toxicol 2005;18:924–933.

Wen Z, Stern ST, Martin DE, Lee K-H, Smith PC. Structural characterization of anti-HIV drug candidate PA-457 [3-*O*-(3′,3′-dimethylsuccinyl)-betulinic acid] and its acyl glucuronides in rat bile and evaluation of *in vitro* stability in human and animal liver microsomes and plasma. Drug Metab Dispos 2006;34:1436–1442.

Wogan GN, Hecht SS, Felton JS, Conney AH, Loeb LA. Environmental and chemical carcinogenesis. Semin Cancer Biol 2004;14:473–486.

Yan Z, Caldwell GW. Stable-isotope trapping and high-throughput screenings of reactive metabolites using the isotope MS signature. Anal Chem 2004;76:6835–6847.

Yang XX, Hu ZP, Chan SY, Zhou SF. Monitoring drug–protein interaction. Clin Chim Acta 2006;365:9–29.

Zhang D, Ogan M, Gedamke R, Roongta V, Dai R, Zhu M, Rinehart JK, Klunk L, Mitroka J. Protein covalent binding of maxipost through a cytochrome P450-mediated orthoquinone methide intermediate in rats. Drug Metab Dispos 2003;31:837–845.

Zhang Z, Chen Q, Li Y, Doss GA, Dean BJ, Ngui JS, Elipe MS, Kim S, Wu JY, DiNinno F, Hammond ML, Stearns RA, Evans DC, Baillie TA, Tang W. In vitro bioactivation of dihydrobenzoxathiin selective estrogen receptor modulators by cytochrome P450 3A4 in human liver microsomes: formation of reactive iminium and quinone-type metabolites. Chem Res Toxicol 2005;18:675–685.

Zhou S. Separation and detection methods for covalent drug–protein adducts. J Chromatogr B 2003;797:63–90.

15

REACTION PHENOTYPING

Susan Hurst, J. Andrew Williams, and Steven Hansel

15.1 INTRODUCTION

Consistent, predictable efficacy, and safety are desired for today's pharmaceutical agents. Since these attributes typically correlate with a drug's systemic exposure, gaining an understanding of contributors to unacceptably risky interindividual pharmacokinetic variability may permit enhanced disease management as well as offer insight into the design of newer, more desirable clinical agents. Sources of potentially significant interindividual pharmacokinetic variation include interacting comedications, polymorphic expression of dispositional metabolizing enzymes or transporters, and dietary influences. This chapter will focus on the practical application of human drug metabolism reaction phenotyping strategies and *in vitro* techniques as they pertain to understanding a drug's metabolic processing via cytochrome P450 (CYPs) and noncytochrome P450 enzyme systems, including conjugation. A brief consideration of the emerging understanding of transporter phenotypic influence upon drug disposition is also included.

Drug metabolism reaction phenotyping by definition is the estimation of the relative contributions of specific enzymes to the metabolism of a test compound (Williams et al., 2005). Such studies may take the form of *in vitro* or *in vivo* assessment using a variety of approaches and tools and can be strategically employed at points along the drug discovery, drug development, and postlaunch continuum. An increased understanding of the clinical significance of how certain drug metabolizing enzymes contribute to

Drug Metabolism in Drug Design and Development, Edited by Donglu Zhang, Mingshe Zhu and W. Griffith Humphreys
Copyright © 2008 John Wiley & Sons, Inc.

interindividual variability in exposure via significant drug–drug interactions or polymorphic expression coupled with the advanced development of tools such as specific chemical or antibody inhibitors of the major P450s, *in vitro–in vivo* scaling techniques, and recombinant enzyme systems (Crespi et al., 1993; Lee et al., 1995; Remmel and Burchell, 1993) have all made reaction phenotyping a staple approach across the pharmaceutical industry. In addition, regulatory expectations now exist whereby new chemical entities (NCEs) are to be sufficiently characterized during development such that potential risks to a patient population can be weighed *en route* to final dosing and labeling (US Department of Health and Human Services et al., 1997, 2001, 2006).

Optimal application of the various reaction phenotyping techniques is dependent upon the scientific issue at hand, required robustness of data to enable decision making, and whether data are being used in compound selection and design versus the full characterization of a single compound. During the earliest stages of the novel drug discovery process, it is generally not known which specific enzymes are responsible for the metabolism of a newly synthesized compound although the role that biotransformation contributes to the total clearance of a compound may be estimated. The extent to which this metabolism is governed by P450s, non-P450s or conjugative enzymes, may be delineated by manipulating *in vitro* systems to assess the impact of (1) eliminating cofactors (e.g., NADPH for P450s, UDPGA for glucuronidation) on metabolic turnover, (2) coincubating with P450 pan-inhibitors (e.g., aminobenzotriazole, sulconazole, SKF-525A), or (3) preboiling preparations to destroy enzymatic activity to establish the contribution of chemical instability. Such evaluations may be incorporated into a general high throughput screening strategy or performed on small subsets of prototypical compounds representative of the template of interest to learn if observed attributes are associated with the structural core or offer the potential to be designed out in a facile manner. Drug discovery strategies often call for the desired development candidate to be eliminated by multiple pathways that may include several P450 enzymes, conjugation or direct elimination into bile or urine. Typically, P450s do play a role in the metabolic processing of small organic molecules so it is often desirable to determine the specific P450 enzyme contribution for mature leads generated during discovery to enable the advancement of lower risk candidates into the clinic. However, such a profile is sometimes not achievable within a chemical series and an overall risk assessment of the clinical prospects will need to be made.

Metabolic reaction phenotyping usually involves the use of enzyme specific chemical inhibitors in fully competent microsomal systems or recombinant enzyme systems individually expressing single enzymes. Once a drug candidate has entered the clinic, efforts will be made to perform a more definitive reaction phenotyping characterization as well as the consideration of clinical probe drug evaluations if patient risk is perceived.

15.2   CYTOCHROME P450 REACTION PHENOTYPING

Many enzyme systems exist that can contribute to the clearance of drugs from the systemic circulation thereby influencing a NCE or its metabolite(s)' systemic exposure and, therefore, its overall efficacy and adverse effect profiles. Most commonly, it is the P450 class of enzymes governing drug clearance via processing within the liver, the most important organ in the elimination of drugs and other xenobiotics. This section will highlight perspectives on identification strategies for determining the contribution that various P450 enzymes may have upon a drug's disposition.

The enzymes in the P450 family are membrane proteins expressed in the endoplasmic reticulum of mammalian cells with their most significant location, based on expression levels, being hepatocytes. More than 50 P450 enzymes are known to exist in the human genome, however, fewer than 10 are known to be expressed sufficiently in human liver to have the potential to play a significant role in xenobiotic metabolism. Two important factors in governing the potential impact that a P450 enzyme has across drug classes is its tissue expression level and substrate specificity. The most important human P450 enzymes involved in drug metabolism are CYPs 1A2, 2C9, 2C19, 2D6, and 3A4 with an emerging understanding that CYPs 2B6, 2C8, and 3A5 may play a role in the metabolism of certain therapeutics. CYPs 2A6 and 2E1 play only a minor role in xenobiotic metabolism. While the CYPs 2D6, 3A4, and 2C9 constitute approximately 50% of total hepatic P450 protein, these three enzyme metabolize nearly 80% of therapeutic drugs (Shimada et al., 1994; Smith et al., 1998). In certain situations larger than usual interindividual variability (greater than twofold) in drug exposure may arise due to the polymorphic expression of enzymes (CYPs 2C9, 2C19, 2D6) that drive a clearance-governing elimination pathway for a drug. For example, clinical studies have shown where CYP2D6 is estimated to contribute to more than 50% of a drug's clearance, reported exposure (area under the curve, AUC) ratios of extensive/poor metabolizers can range as high as 50. For example, systemic exposure to the proton pump inhibitors as expressed by the AUC (area under the plasma level time profiles) is 5–12 times higher in poor metabolizers than in extensive metabolizers (Klotz et al., 2004).

P450 metabolic reaction phenotyping to determine the relative contributions for specific P450 enzymes for metabolic clearance should use (1) chemical inhibition in collaboration with either (2) cDNA expressed systems, and/or (3) the utilization of reaction rate determination across a phenotypically characterized panel of microsomes from multiple individual donors (correlation analysis) (Ring and Wrighton, 2000).

There exist some general considerations across *in vitro* reaction phenotyping studies for P450s and non-P450s that need to be anticipated in the process of defining incubation conditions. While protocol-specific guidance (Method Sheets 1 and 2) is a part of the latter sections of this chapter, it is important to generally consider the concentration of NCE, the nature of the buffer system, the tolerance level of needed organic solvent, and the duration of the incubation experiment. It is generally wise to perform metabolic reaction phenotyping experiments using therapeutically relevant concentrations. Concentration considerations may include plasma and/or microsomal free versus bound drug, C_{max} versus C_{min}, systemic versus portal drug concentrations (oral), and intrahepatic concentration (Ito et al., 1998; Obach, 1997, 1999). Commonly employed buffers include Tris or potassium or sodium phosphate at pH 7.4 to recreate conditions representative of the *in vivo* situation regardless of the differing optimal conditions for individual P450 enzymes. Organic solvents such as methanol or acetonitrile may be needed to aid in compound dissolution, however, sufficient levels of these solvents diminish enzyme activity so it is a good practice to optimize solvent concentrations below 1% for microsomal preparations (Busby et al., 1999; Chauret et al., 1998; Hickman et al., 1998). Other commonly used solvents such as dimethyl sulfoxide (DMSO) should be optimized for the specific enzyme system (generally less than 0.1–0.2%). Experimental duration will depend upon time viability of the enzyme system (45–60 min typical for hepatic microsomes) balanced against the time needed to reliably detect changes in parent drug or metabolite concentration.

Both microsomal and recombinant expressed systems are commercially available and can be used in reaction phenotyping experiments as they contain activities for the major P450s previously highlighted. The NCE is incubated in the desired system and the concentration of parent (substrate depletion) or metabolite (metabolite formation) concentrations determined as a function of time. Linear initial rates, in regard to both enzyme concentration and time, are desired for most appropriate kinetics but these are not always obtained as multiple enzymes may be functioning in parallel or test article may be sufficiently stable to prevent an accurate measure of disappearance. Therefore, the substrate depletion method is most appropriate when sufficiently short *in vitro* half-life compounds are being assessed and kinetics may be confidently derived. A metabolite formation rate approach may be needed when relatively stable compounds are evaluated and parent loss would not be detected with statistical confidence (Obach and Reed-Hagen, 2002). Individual metabolite formation methods work best when authentic metabolite standard or radiolabel is available (again striving for linear initial reaction conditions); this is often not available during the earlier drug discovery phases. The likely contributions of each specific P450 enzyme to human *in vivo* clearance may then be estimated by combining typical hepatic expression levels of the governing P450s and the respective enzyme kinetics of parent; this can then be extended to an *in vivo* clearance prediction (Stormer et al., 2000b). If a bank of phenotyped microsomes is available, the selection of donors representing a range of activities can be used as additional confirmation (Williams et al., 2003).

15.3 NONCYTOCHROME P450 REACTION PHENOTYPING

Similar to P450 reaction phenotyping, designing experimental studies to elucidate the contribution of non-P450 enzymes requires an understanding of the particular enzymes substrate specificity, mechanism of reaction catalysis, tissue expression level, subcellular location, required cofactors/experimental conditions, and selective inhibitors (if identified). The following sections will focus on several examples of non-P450 reaction phenotyping including flavin-containing monooxygenases (FMOs), monoamine oxidases, and esterases.

15.3.1 Flavin-Containing Monooxygenases

Human flavin-containing monooxygenases are NADPH dependent microsomal enzymes that catalyze the oxygenation of many nitrogen, sulfur, selenium, and phosphorous heteroatom-containing chemicals and drugs (Cashman, 1995, 2003; Ziegler, 1993). However, in the top 200 drugs prescribed in the United States, FMOs account for only a small portion of the metabolic pathways when compared to P450s and UDP-glucuronosyltransferases (UGTs) (Williams et al., 2004). Currently there are five functional human FMOs known (Lawton et al., 1994) and FMO3 appears to be the most important FMO present in adult human liver. Based on immunorcactivity studies (Overby et al., 1997), FMO3 is present in adult human liver (Cashman, 2004), in contrast to FMO1 which is not present to any extent in adult liver. However, the expression levels are reversed in fetal livers where FMO1 predominates and FMO3 is not detected (Cashman, 2004; Koukouritaki et al., 2002; Yeung et al., 2000). FMO1 is also located in adult human kidney (Yeung et al., 2000). FMO5 is located in human liver, however, it is more substrate limited than FMO3 (Cashman, 2002; Overby et al., 1995). FMO1, FMO3, and FMO5 are currently available commercially as expressed enzymes. Since FMOs are oxidative enzymes that share a subcellular fraction (microsomes) and cofactor (NADPH) with cytochrome P450 enzymes, both systems will be active in native microsomes under common experimental conditions. Thus, differentiation of FMO and P450 activity is achieved either by selectively inactivating the P450 or the FMO enzymes. To inactivate the P450 component, a pan P450 inhibitor (i.e., aminobenzotriazole) or a detergent can be utilized (as P450 enzymes are more sensitive than FMO enzymes to surfactants). To inactivate the FMO component, either chemical inhibition (i.e., methimazole) or thermal degradation can be used. In comparison to cytochrome P450s, FMOs are less thermally stable and can be inactivated by preincubating for 5 min at 45°C in the absence of NADPH (prior to cooling to 37°C and initiating the reaction). Since FMO is thermolabile microsomal preparation technique can impact FMO activity, therefore when studying FMO dependent mctabolism it is important to use well-characterized microsomes (i.e., microsomes identified to have high functional FMO activity using a probe substrate).

Multiple FMO-3 variants have been identified. The best studied relationship between FMO-3 polymorphisms and enzyme function is the association between *FMO3* genetic variants and trimethylamine (TMA) metabolism (resulting in fish odor syndrome) (Cashman, 2002; Cashman and Zhang, 2002; Hernandez et al., 2003). Currently, FMO-3 (similar to the P450s) can be evaluated utilizing expressed enzyme, correlation analysis, and controlled incubation conditions (i.e., cofactor, chemical inhibitors, and/or thermal degradation). Examples of FMO reaction phenotyping studies include the evaluation of the enzyme system(s) responsible for the *N*-oxidation of clozapine (Tugnait et al., 1997) and the metabolism of sulfinpyrazone sulfide and sulfinpyrazone (He et al., 2001).

The involvement of FMO in the formation of the *N*-oxide of clozapine was determined using chemical inhibition (FMO inhibitor: methimazole), heat inactivation and protection against heat inactivation in the presence of NADPH. In addition clozapine *N*-oxide was catalyzed by purified FMO3 with a K_m similar to that determined using human liver microsomes. In contrast, sulfinpyrazone sulfide and sulfinpyrazone metabolism was determined to be catalyzed by P450s rather than FMO using modified microsomal incubation methods (heat inactivation and surfactant), specific P450 inhibitors with microsomes, and expressed P450s.

15.3.2 Monoamine Oxidases A and B (MAO-A and MAO-B)

Monoamine oxidases are integral outer mitochondrial membrane proteins that catalyze the oxidative deamination of primary and secondary amines as well as some tertiary amines. MAO occurs as two enzymes, MAO-A and MAO-B, which differ in substrate selectivity and inhibitor sensitivity (Abell and Kwan, 2001; Edmondson et al., 2004; Shih et al., 1999). A number of MAO inhibitors have been developed for clinical use as antidepressants and as neuroprotective drugs. Clinically used drug substances include, among others, moclobemide, a relatively selective reversible MAO-A inhibitor, and L-deprenyl, an irreversible selective inhibitor of MAO-B. *In vitro*, clorgyline and L-deprenyl are used as selective irreversible inhibitors of MAO-A and B, respectively. (*Note*: For *in vitro* studies using irreversible inhibitors, preincubation of the irreversible inhibitor with the enzyme prior to initiation of the substrate reaction is required for optimal inhibition.) Expressed MAO-A and MAO-B are not readily available via commercial resources; however, MAO-A and MAO-B have been evaluated and are active in subcellular fractions. While monoamine oxidases are located in the mitochondria, many microsomal preparations are contaminated with monoamine oxidases during the preparation of the microsomal subcellular fraction and thus microsomes are sometimes used to evaluate monoamine oxidase activity in combination with selective inhibitors.

A study comparing monoamine oxidase activity between Japanese and Caucasian livers used rizatriptan as the model substrate (Iwasa et al., 2003). The Japanese livers were evaluated for monoamine oxidase activity using both the mitochondial and microsomal subcellular fractions. Since the oxidative enzymatic activities were highly correlated between the mitochondrial and microsomal subcellular fractions in the Japanese livers, the microsomal subcellular fractions were used for evaluation of the Caucasian livers. For the chemical inhibition component of the study, selective inhibitors were added 5 min prior to addition of the substrate. Rizatriptan was primarily metabolized by MAO-A and there were no apparent differences in MAO-A and MAO-B activity between the Japanese and Caucasian livers.

15.3.3 Esterases

Historically, esterases have been classified according to either the compound metabolized and/or the type of bond cleaved. However, this has changed as technology has advanced regarding the molecular structure and function of esterases, particularly for carboxylesterases (CES) and paraoxonases (PON) (Li et al., 2005; Satoh et al., 2002). CES and cholinesterases both belong to a protein super family designated by the α,β-hydrolase-fold family (Cygler et al., 1993), while PON are unrelated to this protein super family (Draganov and La Du, 2004; Satoh et al., 2002).

CES are the best-characterized esterases in regard to drug metabolism and will be the focus of this section. Based on sequence homology, CES have been classified into four families: CES1, CES2, CES3, and CES4 (Satoh and Hosokawa, 1998). The two major human liver CES enzymes are hCE-1 and hCE-2 that belong to the classes of CES1 and CES2, respectively (Satoh et al., 2002). In mammals, CES is mainly localized in the endoplasmic reticulum of many tissues (Satoh and Hosokawa, 1998; Zhu et al., 2000). Human liver S9 or microsomes are usually used for *in vitro* studies of CES, however, CES is also located in cytosolic fractions (Tabata et al., 2004). In contrast to rat plasma, human plasma contains no (or negligible) CES with esterase activity in human plasma being due to cholinesterases, paroxonase, and/or serum albumin (Li et al., 2005; McCracken et al., 1993). Phenotyping is limited for esterases due to current limited availability of expressed esterases and commercially available purified esterases. Reaction phenotyping for esterases is typically utilized in mechanistic/descriptive studies for specific compounds/series rather than as a broader screen in the drug discovery process like cytochrome P450 reaction phenotyping.

Current examples of reaction phenotyping for CES include the studies conducted describing the bioactivation of the anticancer agent irinotecan (CPT-11, Camptosar)(Humerickhouse et al., 2000; Slatter et al., 1997) and the metabolism of the prodrug capecitabine (Tabata et al., 2004). The bioactivation of irinotecan was evaluated by enzyme kinetic studies using human liver microsomes and potent chemical inhibitors (carboxylesterase inhibitors: physostigmine and bis-nitrophenyphosphate) (Slatter et al., 1997). The enzyme kinetic data in human livers microsomes suggested that two enzymes catalyzed the bioactivation of irinotecan (with low and high affinities). The metabolism of irinotecan was further evaluated by enzyme kinetic studies using purified hCE-1 and hCE-2 and bioactivation studies with purified hCE combined with cytotoxicity studies (Humerickhouse et al., 2000). hCE-1 and hCE-2 were identified as low and high affinity enzymes involved in the bioactivation of irinotecan with the K_m of hCE-2 nearly identical to the K_m of the high affinity enzyme observed previously in microsomes.

The enzyme responsible for the biotransformation of capecitabine to 5′-deoxy-5-fluorocytidine (a precursor to 5-fluorouracil) was evaluated using purified enzyme, cytosol, and microsomes. The purified CES cytosolic enzyme, inhibited by the carboxylesterase inhibitors bis-nitrophenyphosphate and diisopropylfluorophosphate, was identified as belonging to the subgroup CES 1A1 based on the result of the N-terminal amino acid sequence.

15.4 CONJUGATION PHENOTYPING

15.4.1 UGT Reaction Phenotyping

The science behind conjugation reaction phenotyping has advanced more slowly than cytochrome P450 phenotyping, due to the lower incidence and magnitude of drug–drug interactions mediated by enzymes that catalyze conjugation (Williams et al., 2004). Listed in decreasing order of occurrence for the top 200 drugs prescribed in the United States, glucuronidation, sulfation, and conjugation with amino acids or glutathione are possible pathways of metabolism of drugs (Williams et al., 2004). Since amino acid conjugation is rarely a primary or major pathway of metabolism for marketed drugs, reaction phenotyping of glucuronidation catalyzed by UGTs, of sulfation catalyzed by sulfotransferases (SULTs), and of N-acetylation catalyzed by N-acetyltransferases (NATs) will be covered in this section. Targets for (soft) nucleophilic addition of sulfate or glucuronide to drugs occur at similar sites, including hydroxy- and carboxylic acid groups, and nucleophilic nitrogen and sulfur atoms.

Glucuronidation of parent drug is the most commonly listed conjugation pathway of metabolism (Williams et al., 2004). It is important not to confuse glucuronidation as (1) a primary clearance pathway of parent drug (sometimes known as an aglycone), which is the topic of discussion here, with (2) a secondary pathway of clearance of oxidative metabolites mostly catalyzed by cytochrome P450 enzymes. Drugs that are cleared primarily by glucuronidation include propofol, naloxone, zidovudine, and gemcabene (Bauman et al., 2005). The UGT enzymes most commonly listed in catalyzing drug glucuronidation are UGT1A1, UGT1A4, and UGT2B7 (Williams et al., 2004).

As mentioned above, the primary contributors to pharmacokinetic variability of metabolized drugs are clearance by polymorphic enzymes and drug–drug interactions (Williams et al., 2003). However, the magnitude of drug–drug interactions for glucuronidated drugs is typically low (Williams et al., 2004).

Recombinant UGTs

Correlation analyses ⟷ Selective inhibitors

FIGURE 15.1 Three independent approaches for definitive cytochrome P450 phenotyping that can also be applied to UGT phenotyping.

Understanding of genetic influence on pharmacokinetic variability for drugs primarily cleared by glucuronidation has advanced more rapidly in recent years, and convincing evidence of functionally relevant polymorphisms exists only for the *UGT1A1* gene (Miners et al., 2002). The UGT1A1 substrate SN-38 is an active cytotoxic metabolite of irinotecan (Camptosar), indicated for treatment of colon cancer, and *UGT1A1* genotype may therefore offer a useful safety biomarker for lower therapeutic index drugs that are primarily cleared by UGT1A1 (Pharmacia and Upjohn, 2004). Drugs thought to be primarily cleared by UGT1A4 include lamotrigine (Williams et al., 2004), and drugs cleared primarily by UGT2B7 include zidovudine (Williams et al., 2004) and gemcabene (Bauman et al., 2005).

As detailed below, selective chemical inhibition, recombinant UGT enzymes, and correlation analysis are the three independent approaches used for definitive cytochrome P450 phenotyping (Bjornsson et al., 2003) that can also be applied to UGT phenotyping (Fig. 15.1). The cytochrome P450 field is relatively advanced compared to that for UGT reaction phenotyping in that potent and selective chemical inhibitors in addition to probe substrates are well characterized for the major drug metabolizing members of the cytochrome P450 family only (Williams et al., 2005). Flavonoids such as hexamethoxy-flavone (Williams et al., 2004) and tangeretin (Williams et al., 2002a) have been identified as potent inhibitors of UGT1A1, but selectivity is an issue in that these compounds also potently inhibit UGT1A6 and UGT1A9, and possibly other UGT1A enzymes (Goosen et al., 2003).UGT2B7 enzyme activity, however, is not inhibited by flavonoids (Williams et al., 2004). As with cytochrome P450 enzymes, inhibition of UGT enzymes may be substrate selective (Rios and Tephly, 2002). Thus, although flurbiprofen (Bauman et al., 2005) and fluconazole (Uchaipichat et al., forthcoming) demonstrate selectivity in inhibition of UGT2B7-catalyzed fluconazole and gemcabene glucuronidation respectively, it is not necessarily the case that they are suitable for inhibition of all UGT2B7-catalyzed reactions. Until knowledge and understanding of UGT inhibition sufficiently develops, it may therefore be necessary to characterize potency and selectivity of inhibition for glucuronidated drugs using panels of recombinant enzymes where enzyme-selective substrates are absent, before human liver microsomes.

Other major obstacles to quantitative extrapolation of *in vitro* data to predict clearance of UGT substrates in humans include knowledge of relative expression levels of individual UGTs in human tissues and in individual preparations of recombinant enzymes. To date, attempts to quantify protein expression levels in human intestine and liver have not been successful. Quantitative assessment of mRNA levels has been achieved, but the predictive value of this information in anticipating protein levels is unproven. In addition to knowledge of intestinal and hepatic expression levels, measures of relative expression levels of UGTs *in vitro* would context the quantitative contribution of individual UGTs to the clearance of glucuronidated drugs. This is relatively simple for preparations of recombinant P450 enzymes, in that nanomole amounts of P450 per milligram protein can be measured by complexing the P450 enzymes with carbon monoxide and reading absorbance at 450 nm. The absence of validated methods for UGT enzyme quantitation, however, makes measurements of nanomole enzyme per milligram protein for UGTs challenging. The relative abundance *in vitro* to *in vivo* extrapolation approach (Proctor et al., 2004; Rodrigues, 1999) for predicting individual cytochrome P450 enzyme contributions to clearance of P450 substrates could potentially be applied to UGT substrates if the above two challenges of UGT protein quantitation could be met.

A recent example of UGT reaction phenotyping is that for gemcabene, a Pfizer compound indicated for treatment of dyslipidemia, and cleared primarily via a single glucuronide in humans (Bauman et al., 2005). The strategy employed for UGT reaction phenotyping compound is outlined in Fig. 15.2. Although human liver microsomes were assayed in this example, the same experimental approach could also be used for human intestinal microsomes. A brief summary of the approach and key data follows.

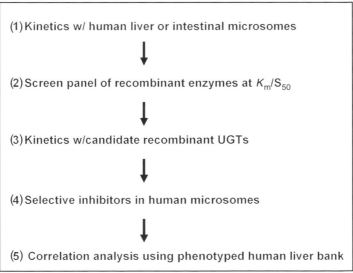

FIGURE 15.2 Strategy for reaction phenotyping.

Preliminary assessment of kinetics in human liver microsomes (Step 1) revealed an S_{50} of approximately 1 mM, which was the concentration used to screen a panel of 12 commercially available recombinant UGTs (Step 2). UGT1A3, UGT2B7, and UGT2B17 demonstrated potential to glucuronidate gemcabene, and kinetics of glucuronidation was assessed using those individual recombinant enzymes to assess similarities/differences in substrate affinity as a potential differentiation factor (Step 3). No differentiation was achieved, however, since the S_{50} values were each approximately equivalent to human liver microsomes 1 mM (Bauman et al., 2005).

Screening activities using candidate selective inhibitors of UGT1A3-, UGT2B7-, or UGT2B17-catalyzed gemcabene glucuronidation revealed 5,6,7,3′,4′,5′-hexamethoxyflavone as an inhibitor of UGT1A3 and UGT2B17 (with four fold greater potency for recombinant UGT1A3) and *S*-flurbiprofen as a selective UGT2B7 inhibitor (Step 4). Hexamethoxyflavone had no effect on gemcabene glucuronidation in human liver microsomes, suggesting that UGT1A3 and UGT2B17 do not significantly contribute to gemcabene glucuronidation. Concordantly, the IC_{50} for *S*-flurbiprofen inhibition of gemcabene glucuronidation in human liver microsomes was similar to that observed for recombinant UGT2B7, indicating that UGT2B7 was the major contributor to gemcabene glucuronidation in human liver microsomes. Further characterization of UGT2B7 as a major contributor to gemcabene metabolism was achieved by using a phenotyped human bank of human liver microsomes (Step 5): Rates of gemcabene glucuronidation correlated with rates of glucuronidation of the UGT2B7 substrate zidovudine, but not with rates of 3-glucuronidation of the UGT1A1 substrate β-estradiol (Bauman et al., 2005).

The approaches described above (Bauman et al., 2005) present a resource-intensive description of definitive UGT reaction phenotyping. Depending on need, one or more of the described steps may be omitted. The area primed for the greatest advance in the near future is the identification of selective glucuronidation inhibitors. Note that competitive substrates for individual enzymes are not necessarily selective inhibitors of those enzymes, since compounds do not need to be substrates of a UGT enzyme to be an inhibitor of that enzyme (Williams et al., 2002a). Recent advances have also been made in *in silico* predictions (Sorich et al., 2002). This promising area should be monitored for significant advances.

15.4.2 N-Acetylation Reaction Phenotyping

There are two human N-acetyltransferases, (NAT1 and NAT2). Substrates of NAT1 include sulfamethoxazole (Nakamura et al., 1995), an ingredient of the antibiotic combination Bactrim™, whereas isoniazid is a substrate of NAT2 (Kinzig-Schippers et al., 2005). The polymorphic nature of the *NAT1* gene is equivocal, whereas convincing evidence exists for functional polymorphisms in the *NAT2* gene: genotype for *NAT2* explains 88% of the variability in pharmacokinetics observed for isoniazid. Recombinant forms of these enzymes are commercially available, and human liver cytosol serves as another source. Acetyl coA is the required cofactor for this enzyme family.

15.4.3 Sulfation Reaction Phenotyping

Sulfation, using 3'-phosphoadenosine-5'-phosphosulphate (PAPS) as cofactor, is rarely a primary clearance mechanism for drugs relative to glucuronidated drugs or those cleared by cytochrome P450 enzymes. Exceptions include minoxidil, acetaminophen at therapeutic doses, and ethinylestradiol. Recombinant enzymes are commercially available for sulfotransferase phenotyping (Schneider and Glatt, 2004), but selective inhibitors have yet to be identified.

15.5 TRANSPORTER PHENOTYPING

Transporters have the potential to contribute to clearance. Strictly speaking, a reaction is not taking place, but since the consequence of transporter action may be the same as with a drug metabolizing enzyme (removal of unchanged drug from the plasma), inclusion in this chapter is warranted. A good example of the developing field of transporter phenotyping is that for hepatic uptake of pitivastatin, which modeled the "relative activity factor" approach previously taken for cytochrome P450 enzymes (Hirano et al., 2004). The kinetics of pitivastatin uptake was monitored in cells transfected with the hepatic uptake transporters OATP1B1 and OATP1B3. The hepatic uptake of pitivastatin was also modeled using the "relative abundance" approach with protein expression levels of OATP1B1 and OATP1B3 measured in transfectants and human hepatocytes. Both methods predicted approximately 90 and 10% contributions of OATP1B1 and OATP1B3 to the hepatic uptake of pitivastatin in human liver, respectively (Hirano et al., 2004). However the overall hepatic uptake in hepatocytes using the "relative abundance" approach overestimated the observed hepatic clearance. This over estimation may be due to (1) differences in the recovery of the transporter protein in samples for Western Blot analysis for transfectants and hepatocytes or (2) the total amount of protein in the whole-cell crude membrane may not be reflective of the functional transporter on the cell surface. The pitivastatin example provides a significant advance over the previously applied approach, where compounds were screened in cells transfected with various recombinant transporters. Although this formerly applied approach provided assessment of potential contribution of individual transporters, it could not offer an assessment of relative contributions, unlike the relative activity factor approach (Hirano et al., 2004).

15.6 NONRADIOLABELED REACTION PHENOTYPING

For the purposes of definition, nonradiolabeled reaction phenotyping is that conducted ahead of data from radiolabel studies in humans, and is typically conducted before first dosing of humans (Williams et al., 2003). This differs from reaction phenotyping later in development, which in the presence of radiolabel or metabolite standards is more definitive in nature (Bjornsson et al., 2003). Early reaction phenotyping may employ one or more of the independent approaches described in Fig. 15.1, whereas late (definitive) reaction phenotyping is likely to use all three. Early reaction phenotyping has been covered previously (Williams et al., 2004, 2005), and the essential elements will be presented in the following sections, with relevant updates.

Perhaps the most important consideration ahead of conduct of experiments is that of downstream value of the data, with particular regard to drug safety. Compounds that are substrates of cytochrome P450 enzymes are most likely to exhibit interindividual variability in pharmacokinetics due to drug–drug interactions, which would justify a focused effort on cytochrome P450 reaction phenotyping in the early stages of drug discovery (Williams et al., 2004). Compounds that appear to be substrates of other enzyme systems or families, such as UGTs (Williams et al., 2004), or esterases are less likely to be of concerns to safety in patients, and therefore would not necessarily justify in depth phenotyping of the individual enzymes, especially in the early stages.

15.6.1 Objective

The objective of early reaction phenotyping is to make a semiquantitative or in some cases quantitative prediction of the relative contributions of individual enzymes to the clearance of a compound in humans.

15.6.2 Selection of Appropriate Experimental Systems

15.6.2.1 In silico Advances continue to be made in the prediction of clearance mechanisms using *in silico* approaches (Ekins et al., 2001), although the status of the technology at this time cannot replace the conduct of experiments in the laboratory.

15.6.2.2 Expressed Enzymes Recombinant P450 enzymes are now available from multiple vendors in different formats. One major advantage of recombinant enzymes is that in many preparations the metabolic competence is significantly higher than in human liver microsomes per milligram protein, so that the potential for each of the five major cytochrome P450s (CYP1A2, CYP2C9, CYP2C19, CYP2D6 or CYP3A4) or others to contribute to the metabolism of the test compound of interest can be rapidly assessed (Williams et al., 2005). Other advantages include the ability to compare rates of metabolic turnover between individual enzymes, and the identification of metabolites formed by individual enzymes. Additionally, for P450 enzymes, an advantage is that the amounts of enzyme on a nanomolar basis within each incubation can be controlled, which in addition to the relative amounts in the liver, allow for quantitative predictions to clearance in humans by each P450. This forms the basis of the "relative abundance *in vitro* to *in vivo* extrapolation approach" (Rodrigues, 1999). As stated above (Section 15.4.1), similar data for UGT enzymes is lacking.

Perhaps the most important contribution to error in use of expressed cytochrome P450s is in the interpretation and extrapolation of generated data. The increased metabolic rate per milligram protein or per nanomole P450 relative to human liver microsomes can be overinterpreted in some cases where experimental conditions are not appropriately controlled. For example, there is potential for false positive results in some cases, where the rates of metabolic turnover for one enzyme may overestimate the contribution predicted from a single enzyme, based on subsequent data from other experimental systems such as human liver microsomes with selective chemical inhibitors.

15.6.2.3 Subcellular Fractions P450s are located in the endoplasmic reticulum of the cell, thus the subcellular fraction utilized for P450 reaction phenotyping is microsomes. Selective P450 inhibitors allow semiquantitation of the relative contributions of each enzyme to the *in vitro* metabolism of a test substance (Bjornsson et al., 2003). Table 15.1 shows a list of preferred inhibitors of the major human cytochrome P450s—(Tucker et al., 2001). Other enzymes expressed in human liver microsomes include flavin monooxygenase, UGTs, glutathione-*S*-transferases, esterases, microsomal epoxide hydrolase, and via impurity, measurable levels of monoamine oxidases.

TABLE 15.1   Preferred and accepted inhibitors for P450s (adapted from Tucker et al., 2001).

P450	Preferred	Acceptable
CYP1A2	Furafylline ($1-10\ \mu M$)	α-Napthoflavone (can also activate and inhibit CYP3A4)
CYP2A6	8-methoxypsoralen ($20\ \mu M$)	Coumarin, Sertraline (also inhibits CYP2D6)
CYP2C8	Montelukast ($0.2\ \mu M$)	
CYP2C9	Sulfaphenazole ($2.5-25\ \mu M$)	Ticlopidine (also inhibits CYP2D6) Nootkatone (also inhibits CYP2A6)
CYP2C19		
CYP2D6	Quinidine ($2.5-25\mu M$)	
CYP3A4	Ketoconazole ($1-2.5\ \mu M$)	
	Troleandomycin	Cyclosporine

Hepatic cytosol is a useful source of sulfotransferases (Pacifici, 2004) and aldehyde oxidase (Lake et al., 2002). Compared with cytochrome P450s, these enzymes are rarely utilized in early drug metabolism screening and thus may be limited to specific mechanistic studies or later stages of the drug development process.

15.6.2.4 Whole-Cell Systems

Hepatocytes offer the advantage over liver microsomes of an integrated metabolic system, closer to the *in vivo* situation. For prediction of metabolites formed *in vivo* hepatocytes therefore offer a useful system. For reaction phenotyping efforts the intact cell membrane presents a barrier to entry into the cell for both the substrate and any selective enzyme inhibitor that may be used. However, the sequential metabolism of substrates that would be expected to occur in hepatocytes may not generate sufficient clarity on which enzyme is the rate-limiting step determining *in vitro* clearance. Thus, subcellular fractions may be optimal systems for mechanistic reaction phenotyping.

One significant advantage of hepatocytes over liver microsomes is that the contribution of cytochrome P450 enzymes versus other cytosolic enzyme systems can be assessed, using pan-cytochrome P450 inhibitors. Aminobenzotriazole is a mechanism-based inhibitor of cytochrome P450 enzymes, and inhibits all of the major cytochrome P450s in hepatocytes except CYP2C9 (using diclofenac as substrate), without affecting glucuronidation or sulfation activities (Takahashi and Bauman, Pfizer Global Research and Development, personal communication). Imidazole containing compounds such as sulconazole may be more suitable compounds, since unlike aminobenzotriazole no preincubation is required and CYP2C9 (using diclofenac as substrate) is also inhibited (Takahashi and Bauman, Pfizer, personal communication).

15.6.3 Experimental Approach Considerations

This topic has been covered in some detail in a previous publication (Williams et al., 2003), and the key points are summarized below.

15.6.3.1 Selection of Substrate and Enzyme Concentration

Substrate concentration is a key consideration for experimental design, since enzyme-catalyzed metabolism is a saturable process, and the potential exists for differing conclusions at different substrate concentrations when multiple enzymes contribute to *in vitro* turnover. A good example is that of amitryptyline, where in human liver microsomes CYP3A4 predominates at high concentrations, whereas CYP2D6 predominates at low concentrations (Ghahramani et al., 1997). In cases where the predicted therapeutic index of the compound is low, a determination of K_m value (substrate concentration at V_{max}—half-maximal reaction velocity) may be valuable in understanding the potential for nonlinear pharmacokinetics *in vivo*. Compounds with very low K_m (e.g., below 1 µM) would have a higher chance of saturating the enzyme(s) responsible for clearance than compounds with much higher K_ms. Compounds primarily cleared by glucuronidation typically have high K_ms, and therefore rarely saturate their clearance mechanism (Williams et al., 2004). In the absence of knowledge of K_m, a substrate concentration of 1 μM can be chosen for the "substrate depletion" approach, as described below. A recent report describes a method for determination of both V_{max} and K_m using the substrate depletion approach (Nath and Atkins, 2006). In later development, the substrate concentration can be optimized to the therapeutic concentration.

Although the aim is to make a semiquantitative assessment that should approximate the *in vivo* concentration, accumulating evidence suggests that plasma concentrations *in vivo* do not necessarily equal liver concentrations. For example, as mentioned above, pitivastatin is actively taken up into the liver by the hepatic uptake transporters OATP1B1 and OATP1B3 (Hirano et al., 2004), which likely results in a high liver:plasma ratio.

Additionally, the role of protein binding to plasma proteins in defining drug concentrations available to drug metabolizing enzymes in liver microsomes *in vitro* is not well understood. Advances in measurement and prediction of protein binding *in vitro* offer the potential to fill this gap (Obach, 1997; Wring et al., 2002). As mentioned above, understanding of liver/plasma ratios *in vivo* is still an area for developing further understanding.

Enzyme concentrations in *in vitro* incubations will determine rate of turnover. Typical concentrations for human liver microsomes range between 0.5 and 1.0 mg/mL protein. At concentrations above 2 mg/mL the nonspecific binding of parent drug and metabolites may be too great to yield useful information. Typical cell concentrations for hepatocytes range from 0.5 to 1.0 million cells/mL (Williams et al., 2003).

15.6.3.2 Substrate Depletion Versus Metabolite Formation   In many cases multiple enzymes will catalyze the formation of multiple metabolites from the compound of interest, which presents a level of complexity for analysis of metabolite formation that may not be justifiably resourced in the earlier stages of development compared to the later stages. The substrate depletion approach provides a method to determine relative contributions of individual enzymes using a single analytical assay and appropriate calibration curves for the parent compound (Williams et al., 2003). This would therefore offer an opportunity to understand potential intersubject variability in pharmacokinetics of the parent compound as opposed to variability in the individual clearance pathways. It also provides a method to determine the K_m–V_{max}, and therefore the potential for nonlinear pharmacokinetics (Obach and Reed-Hagen, 2002). A key assumption of the substrate depletion approach is that the substrate concentration is well below K_m. Therefore, 1 μM is an appropriate default concentration.

Compounds most suited to the substrate depletion approach are those that are turned over rapidly. For compounds that are turned over more slowly, for example those where the depletion of parent is less than 20% over the incubation period, it will not be possible to define the relative contributions of each enzyme. This presents a decision point on whether to advance the compound forward to later stages in development with unknown enzymology, or to proceed with the "metabolite formation approach", which requires chromatography to separate metabolites (Williams et al., 2005). In early stages, radiolabeled compounds and authentic standards are not available, and thus peak responses should be considered with care with regard to interpretation of relative amounts.

Quantitation of metabolites is challenging in the absence of radiolabel or authentic standards. The most commonly used methods for metabolite monitoring are detection by ultraviolet (UV) absorbance and by mass spectrometry. When UV absorbance is being used for detection, the peak response will be approximately proportional to the amount of metabolite if the chromophore is not altered, such as hydroxylation of an aliphatic side group. However, if the molecule is cleaved, altering conjugated systems, relative UV peak responses may not accurately represent the relative amounts of parent compound and each metabolite. Mass spectrometry is even less reliable than UV absorbance in the absence of authentic standards, as ionization efficiency for each detected molecule will differ. In practice, a combination of UV and mass spectrometry is most appropriate. However, unlike the substrate depletion approach, the possibility would still exist of "missing" metabolites if the detection method does not pick up the metabolite of interest.

15.6.3.3   Reaction Velocity Linearity   To achieve the most accurate assessment of relative contributions of various enzymes to metabolism of a test compound using the substrate depletion approach, the reaction must be monitored where the rate of depletion of parent is proportional to its concentration (a first order reaction). This justifies the selection of low substrate concentrations (e.g., 1 μM) in the incubation. This relationship would appear as curved line using a linear scale (Fig. 15.3a) or a straight line using a log-linear plot. For the metabolite formation approach, the generation of metabolite would need to be linear with time and with protein, and would need to be conducted under conditions where the concentration of parent compound was not significantly (>15%) depleted. Figures 15.3b and c respectively show examples of a compound where metabolite formation is linear to 15 min incubation time (but not past that time period) and up to 1 mg/mL protein (but not above that concentration). In general, reactions catalyzed by cytochrome P450 enzymes do not maintain linearity past 45 min incubation time. In contrast, reactions catalyzed by UGT enzymes can demonstrate linearity of glucuronide production for up to 4 h.

15.6.4   Selection of Appropriate Experimental Designs

15.6.4.1   Expressed Enzymes   In the early stages of drug development it is appropriate to focus efforts on the five major cytochrome P450 enzymes (CYP1A2, CYP2C9, CYP2C19, CYP2D6, and CYP3A4/5). For the purposes of consistency, amounts of enzyme should be normalized to equivalent amounts of nanomole P450 per incubation (Williams et al., 2002b). Suppliers often provide ready made buffers, or these can be made quite simply in the laboratory. Preincubation of enzyme, substrate, and buffer at 37°C with shaking should ensure the initial reaction velocity is near its theoretical maximum when the reaction is initiated by the addition of cofactor

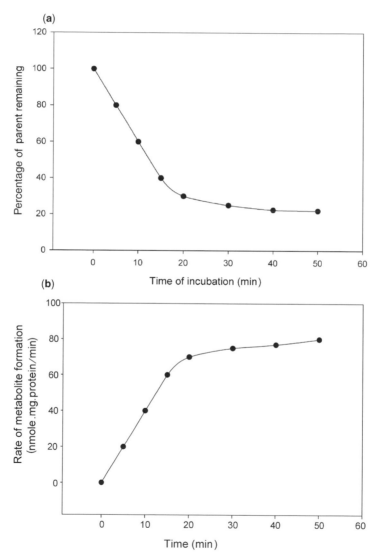

FIGURE 15.3 Assessment of time linearity using (**a**) substrate depletion and (**b**) metabolite formation approaches and (**c**) confirmation of linearity with respect to protein for the metabolite formation approach.

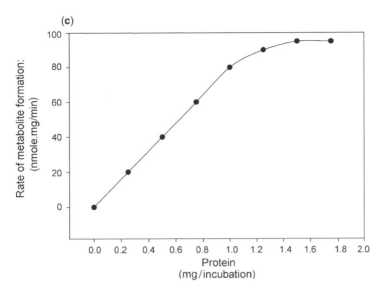

FIGURE 15.3 (*Continued*)

(e.g., NADPH). The concentration of solvents, particularly methanol and DMSO, should be kept to a minimum (see Section 15.2). Reactions are typically stopped by the addition of an ice-cold volume of ice-cold protein precipitant (acetonitrile and trichloroacetic acid are common choices), and spinning for precipitation will reduce the amount of unwanted protein injected onto the HPLC column. Analysis of a compound is most easily achieved by direct injection of an aliquot of the supernatant, rather than extraction, and the former approach is most commonly used.

For substrate depletion approaches, the data can be plotted on a graph (Fig. 15.4a) using simple visual inspection of the data. If only one recombinant enzyme depletes parent drug, then it can be concluded that the enzyme in question is the major contributor to metabolism *in vitro* (Figure 15.4a—recombinant CYP1A2 is the only P450 contributing to parent depletion in this example). If more than one enzyme appears to be depleting parent, then the relative contributions of each can be estimated using a combination of the slopes of (Fig 15.4b—both recombinant CYP1A2 and CYP2C9 are shown to deplete parent drug over time), and the relative hepatic expression levels of the metabolizing enzymes.

15.6.4.2 Chemical/Antibody Inhibition The selectivity of chemical inhibition for cytochrome P450 inhibitors is often dependent on inhibitor concentration. For example, ketoconazole is a potent and selective inhibitor of CYP3A enzymes at 1 μM concentration, whereas at higher concentrations it also inhibits CYP2C8 and other enzymes. A list of selective chemical cytochrome P450 inhibitors is shown in Table 15.1 (Tucker et al., 2001). Note that in

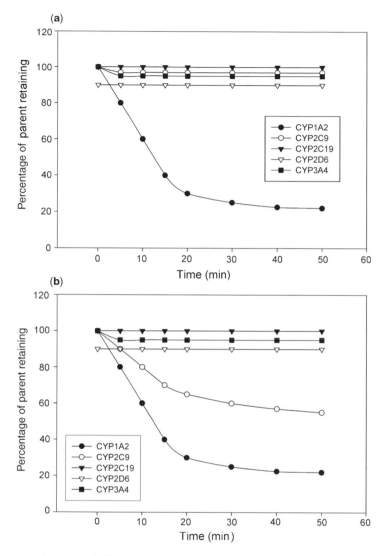

FIGURE 15.4 Use of the substrate depletion approach for assessing potential contribution to clearance. (a) Recombinant CYP1A2 is depleting parent drug and (b) both recombinant CYP1A2 and CYP2C9 are depleting parent drug.

addition to the five major listed cytochrome P450s (CYP1A2, CYP2C9, CYP2C19, CYP2D6, and CYP3A), the other listed inhibitors include CYP2C8 (substrates include rosiglitazone and some oncology compounds) and CYP2A6 (substrates include coumarin type compounds).

Cytochrome P450-selective antibodies are also commercially available. They offer increased selectivity over some of the chemical inhibitors, however cost may prohibit their routine use at earlier stages of drug discovery.

15.6.4.3 Correlation Analysis   A human liver bank phenotyped for cyto-chrome P450 activities (and potentially UGT activities) can provide further confidence on the major pathways contributing to *in vitro* clearance. Although more commonly associated with definitive reaction phenotyping where more than 20 livers are used to assess rates of metabolism of the compound of interest and compared to rates of metabolism, of the probe substrate (Hyland et al., 2003), selection of a smaller number of livers (e.g., six, comprised of three with high activity and three with low activity) in the earlier stages may also be informative and could add further confidence. In the earlier stages of development, correlation analysis is not commonly used. If resources are constrained to a single approach, chemical inhibition is recommended.

15.6.4.4 Quantitative Reaction Phenotyping Using Metabolite Standard   In the presence of authentic standard of the major metabolite(s), it is first necessary to develop an analytical assay using an appropriate detection method (most commonly mass spectrometry). Secondly, it is important to define the kinetics of metabolite formation using human liver microsomes. For reaction phenotyping using authentic metabolite standard(s), an example of the utility of recombinant enzymes and human liver microsomes with inhibitors is that of sildenafil (Hyland et al., 2001). The major circulating metabolite for this compound is formed via the N-demethylation of the piperazine ring. Assessment of the kinetics of sildenafil N-demethylation in human liver microsomes showed that at least two P450 enzymes were involved. At 250 μM sildenafil concentration N-demethylation was primarily mediated through a low affinity high K_m (81 μM) enzyme, whereas at 2.5 μM there was a greater role for the high affinity, low K_m (6 μM) enzyme (61%). Since ketoconazole strongly inhibited metabolism at both substrate concentrations, it was concluded that 75% or more of sildenafil N-demethylation activity is probably attributable to CYP3A4 (Fig. 15.5a). Experiments with recombinant enzymes indicated that CYP3A4 and to a lesser extent CYP2C9 contributed to sildenafil *N*-demethylation (Fig. 15.5b). In addition, multivariate analysis of correlation data across a human liver bank showed that the rate of sildenafil N-demethylation strongly correlated with the rate of both CYP2C9 (phenytoin 4-hydroxylation) and CYP3A4 (testosterone 6β-hydroxylation) mediated metabolism.

15.6.5 Quantitative Reaction Phenotyping: Expressed or Purified Enzyme Systems

Traditionally there are two approaches to quantitative reaction phenotyping using either expressed or purified enzymes. One approach, the relative activity factor (RAF) approach (Crespi and Miller, 1999; Stormer et al., 2000a; Venkatakrishnan et al., 2001), is being increasingly used in estimating the quantitative scaling of single enzyme data to multienzyme systems. The relative activity factor is a methodology that normalizes the activity in expressed or

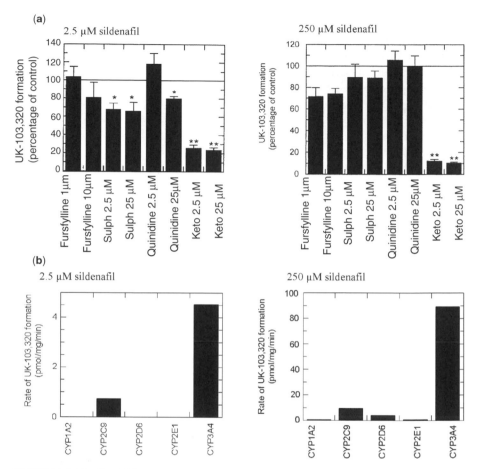

FIGURE 15.5   Definitive cytochrome P450 reaction phenotyping for *N*-demethylation of sildenafil – **(a)** Use of selective enzyme inhibitors **(b)** Use of recombinant enzymes. Reprinted from Hyland et al. (2001) by permission of Blackwell Publishing.

purified enzyme systems to multienzyme (i.e., microsomes or hepatocytes) systems. To utilize this approach, a standard probe substrate activity is used to normalize expressed enzyme system activity to human liver microsomes. RAFs are usually calculated as the ratio of V_{max} (reaction velocity at saturating substrate concentrations) in human liver microsomes of the enzyme-specific index reaction divided by the V_{max} of the reaction catalyzed by the cDNA-expressed enzyme. Preferably the probe substrate activity should be conducted in the same laboratory for both the expressed and microsomal systems. Once the RAF for the enzyme of interest has been established using an index substrate, the recombinant enzyme data can be converted to a microsomal equivalent. Appropriate care should be used in the choice of probe substrate(s) for the calculation of the RAF value(s) particularly for enzymes that demonstrate substrate dependent kinetic relationships (i.e., CYP3A4). Other potential confounding factors using the RAF approach are nonspecific protein binding and potential differences between expressed enzyme systems. Since higher protein concentrations are usually used with microsomal incubations, compared to expressed enzyme incubations, nonspecific protein binding may be different for the same substrate across the two matrices, especially for lipophilic amines. Also in consideration of P450 scaling, expressed systems often contain excess cytochrome b5, over expressed P450 reductase, or a different lipid environment when compared to microsomes. Therefore, an RAF value calculated for one expression system or lot of expressed enzyme may not be relevant for a different expression system or lot of expressed enzyme.

Another quantitative approach, scales the expressed enzyme activity using the relative abundance of the enzymes of interest. The recombinant or purified enzyme reaction rate is scaled utilizing the relative abundance of the particular enzyme in microsomes, hepatocytes, or liver (Proctor et al., 2004; Rodrigues, 1999). A modified approach that has been utilized for P450s incorporates a scaling factor (Inter System Extrapolation factor: ISEF) to account for the potential differences in intrinsic activity per unit enzyme between expressed enzyme and human liver microsomes (Proctor et al., 2004). These approaches based on enzyme abundance require accurate immunohistochemical data regarding the enzyme of interest. A comparison of the relative abundance approach, RAF based on V_{max}, and RAF based on CL_{int} has also been conducted utilizing the substrate depletion approach in early discovery; where the most accurate prediction method was determined to be RAF based on CL_{int} (Emoto et al., 2006). Currently, scaling by relative abundance for metabolic enzymes has only been conducted for P450s.

15.7   RADIOLABELED REACTION PHENOTYPING

Radiolabeled reaction phenotyping studies are usually conducted in the later stages of drug development due to the timing of the availability of the radiolabeled compound. However the timing of radiolabeled *in vitro* studies (and preclinical *in vivo* studies) can be adjusted due to the development needs of a NCE. Tritium is often used instead of ^{14}C in early drug discovery studies depending on the particular isotopic challenges in the radiosynthesis of a compound. Isotopic stability (loss of the tritium) is a particular problem with tritium labeled compounds and should be evaluated prior to the conduct of reaction phenotyping studies. As a compound moves through the drug discovery–drug development continuum, the evaluation of its metabolic processes and the subsequent evaluations (if risk is projected) of interindividual variability (due to polymorphic enzymes) and drug–drug interactions become more definitive.

15.7.1 Quantitative *In vitro* Radiolabeled Reaction Phenotyping Studies

Experimental methodologies utilizing radiolabeled compound are similar to those where authentic standard is available (see Section 15.6.3). Radiolabeled studies can be conducted using the following: expressed enzyme and/or human liver microsomes for the evaluation of K_m–V_{max} chemical inhibition, and/or correlation analysis using a characterized bank of human liver microsomes (Tuvesson et al., 2005; Zhu et al., 2005; Zhang et al., 2007).

Often in the early stages of drug development the relative abundance of a metabolite (in the absence of a synthesized standard or radiolabeled material) is estimated utilizing multiple detection methods. These methods include comparing either the ionization or the UV absorbance of a metabolite to the parent drug utilizing mass spectroscopy or UV detection (within the limits outlined above in Section 15.6.3.2), respectively. However, once radiolabeled material becomes available the relative abundance of a metabolite in an *in vitro* reaction phenotyping study can be determined by utilizing HPLC—radiometric detection via a comparison of the characterized radiometric peaks (Fig. 15.6). This assumes however that the metabolite of interest contains either the radiolabeled moiety or there is a clear and well-understood relationship between a radiolabeled component and the metabolite of interest (i.e., single cleavage product).

To estimate the total abundance of a particular metabolite in an *in vitro* reaction phenotyping sample using a single ^{14}C label and HPLC-radiometric detection, the amount of the metabolite in a given sample is estimated by multiplying the percent abundance of the metabolite obtained from a radiometric chromatogram of the sample (Fig. 15.6) to the molar amount of the initial radiolabeled parent drug present in the sample (assuming overall radioactive recovery is greater than 90%: taking into account incubation recovery, sample extraction efficiency, and column recovery). If the overall radioactive recovery is poor, then the metabolic profile obtained may not

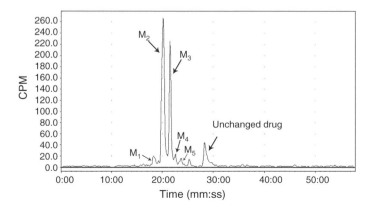

FIGURE 15.6 Example radiochromatogram.

accurately reflect the true metabolic profile of the compound and may reflect experimental or analytical issues (i.e., the radiolabel located on a metabolically unstable position, specific metabolites not extracting or being retained on the HPLC column). This calculation can only be conducted for radiolabeled components that are present on the radiochromatogram and it requires well-defined chromatographic separation of the metabolic peaks of interest. Radiolabeled components that are only observed utilizing mass spectroscopy would be considered trace amounts.

15.7.2   *In vivo* Quantitative ADME Studies

The extent of formation of a particular metabolite and its impact on overall drug elimination is ultimately evaluated in a radiolabeled *in vivo* absorption, distribution, metabolism, and excretion (ADME) study. Human radiolabeled ADME studies are often conducted to determine overall elimination pathways of a drug of interest. If a polymorphic enzyme primarily metabolizes the compound of interest, the subjects may be genotyped for this enzyme and a poor metabolizer cohort added to the study. Essentially the concentration time profiles of a drug and its metabolites as well as the final routes of elimination of an administered dose of drug are determined in a radiolabeled ADME study. The matrices that are commonly collected in human radiolabeled ADME studies include blood, plasma, urine, and feces with additional matrices collected for metabolite characterization if needed and available (i.e., bile). The total amount of the radioactive dose eliminated is determined for each excretory matrix (i.e., urine and feces). The excretory matrices are then profiled utilizing HPLC-radiometric detection and the observed radiometric peaks are then characterized utilizing LC/MS/MS and NMR (if regiochemical identification is required).

The % dose (or fraction metabolized) for each metabolite is calculated by summing the % dose calculated for each excretion pathway (urine or feces) of the metabolite and any subsequent secondary metabolites. The % dose for each excretion pathway is calculated by multiplying the percent abundance of a characterized metabolite on a radiometric chromatogram by the total radioactivity present in the evaluated sample (once extraction and column efficiencies have been taken into account). More detailed analysis would be required to calculate the fraction metabolized if secondary metabolites were formed from more than one primary metabolite (Bu et al., 2005, 2006).

The systemic exposure of circulating metabolite is calculated by estimating the area under the curve for a mean radioactivity versus time plot (Fig. 15.7). The fraction of the systemic exposure contributed by a particular metabolite is calculated by dividing the AUC estimated for a particular metabolite by the AUC of the total radioactivity. Alternatively, the metabolite concentration time profile and metabolite concentrations can be constructed using nonradiometric bioanalytical techniques if metabolite standards are available.

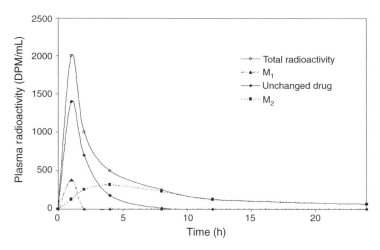

FIGURE 15.7 Example plasma radioactivity—time profile for radiolabeled ADME studies.

The pharmacokinetic and metabolic evaluation of a single oral and IV dose of sildenafil (see Section 15.6.4.4 for *in vitro* reaction phenotyping data) in humans is an example of a quantitative radiolabeled ADME study (Muirhead et al., 2002). No unchanged drug was recovered from the urine or feces indicating that metabolism is the major route of elimination for sildenafil. The principal routes of metabolism were identified (*N*-demethylation, multiple oxidation, aliphatic hydroxylation, and the loss of two carbon fragment from the piperazine ring). The major circulating component in plasma was identified as sildenafil accounting for 60 and 32% of the total radioactivity (AUC ratio) for IV and oral administration, respectively. The corresponding major circulating sildenafil metabolite was UK-103,320 (formed by piperazine N-demethylation) accounting for 8.7% and 17% following IV and oral administration, respectively. As mentioned in Section 15.6.4.4, *in vitro* studies indicate that sildenafil is metabolized to UK-103,320 by CYP2C9 and CYP3A4 (Hyland et al., 2001; Warrington et al., 2000).

15.7.3 Drug–Drug Interaction Potential

When projecting the drug–drug interaction potential of a particular drug, the extent to which the drug is eliminated by the inhibited enzyme is key to understanding the overall impact of the inhibition. Once the fraction metabolized (% dose) for the major metabolic pathways of a compound have been estimated (see Section 15.7.2) and the enzymes responsible for those pathways have been identified and the contribution quantitated, the overall fraction metabolized by a specific enzyme can be estimated ($f_{m(enz)}$). Prior to the availability of the radiolabeled human ADME study, $f_{m(enz)}$ can be projected as a product of the predictions for overall fraction metabolized, fraction metabolized via a specific enzyme system, and fraction metabolized by the specific enzyme. It is important when making projections to keep in mind the overall contribution of an enzyme. For example, expressed P450 enzyme and microsomal chemical inhibition studies may indicate that CYP3A4 is the only P450 involved in the metabolism of a compound. However, if the compound is estimated to be 50% cleared by renal excretion and 20% of the fraction metabolized is projected to be via N-acetylation, then the overall fraction metabolized via CYP3A4 would be projected to be no greater than 40% (i.e., $f_{m(CYP3A4)} \leq 0.5 \times 0.8 \times 1$).

The estimated $f_{m(enz)}$ value is utilized in the estimation of the drug–drug interaction potential of the compound of interest (Eq. 15.1) (Ito et al., 2005; Rowland and Matin, 1973). For Equation 15.1, the ratio of the systemic exposures of the inhibited compound in its inhibited versus noninhibited state (AUC_i/AUC) is a function of the *in vivo* inhibitor concentration [I], the binding affinity of the inhibitor for the inhibited enzyme (K_i), and the fraction of the drug of interest that is metabolized by the inhibited pathway ($f_{m(enz)}$). For orally administered drugs where the inhibitor also impacts intestinal metabolism, the effect of the inhibitor on the ratio of the systemic exposures of the inhibited compound is more complicated (see Chapter 5). However, $f_{m(enz)}$ for a particular drug by the inhibited metabolic pathway is also a key parameter for this projection as well. Essentially, the more the drug of interest is eliminated by the inhibited metabolic pathway the larger the potential drug–drug interaction (for a given [I] and K_i) (Fig. 15.8).

$$\frac{AUC_i}{AUC} = \frac{1}{\frac{f_{m(enz)}}{\left(1+\frac{[I]}{K_i}\right)} + \left(1 - f_{m(enz)}\right)} \tag{15.1}$$

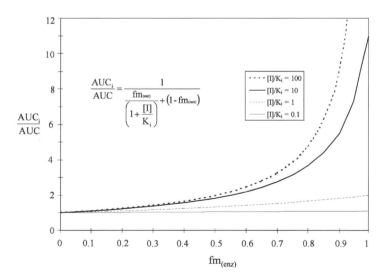

FIGURE 15.8 Simulation of changes in the AUC_i/AUC ratio of a drug as a function of $f_{m(enz)}$ using Equation 15.1.

15.7.4 Specialized Clinical Studies

Clinical studies may be conducted to further evaluate the impact of metabolism via a polymorphic enzyme or to establish a clinical drug–drug interaction risk assessment.

The venlafaxine study described below provides a recent example of polymorphic reaction phenoyping. A major route of venlafaxine metabolism (O-demethylation) was identified *in vitro* as being mediated by CYP2D6 using microsomes from both poor and extensive metabolizers of CYP2D6 (Otton et al., 1996). Subsequently, a clinical study was conducted to evaluate the metabolism of venlafaxine in poor and extensive metabolizers of CYP2D6 with and without a low dose of quinidine (CYP2D6 inhibitor) (Lessard et al., 1999). In the absence of quinidine, the oral clearance of venlafaxine in the poor CYP2D6 metabolizers was fourfold less than for the extensive metabolizers. In the presence of quinidine, the oral clearance of venlafaxine in the extensive CYP2D6 metabolizers was approximately fivefold less than in the absence of quinidine with a decrease in the formation of *O*-desmethyl-venlafaxine (CYP2D6 mediated pathway). In poor CYP2D6 metabolizers quinidine did not affect the oral clearance or the metabolic clearance of venlafaxine to its metabolites. Clinical drug–drug interactions studies will be discussed in Chapter 5.

15.8 SUMMARY AND FUTURE DIRECTIONS

Reaction phenotyping for P450 metabolism is well established and has become a routine technique in the early discovery evaluation for NCEs. For non-P450 enzymes, reaction phenotyping is usually focused on answering specific mechanistic questions that arise during the drug development process. Designing effective reaction phenotyping studies requires a thorough understanding of the enzyme of interest (i.e., the enzyme's substrate specificity, mechanism of reaction catalysis, tissue expression level, subcellular location, required cofactors/experimental conditions, and selective inhibitors) in addition to the "appropriate tools."

As technology continues to progress in the science of reaction phenotyping, particularly for the non-P450 enzymes, it will enhance our ability to move compounds forward in the drug development process with the confidence that clinically significant interindividual variability in pharmacokinetics due to either significant polymorphic metabolism or drug–drug interactions can be avoided or hopefully minimized.

ACKNOWLEDGMENTS

The authors would like to acknowledge the kind assistance of Ms. Theresa Davis and Mr. Michael Zientek in the preparation of this manuscript. Figure 15.5 was graciously supplied by Dr. Ruth Hyland and Blackwell Publishing.

REFERENCES

Abell CW, Kwan SW. Molecular characterization of monoamine oxidases A and B. Prog Nucleic Acid Res Mol Biol 2001;65:129–156.

Bauman JN, Goosen TC, Tugnait M, Peterkin V, Hurst SI, Menning LC, Milad M, Court MH, Williams JA. UDP-Glucuronosyltransferase 2B7 is the major enzyme responsible for gemcabene glucuronidation in human liver microsomes. Drug Metab Dispos 2005;33:1349–1354.

Bjornsson TD, Callaghan JT, Einolf HJ, Fischer V, Gan L, Grimm S, Kao J, King SP, Miwa G, Ni L, Kumar G, McLeod J, Obach RS, Roberts S, Roe A, Shah A, Snikeris F, Sullivan JT, Tweedie D, Vega JM, Walsh J, Wrighton SA. The conduct of *in vitro* and *in vivo* drug–drug interaction studies: a pharmaceutical research and manufacturers of America (PhRMA) perspective. Drug Metab Dispos 2003;31:815–832.

Bu HZ, Kang P, Zhao P, Pool WF, Wu EY. A simple sequential incubation method for deconvoluting the complicated sequential metabolism of capravirine in humans. Drug meta dispos: the biological fate of chemicals 2005;33:1438–1445.

Bu HZ, Zhao P, Kang P, Pool WF, Wu EY. Identification of enzymes responsible for primary and sequential oxygenation reactions of capravirine in human liver microsomes. Drug metab dispos: the biological fate of chemicals 2006;34:1798–1802.

Busby WF, Jr, Ackermann JM, Crespi CL. Effect of methanol, ethanol, dimethyl sulfoxide, and acetonitrile on *in vitro* activities of cDNA-expressed human cytochromes P-450. Drug Metab Dispos 1999;27:246–249.

Cashman JR. Structural and catalytic properties of the mammalian flavin-containing monooxygenase. Chem Res Toxicol 1995;8:166–181.

Cashman JR. Human flavin-containing monooxygenase (form 3): polymorphisms and variations in chemical metabolism. Pharmacogenomics 2002;3:325–339.

Cashman JR. The role of flavin-containing monooxygenases in drug metabolism and development. Curr Opin Drug Discov Devel 2003;6:486–493.

Cashman JR. The implications of polymorphisms in mammalian flavin-containing monooxygenases in drug discovery and development. Drug Discov Today 2004;9:574–581.

Cashman JR, Zhang J. Interindividual differences of human flavin-containing monooxygenase 3: genetic polymorphisms and functional variation. Drug Metabol Dispos 2002;30:1043–1052.

Chauret N, Gauthier A, Nicoll-Griffith DA. Effect of common organic solvents on *in vitro* cytochrome P450-mediated metabolic activities in human liver microsomes. Drug Metab Dispos 1998;26:1–4.

Crespi CL, Langenbach R, Penman BW. Human cell lines, derived from AHH-1 TK +/− human lymphoblasts, genetically engineered for expression of cytochromes P450. Toxicology 1993;82:89–104.

Crespi CL, Miller VP. The use of heterologously expressed drug metabolizing enzymes—state of the art and prospects for the future. Pharmacol Ther 1999;84:121–131.

Cygler M, Schrag JD, Sussman JL, Harel M, Silman I, Gentry MK, Doctor BP. Relationship between sequence conservation and three-dimensional structure in a large family of esterases, lipases, and related proteins. Protein Sci 1993;2:366–382.

Draganov DI, La Du BN. Pharmacogenetics of paraoxonases: a brief review. Naunyn-Schmiedebergs Arch Pharmacol 2004;369:78–88.

Edmondson DE, Mattevi A, Binda C, Li M, Hubalek F. Structure and mechanism of monoamine oxidase. Curr Med Chem 2004;11:1983–1993.

Ekins S, de Groot MJ, Jones JP. Pharmacophore and three-dimensional quantitative structure activity relationship methods for modeling cytochrome p450 active sites. Drug Metab Dispos 2001;29:936–944.

Emoto C, Murase S, Iwasaki K. Approach to the prediction of the contribution of major cytochrome P450 enzymes to drug metabolism in the early drug-discovery stage. Xenobiotica; the fate of foreign compounds in biological systems 2006;36:671–683.

Ghahramani P, Ellis SW, Lennard MS, Ramsay LE, Tucker GT. Cytochromes P450 mediating the N-demethylation of amitriptyline. Br J Clin Pharmacol 1997;43:137–144.

Goosen TC, Arimoto R, Gifford E, Ball SE, Hurst SI, Tugnait M, Hollenberg PF, Williams JA. A computational model based on UGT1A inhibition in human liver microsomes predicts the tacrolimus-mycophenolic acid metabolic drug interaction. Drug Metab Rev 2003;35:58.

He M, Rettie AE, Neal J, Trager WF. Metabolism of sulfinpyrazone sulfide and sulfinpyrazone by human liver microsomes and cDNA-expressed cytochrome P450s. Drug Metab Dispos 2001;29:701–711.

Hernandez D, Addou S, Lee D, Orengo C, Shephard EA, Phillips IR. Trimethylaminuria and a human FMO3 mutation database. Hum Mutat 2003;22:209–213.

Hickman D, Wang JP, Wang Y, Unadkat JD. Evaluation of the selectivity of *in vitro* probes and suitability of organic solvents for the measurement of human cytochrome P450 monooxygenase activities. Drug Metab Dispos 1998;26:207–215.

Hirano M, Maeda K, Shitara Y, Sugiyama Y. Contribution of OATP2 (OATP1B1) and OATP8 (OATP1B3) to the hepatic uptake of pitavastatin in humans. J Pharmacol Exp Ther 2004;311:139–146.

Humerickhouse R, Lohrbach K, Li L, Bosron WF, Dolan ME. Characterization of CPT-11 hydrolysis by human liver carboxylesterase isoforms hCE-1 and hCE-2. Cancer Res 2000;60:1189–1192.

Hyland R, Jones BC, Smith DA. Identification of the cytochrome P450 enzymes involved in the N-oxidation of voriconazole. Drug Metab Dispos 2003;31:540–547.

Hyland R, Roe EG, Jones BC, Smith DA. Identification of the cytochrome P450 enzymes involved in the N-demethylation of sildenafil. Br J Clin Pharmacol 2001;51:239–248.

Ito K, Hallifax D, Obach RS, Houston JB. Impact of parallel pathways of drug elimination and multiple cytochrome P450 involvement on drug-drug interactions: CYP2D6 paradigm. Drug Metab Dispos 2005;33:837–844.

Ito K, Iwatsubo T, Kanamitsu S, Nakajima Y, Sugiyama Y. Quantitative prediction of *in vivo* drug clearance and drug interactions from *in vitro* data on metabolism, together with binding and transport. Annu Rev Pharmacol Toxicol 1998;38:461–499.

Iwasa T, Sano H, Sugiura A, Uchiyama N, Hara K, Okochi H, Nakagawa K, Yasumori T, Ishizaki T. An *in vitro* interethnic comparison of monoamine oxidase activities between Japanese and Caucasian livers using rizatriptan, a serotonin receptor 1B/1D agonist, as a model drug. Br J Clin Pharmacol 2003;56:537–544.

Kinzig-Schippers M, Tomalik-Scharte D, Jetter A, Scheidel B, Jakob V, Rodamer M, Cascorbi I, Doroshyenko O, Sorgel F, Fuhr U. Should we use N-acetyltransferase type 2 genotyping to personalize isoniazid doses?Antimicrob Agents Chemother 2005;49:1733–1738.

Klotz U, Schwab M, Treiber G. CYP2C19 polymorphism and proton pump inhibitors. Basic Clin Pharmacol Toxicol 2004;95:2–8.

Koukouritaki SB, Simpson P, Yeung CK, Rettie AE, Hines RN. Human hepatic flavin-containing monooxygenases 1 (FMO1) and 3 (FMO3) developmental expression. Pediatric Res 2002;51:236–243.

Lake BG, Ball SE, Kao J, Renwick AB, Price RJ, Scatina JA. Metabolism of zaleplon by human liver: evidence for involvement of aldehyde oxidase. Xenobiotica 2002;32:835–847.

Lawton MP, Cashman JR, Cresteil T, Dolphin CT, Elfarra AA, Hines RN, Hodgson E, Kimura T, Ozols J, Phillips IR. A nomenclature for the mammalian flavin-containing monooxygenase gene family based on amino acid sequence identities. Arch Bioch Biophys 1994;308:254–257.

Lee CA, Kadwell SH, Kost TA, Serabjit-Singh CJ. CYP3A4 expressed by insect cells infected with a recombinant baculovirus containing both CYP3A4 and human NADPH-cytochrome P450 reductase is catalytically similar to human liver microsomal CYP3 A4. Arch Biochem Biophys 1995;319:157–167.

Lessard E, Yessene MA, Hamelin BA, O'Hara G, LaBlanc J, Turgeon J. Influence of CYP2D6 on the Disposition and Cardiovascular Toxicity of Antidepressent Agent Venlafaxine in Humans. Pharmacogenetics 1999;9:435–443.

Li B, Sedlacek M, Manoharan I, Boopathy R, Duysen EG, Masson P, Lockridge O. Butyrylcholinesterase, paraoxonase, and albumin esterase, but not carboxylesterase, are present in human plasma. Biochem Pharmacol 2005;70:1673–1684.

McCracken NW, Blain PG, Williams FM. Human xenobiotic metabolizing esterases in liver and blood. Biochem Pharmacol 1993;46:1125–1129.

Miners JO, McKinnon RA, Mackenzie PI. Genetic polymorphisms of UDP-glucuronosyltransferases and their functional significance. Toxicology 2002;181–182:453–456.

Muirhead GJ, Rance DJ, Walker DK, Wastall P. Comparative human pharmacokinetics and metabolism of single-dose oral and intravenous sildenafil. Br J Clin Pharmacol 2002;53 (Suppl 1):13S–20S.

Nakamura H, Uetrecht J, Cribb AE, Miller MA, Zahid N, Hill J, Josephy PD, Grant DM, Spielberg SP. In vitro formation, disposition and toxicity of N-acetoxy-sulfamethoxazole, a potential mediator of sulfamethoxazole toxicity. J Pharmacol Exp Ther 1995;274:1099–1104.

Obach RS. Nonspecific binding to microsomes: impact on scale-up of in vitro intrinsic clearance to hepatic clearance as assessed through examination of warfarin, imipramine, and propranolol. Drug Metab Dispos 1997;25:1359–1369.

Obach RS. Prediction of human clearance of twenty-nine drugs from hepatic microsomal intrinsic clearance data: An examination of in vitro half-life approach and nonspecific binding to microsomes. Drug Metab Dispos 1999;27:1350–1359.

Obach RS, Reed-Hagen AE. Measurement of Michaelis constants for cytochrome P450-mediated biotransformation reactions using a substrate depletion approach. Drug Metab Dispos 2002;30:831–837.

Otton SV, Ball SE, Cheung SW, Inaba T, Rudolph RL, Sellers EM. Venlafaxine oxidation *in vitro* is catalysed by CYP2 D6. Br J Clin Pharmacol 1996;41:149–156.

Overby LH, Buckpitt AR, Lawton MP, Atta-Asafo-Adjei E, Schulze J, Philpot RM. Characterization of flavin-containing monooxygenase 5 (FMO5) cloned from human and guinea pig: evidence that the unique catalytic properties of FMO5 are not confined to the rabbit ortholog. Arch Biochem Biophys 1995;317: 275–284.

Overby LH, Carver GC, Philpot RM. Quantitation and kinetic properties of hepatic microsomal and recombinant flavin-containing monooxygenases 3 and 5 from humans. Chem Biol Interact 1997;106:29–45.

Pacifici G. Inhibition of human liver and duodenum sulfotransferases by drugs and dietary chemicals: a review of the literature. Int J Clin Pharmacol Ther 2004;42: 488–495.

PHARMACIA,UPJOHN. 2004. Drug label approved on 07/21/2005 for CAMPTOSAR, NDA no. 020571, U.S. Food and Drug Administration. Avaliable at.http://www.accessdata.fda.gov/scripts/cder/drugsatfda/.

Proctor NJ, Tucker GT, Rostami-Hodjegan A. Predicting drug clearance from recombinantly expressed CYPs: intersystem extrapolation factors. Xenobiotica 2004;34:151–178.

Remmel RP, Burchell B. Validation and use of cloned, expressed human drug-metabolizing enzymes in heterologous cells for analysis of drug metabolism and drug-drug interactions. Biochem Pharmacol 1993;46:559–566.

Ring BJ, Wrighton SA. Industrial Viewpoint: Application of *In Vitro* Drug Metabolism in Various Phases of Drug Development. Philidelphia:Lippincott Williams & Wilkins;2000.

Rios GR, Tephly TR. Inhibition and active sites of UDP-glucuronosyltransferases 2B7 and 1 A1. Drug Metab Dispos 2002;30:1364–1367.

Rodrigues AD. Integrated cytochrome P450 reaction phenotyping: attempting to bridge the gap between cDNA-expressed cytochromes P450 and native human liver microsomes. Biochem Pharmacol 1999;57:465–480.

Rowland M, Matin S. Kinetics of drug-drug interactions. J Pharmacokinet Biopharm 1973;1:553–567.

Satoh T, Hosokawa M. The mammalian carboxylesterases: from molecules to functions. Annu Rev Pharmacol Toxicol 1998;38:257–288.

Satoh T, Taylor P, Bosron WF, Sanghani SP, Hosokawa M, La Du BN. Current progress on esterases: from molecular structure to function. Drug Metab Dispos 2002;30:488–493.

Schneider H, Glatt H. Sulpho-conjugation of ethanol in humans *in vivo* and by individual sulphotransferase forms *in vitro*. Biochem J 2004;383:543–549.

Shih JC, Chen K, Ridd MJ. Monoamine oxidase: from genes to behavior. Annu Rev Neurosci 1999;22:197–217.

Shimada T, Yamazaki H, Mimura M, Inui Y, Guengerich FP. Interindividual variations in human liver cytochrome P-450 enzymes involved in the oxidation of drugs, carcinogens and toxic chemicals: studies with liver microsomes of 30 Japanese and 30 Caucasians. J Pharmacol Exp Ther 1994;270:414–423.

Slatter JG, Su P, Sams JP, Schaaf LJ, Wienkers LC. Bioactivation of the anticancer agent CPT-11 to SN-38 by human hepatic microsomal carboxylesterases and the *in vitro* assessment of potential drug interactions. Drug Metab Dispos 1997;25:1157–1164.

Smith DA, Abel SM, Hyland R, Jones BC. Human cytochrome P450s: selectivity and measurement *in vivo*. Xenobiotica 1998;28:1095–1128.

Sorich MJ, Smith PA, McKinnon RA, Miners JO. Pharmacophore and quantitative structure activity relationship modelling of UDP-glucuronosyltransferase 1A1 (UGT1A1) substrates. Pharmacogenetics 2002;12:635–645.

Stormer E, von Moltke LL, Greenblatt DJ. Scaling drug biotransformation data from cDNA-expressed cytochrome P-450 to human liver: a comparison of relative activity factors and human liver abundance in studies of mirtazapine metabolism. J Pharmacol Exp Ther 2000a;295:793–801.

Stormer E, von Moltke LL, Shader RI, Greenblatt DJ. Metabolism of the antidepressant mirtazapine *in vitro*: contribution of cytochromes P-450 1A2, 2D6, and 3A4. Drug Metab Dispos 2000b;28:1168–1175.

Tabata T, Katoh M, Tokudome S, Nakajima M, Yokoi T. Identification of the cytosolic carboxylesterase catalyzing the $5'$-deoxy-5-fluorocytidine formation from capecitabine in human liver. Drug Metab Dispos 2004;32:1103–1110.

Tucker GT, Houston JB, Huang SM. Optimizing drug development: strategies to assess drug metabolism/transporter interaction potential–towards a consensus. British J Clin Pharmacol 2001;52:107–117.

Tugnait M, Hawes EM, McKay G, Rettie AE, Haining RL, Midha KK. N-oxygenation of clozapine by flavin-containing monooxygenase. Drug Metab Dispos 1997;25: 524–527.

Tuvesson H, Hallin I, Persson R, Sparre B, Gunnarsson PO, Seidegård J. Cytochrome P450 3A4 is the major enzyme responsible for the metabolism of laquinimod, a novel immunomodulator. Drug metab dispos: the biological fate of chemicals 2005;33: 866–872.

U.S. Department of Health and Human Services,U.S. Food and Drug Administration, Center for Drug Evaluation and Research and Research (CDER).1997. Guidance for industry - drug metabolism/drug interaction studies in the drug development process: studies *in vitro*. Available at www.fda.gov/cder/guidance/clin3.pdf.

U.S. Department of Health and Human Services,U.S. Food and Drug Administration, Center for Drug Evaluation and Research (CDER),Center for Biologics Evaluation and Research (CBER). 2001. Guidance for industry: *in vivo* drug metabolism/drug interaction studies - study design, data analysis, and recommendations for dosing and labeling. Available at www.fda.gov/cder/guidance/2635fnl.htm.

US Department of Health and Human Services,Food and Drug Administration, Center for Drug Evaluation and Research (CDER), Center for Biologics Evaluation and Research (CBER). 2006. Draft guidance for industry: drug interaction studies - study design, data analysis, and implications for dosing and labeling. Available at http:// www.fda.gov/cder/guidance/6695dft.pdf.

Uchaipichat V, Winner LK, Mackenzie PI, Elliot DJ, Williams JA, Miners JO. Quantitative prediction of *in vivo* inhibitory interactions involving glucuronidated drugs from *in vitro* data: the effect of fluconazole on zidovudine glucuronidation. Br J Clin Pharmacol. Forthcoming.

Venkatakrishnan K, von Moltke LL, Greenblatt DJ. Application of the relative activity factor approach in scaling from heterologously expressed cytochromes p450 to human liver microsomes: studies on amitriptyline as a model substrate. J Pharmacol Exp Ther 2001;297:326–337.

Warrington JS, Shader RI, von Moltke LL, Greenblatt DJ. *In vitro* biotransformation of sildenafil (Viagra): identification of human cytochromes and potential drug interactions. Drug Metab Dispos 2000;28:392–397.

Williams JA, Bauman J, Cai H, Conlon K, Hansel S, Hurst S, Sadagopan N, Tugnait M, Zhang L, Sahi J. *In vitro* ADME phenotyping in drug discovery: current challenges and future solutions. Curr Opin Drug Discov Devel 2005;8:78–88.

Williams JA, Hurst SI, Bauman J, Jones BC, Hyland R, Gibbs JP, Obach RS, Ball SE. Reaction phenotyping in drug discovery: moving forward with confidence? Curr Drug Metab 2003;4:527–534.

Williams JA, Hyland R, Jones BC, Smith DA, Hurst S, Goosen TC, Peterkin V, Koup JR, Ball SE. Drug-drug interactions for UDP-glucuronosyltransferase substrates: a pharmacokinetic explanation for typically observed low exposure (AUCi/AUC) ratios. Drug Metab Dispos 2004;32:1201–1208.

Williams JA, Ring BJ, Cantrell VE, Campanale K, Jones DR, Hall SD, Wrighton SA. Differential modulation of UDP-glucuronosyltransferase 1A1 (UGT1A1)-catalyzed estradiol-3-glucuronidation by the addition of UGT1A1 substrates and other compounds to human liver microsomes. Drug Metab Dispos 2002a;30:1266–1273.

Williams JA, Ring BJ, Cantrell VE, Jones DR, Eckstein J, Ruterbories K, Hamman MA, Hall SD, Wrighton SA. Comparative metabolic capabilities of CYP3A4, CYP3A5, and CYP3 A7. Drug Metab Dispos 2002b;30:883–891.

Wring SA, Silver IS, Serabjit-Singh CJ. Automated quantitative and qualitative analysis of metabolic stability: A process for compound selection during drug discovery. Methods Enzymol 2002;357:285–296.

Yeung CK, Lang DH, Thummel KE, Rettie AE. Immunoquantitation of FMO1 in human liver, kidney, and intestine. Drug Metab Dispos 2000;28:1107–1111.

Zhang D, Wang L, Chandrasena G, Ma L, Zhu M, Zhang H, Davis CD, Humphreys WG. Involvement of multiple cytochrome P450 and UDP-glucuronosyltransferase enzymes in the *in vitro* metabolism of muraglitazar. Drug Metab Dispos 2007; 35:139–149.

Zhu M, Zhao W, Jimenez H, Zhang D, Yeola S, Dai R, Vachharajani N, Mitroka J. Cytochrome P450 3A-mediated metabolism of buspirone in human liver microsomes. Drug metab dispos: the biological fate of chemicals 2005;33:500–507.

Zhu W, Song L, Zhang H, Matoney L, LeCluyse E, Yan B. Dexamethasone differentially regulates expression of carboxylesterase genes in humans and rats. Drug Metab Dispos 2000;28:186–191.

Ziegler DM. Recent studies on the structure and function of multisubstrate flavin-containing monooxygenases. Annu Rev Pharmacol Toxicol 1993;33:179–199.

APPENDIX A: REACTION PHENOTYPING—EXPRESSED cDNA ENZYME INCUBATION METHOD SHEET

I. Objective
   A. To determine whether individually expressed human cDNA P450 enzymes are capable of contributing to metabolism by measuring the disappearance of parent compound.

II. Materials
   A. Appropriate buffer pH 7.4 (i.e., 50-mM potassium phosphate (KP)).
   B. β-NADPH (1 mM final concentration).
   C. P450 expressed enzyme
      1. Recombinant human P450 and NADPH-P450 reductase.
      2. P450 control for expression system
   D. Drug stock solutions
      1. Prepare a stock solution that can provide a final incubate drug concentration of 1 μM or less if K_m known (0.1–1.0 mM or 100–1000 × final drug concentration).
      2. Organic solvent content added must be less than 1% of the total incubation volume and the content of DMSO is less than 0.1% of the total incubation volume.

III. Methods
   A. Drug incubates
      1. Drug incubations performed in triplicate (N = 3) for each inhibitor.
      2. On ice, add all components of the incubation except β-NADPH.
      3. Preincubate samples at 37°C in shaking water bath for at least 3 min.
      4. Add prewarmed β-NADPH and swirl gently to start reaction, keep covered throughout incubation (for minus β-NADPH controls buffer is substituted for β-NADPH).
      5. Sample quench
         a. Remove aliquots at predetermined times and transfer to Eppendorf tubes or 96-well plate containing quench solvent.
      6. Samples may be stored at −80°C until analysis.

APPENDIX B: REACTION PHENOTYPING—MICROSOMAL CHEMICAL INHIBITION

I. Objective
   A. To specifically inhibit CYP450s in human liver microsomes to determine which enzymes are contributing to metabolism by observing the changes in the disappearance of parent compound.

II. Materials
   A. Appropriate buffer pH 7.4 (i.e, 50-mM potassium phosphate buffer (KP)).
   B. β-NADPH (1 mM final concentration).
   C. Human liver microsomes.
   D. Stock Solutions
      1. Compound
         a. Prepare a stock solution that can provide a final incubate drug concentration of 1 μM or less if K_m known (0.1–1.0 mM or 100–1000 × final drug concentration).
      2. Inhibitors
         a. Dissolve primary stocks in DMSO.
         b. Dilute primary stock with ACN so final stock is 75:25, %v/v (ACN:DMSO).
      3. Organic solvent content added must be less than 1% of the total incubation volume and the content of DMSO is less than 0.1% of the total incubation volume.

III. Methods
   A. Drug incubates
      1. Drug incubations performed in triplicate ($N = 3$) for each inhibitor.
      2. On ice, add all components of the incubation except β-NADPH.
      3. Preincubate samples at 37°C in shaking (80 rpm) water bath for at least 3 min.
      4. Add prewarmed β-NADPH and swirl gently to start reaction, keep covered throughout incubation.
      5. Sample quench
         a. Remove aliquots at predetermined times and transfer to Eppendorf tubes or 96-well plate containing quench solvent.
      6. Samples may be stored at −80°C until analysis.

16

ANALYSIS OF *IN VITRO* CYTOCHROME P450 INHIBITION IN DRUG DISCOVERY AND DEVELOPMENT

Magang Shou and Renke Dai

16.1 INTRODUCTION

Investigation of the inhibition potential of drugs or drug candidates toward human drug metabolizing enzymes (DMEs), particularly hepatic cytochromes P450 (CYPs) is of great interest to pharmaceutical scientists and clinicians. In the pharmaceutical industry, *in vitro* screening of CYP enzyme inhibition of new chemical entities (NCEs) provides critical information for lead finding, optimization of drug candidates and evaluation of drug–drug interaction potential in humans (Bjornsson et al., 2003; Ito et al., 1998; Lin, 2000; Lin and Lu, 2001; Obach 2003; Obach et al., 2005; Obach et al., 2006; Shou 2005). A metabolism-based drug interaction implies that one drug causes a change in the metabolic clearance of another drug, in turn either decreasing or increasing concentrations of the drug in plasma, and presumably also causes a change at the site of action, that has led and can lead to either pharmacodynamic (PD) inhibition or enhancement of the clinical effects or serious toxicological consequences of another drug in human (Backman et al., 1994; Gomez et al. 1995; Gorski et al., 1998; Greenblatt et al., 1998a, 1998b, 1999; Heerey et al., 2000, 1993a, 1993b, 1993c; Honig and Cantilena 1994; Olkkola et al., 1994;

Drug Metabolism in Drug Design and Development, Edited by Donglu Zhang, Mingshe Zhu and W. Griffith Humphreys
Copyright © 2008 John Wiley & Sons, Inc.

Thummel and Wilkinson 1998). The majority of drug interactions of clinical significance have occurred through interactions at the level of CYPs. Clinical pharmacokinetic (PK) drug–drug interaction (DDI) studies have high cost and are time consuming. As a consequence, there is a realistic limit to the number and scope of clinical drug interaction studies that can be performed. The search for alternative approaches to study drug interactions at early stage of drug discovery has been encouraged. Thus, screening for the DDI using *in vitro* systems has become increasingly important for DDI predictions in man and for more informed planning of clinical drug interaction studies.

For the inhibition assessment, a common strategy is to monitor effect of test compound on the metabolism of CYP probe substrate using human liver microsomes (HLM). Generally, most CYP-mediated reactions follow simple Michaelis–Menten kinetics and their kinetic constants (K_m and V_{max}) are easily derived. If the addition of a CYP inhibitor results in inhibition of the enzyme reaction, a value for K_i (or IC_{50}) for reversible inhibition or K_I and k_{inact} for irreversible inhibition can be determined with suitable kinetic models. It is accepted that the kinetic parameters can be used to predict and understand *in vivo* PK and PD consequences caused by exposure to one or multiple drugs. Among numerous CYP enzymes identified to date, six human hepatic CYP isoforms (CYP1A2, 2C8, 2C9, 2C19, 2D6, and 3A4) play dominant roles in the metabolism of clinical drugs, and these enzymes are commonly employed to evaluate NCEs as inhibitors of the enzymes (Food and Drug Administration, 1997 and 1999). Many isoform-specific probe substrates have been recommended by US regulatory agency (FDA) and Pharmaceutical Research and Manufactures of America (PhRMA) (Bjornsson et al., 2003) and wildly used in pharmaceutical industry.

In practice, enzyme inhibition is one of the easiest phenomena to measure in a high throughput manner, as a result of experimental simplicity (monitoring the effect of a test compound on the formation of metabolite(s) formed in reaction of the CYP probe substrate with HLM, cDNA-expressed CYPs, or hepatocytes). CYP inhibition evaluation is performed for one single isozyme or multiple enzymes at a time. Various attempts have been made to increase the throughput of these assays. Fluorescence-based assays for determining the inhibition of CYP activity are available for a number of isozymes and are being used in high throughput screening (HTS) of small molecules (Cohen et al., 2003; Crespi and Stresser, 2000; Donato et al., 2004; Stresser et al., 2002;) In these assays, the nonfluorescent substrate, upon enzymatic conversion, generates a fluorescent metabolite that is detected in the fluorescence plate reader. The assay is often performed with single endpoint measurement of the reaction to make it easy to adapt to robotic systems. The most serious drawback of this method is the interference from the intrinsic fluorescence or possible fluorescence-quenching effects of compounds with the detection of substrate metabolite. In addition, these compounds also lack the required specificity as substrates for these enzymes, and hence cannot be used with HLM but have to be used with individually expressed CYP enzymes.

Radioisotope-based inhibition assays have been also developed for the CYP inhibition using probe substrates, in which the O- or N-alkyl group contained either tritium or ^{14}C-labeled (Draper et al., 1998a, 1998b; Di Marco et al., 2005; Moody et al., 1999; Riley and Howbrook, 1997). The assays do offer some advantages over the fluorometric assays including substrate selectivity, better solubility of substrates, and higher rates of substrate turnover. However, the major disadvantage of the assays is that they require a separation procedure, such as solid-phase extraction, for separating metabolites from the parent compound before analysis, which limits these approaches inadequate for HTS. An LC/MS method with specific drug probes appears to overcome many of these limitations by above two methods (Cohen et al., 2003; Bu et al., 2001a, 2001b; Nomeir et al., 2001; Yin et al., 2000; Zhang et al., 2002). It can provide sufficient throughput (automated 96-well plate and fast direct online injection to LC/MS/MS), flexibility (variety of probe substrate and enzyme source, i.e., recombinant CYPs and HLM), and instrument selectivity, sensitivity, and speed (rapid gradient LC and simultaneous detection of multiple metabolites). For example, the LC/MS uses both ballistic gradient liquid chromatography and mass spectrometry for quantification of the drug and the metabolite(s). Mass-spectrometric analysis using single ion monitoring (SIM) provides high sensitivity and eliminates the need for long gradient for the resolution of the peaks during liquid chromatography. In addition, a new and robust method for the simultaneous evaluation of the multiple CYP enzyme activities in HLM has been also accomplished, using rapid gradient LC/MS with SRM of specific metabolites (Bu et al., 2001a, 2001b; Dierks et al., 2001). In addition to the above CYP inhibition assays, automation and validation of the assays following good laboratory practice (GLP) level have been largely dealt and developed, thereby leading to data with high quality and precise (Jenkins et al., 2004; Kremers 2002; Walsky and Obach, 2004; Yao et al., 2007). The focus of this chapter is to provide theoretical background for enzyme inhibition kinetics (reversible and irreversible), general methods for determining the inhibition kinetic parameters, inhibition assays specific to individual CYPs with LC/MS/MS approaches, and available tools and guidelines for prediction of DDIs in human from the *in vitro* inhibition data.

16.2 REVERSIBLE INHIBITION

When an inhibitor is added to the enzyme reaction, the reaction mixture may comprise more than one enzyme complex, namely ES, EI, and/or ESI (Scheme 16.1) (Segel, 1987; Shou et al., 2001). Since the ES concentration ([ES]) decreases with an increase in [I], the rate of product formation (k_p[ES]) can decline ($0 \leq \beta < 1$). Scheme 16.1 depicts a general kinetic model used to describe the interaction between substrate (S), inhibitor (I) and enzyme (E). Based on nature of inhibition, inhibition kinetics can be categorized to competitive, noncompetitive, uncompetitive and mixed type inhibitions.

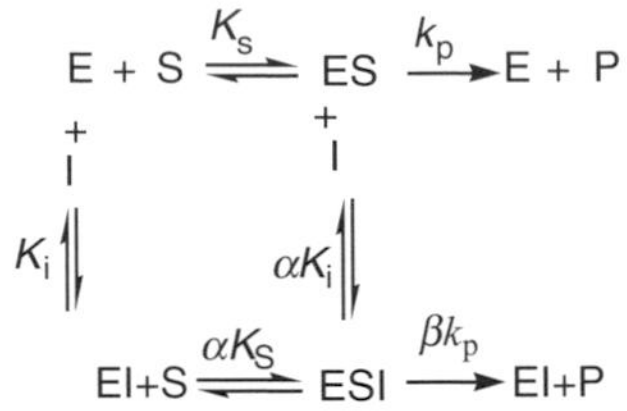

$$E + S \xrightleftharpoons{K_s} ES \xrightarrow{k_p} E + P$$

$$+\qquad\qquad +$$
$$I\qquad\qquad I$$

$$K_i \Big\| \qquad\qquad \alpha K_i \Big\|$$

$$EI + S \xrightleftharpoons{\alpha K_S} ESI \xrightarrow{\beta k_p} EI + P$$

SCHEME 16.1 General kinetics for competitive ([ESI] = 0), noncompetitive (α = 1 and β = 0), uncompetitive (α = 1, β = 0 and [EI] = 0), and mixed type inhibition ($\alpha \neq 1$ and $0 < \beta < 1$). K_S = dissociation constant for substrate (S); K_i = dissociation constant for inhibitor (I); k_p = rate constant for the formation of product; α and β are factors that change K_S or K_i and k_p when the second molecule (either I or S) is bound. Rate equations for above inhibition kinetics are given in Table 16.4 (competitive, noncompetitive, uncompetitive and mixed type).

16.2.1 Materials and Reagents

Chemicals available from the commercial sources below: phenacetin, acetaminophen, diclofenac, cortisone, flufenamic acid, propranolol, 4'-hydroxy-butyranilide, sulfaphenazole, quinidine, testosterone, ketoconazole, 6β-OH-testosterone, taxol, quercetin, baccatin, 4'-hydroxydiclofenac, 4-hydroxytria-zolam, (R)-(+)-propranolol, phenytoin, dextromethorphan, testosterone, dextrorphan, α-naphthoflavone, and NADPH from Sigma (St Louis, MO); fluvoxamine maleate from Tocris (Ballvin, MO); midazolam, 1-OH-midazo-lam, (S)-mephenytoin, (+)-N-3-benzylnirvanol, bufuralol, 1'-OH-bufurarol, 4'-OH-mephenytoin from BD Biosciences (Woburn, MA, USA); (R)-N-3-benzyl-phenobarbital, and L-000706631 from Merck compound library; and 6β-OH-progesterone from Steraloids, Inc. (Newport, RI, USA). Pooled HLM were purchased from Tissue Transformation Technologies (Exton, PA) and from Xenotech (Kansas, KS), respectively. Water was purified by a Mill-Q-System from Millipore Corp. (Milford, MA). Formic acid (analytical grade) was from J.T. Baker (Phillipsburg, NJ). All chemical inhibitors and substrates were dissolved in 50% (v/v) acetonitrile/water or DMSO. The final volume of acetonitrile or DMSO in the reaction mixture was less than 2% (v/v) or 0.2%, respectively. Activity in the presence of the solvent alone was assigned as "control" (100%). Reaction 96-well plates (300 µL), preparation 96-well plates (1.5 mL), and 96-well plate receivers (1.5 mL) were purchased from VWR (West Chester, PA). The reaction and preparation plates were treated with acetonitrile, reconditioned with water and dried by centrifugation prior to use. Hydrophobic and hydrophilic 96-well filtration plates (0.45 µm polytetrafluor-oethylene) were purchased from Millipore Corp (Billerica, MA).

16.2.2 Instrument

A TECAN Genesis RSP 200 liquid handling workstation equipped with eight tips, shaking and temperature control (heating block) was used for sample

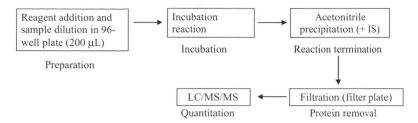

SCHEME 16.2 High throughput screening of CYP inhibition by automated TECAN.

transfer, dilution, and incubation. Gemini software version 4.0 was used for programming the operation. A general instruction of the automated assay is depicted in Scheme 16.2 (Yao et al., 2007). All analytical methods were conducted by LC/MS/MS. The LC/MS/MS consists of a triple quadrupole mass spectrometer (Sciex, API-365, Biosystems MD, Toronto, Canada) equipped with a turbo ion spray ionization source, two Shimadzu LC-10ADvp Solvent Delivery Modules (pumps), an SCL-10ADvp controller (Shimadzu, Columbia, MD), a DGU-14 solvent degasser (Shimadzu, Columbia, MD), and an autosampler (Perkin–Elmer serious 200, Norwalk, CT). Data were collected and processed using Sciex Analyst 1.1 data collection and integration software on an IBM compatible computer. A centrifuge equipped with a 96-well plate rotor was used to conduct the filtration process to separate precipitated proteins. Table 16.1 shows the LC/MS/MS analyses of individual CYP marker substrates in pooled human liver microsomes.

16.2.3 Optimization of Kinetic Reaction

Phosphate buffer (100 mM, pH 7.4) containing 1 mM EDTA was prepared by a dilution of 400 mM mono and dibasic potassium phosphate stock solutions (stored at 4°C) and stored at ambient temperature. Frozen stocks of HLM were used and the remaining was discarded after the first thawing. NADPH stock, at 10 mM in phosphate buffer was made fresh daily. Stock solutions of analytes (i.e., metabolites) were prepared in solvent and stored at -20°C or 4°C. Internal standards were dissolved in either DMSO or 50% acetonitrile in water.

To measure accurate kinetic parameters, all reaction rates should be linear to protein concentrations (0.1–0.25 mg/mL) and incubation times (up to 20 min). Substrate consumption was limited to less than 20%. Because the HLM contains multiple CYPs an isoform-mediated metabolite(s) is (are) monitored to measure the enzyme activity. If the metabolite detection is highly sensitive, a low protein concentration is recommended for the highly protein bound compounds to minimize protein binding. To obtain a K_m value eight substrate concentrations are routinely chosen, which are set around the K_m (0.1–3 K_m). K_m is determined by a nonlinear regression on a velocity versus

TABLE 16.1 LC–MS/MS analyses of individual CYP marker substrates in pooled human liver microsomes.

CYP involved	Substrate	LC column	LC gradient	Metabolite $Q1/Q3$ mass (m/z)	Internal standard $Q1/Q3$ mass (m/z)
CYP1A2	Phenacetin	AquaSep, 5 μm, 2.0 × 50 mm	5–50% B 1.4 min linear	Acetaminophen 151.9/110.1	4′-OH-Butyranilide 180.2/71.0
CYP2A6	Coumarin	Zorbax SBC8, 3.5 μm, 4.6 × 75 mm	15–65% 7 min linear	7-OH-Coumarin HPLC	
CYP2B6	Bupropion	BDS Hypersil C8, 5 μm, 2 × 50 mm	20–80% B 1.9 min linear	OH-Bupropion 256.2/184.2	dl-Propranolol 260.2/155.1
CYP2C8	Diazepam	Extend-C18, 3.5 μm, 4.6 × 50 mm	30–70% B 7.0 min linear	Nordiazepam 300.9/255.2	Lorazepam 320.9/275.2
CYP2C8	Taxol	Zorbax 300 Extend C18, 5 μm, 4.6 × 50 mm	20–80% B 2.0 min linear	6α-OH Taxol 870.3/525.2	Baccatin III 587.2/405.2
CYP2C9	Diclofenac	BDS Hypersil C8, 5 μm, 2 × 50 mm	10–70% B 1.9 min linear	4′-OH-Diclofenac 312/231.1	Flufenamic acid 282.1/264.1
CYP2C19	(S)-Mephenytoin	Zorbax SB-Aq, 5 μm, 4.6 × 50 mm	20–80% B 1.9 min linear	4′-OH-Mephenytoin 235.1/150.1	Phenytoin 253.2/182.2
CYP2D6	Dextromethorphan	XDB C8, 5 μm, 2.1 × 50 mm Agilent	0–60% B 2.1 min linear	Dextrorphan 258.2/157.0	Levallorphan 284.1/199.2

Enzyme	Substrate	Column	Gradient	Metabolite	Internal standard
	Bufuralol	BDS Hypersil C8, 5 μm. 2 × 50 mm	0–50% B 1.9 min linear	1'-OH-Bufurarol 278.1/186.1	DL-Propranolol 260.2/155.1
CYP2E1	Chlorzoxazone	Synergy Polar-RP 4 μm, 50 × 4.6 mm	20–80% B 2 min linear	6-OH-Chlorzoxazone 168.1/132.0	Cpd A 202.2/155.1
CYP3A4	Nifedipine	BDS Hypersil C8, 5 μm, 2 × 50 mm	0–50% B 1.9 min linear	Oxidized nifedipine 344.9/284.1	Nimodipine 419.1/343.2
CYP3A4	Testosterone	BDS Hypersil C8, 5 μm, 2 × 50 mm	0–50% B 1.9 min linear	6β-OH-Testosterone 305.5/269	Cortisone 361.0/163.2
	Dextromethorphan	XDB C8, 5 μm, 2.1 × 50 mm Agilent	0–60% B 2.1 min linear	3-Methoxymorphinan 258.2/171.1	Levallorphan 284.1/199.2
	Diazepam	Extend-C18, 3.5 μm, 4.6 × 50 mm	30–70% B 7.0 min linear	Temazepam 270.9/140.0	Lorazepam 320.9/275.2
	Nordiazepam	Extend-C18, 3.5 μm, 4.6 × 50 mm	30–70% B 7.0 min linear	Oxazepam 292.0/246.2	Lorazepam 320.9/275.2
	Midazolam	ACE 5 C18-A3654 5 μm, 4.6 × 50 mm	10–80% B 2.5 min linear	1'-OH-Midazolam 342.1/203.1	Triazolam 343.1/308.1

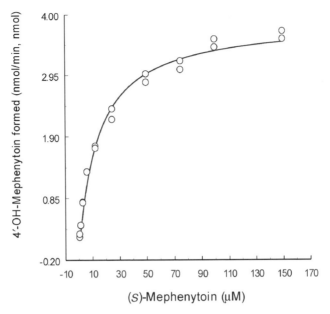

FIGURE 16.1   Hyperbolic saturation curve of CYP2C19-catalyzed (*S*)-mephenytoin 4′-hydroxylation in human liver microsomes. K_m and V_{max} values were calculated with Equation 16.1 ($K_m = 23.7 \pm 2.15\ \mu M$ and $V_{max} = 0.36 \pm 0.01\ nmol/min/mg$). Data were fitted by nonlinear regression (Eq. 16.1, SigmaPlot 9.0).

substrate concentration profile using Michaelis–Menten equation (Eq. 16.1). Figure 16.1 shows the typical M–M kinetics of (*S*)-mephenytoin 4′-hydroxylation. Table 16.2 shows K_m values of the various substrates for the individual human CYP activities in HLM.

$$v = \frac{V_{max}[S]}{K_m + [S]} \tag{16.1}$$

16.2.4   LC/MS/MS Analysis

For all 16 assays in Table 16.1, HPLC separation of metabolite(s) formed from each substrate was achieved using LC columns eluted with the following mobile phase gradient at a flow rate of 1.5 mL/min. The mobile phase solvents were used for CYP1A2 assay (A: 0.05% formic acid in water; B: 0.05% formic acid in acetonitrile) and for all other CYP assays (A: 0.05% formic acid in water:methanol 90:10; B: 0.05% formic acid in acetonitrile) as indicated in Table 16.1. The columns were re-equilibrated to starting conditions before the next set of samples was injected. HPLC flow was diverted from the mass spectrometer to waste for the first and last minute of the gradient.

TABLE 16.2 K_m values for individual CYPs in pooled human liver microsomes.

Enzyme involved	Substrate	Metabolite	Internal standard	K_m (μM)
CYP1A2	Phenacetin	Acetaminophen	4'-OH-Butyranilide	87.7
CYP2A6	Coumarin	7-OH-Coumarin	None	97.8
CYP2B6	Bupropion	OH-Bupropion	DL-Propranolol	98.0
CYP2C8	Taxol	6α-OH-Taxol	Baccatin III	14.4
CYP2C9	Diclofenac	4'-OH-Diclofenac	Flufenamic acid	11.2
CYP2C19	(S)-Mephenytoin	4'-Mephenytoin	Phenytoin	77.9
CYP2D6	Bufuralol	1'-OH-Bufuralol	DL-Propranolol	14.0
CYP2D6	Dextromethorphan	Dextrorphan	Levallorphan	11.7
CYP2E1	Chlorzoxazone	6-OH-Chlorzoxazone	Cpd A	274
CYP3A4	Testosterone	6β-OH-Testosterone	Cortisone	47.2
CYP3A4	Nifedipine	Oxidized nifedipine	Nimodipine	13.0
CYP3A4	Midazolam	1'-OH-Midazolam	1'-OH-Triazolam	2.98
CYP3A4	Midazolam	4-OH-Midazolam	4-OH-Triazolam	24.4

During the enzyme reaction of a substrate, multiple metabolite isomers may be formed. Thus, a baseline LC separation of the metabolites is desired. For example, more than five hydroxylated testosterone isomers (2-, 6-, 15-, and 16-positions) were seen after an incubation of testosterone with HLM. BDS Hypersil C8 was successfuly used for a separation of the five isomers (Table 16.1). Majority of the metabolites (>82%) was due to the 6β-OH-testosterone. Similarly, two hydroxylated isomers of midazolam at 1 and 4'-positions were also isolated with the given HPLC condition.

Mass to charge value for each metabolite formed is shown in Table 16.1. For quantitation of the metabolites, the mass spectrometer was operated in the multiple reaction-monitoring (MRM) modes to monitor the metabolite(s) and internal standard with dwell time set to 150 ms for each reaction. After optimization, heated nebulizer parameters were set as follows (arbitrary units): NEB: 10, CUR: 8, TEMP: 300°C, and NC: 5. The flow rate of heated gas (gas 2) was operated at 5 L/min.

16.2.5 Automated Sample Preparation and Incubation

An automation procedure for sample preparation and incubation has been described previously (Yao et al., 2007). A working solution of HLM was prepared with phosphate buffer to form 0.1 to 0.25 mg/mL for the assays. Metabolite standards were manually prepared with 50% acetonitrile in water as stock solution, and further diluted with HLM for quantitation of the metabolites formed. Individual probe substrates were prepared at designated concentrations around their K_m values. Positive controls (inhibitors) or test compounds were prepared in DMSO (or 50% acetonitrile) as stock solution.

2.5 µL of the stock was dissolved in HLM working solution to appropriate concentrations. All working solutions were on ice before TECAN transfer and dilution. Serial dilutions were conducted by TECAN for all samples. Seven concentrations for a standard curve and four concentrations (3 × LLQ, low, median, and high) for QC were prepared. Eight concentrations of each positive inhibitor or test compound were used for determination of IC_{50}.

The assay was designed to run six compounds (five test compounds and one positive control) for each CYP enzyme. Two plates were used to determine IC_{50} values for five test compounds. A standard, a positive control and two test compounds at the highest concentration were manually prepared and spiked into the last row (H) of each column (1, 3, 5, and 7) in a 2-mL 96-well preparation plate for serial dilution by TECAN. The HLM solution was transferred to column 1, 3, 5, and 7 except the row H. The individual test compounds were diluted serially with HLM to eight concentrations. DMSO was adjusted to ~0.16% (v/v) as a final concentration for each diluted sample. The tip used for dilution was rotated after each transfer of the sample, aspirated and dispensed three times before use to ensure the sample was well mixed in each dilution.

After serial dilution by TECAN, 180 µL of the mixture in column 1, 3, 5, and 7 was transferred to an incubation plate in triplicate. After preincubation at 37°C in a 96-well temperature-controlled heater block for 5 min, 20 µL NADPH (10 mM in 100 mM phosphate buffer) was added to the reaction plate to a final volume of 200 µL to initiate the reaction. During the course of reaction, 240 µL of acetonitrile containing internal standard was prepared and added into a filter plate. After the incubation, 120 µL reaction mixture in the well was transferred into the filter plate for termination of the reaction. One hundred and eight microliters mixture in the well containing standard sample and 12 µL of NADPH were also transferred into a separate filter plate as a need for standard calibration and QC. Thus, the filter plates containing terminated incubation mixtures, standard and QC samples, respectively, was stacked on a 2 mL 96-well receiver plate, vortexed for 30 s, and the mixtures were filtered (0.45 µm, hydrophobic or hydrophilic PTFE membranes) by a centrifuge for 5 min at 2000×g, into the receiver plate. The receiver plate was vortexed and sealed with a polypropylene film, and 10–25 µL of each sample was injected to LC/MS/MS for analysis. Filtration technology was used to separate soluble metabolites from matrix. A newly 96-well filtration plate with chemically resistant 0.45 µm polytetrafluoroethylene (PTFE) membrane (Millipore Inc.) is recommended, which processes up to 1.8 mL of aqueous organic phase and allows for good separation of the metabolites from precipitated protein. Samples in the 96-well filtration plate can be quantitatively transferred into a common 96-well collection plate by either vacuum or centrifuge (5 min) equipped with a plate carrier. The filtered samples are clean and directly injected into LC/MS from the collection plate with no clog in tubing and needle sprayer. Thus, the sample variation is low.

16.2.6 Data Analysis

Nonlinear curve fitting of enzyme kinetic data was accomplished with the Enzyme Kinetics module of either Grafit version 5.0 (Erithacus Software Ltd, Horley Surrey, UK) or Sigmaplot version 9.0 (SPSS, Inc., Chicago, IL).

When the behavior of inhibition is identified, appropriate equations (velocity reduction or % inhibition) can be used for quantitative inhibition of a metabolite formed in the presence of the inhibitor (Table 16.3). Figure 16.2 shows plots for each type of inhibition and graphic determinations for their K_i. IC_{50} is an estimate to evaluate K_i when the substrate concentration approaches K_m. IC_{50} equals to two units of K_i for both competitive and uncompetive inhibitions, and to one unit of K_i for noncompetitive inhibition. However, the IC_{50} value cannot be used as an estimate of K_i in the mixed type inhibition, due to the interference of factors α (Scheme 16.1), which is a change in binding affinity of inhibitor to enzyme (Segel, 1997; Shou et al., 2001).

For determination of enzyme kinetic parameters, replicates of $n = 3$ were run at eight substrate concentrations to obtain K_m and replicates of $n = 3$ were run at eight inhibitor concentrations in the presence of substrate concentration approaching K_m to generate IC_{50} value. Known inhibitors were usually run in parallel as positive controls. IC_{50} parameters were obtained by curve fitting of nonlinear regression to observed data using Equation 16.2, where V_o is the uninhibited velocity, v is the observed velocity, s is the slope factor, and [I] is the inhibitor concentration. Alternatively, the inhibition of enzyme activity can be expressed as percentage of control activity. Figure 16.3 and Table 16.4 show the curve fitting of inhibition of six individual CYP activities

TABLE 16.3 Enzyme inhibition kinetics from the kinetic model in Scheme 16.1: types of enzyme inhibition, velocity and % of enzyme activity (% Control) for the metabolism of substrate (S) in the presence of an inhibitor (I), and relationship between [S] and [I] in different types of inhibition as $[S] = K_m$.

Inhibition type	$\dot{v}$ (velocity)	Percentage of Control	Relationship $(IC_{50}$ and $K_i)$ $([S] = K_m)$
Competitive ([ESI] = 0)	$\dfrac{V_{max}[S]}{K_m\left(1+\frac{[I]}{K_i}\right)+[S]}$	$\dfrac{K_m+[S]}{K_m\left(1+\frac{[I]}{K_i}\right)+[S]}$	$IC_{50} = 2K_i$
Noncompetitive ($\alpha = 1$)	$\dfrac{V_{max}[S]}{K_m\left(1+\frac{[I]}{K_i}\right)+[S]\left(1+\frac{[I]}{K_i}\right)}$	$\dfrac{K_m+[S]}{(K_m+[S])\left(1+\frac{[I]}{K_i}\right)}$	$IC_{50} = K_i$
Uncompetitive ([EI] = 0, $\alpha = 1$)	$\dfrac{V_{max}[S]}{K_m+[S]\left(1+\frac{[I]}{K_i}\right)}$	$\dfrac{K_m+[S]}{K_m+[S]\left(1+\frac{[I]}{K_i}\right)}$	$IC_{50} = 2K_i$
Mixed type ($\alpha \neq 1$)	$\dfrac{V_{max}[S]}{K_m\left(1+\frac{[I]}{K_i}\right)+[S]\left(1+\frac{[I]}{\alpha K_i}\right)}$	$\dfrac{K_m+[S]}{K_m\left(1+\frac{[I]}{K_i}\right)+[S]\left(1+\frac{[I]}{\alpha K_i}\right)}$	$IC_{50} = \dfrac{2\alpha K_i}{(\alpha+1)}$

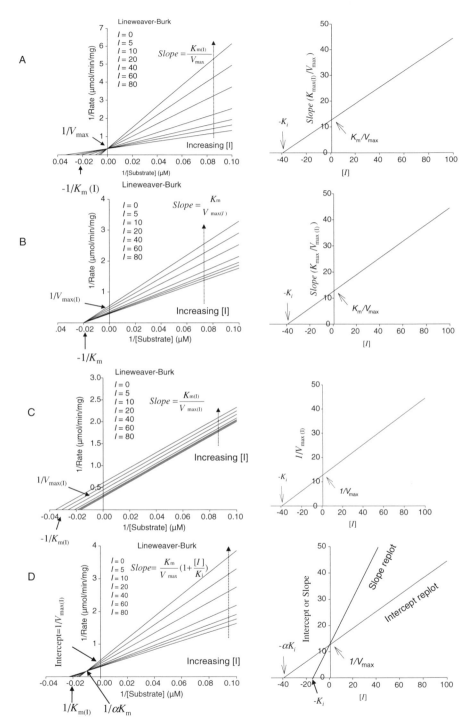

FIGURE 16.2 Methods to measure K_i values of competitive (**a**), noncompetitive (**b**), uncompetitive (**c**), and mixed type inhibition (**d**).

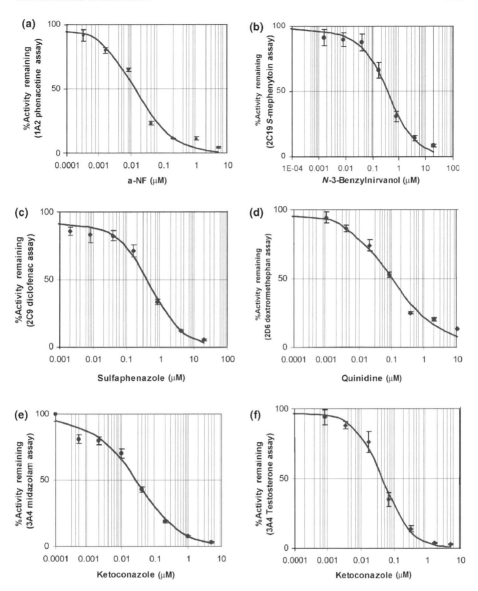

FIGURE 16.3 CYP inhibition curves of six known chemical inhibitors in the metabolism of marker substrates at their K_m values by human liver microsomes using Equation 16.2 (Yao et al., 2007).

and the resulting IC_{50} values, respectively, which were consistent with those in literature.

$$v = \frac{V_o}{1 + \left(\dfrac{[I]}{IC_{50}}\right)^s} \qquad (16.2)$$

TABLE 16.4   IC$_{50}$ Values of selective chemical inhibitors for individual CYPs in pooled human liver microsomes.

CYP involved[a]	Reaction[b]	Inhibitor	Concentration range (μM)	IC$_{50}$ (μM)
CYP1A2	Phenacetin O-deethylation	Fluvoxamine	0.0046–10	0.4
		α-Naphthoflavone	0.0004–5	0.014
CYP2A6	Coumarin 7-hydroxylation	Tranylcypromine	0.0046–10	0.15
CYP2B6	Bupropion hydroxylation	MBA[c]	0.0046–10	0.05
CYP2C8	Taxol 6α-hydroxylation	Montelukast	0.0023–5	0.27
CYP2C9	Diclofenac 4′-hydroxylation	Sulfaphenazole	0.0046–10	0.7
CYP2C19	(S)-Mephenytoin 4′-hydroxylation	(R)-N-3-benzyl-phenobarbital	0.0046–10	0.4
CYP2D6	Bufuralol 1′-hydroxylation	Quinidine	0.0046–10	0.1
	Dextromethorphan	Quinidine	0.001–10	0.13
CYP2E1	Chlorzoxazone 6-hydroxylation	4-Methylpyrazole	0.09–10	0.74
CYP3A4	Testosterone 6β-hydroxylation	Ketoconazole	0.0046–10	0.03
	Midazolam 1′-hydroxylation	Ketoconazole	0.0005–5	0.03

[a]CYP involved in human liver microsomal incubation reaction (Yao et al., 2007).
[b]Substrate concentration equal to K_m.
[c]MBA = N-(α-methylbenzyl)-1-aminobenzotriazole.

For K_i determination, six various concentrations of the given substrate (spread around K_m) and test compound (or known inhibitor, spread around K_i), respectively, were prepared by serial dilution with HLM (6 × 6 = 36 experimental points). The kinetic model (i.e., competitive inhibition, Table 16.3) was used to fit observed data (scatter points) and the fitting of the data were plotted in Fig. 16.4 (2D plot) and Fig. 16.5 (surface plot, SigmaPlot 9.0). K_i values are calculated accordingly.

16.3   IRREVERSIBLE INHIBITION

Irreversible inhibition (mechanism-based inhibition, MBI) is among the most specific enzyme inhibitions, which includes CYP suicide inactivation process (the more widely studied process) and metabolite–intermediate complex (MI) formation (Silverman, 1995; Waley, 1980). The former involves metabolism of drugs to products that denature the CYP. In this case, the inactivator for the

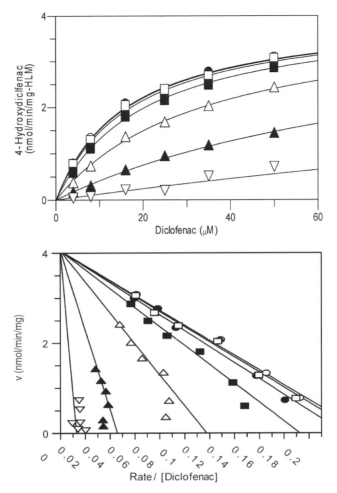

FIGURE 16.4 Competitive inhibition of sulfaphenazole in the metabolism of diclofenac by human liver microsomes. K_m (16.7 ± 0.07 μM), V_{max} (4.1 ± 0.77 nmol/min/mg) and K_i (0.2840 ± 0.01 μM) were generated using the equantion of competitive inhibition in Table 16.3. (Yao et al., 2007).

target enzyme is the substrate that in the process of catalytic turnover is metabolized to a reactive intermediate that inactivates the enzyme irreversibly (suicide inhibition). The latter is MI complexation in which CYP produces a metabolite with the capacity to bind tightly to the CYP heme (Pessayre et al., 1982; Mayhew et al., 2000). The distinction between CYP inactivation and MI complexation is that, in the latter, the hemoprotein is not actually destroyed, even though it is rendered catalytically inert. There are most three pathways for inactivation of the enzyme by the reactive intermediate: covalent modifications on the apoprotein and the heme, and crosslink between apoprotein and heme.

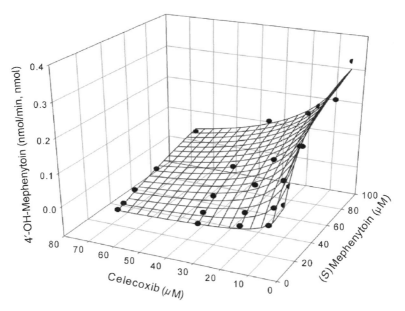

FIGURE 16.5 Celecoxib inhibition of CYP2C19-catalyzed (*S*)-mephenytoin 4′-hydroxylation in human liver microsomes: $V_{max} = 0.40 \pm 0.02$ nmol/min/mg, $K_m = 13.8 \pm 1.9\ \mu M$, $K_i = 3.2 \pm 0.4\ \mu M$, RSS = 0.002, and $R^2 = 0.981$, respectively. All kinetic values were determined by multiple nonlinear regression (3-dimension) using the velocity equation of competitive inhibition in Table 16.3.

In comparison with the reversible enzyme inhibition, the MBI can be characterized to be (1) dose-dependent; (2) preincubation time-dependent; (3) bioactivation required for inactivation of the target enzyme; (4) *de novo* protein synthesis to return metabolic capacity; and (5) slow onset of the effects but more profound than the reversible inhibition.

16.3.1 Kinetic Model for Mechanism-Based Inhibition

Kinetic model for mechanism-based inhibition is proposed in Scheme 16.3 (Waley, 1980; Walsh et al., 1978). Inactivation of the enzyme is an irreversible process over the time scale of the experiment. At the given concentrations of inhibitor and enzyme, the reactions indicated in Scheme 16.3 are governed by the first-order rate constants k_1, k_{-1}, k_2, k_3 and k_4, respectively. The rate of enzyme inactivation can be introduced by Equation 16.3 (Jung and Metcalf, 1975; Kitz and Wilson, 1962).

$$\ln \frac{[E]}{[E]_{tot}}(\%\ \text{activity remaining}) = \frac{-k_{inact}[I]t}{K_I + [I]} \tag{16.3}$$

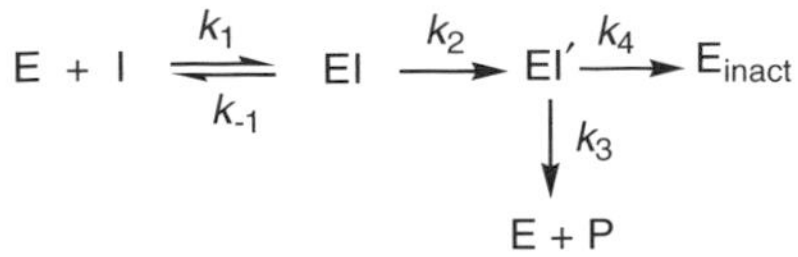

$$E + I \underset{k_{-1}}{\overset{k_1}{\rightleftharpoons}} EI \xrightarrow{k_2} EI' \xrightarrow{k_4} E_{inact}$$

$$\downarrow k_3$$

$$E + P$$

SCHEME 16.3 Proposed kinetic scheme for mechanism-based enzyme inhibition. E, I and P stand for enzyme, inhibitor and product, respectively, and EI is the initial binding of inhibitor to enzyme (the enzyme–inhibitor complex) and EI' is the active form of the complex in which the inhibitor is catalyzed to intermediate (reactive metabolite). E_{inact} is the inactivated enzyme by the reactive metabolite formed. Inactivation of the enzyme is an irreversible process over the time scale of the experiment. At the given concentrations of inhibitor and enzyme, the reactions are governed by the first-order rate constants k_1, k_{-1}, k_2, k_3 and k_4, respectively.

where [E] is the active enzyme concentration at time t; [I] the inhibitor concentration, k_{inact} the inactivation rate constant, and K_I the inactivator concentration that produces half-maximal rate of inactivation. The potency of an inhibitor can be described by the partitioning ratio of the rate for metabolite formation to that for enzyme inactivation (Eq. 16.4). Partition ratio is a measure of the efficiency of inhibitor (average number of cycles the enzyme can traverse before it is inactivated). It can vary from 0 to several thousands. The value of less than 10 is usually defined as a potent inhibitor. The ratio is dependent on the reactivity of the reactive intermediate, rate of release of the reactive intermediate from active site (k_3), rate of enzyme inactivation (k_4) and the proximity of an appropriate target molecule(s) for the covalent bond formation (Jushchyshyn et al., 2003; Walsh et al., 1978).

$$\text{Partioning} = \frac{k_{cat}}{k_{inact}} = \frac{k_3}{k_4} \tag{16.4}$$

16.3.2 Measurements of Kinetic Parameters

Evaluation of compounds as mechanism-based inhibitors of major CYP isoforms (i.e., CYP3A4, CYP1A2, CYP2C8, CYP2C9, CYP2C19, and CYP2D6) in a high throughput mode has been implemented with the marker substrates at early stage of drug discovery in pharmaceutical industry. The analytical assays are similar to those described in the reversible enzyme inhibition. The MBI is commonly performed in preincubation of an inhibitor at a single concentration and variable incubation times. Liver microsomes (0.5–2 mg/mL) from different species can be used to serve as enzyme sources. The inactivation (activity remaining) is monitored by a reaction with each of individual isoform-specific substrates after preincubation time with the inhibitor of test. At the discovery stage, screening for mechanism-based inhibition potential is generally conducted at 10 or 50 μM of test compound, using a preincubation time period of 30 min. A decrease of enzyme activity in 30 min (first order kinetics) is measured and apparent rate constant (k_{obs}) can be obtained by an initial slope of the linear regression lines of semilogarithmic plots (Fig. 16.6a). A comprehensive evaluation of compounds on MBI is perfomred to generate its kinetic parameters (K_I and k_{inact}) (Fig. 16.6b and c). As reported in literature, K_I and k_{inact} values for most

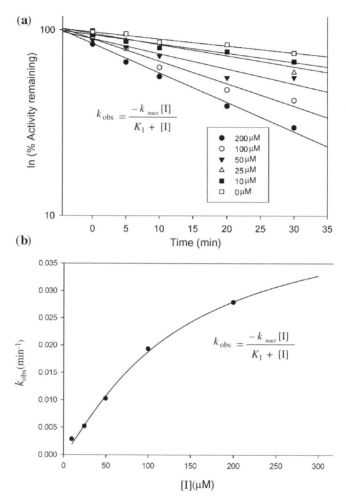

FIGURE 16.6 Methods to determine K_I and k_{inact} of Zileuton in human liver microsomes. (**a**) Aliquots were removed from the primary reaction mixture at the indicated time points and were assayed for residual marker activity (phenacetin O-deethylation). Percent activity remaining (related to time zero in the presence of solvent alone) was plotted in the logarithmic scale determined from a single experiment. The slopes in linear ranges at various inhibitor concentrations were defined as k_{obs}. (**b**) Determination of k_{inact} (0.045 min^{-1}) and K_I (131.7 μM) using nonlinear regression (Eq. 16.5). (**c**) Determination of k_{inact} (0.035 min^{-1}) and K_I (117 μM) using Linewever–Burk plot (Eq. 16.6).

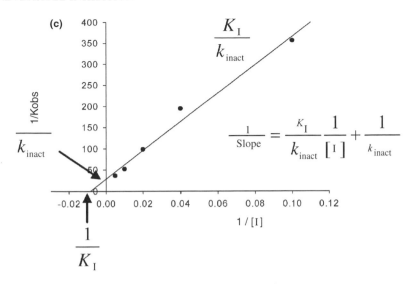

FIGURE 16.6   (*Continued*)

known mechanism-based inhibitors ranged from 0.04 (ritonavir) to 108 μM (rutaecarpine), and from 0.01 (diltiazem) to 1.62 min^{-1} (L-754394), respectively (Atkinson et al., 2005; Chiba et al., 1995; Iwata et al., 2005; Luo et al., 2003).

16.3.3   General Incubation Procedure and Sample Preparation

A general experimental procedure for an enzyme inactivation reaction has been described previously (Lu et al., 2003). The incubation mixtures contained liver microsomal proteins (0.5–2 mg/mL final concentration), 1 mM NADPH and 100 mM potassium phosphate buffer (pH 7.4). The samples were incubated at 37°C (up to 30 min) in the presence of solvent alone, or an inhibitor at varying concentrations (eight concentration points assigned between 0.25 and 4 K_I) in a shaking water bath (final volume of 0.2 mL). All chemical inhibitors and substrates were dissolved in 50% (v/v) acetonitrile/water. The final volume of acetonitrile in the incubation was less than 1% (v/v). At the appropriate times (0, 5, 10, 20, and 30 min), aliquots (25 μL) of the incubate were removed and added to separate vials containing 225 μL (10-fold dilution) of fresh buffer preheated at 37°C. A marker substrate (Table 16.1) and NADPH (1.0 mM) were added and the reaction was allowed to proceed for additional 10 min. The reaction was terminated with acetonitrile (two volumes) containing an appropriate amount of internal standard. The samples were vortexed, centrifuged (3500 rpm for 10 min) and the supernatant was separated and diluted with an equal volume of 0.05% formic acid in water. Aliquots (20 μL) were subjected to LC/MS/MS analysis. Incubations carried out in the presence of solvent alone (time zero) were designated controls (100%). HPLC flow was diverted from the mass spectrometer to waste for the first and last minute of the gradient to remove nonvolatile salts. Mass to charge value for each metabolite formed is shown in Table 16.1. For quantitation of the metabolites, the mass spectrometer was operated in the MRM modes to monitor area ratios of metabolite(s) to internal standard.

16.3.4 Data Analysis

The observed rates of individual CYP inactivations (k_{obs}) are calculated from the initial slopes of the linear regression lines of semilogarithmic plots (natural logarithm of remaining activity versus preincubation time, Fig. 16.6a). k_{inact} and K_I can be obtained by nonlinear regression using Equation 16.5 (Fig. 16.6b). In most cases, the parameters are also generated by the double reciprocal Lineweaver–Burk plot ($1/k_{obs}$ versus $1/[I]$) as depicted in Fig. 16.6c (Eq. 16.6), which shows k_{inact} estimate at reciprocal of the y-intercept and K_I at negative reciprocal of the x-intercept.

$$k_{obs} = \frac{k_{inact}[I]}{K_I + [I]} \tag{16.5}$$

$$\frac{1}{k_{obs}} = \frac{K_I}{k_{inact}} \times \frac{1}{[I]} + \frac{1}{k_{inact}} \tag{16.6}$$

16.4 FLUORESCENT ASSAY

The assays with fluorescence detection in a HTS manner have been employed in pharmaceutical companies to determine the CYP inhibition potential of the drug candidates. These assays are usually applied for major human CYPs, such as 1A2, 2C9, 2C19, 2D6, and 3A4, in which a fluoroprobe, through a CYP-mediated biotransformation, becomes fluorescent (Crespi et al., 2000; Donato et al., 1998; Kennedy and Jones, 1994). These fluoroprobes may not be selective for an individual CYP, and thus the recombinant cDNA-expressed CYPs in microsomal preparations have to be used (Crespi et al., 2000; Kariv et al., 2001; Trubetskoy et al., 2005). The fluorescent HTS has been well developed into 96-well, 384-well, or even higher microplate format (Crespi et al., 2000; Kariv et al., 2001; Kennedy and Jones, 1994; Trubetskoy et al., 2005). The technology has been further developed into cell-based investigation, where CYPs are transfected into and expressed in cell lines (Crespi and Stresser, 2000; Donato et al., 2004). Microarray technology, combining solidified enzymes and fluorescence, has been recently reported for HTS (Sakai-Kato et al., 2005). The major fluorescence HTS assays for CYP inhibition are summarized on Table 16.5.

The better fluoroprobes would produce strong fluorescent metabolite with higher excitation (>400 nm) and emission (>500 nm) wavelengths to reduce

TABLE 16.5 Fluoroprobes and fluorescent metabolites for individual CYPs.

Substrates→products	CYPs involved	Fluorescent detection: excitation (nm)/emission (nm)
AMMC→AHMC	2D6	390/460
BFC→HFC	3A4, 3A5, 3A7	409/530
BOMFC→HFC	3A4, 3A5, 2B6	400/502
BOMCC→CHC	3A4, 3A5, 2B6, 2C9	409/460
BOMF→fluorescein	2C9	485/530
BOMR→resorufin	3A4	530/585
7BQ→7-HQ	3A4	409/530
CEC→CHC	1A2, 2C19	409/460
BzRes→resorufin	1A1, 1B1, 3A4	530/590
Coumarin→7-HC	2A6	390/460
DBF→fluorescein	2C8, 2C9, 2C19, 3A4, 19	485/538
DBOMF→fluorescein	3A4, 3A5	485/530
EFC→HFC	2B6	409/530
EOMCC→CHC	2D6, 1A2, 2C19, 2E1	409/460
7-MFC→HFC	2C9, 2E1	409/530
MAMC→HAMC	2D6	390/460
MOBFC→HFC	2D6	400/502
OMF→fluorescein	2C19	485/538
OOMR→resorufin	2C9	530/585

AMMC, 3-[2-(N,N-diethyl-N-methylamino)ethyl]-7-methoxy-4-methylcoumarin; AHMC, 3-[2-(N,N-diethylamino)ethyl]-7-hydroxy-4-methylcoumarin hydrochloride; BFC, 7-benzyloxy-4-(trifluoromethyl)-coumarin; BOMFC, 7-benzyloxymethyloxy-4-trifluoromethyl coumarian; BOMCC, 7-benzyloxymethyloxy-3-cyanocoumarian; BOMF, benzyloxymethylfluorescein; BOMR, benzyloxymethylresorufin; 7BQ, 7-benzyloxyquinoline; BzRes, resorufin benzyl ether. CEC, 3-cyano-7-ethoxycoumarin; CHC, 3-cyano-7-hydroxycoumarin; DBF, dibenzylfluorescein; DBOMF, dibenzyloxcymethylfluorescein EFC, 7-ethoxy-4-trifluoromethylcoumarin; EOMCC, ethyloxymethyloxy-3-cyanocoumarian; 7-HQ, 7-hydroxyquinoline; HFC, 7-hydroxy-4-trifluoromethylcoumarin; 7-HC, 7-hydroxycoumarin; HAMC, 7-hydroxy-4-(aminomethyl)coumarin; 7-MFC, 7-methoxy-4-trifluoromethylcoumarin; MAMC, 7-methoxy-4-(aminomethyl)coumarin; MOBFC, 7-(p-methoxybenzyloxy-4-trifluorocoumarin; OMF, 3-O-methylfluorescein; OOMR, n-Octyloxymethylresorufin.

interference. The fluorescent HTS comes with great sensitivity, simple operation, fast detection, lower expenditure, and quick turnaround time (Crespi et al., 2000; Donato et al., 2004; Kariv et al., 2001; Sakai-Kato et al., 1998; Trubetskoy et al., 2005). However, the fluorescence nature of the test compounds and/or NADPH would interfere with the fluorescence detection. In addition, most fluoroprobes are lacked of selectivity to individual CYPs and are not the *in vivo* probe substrates for CYPs. Poor correlations of fluorescence HTS with LC/MS/MS assays using substrates selective to individual human CYPs have been reported (Bjornsson et al., 2003; Cohen et al., 2003; Crespi and Stresser, 2000; Donato et al., 2004; Stresser et al., 2002). It has been suggested that the fluorescence results are not recommended for regulatory submission (Bjornsson et al., 2003).

16.5 PREDICTION OF HUMAN DRUG–DRUG INTERACTIONS FROM *IN VITRO* CYP INHIBITION DATA

When an enzyme inhibitor is identified to change biotransformation profiles of a victim drug or drug candidate, drug–drug interactions probably occur, largely due to the change in enzyme activity. It may lead to severe clinically adverse effects associated with metabolism-based DDIs. Thus, evaluation of the compounds as inhibitors of the DMEs and prediction of the probable *in vivo* DDI potentials are crucial at the discovery stage. *In vitro* kinetic data from enzyme inhibition are typically required to assess the degree of the inhibition (IC_{50} and K_i for reversible enzyme inhibition and K_I and k_{inact} for mechanism-based inhibition). Apparent intrinsic clearance ($CL_{int,[I]}$) in the presence of inhibitor can be calculated. Ratios of systemic exposure in the presence and absence of inhibitor can be determined according to appropriate *in vitro* approaches (Azie et al., 1998; Bjornsson et al., 2003; Ito et al., 1998a, 1998b, 2004; Kanamitsu et al., 2000; Lin and Lu, 2001; Mayhew et al., 2000; Obach, 2003; Obach et al., 2005, 2006; Shou, 2005; Thummel and Wilkinson, 1998; Wang et al., 2000, 2004).

16.5.1 Reversible CYP Inhibition

Since enzyme inhibition involves reversible mechanisms, $CL_{int,[I]}$ may vary with regard to the type and concentration of inhibitor. The concentrations of an inhibitor (or drug) that are relevant to clinical application can be approached for the prediction in the *in vivo* situation. In practice, a ratio in AUC, hepatic clearance (CL_{hept}), plasma concentration at steady state (C_{ss}), or intrinsic clearance (CL_{int}) caused by metabolism-based DDIs is commonly used to assess the degree of metabolism inhibition *in vivo* (Eq. 16.7). If a drug is eliminated due to both metabolism and renal excretion, the fraction of the drug metabolized by the inhibited enzyme (f_m) should be introduced to the prediction. With inclusion of f_m, the ratio change in AUC in the presence and absence of an inhibitor can be expressed for competitive and noncompetitive (Eq. 16.8).

$$\frac{AUC_{[I]}}{AUC_{ctr}} = \frac{C_{ss,[I]}}{C_{ss}} = \frac{CL_{hept}}{CL_{hept,[I]}} = \frac{CL_{int}}{CL_{int,[I]}} \tag{16.7}$$

$$\frac{AUC_{[I]}}{AUC_{ctr}} = \frac{1}{\dfrac{f_m}{\left(1 + \dfrac{f_u[I]}{K_i}\right)} + (1 - f_m)} \tag{16.8}$$

where f_m represents the fraction of CYP-dependent metabolism catalyzed by the inhibited CYP; K_i the inhibitor dissociation constant; f_u the unbound fraction of inhibitor in plasma and [I] the projected plasma concentration of

TABLE 16.6 Fractions of metabolism of substrates by individual CYPs reported in literature.

Substrate	f_m	Substrate	f_m
Alprazolam	0.8	Midazolam	0.99 (0.94)
Buspirone	0.99	Nifedipine	0.71
Carbamazepine	0.6	Nisoldipine	0.99
Cisapride	0.95	Pimozide	0.4
Cyclosporine	0.71	Quinidine	0.76
Diazepam	0.8	Simvastatin	0.99
Felodipine	0.99 (0.81)	Taxol	0.7–0.9
Loratadine	0.6	Terfenadine	0.74
Lovastatin	0.99	Triazolam	0.98

inhibitor (usually $[I]_{max,ss}$). If the drug elimination is due largely to a CYP ($f_m = 1$), Equation 16.8 can be simplified to Equation 16.9.

$$\frac{AUC_{[I]}}{AUC_{ctr}} = 1 + \frac{f_{u,p}[I]}{K_i} \tag{16.9}$$

The clinical implications of CYP inhibition by inhibitors are dependent on the *in vivo* concentration of the inhibitor and the role of that CYP in the metabolism of the coadministered drug (f_m) (Table 16.6). Clinical relevance of competitive CYP inhibition to human DDI prediction is given in Table 16.7 (Bjornsson et al., 2003a, 2003b). The equations are used to quantitatively predict DDI potential in human from *in vitro* competitive, noncompetitive and mixed type inhibition. As a conservative approach, the inhibitor $[I]_{max}$ at steady-state and at the highest clinical dose expected should be used in the estimation of AUC change. It was found that a DDI would likely occur if the ratio of inhibitor $[I]_{max}/K_i$ were greater than 1 (Table 16.7). DDIs at the ratios between 1 and 0.1 or below 0.1 are possible or remote.

16.5.2 Prediction of Human Drug–Drug Interactions from Mechanism-Based CYP Inhibition

The ability to accurately predict DDI secondary to MBI of a CYP(s) has been greatly enhanced by the knowledge of the specific CYP-mediated reactions,

TABLE 16.7 Prediction of clinical relevance of competitive CYP inhibition (Bjornsson et al., 2003a, 2003b).

$[I]/K_i$	Prediction
$[I]_{max}/K_i > 1$	Likely
$1 > [I]_{max}/K_i > 0.1$	Possible
$0.1 > [I]_{max}/K_i$	Remote

although the prediction of MBI-mediated DDIs is more complicated than that from reversible CYP inhibition (Brown et al., 2005; Galetin et al., 2005, 2006; Ito et al., 1998; Kenworthy et al., 1999; Mayhew et al., 2000; Venkatakrishnan et al., 2001; von Moltke et al., 1998; Yao and Levy, 2002). Equation 16.10 is a commonly used model for the quantitative prediction of human DDIs (AUC ratio change in the presence and absence of an inhibitor) from *in vitro* MBI (Brown et al., 2005; Galetin et al., 2006; Mayhew et al., 2000; Wang et al., 2004).

$$\frac{AUC_{[I]}}{AUC_{[contr]}} = \frac{1}{\left(\dfrac{f_m}{1 + \left(\dfrac{k_{inact}f_u[I]}{(K_I + f_u[I])k_E}\right)}\right) + (1 - f_m)} \tag{16.10}$$

where f_m represents the fraction of P450-dependent metabolism catalyzed by the inhibited CYP; k_E the rate constant for enzyme degradation; [I] the projected plasma concentration of inhibitor (usually $C_{max,ss}$); f_u the unbound fraction of inhibitor in plasma. The equation can be modified based on several situations below. If the drug elimination is due largely to the CYP ($f_m = 1$), Equation 16.10 can be simplified with Equation 16.11. Table 16.8 shows examples of prediction for *in vivo* DDI potential from *in vitro* MBI data and their comparisons with

TABLE 16.8   Prediction of human DDIs from *in vitro* mechanism-based CYP3A4 inhibition.

Inhibitor	Substrate	$[I]_{p,ss}$[a] (µM)	$f_{u,p}$[b]	f_m	K_I (µM)[c]	k_{inact} (min^{-1})	AUC (pred)[d]	AUC (obs)[e]
	Triazolam	4.28	0.3	0.98	5.5	0.07	7.6	5.3
Clarithromycin	Cisapride	0.9	0.3	0.95	5.5	0.07	5.8	3.2
	Midazolam	4.28	0.3	0.94	5.5	0.07	10.7	8.4
	Simvastatin	4.1	0.16	0.99	10.9	0.05	5.9	6.2
Erythromycin	Buspirone	4.1	0.16	0.99	10.9	0.05	5.9	5.9
	Midazolam	4.1	0.16	0.94	10.9	0.05	5.6	4.4
Mibefradil	Midazolam	10.2	0.005	0.94	2.3	0.4	9.0	8.9
Nelfinavir	Simvastatin	7.0	0.015	0.99	1	0.22	8.3	6.1
Ritonavir	Triazolam	3.15	0.015	0.98	0.17	0.4	39.1	20.3
Saquinavir	Midazolam	1.12	0.02	0.94	0.65	0.26	9.0	5.2
	Simvastatin	0.91	0.1	0.99	4.2	0.09	4.7	4.2
Verapamil	Buspirone	0.48	0.1	0.99	4.2	0.09	3.0	3.5
	Midazolam	0.48	0.1	0.94	4.2	0.09	2.7	2.9

[a]Maximum plasma concentrations of the inhibitors at steady state reported in literature.
[b]Unbound fractions of the inhibitors in human plasma.
[c]K_I values used were not corrected with unbound fraction of the inhibitors in liver microsomes.
[d]Predicted AUC ratios from *in vitro* data using Equation 16.10. A half life of (23 hours) for CYP3A4 degradation was used in the prediction.
[e]Clinical observed AUC ratios reported in literature.

clinical observed DDIs.

$$\frac{AUC_{[I]}}{AUC_{ctr}} = 1 + \frac{k_{inact}f_u[I]}{(K_I + f_u[I])k_E}$$ (16.11)

If the CYP in gut contributes to the metabolism of a drug and is inhibited by a mechanism-based inhibitor, a change of bioavailability of the drug in gut should be taken into account. The ratio of $F_{g,[I]}$ to $F_{g,ctr}$ (the intestinal wall availability in the presence and absence of inhibitor) can be incorporated into the model (Equation 16.12). The ratio of $F_{g,[I]}$ to $F_{g,ctr}$ is estimated from the relative change in $CL_{int,g}$ (gut intrinsic clearance) caused by the inhibitor (Equation 16.13).

$$\frac{AUC_{[I]}}{AUC_{ctr}} = \frac{F_{g,[I]}}{F_{g,ctr}} \frac{1}{\left(\dfrac{f_m}{1 + \dfrac{k_{inact}f_u[I]}{(K_I + f_u[I])k_E}} + (1 - f_m) \right)}$$ (16.12)

$$\frac{F_{g,[I]}}{F_{g,ctr}} = \frac{1}{F_{g,ctr} + (1 - F_{g,ctr})\left(\dfrac{CL_{int,g,[I]}}{CL_{int,g}} \right)}$$ (16.13)

16.5.3 Factors Affecting the Prediction of Drug–Drug Interactions

Although research on the prediction of DDI from *in vitro* data achieved some extent in success, many unresolved issues that affect accurate prediction need to be addressed and can complicate the prediction. These factors include (1) choice of steady state plasma concentration ($[I]_{max}$, $[I]_{ave}$ or $[I]_{trough}$), or concentration in total hepatic circulation ($[I]_{hept,inlet}$) that approaches that at active site of enzyme (hepatic intracellular concentration); (2) impact of unbound fraction of the drug (f_u) in plasma and liver microsomes (Austin et al., 2002; Giuliano et al., 2005), particularly for highly bound inhibitors; (3) accurate *in vitro* assessment on f_m of individual victim drugs (interacting drugs) for the inhibited enzyme. If a drug is metabolized by multiple enzymes that are inhibited by the inhibitor, f_m value of the drug is contributed by multiple enzymes ($f_m = f_{m, CYP3A4} + f_{m, CYP2C8} \cdots$) (Rodrigues, 2005); (4) the enzyme inhibition by metabolite(s) formed; (5) enzyme inhibition occurring in intestine ($CL_{int,g}$), which may underestimate the prediction; (6) concurrence in both reversible and mechanism-based inhibition; (6) concurrence in induction and inhibition of the enzyme by the inhibitor (Greenblatt et al., 1999); (7) effect of transporters or poor permeability on intracellular concentration of the inhibitor (Zhang et al., 2006); (8) effect of non-CYP enzymes (i.e., elimination of parent substrates by conjugation pathways) on f_m; (9) choice of K_E value for MBI. Varying half-lives for CYP3A4 recovery have been reported in literature ($t_{1/2} = 9$–144 h, $K_E = 0.693/t_{1/2}$) (Correia, 1991; Fromm et al., 1996; Greenblatt et al., 2003; Hsu et al., 1997; Lai et al., 1978;

Renwick et al., 2000); (10) Interindividual variability of enzymes in human; (11) potential complication due to renal or hepatic impairment; and (12) interactions with dietary ingredients.

16.6 CONCLUSION

In vitro evaluation of NCEs as inhibitors of CYP activity at early stages of drug development is crucial for prediction of DDI potential in human. The chapter provides theoretic background of enzyme inhibition kinetics, experimental approaches for determination of kinetic parameters for both reversible and mechanism-based CYP inhibition, and individual CYP inhibition assays, particularly with LC/MS/MS analysis from a perspective of pharmaceutical industry. These allow adaptation of the system to determine detailed enzyme kinetics and inhibition parameters in an HTS fashion for drug metabolism scientists. The enzyme kinetic values obtained from the assays can be used to assess the potency of the inhibitors, to predict DDI potential in clinic, to drive decision-making on drug development at early stages and to direct informed planning of clinical DDI studies. In addition to the prediction from *in vitro* inhibition data, drug interaction studies in animals, when coupled with the corresponding *in vitro* studies, can be of great benefit in understanding the relationships between *in vitro* inhibition and *in vivo* PK-based DDIs and provide accuracy of the prediction.

ACKNOWLEDGMENT

The authors would like to thank Ping Lu, Regina Wang, Chris Kochansky, Ryan Norcross, Rich Edom, Ken Korzekwa, Juinn Lin, Tom Baillie (Merck Research Laboratories, West Point, PA19486, USA), Ming Yao (Bristol-Myers Squibb Pharmaceutical Research Institute, Princeton, NJ 08543, USA) for their contribution and comments.

REFERENCES

Atkinson A, Kenny JR, Grime K. Automated assessment of time-dependent inhibition of human cytochrome P450 enzymes using liquid chromatography-tandem mass spectrometry analysis. Drug Metab Dispos 2005;33:1637–1647.

Austin RP, Barton P, Cockroft SL, Wenlock MC, Riley RJ. The influence of nonspecific microsomal binding on apparent intrinsic clearance, and its prediction from physicochemical properties. Drug Metab Dispos 2002;30:1497–1503.

Azie NE, Brater DC, Becker PA, Jones DR, Hall SD. The interaction of diltiazem with lovastatin and pravastatin. Clin Pharmacol Ther 1998;64:369–377.

Backman JT, Olkkola KT, Aranko K, Himberg JJ, Neuvonen PJ. Dose of midazolam should be reduced during diltiazem and verapamil treatments. Br J Clin Pharmacol 1994;37:221–225.

Bjornsson TD, Callaghan JT, Einolf HJ, Fischer V, Gan L, Grimm S, Kao J, King SP, Miwa G, Ni L, Kumar G, McLeod J, Obach RS, Roberts S, Roe A, Shah A, Snikeris F, Sullivan JT, Tweedie D, Vega JM, Walsh J, Wrighton SA. The conduct of *in vitro* and *in vivo* drug–drug interaction studies: a Pharmaceutical Research and Manufacturers of America (PhRMA) perspective. Drug Metab Dispos 2003a;31: 815–832.

Bjornsson TD, Callaghan JT, Einolf HJ, Fischer V, Gan L, Grimm S, Kao J, King SP, Miwa G, Ni L, Kumar G, McLeod J, Obach SR, Roberts S, Roe A, Shah A, Snikeris F, Sullivan JT, Tweedie D, Vega JM, Walsh J, Wrighton SA. The conduct of *in vitro* and *in vivo* drug–drug interaction studies: a PhRMA perspective. J Clin Pharmacol 2003b;43:443–469.

Brown HS, Ito K, Galetin A, Houston JB. Prediction of *in vivo* drug–drug interactions from *in vitro* data: impact of incorporating parallel pathways of drug elimination and inhibitor absorption rate constant. Br J Clin Pharmacol 2005;60:508–518.

Bu HZ, Knuth K, Magis L, Teitelbaum P. High-throughput cytochrome P450 (CYP) inhibition screening via cassette probe-dosing strategy: III. Validation of a direct injection/on-line guard cartridge extraction-tandem mass spectrometry method for CYP2C19 inhibition evaluation. J Pharm Biomed Anal 2001a;25:437–442.

Bu HZ, Magis L, Knuth K, Teitelbaum P. High-throughput cytochrome P450 (CYP) inhibition screening via a cassette probe-dosing strategy. VI. Simultaneous evaluation of inhibition potential of drugs on human hepatic isozymes CYP2A6, 3A4, 2C9, 2D6 and 2E1. Rapid Commun Mass Spectrom 2001b;15:741–748.

Chiba M, Nishime JA, Lin JH. Potent and selective inactivation of human liver microsomal cytochrome P-450 isoforms by L-754,394, an investigational human immune deficiency virus protease inhibitor. J Pharmacol Exp Ther 1995;275:1527–1534.

Cohen LH, Remley MJ, Raunig D, Vaz ADN. *In vitro* drug interactions of cytochrome p450: an evaluation of fluorogenic to conventional substrates. Drug Metab Dispos 2003;31:1005–1015.

Correia MA. Cytochrome P450 turnover. Methods Enzymol 1991;206:315–325.

Crespi CL, Stresser DM. Fluorometric screening for metabolism-based drug–drug interactions. J Pharmacol Toxicol Methods 2000;44:325–331.

DiMarco A, Marcucci I, Verdirame M, Perez J, Sanchez M, Pelaez F, Chaudhary A, Laufer R. Development and validation of a high-throughput radiometric CYP3A4/5 inhibition assay using tritiated testosterone. Drug Metab Dispos 2005;33:349–358.

Dierks EA, Stams KR, Lim HK, Cornelius G, Zhang H, Ball SE. A method for the simultaneous evaluation of the activities of seven major human drug-metabolizing cytochrome P450s using an *in vitro* cocktail of probe substrates and fast gradient liquid chromatography tandem mass spectrometry. Drug Metab Dispos 2001;29:23–29.

Donato MT, Jimenez N, Castell JV, Gomez-Lechon MJ. Fluorescence-based assays for screening nine cytochrome P450 (P450) activities in intact cells expressing individual human P450 enzymes. Drug Metab Dispos 2004;32:699–706.

Draper AJ, Madan A, Latham J, Parkinson A. Development of a non-high pressure liquid chromatography assay to determine [^{14}C]chlorzoxazone 6-hydroxylase (CYP2E1) activity in human liver microsomes. Drug Metab Dispos 1998a;26:305–312.

Draper AJ, Madan A, Smith K, Parkinson A. Development of a non-high pressure liquid chromatography assay to determine testosterone hydroxylase (CYP3A) activity in human liver microsomes. Drug Metab Dispos 1998b;26:299–304.

Fromm MF, Busse D, Kroemer HK, Eichelbaum M. Differential induction of prehepatic and hepatic metabolism of verapamil by rifampin. Hepatology 1996;24: 796–801.

Galetin A, Burt H, Gibbons L, Houston JB. Prediction of time-dependent CYP3A4 drug–drug interactions: impact of enzyme degradation, parallel elimination pathways, and intestinal inhibition. Drug Metab Dispos 2006;34:166–175.

Galetin A, Ito K, Hallifax D, Houston JB. CYP3A4 substrate selection and substitution in the prediction of potential drug–drug interactions. J Pharmacol Exp Ther 2005;314:180–190.

Giuliano C, Jairaj M, Zafiu CM, Laufer R. Direct determination of unbound intrinsic drug clearance in the microsomal stability assay. Drug Metab Dispos 2005;33:1319–1324.

Gomez DY, Wacher VJ, Tomlanovich SJ, Hebert MF, Benet LZ. The effects of ketoconazole on the intestinal metabolism and bioavailability of cyclosporine. Clin Pharmacol Ther 1995;58:15–19.

Gorski JC, Jones DR, Haehner-Daniels BD, Hamman MA, O'Mara EMJ, Hall SD. The contribution of intestinal and hepatic CYP3A to the interaction between midazolam and clarithromycin. Clin Pharmacol Ther 1998;64:133–143.

Greenblatt DJ, von Moltke LL, Daily JP, Harmatz JS, Shader RI. Extensive impairment of triazolam and alprazolam clearance by short-term low-dose ritonavir: the clinical dilemma of concurrent inhibition and induction. J Clin Psychopharmacol 1999;19:293–296.

Greenblatt DJ, von Moltke LL, Harmatz JS, Chen G, Weemhoff JL, Jen C, Kelley CJ, LeDuc BW, Zinny MA. Time course of recovery of cytochrome p450 3A function after single doses of grapefruit juice. Clin Pharmacol Ther 2003;74:121–129.

Greenblatt DJ, von Moltke LL, Harmatz JS, Mertzanis P, Graf JA, Durol AL, Counihan M, Roth-Schechter B, Shader RI. Kinetic and dynamic interaction study of zolpidem with ketoconazole, itraconazole, and fluconazole. Clin Pharmacol Ther 1998a;64:661–671.

Greenblatt DJ, Wright CE, von Moltke LL, Harmatz JS, Ehrenberg BL, Harrel LM, Corbett K, Counihan M, Tobias S, Shader RI. Ketoconazole inhibition of triazolam and alprazolam clearance: differential kinetic and dynamic consequences. Clin Pharmacol Ther 1998b;64:237–247.

Heerey A, Barry M, Ryan M, Kelly A. The potential for drug interactions with statin therapy in Ireland. Ir J Med Sci 2000;169:176–179.

Honig PK, Cantilena LR. Polypharmacy. Pharmacokinetic perspectives. Clin Pharmacokinet 1994;26:85–90.

Honig PK, Woosley RL, Zamani K, Conner DP, Cantilena LRJ. Changes in the pharmacokinetics and electrocardiographic pharmacodynamics of terfenadine with concomitant administration of erythromycin. Clin Pharmacol Ther 1992;52:231–238.

Honig PK, Worham DC, Zamani K, Mullin JC, Conner DP, Cantilena LR. The effect of fluconazole on the steady-state pharmacokinetics and electrocardiographic pharmacodynamics of terfenadine in humans. Clin Pharmacol Ther 1993a;53:630–636.

Honig PK, Wortham DC, Hull R, Zamani K, Smith JE, Cantilena LR. Itraconazole affects single-dose terfenadine pharmacokinetics and cardiac repolarization pharmacodynamics. J Clin Pharmacol 1993b;33:1201–1206.

Honig PK, Wortham DC, Zamani K, Conner DP, Mullin JC, Cantilena LR. Terfenadine-ketoconazole interaction. Pharmacokinetic and electrocardiographic consequences. JAMA 1993c;269:1513–1518.

Hsu A, Granneman GR, Witt G, Locke C, Denissen J, Molla A, Valdes J, Smith J, Erdman K, Lyons N, Niu P, Decourt JP, Fourtillan JB, Girault J, Leonard JM. Multiple-dose pharmacokinetics of ritonavir in human immunodeficiency virus-infected subjects. Antimicrob Agents Chemother 1997;41:898–905.

Ito K, Brown HS, Houston JB. Database analyses for the prediction of *in vivo* drug–drug interactions from *in vitro* data. Br J Clin Pharmacol 2004;57:473–486.

Ito K, Iwatsubo T, Kanamitsu S, Nakajima Y, Sugiyama Y. Quantitative prediction of *in vivo* drug clearance and drug interactions from *in vitro* data on metabolism, together with binding and transport. Annu Rev Pharmacol Toxicol 1998a;38:461–499.

Ito K, Iwatsubo T, Kanamitsu S, Ueda K, Suzuki H, Sugiyama Y. Prediction of pharmacokinetic alterations caused by drug–drug interactions: metabolic interaction in the liver. Pharmacol Rev 1998b;50:387–412.

Ito K, Iwatsubo T, Kanamitsu S, Ueda K, Suzuki H, Sugiyama Y. Prediction of pharmacokinetic alterations caused by drug–drug interactions: metabolic interaction in the liver. Pharmacol Rev 1998c;50:387–412.

Iwata H, Tezuka Y, Kadota S, Hiratsuka A, Watabe T. Mechanism-based inactivation of human liver microsomal CYP3A4 by rutaecarpine and limonin from Evodia fruit extract. Drug Metab Pharmacokinet 2005;20:34–45.

Jenkins KM, Angeles R, Quintos MT, Xu R, Kassel DB, Rourick RA. Automated high throughput ADME assays for metabolic stability and cytochrome P450 inhibition profiling of combinatorial libraries. J Pharm Biomed Anal 2004;34:989–1004.

Jung MJ, Metcalf BW. Catalytic inhibition of gamma-aminobutyric acid - alpha-ketoglutarate transaminase of bacterial origin by 4-aminohex-5-ynoic acid, a substrate analog. Biochem Biophys Res Commun 1975;67:301–306.

Jushchyshyn MI, Kent UM, Hollenberg PF. The mechanism-based inactivation of human cytochrome P450 2B6 by phencyclidine. Drug Metab Dispos 2003;31:46–52.

Kanamitsu S, Ito K, Sugiyama Y. Quantitative prediction of *in vivo* drug–drug interactions from *in vitro* data based on physiological pharmacokinetics: use of maximum unbound concentration of inhibitor at the inlet to the liver. Pharm Res 2000;17:336–343.

Kariv I, Fereshteh MP, Oldenburg KR. Development of a miniaturized 384-well high throughput screen for the detection of substrates of cytochrome P450 2D6 and 3A4 metabolism. J Biomol Screen 2001;6:91–99.

Kennedy SW, Jones SP. Simultaneous measurement of cytochrome P4501A catalytic activity and total protein concentration with a fluorescence plate reader. Anal Biochem 1994;222:217–223.

Kenworthy KE, Bloomer JC, Clarke SE, Houston JB. CYP3A4 drug interactions: correlation of 10 *in vitro* probe substrates. Br J Clin Pharmacol 1999;48:716–727.

Kitz R, Wilson IB. Esters of methanesulfonic acid as irreversible inhibitors of acetylcholinesterase. J Biol Chem 1962;237:3245–3249.

Kremers P. *In vitro* tests for predicting drug–drug interactions: the need for validated procedures. Pharmacol Toxicol 2002;91:209–217.

Lai AA, Levy RH, Cutler RE. Time-course of interaction between carbamazepine and clonazepam in normal man. Clin Pharmacol Ther 1978;24:316–323.

Lin JH. Sense and nonsense in the prediction of drug–drug interactions. Curr Drug Metab 2000;1:305–331.

Lin JH, Lu AY. Interindividual variability in inhibition and induction of cytochrome P450 enzymes. Annu Rev Pharmacol Toxicol 2001;41:535–567.

Lu P, Schrag ML, Slaughter DE, Raab CE, Shou M, Rodrigues AD. Mechanism-based inhibition of human liver microsomal cytochrome P450 1A2 by zileuton, a 5-lipoxygenase inhibitor. Drug Metab Dispos 2003;31:1352–1360.

Luo G, Lin J, Fiske WD, Dai R, Yang TJ, Kim S, Sinz M, LeCluyse E, Solon E, Brennan JM, Benedek IH, Jolley S, Gilbert D, Wang L, Lee FW, Gan LS. Concurrent induction and mechanism-based inactivation of CYP3A4 by an L-valinamide derivative. Drug Metab Dispos 2003;31:1170–1175.

Mayhew BS, Jones DR, Hall SD. An *in vitro* model for predicting *in vivo* inhibition of cytochrome P450 3A4 by metabolic intermediate complex formation. Drug Metab Dispos 2000;28:1031–1037.

Moody GC, Griffin SJ, Mather AN, McGinnity DF, Riley RJ. Fully automated analysis of activities catalysed by the major human liver cytochrome P450 (CYP) enzymes: assessment of human CYP inhibition potential. Xenobiotica 1999;29:53–75.

Nomeir AA, Ruegg C, Shoemaker M, Favreau LV, Palamanda JR, Silber P, Lin CC. Inhibition of CYP3A4 in a rapid microtiter plate assay using recombinant enzyme and in human liver microsomes using conventional substrates. Drug Metab Dispos 2001;29:748–753.

Obach RS. Drug–drug interactions: an important negative attribute in drugs. Drugs Today (Barc) 2003;39:301–338.

Obach RS, Walsky RL, Venkatakrishnan K, Gaman EA, Houston JB, Tremaine LM. The utility of *in vitro* cytochrome p450 inhibition data in the prediction of drug–drug interactions. J Pharmacol Exp Ther 2006;316:336–348.

Obach RS, Walsky RL, Venkatakrishnan K, Houston JB, Tremaine LM. *In vitro* cytochrome P450 inhibition data and the prediction of drug–drug interactions: qualitative relationships, quantitative predictions, and the rank-order approach. Clin Pharmacol Ther 2005;78:582–592.

Olkkola KT, Backman JT, Neuvonen PJ. Midazolam should be avoided in patients receiving the systemic antimycotics ketoconazole or itraconazole. Clin Pharmacol Ther 1994;55:481–485.

Pessayre D, Larrey D, Vitaux J, Breil P, Belghiti J, Benhamou JP. Formation of an inactive cytochrome P-450 Fe(II)-metabolite complex after administration of troleandomycin in humans. Biochem Pharmacol 1982;31:1699–1704.

Renwick AB, Watts PS, Edwards RJ, Barton PT, Guyonnet I, Price RJ, Tredger JM, Pelkonen O, Boobis AR, Lake BG. Differential maintenance of cytochrome P450 enzymes in cultured precision-cut human liver slices. Drug Metab Dispos 2000;28:1202–1209.

Riley RJ, Howbrook D. *In vitro* analysis of the activity of the major human hepatic CYP enzyme (CYP3A4) using [*N*-methyl-^{14}C]-erythromycin. J Pharmacol Toxicol Methods 1997;38:189–193.

Rodrigues AD. Impact of CYP2C9 genotype on pharmacokinetics: are all cyclooxygenase inhibitors the same? Drug Metab Dispos 2005;33:1567–575.

Sakai-Kato K, Kato M, Homma H, Toyo'oka T, Utsunomiya-Tate N. Creation of a P450 array toward high-throughput analysis. Anal Chem 2005;77:7080–7083.

Segel IH. Citation-classic - enzyme-kinetics - behavior and analysis of rapid equilibrium and steady-state enzyme-systems. Curr Contents/Life Sci 1987;14.

Shou M. Prediction of pharmacokinetics and drug–drug interactions from *in vitro* metabolism data. Curr Opin Drug Discov Devel 2005;8:66–77.

Shou M, Lin Y, Lu P, Tang C, Mei Q, Cui D, Tang W, Ngui JS, Lin CC, Singh R, Wong BK, Yergey JA, Lin JH, Pearson PG, Baillie TA, Rodrigues AD, Rushmore TH. Enzyme kinetics of cytochrome P450-mediated reactions. Curr Drug Metab 2001;2:17–36.

Silverman RB. Mechanism-based enzyme inactivators. Methods Enzymol 1995;249: 240–283.

Stresser DM, Turner SD, Blanchard AP, Miller VP, Crespi CL. Cytochrome P450 fluorometric substrates: identification of isoform-selective probes for rat CYP2D2 and human CYP3A4. Drug Metab Dispos 2002;30:845–852.

Thummel KE, Wilkinson GR. *In vitro* and *in vivo* drug interactions involving human CYP3A. Annu Rev Pharmacol Toxicol 1998;38:389–430.

Trubetskoy OV, Gibson JR, Marks BD. Highly miniaturized formats for *in vitro* drug metabolism assays using vivid fluorescent substrates and recombinant human cytochrome P450 enzymes. J Biomol Screen 2005;10:56–66.

Venkatakrishnan K, von Moltke LL, Greenblatt DJ. Human drug metabolism and the cytochromes P450: application and relevance of *in vitro* models. J Clin Pharmacol 2001;41:1149–1179.

Von Moltke LL, Greenblatt DJ, Schmider J, Wright CE, Harmatz JS, Shader RI. *In vitro* approaches to predicting drug interactions *in vivo*. Biochem Pharmacol 1998;55:113–122.

Waley SG. Kinetics of suicide substrates. Biochem J 1980;185:771–773.

Walsh C, Cromartie T, Marcotte P, Spencer R. Suicide substrates for flavoprotein enzymes. Methods Enzymol 1978;53:437–448.

Walsky RL, Obach RS. Validated assays for human cytochrome P450 activities. Drug Metab Dispos 2004;32:647–660.

Wang RW, Newton DJ, Liu N, Atkins WM, Lu AY. Human cytochrome P-450 3A4: *in vitro* drug–drug interaction patterns are substrate-dependent. Drug Metab Dispos 2000;28:360–366.

Wang YH, Jones DR, Hall SD. Prediction of cytochrome P450 3A inhibition by verapamil enantiomers and their metabolites. Drug Metab Dispos 2004;32:259–266.

Yao C, Levy RH. Inhibition-based metabolic drug–drug interactions: predictions from *in vitro* data. J Pharm Sci 2002;91:1923–1935.

Yao M, Zhu M, Sinz MW, Zhang H, Humphreys WG, Rodrigues AD, Dai R. Development and full validation of six inhibition assays for the five major cytochrome P450 enzymes in human liver microsomes using an automated 96-well microplate incubation format and LC-MS/MS analysis. J Pharm Biomed Anal 2007;44:211–223.

Yin H, Racha J, Li SY, Olejnik N, Satoh H, Moore D. Automated high throughput human CYP isoform activity assay using SPE-LC/MS method: application in CYP inhibition evaluation. Xenobiotica 2000;30:141–154.

Zhang L, Strong JM, Qiu W, Lesko LJ, Huang SM. Scientific perspectives on drug transporters and their role in drug interactionst. Mol Pharm 2006;3:62–69.

Zhang T, Zhu Y, Gunaratna C. Rapid and quantitative determination of metabolites from multiple cytochrome P450 probe substrates by gradient liquid chromatography-electrospray ionization-ion trap mass spectrometry. J Chromatogr B Analyt Technol Biomed Life Sci 2002;780:371–379.

17

TESTING DRUG CANDIDATES FOR CYP3A4 INDUCTION

GANG LUO, LIANG-SHANG GAN, AND THOMAS M. GUENTHNER

17.1 INTRODUCTION

The primary objective of preclinical drug evaluation is characterization of the direct pharmacological and toxicological activities of the candidate molecules. However, a major concern of preclinical drug development is also the prediction of eventual interactions between the drug candidate and other drugs or xenobiotics that may be simultaneously present. Drug–drug interactions are normally understood to occur when the effects of two or more drugs are qualitatively or quantitatively different from the simple sum of the effects observed when the same doses of the drugs are present separately. Drug–drug interactions involving drug metabolism are certainly among the most prevalently encountered therapeutically, and elucidation of the potential of candidate drug molecules to either inhibit or induce drug metabolism is among the most important objectives of the drug development process.

A prevalent type of metabolism-dependent drug–drug interaction occurs when one drug increases the clearance of a second, by increasing its rate of metabolism, usually through a mechanism of action known as enzyme induction. Although this type of drug–drug interaction is probably less common than that involving inhibition of drug metabolism, and is less likely to jeopardize patient safety, it nevertheless can produce significant clinical consequences, manifested primarily as lower blood levels, decreased half-life, and possible therapeutic failure. A well-documented historical example of this

Drug Metabolism in Drug Design and Development, Edited by Donglu Zhang, Mingshe Zhu and W. Griffith Humphreys
Copyright © 2008 John Wiley & Sons, Inc.

type of drug–drug interaction is the increased incidence of failure of oral contraceptives occurring in women who were concurrently treated with rifampicin. A 42% decrease in the bioavailability of both ethinyl estradiol and norethistereone, resulting in unexpected failure of contraception, was observed in these patients, due to the induction by rifampicin of both Phase I (primarily CYP3A4), and Phase II (UDP-glucuronosyltransferase) drug metabolizing enzymes (LeBel et al., 1998).

The term induction, in the context of drug metabolism enzymes, is understood to indicate the increased expression of a drug metabolizing enzyme or enzymes in response to exposure of the cell or organism to an exogenous or endogenous molecular inducing agent. This increased expression of the induced enzyme protein is associated with enhanced drug metabolism, and is usually (but not always) the result of enhanced transcription of the associated gene (Ronis and Ingelman-Sundberg, 1999). The phenomenon of induction of xenobiotic metabolism has been recognized for over 40 years, and the effects of induction on the intensity and duration of the action of human pharmacotherapeutic agents has been an important area of pharmacological research for equally as long (Okey, 1990; Sueyoshi and Negishi, 2001). Although drugs may induce their own metabolism, a phenomenon often referred to as "autoinduction," induction of the metabolism of one drug may also occur through the action of a concomitantly administered second drug. This second type of induction constitutes one of the most common types of drug–drug interaction, whereby one drug decreases the activity of a second drug by enhancing its metabolism. By inducing one or more drug metabolizing enzymes, particularly those that are relatively substrate nonspecific, a single drug or drug candidate has the potential not only to enhance its own clearance, but to produce drug–drug interactions with hundreds of other therapeutic agents. The potential of a molecule to produce such drug–drug interactions will ultimately be a major factor in determining its therapeutic usefulness, and this potential should ideally be determined early on in the drug discovery process.

Probably the most well studied paradigm for induction-based drug–drug interactions involves the induction of one or more forms of cytochrome P450 (CYP), the superfamily of heme-thiolate proteins whose members play a central role in hepatic and extrahepatic drug metabolism. Of all the members of the human CYP superfamily, those of the CYP3A family, and in particular CYP3A4, are commonly considered to have the greatest overall impact on human pharmacotherapy. The CYP3A subfamily is the most abundantly expressed CYP subfamily in the human liver, and CYP3A4 is known to metabolize approximately 50–60% of all known therapeutic drugs (Luo et al., 2004; Thummel and Wilkinson, 1998; Wrighton and Stevens, 1992). Perhaps most important to the discussion in this chapter, the expression of CYP3A4 in human liver can be enhanced or induced by a large number of therapeutic drugs and other xenobiotics (Luo et al., 2004; Ronis and Ingelman-Sundberg, 1999; Sueyoshi and Negishi, 2001). CYP3A4 is not the only inducible CYP isoform that plays an important role in the metabolism of drugs and xenobiotics in humans. In addition to the noninducible isoform CYP2D6, which is capable of metabolizing approximately 25% of known prescription drugs in humans, other inducible forms of CYP that are important in human drug metabolism include CYP1A2, CYP2E1, members of the CYP2C family, and CYP2B6. Limitations on the length of this chapter preclude an in-depth discussion of the experimental approaches to investigating the potential induction of all of these important CYP isoforms, or of the induction of equally important human Phase II enzymes such as the UDP-glucuronosyl transferases. This chapter will therefore focus on methods for routine screening and initial determination of the potential of drugs or drug candidates to induce what is arguably the most important CYP isoform, in terms of overall impact on human therapeutics, namely, CYP3A4. It should be recognized that in the case of CYP2C and CYP2B6, molecular pathways that mediate their induction in humans are similar, but not identical to, those that mediate CYP3A4 induction (Eloranta et al., 2005; Pascussi et al., 2003). Therefore many of the approaches that are used to examine the potential for CYP3A4 induction might theoretically be adaptable to investigating potential induction of these other isoforms.

In this chapter, three general approaches for screening for potential inducers of CYP3A4 will be discussed, and specific methodologies will be described. (1) Intact animals. In the past, prediction of the potential of a drug candidate to induce CYP3A4 in humans was often based on animal studies. Studies using common animal species are now known to be often poorly predictive of human CYP3A4 induction, due to well documented species-dependent differences in CYP3A induction between humans and common laboratory animal species. More recently, transgenic "humanized" mice have been developed, which express the human PXR receptor. These strains appear to provide a much more accurate intact mouse model for predicting CYP3A4 induction in humans. (2) *In vitro* models. *In vitro* models include primary human hepatocyte cultures, immortalized and cryopreserved human hepatocytes, and the PXR reporter gene assay. Primary human hepatocyte cultures provide an integrated model of intact human liver, have the advantage of providing easily controlled exposure to potential inducing agents. This model has major disadvantages of expense and lack of ready availability. The PXR reporter gene assay, employing cultured cells transfected with human signaling genes, appears to currently provide the best balance between economy and efficiency in screening on one hand, and accurate prediction of CYP3A4 induction in humans on the other. These methods measure binding of putative inducers with human signaling molecules in transfected cultured cells, and subsequent activation of gene expression through pathways responsible for activation of CYP3A4 expression. These *in vitro* methods hold great promise for rapid, economical screening, and accurate prediction of the potential of a drug candidate for CYP3A4 induction in humans. (3) Direct measurement of induction in humans. Potential inducing agents can be directly administered to volunteer human subjects, and induction can be assessed by measuring the metabolism or clearance of model substrates for the drug metabolizing enzymes of interest. This method has the advantage of measuring directly in humans the end result of induction, that is, enhanced drug metabolizing activity, and is therefore undoubtedly the most relevant to actual clinical exposure. The disadvantage of this model lies in its expense, and its potential hazard to experimental subjects who derive no therapeutic benefit from exposure to the agent being tested. These three approaches and their relative advantages and disadvantages will be discussed.

17.2   ASSESSMENTS

17.2.1   Assessment of Induction Potential Using Intact Animal Models

Intact animal models have long been used to predict drug–drug interactions in humans, and have historically provided the primary model for investigating the cellular events and mechanisms that mediate induction of drug metabolizing enzymes by xenobiotics. Although we now recognize that the value of preclinical studies in animals for predicting induction of drug metabolism in humans may be limited, animal models have historically been very valuable in both identifying and characterizing potential drug metabolism inducing agents. They provide an intact, physiologically relevant system for presentation of the putative inducer to the signaling system, as well as an intact signaling system itself, complete with all intracellular and extracellular signaling modulators. Furthermore, animal models provide expression of a complete drug metabolism system, which potentially provides the expression of multiple drug metabolizing enzymes that may be responsive to the inducing agent. This complete system can therefore be used in carefully selected induction studies, from overall *in vivo* clearance measurements to subcellular or molecular mechanism studies, all in the same single integrated model.

However, intact animal models also have important limitations that may diminish their efficiency and usefulness in screening for human induction potential. They are a more expensive, lower throughput system than *in vitro* systems or cell culture. But certainly the primary drawback lies in the fact that the molecular mediators of induction, including receptors and signaling factors, that are present in animal species, may be very different, both qualitatively and quantitatively, from those present in humans. The existence of interspecies differences in the structure and function of receptors, and other signaling molecules involved in induction, means that studies using common laboratory animal species may poorly predict the ability of a drug candidate to induce the same enzymes in humans. Considering the cost involved in large scale animal studies, and their potential lack of predictive value, their efficiency and usefulness in screening drug candidates for potential inducing activity must now be questioned, especially in the light of newer well developed and readily available *in vitro* or cell culture screening systems. Animal-based screens may only be useful if the signaling mechanisms responsible for inducing the specific drug metabolizing enzymes targeted by the screening procedure have similar structural and mechanistic properties. In other words, the relative predictive usefulness of animal studies in screening for inducing potential of a candidate compound will depend almost entirely upon the particular drug metabolizing enzyme(s) whose potential induction is being investigated.

For example, members of the CYP1A family are induced by similar mechanisms in humans and common laboratory animal species, and the signaling molecules involved appear not to exhibit major interspecies structural variation (Hankinson, 1995; Nebert et al., 1993; Okey et al., 1994). The number of false negative and false positive results that might be generated by screening candidate compounds at sufficiently high doses in animals for induction of human CYP1A enzymes would appear to be acceptably low. In fact, animal studies might be preferable to direct studies in humans for many of these compounds, because of the inherent potential toxicity and genotoxicity of many planar aromatic Ah receptor ligands.

On the contrary, using common laboratory animal models to screen candidate compounds for their potential to induce CYP3A family members is much more problematic. Because of the well-recognized and well-documented interspecies differences in the structure and function of molecules mediating the CYP3A induction signaling cascade, particularly those differences in PXR receptor structure, such animal screening studies are likely to be poorly predictive of induction in humans (Bertilsson et al., 1998; Blumberg et al., 1998; Jones et al., 2000; Kliewer et al., 1998; Lehmann et al., 1998; Luo et al., 2004; Moore et al., 2002 Zhang et al., 1999). Whereas the marked species differences in induction of metabolic activity observed in many laboratories (Jones et al., 2000; Kocarek et al., 1995; LeCluyse, 2001a; Lu and Li, 2001; Moore et al., 2003) cannot be attributed to the relatively small variations in sequence homology observed in the DNA binding domain (DBD) region of a species-specific PXR receptor, the more extensive deviations observed in the ligand-binding domain (LBD) region of the receptor provide a more rational basis for the observed species differences (Goodwin et al., 2002; Kliewer and Willson, 2002; LeCluyse, 2001a; Moore et al., 2002). Correlations between the activation profiles of human, rat, and rabbit PXRs, and CYP3A4 induction profiles obtained from *in vitro* and *in vivo* experiments for the same species, confirm this hypothesis. Notably, mouse and rat PXRs share 97% identity in the LBD region, while human and rhesus monkey PXRs share 95% identity in the same region. On the contrary, sequence identity between the human LBD region and that of the rat or mouse is 76–77% (Luo et al., 2004; Moore et al., 2002). These structural data provide a molecular explanation for the numerous observations that CYP3A induction profiles for many compounds vary greatly between humans and rats or mice, yet CYP3A induction profiles for mice and rats are very similar to each other, as are those for humans and rhesus monkey (Jones et al., 2000; Kocarek et al., 1995; LeCluyse, 2001a; Lu and Li, 2001; Moore et al., 2003).

It is evident, therefore, that while animal studies may offer valuable information regarding the potential of compounds for inducing those drug-metabolizing enzymes whose mechanisms and molecular mediators are qualitatively and quantitatively similar in preclinical species and in humans (e.g., CYP1A family members), and they are expected to be poorly predictive and marginally useful in cases where major species differences exist (e.g., CYP3A family members). However, recent technological developments offer a third possibility. The generation of transgenic mouse models, which incorporate genes that code for human-specific signaling molecules, can result in humanized mice, which respond to potential inducers that interact with those molecules in a manner very similar to that of humans. The main determining factor for the species-specific pattern of CYP3A inducibility is the interaction of the inducer with the species-specific PXR receptor, rather than in the interaction of the activated PXR with the xenobiotic response element (XRE) in the regulatory region of the CYP3A4 gene (Luo et al., 2004; Xie et al., 2000). Xie et al. have generated a transgenic mouse strain that lacks the mouse PXR gene, but expresses the human PXR gene. This animal model appears to provide a reliable assessment of the CYP3A induction potential of candidate molecules in an intact mouse whose response to xenobiotic PXR ligands mirrors that of humans (Xie et al., 2000; Xie and Evans, 2002). These humanized mice offer a potentially powerful tool for drug discovery and development that will reliably predict the ability of candidate molecules to induce CYP3A enzymes in humans. Furthermore, because this is a whole-animal system, the effects of CYP3A induction (as well as the activation of other PXR-mediated pathways) on the pharmacodynamics, pharmacokinetics, and toxicological profiles of test compounds can be directly determined.

We see then, that while the results of testing for inducers of drug metabolizing enzymes in preclinical species cannot be automatically assumed to provide a reliable assessment of their potential for producing similar effects in humans, in certain cases, animal models may serve as a valuable bridge between *in vitro* and cell culture studies on one hand, and clinical studies on the other. When the potential predictive value of animal studies warrant their consideration, some general principles related to study design should be kept in mind. Animals are normally treated from one to several days with multiple bolus oral, IP, or IV doses of potential inducing agents, or exposed for longer periods of time via the diet. The length of treatment will depend on the rate of clearance of the inducing agent and the rate of turnover of the target protein. Normally, 4–5 days of single dose treatment is required to achieve maximal induction. The dose–response curve for induction is relatively steep, and usually, only a maximal response is sought. Therefore, a complete dose–response curve does not need to be generated, and animals can be treated at or near the maximally tolerated dose, as determined by previous acute toxicity studies. In the case of humanized transgenic mice, expense and availability may be an issue. Numbers of animals may, therefore, be limited to three or four per treatment group. The route of administration must also be considered. The oral route of administration is preferred for preclinical studies, either by gavage or as an admixture in the animal diet, to provide a more relevant route of exposure for compounds that will eventually be administered p.o. in humans, especially those compounds that undergo significant first pass metabolism in the gut epithelium or liver, or compounds that may require metabolic activation to produce a metabolite that is an active inducer. Of course the proper control groups must be included. These should include vehicle controls for injection, or purified diet controls for dietary administration. Positive control groups dosed with a known enzyme-specific inducer are also recommended. For example, positive controls for CYP3A induction in humanized mice should be administered known effective doses of a human CYP3A-specific inducer such as rifampicin. In the case of humanized mice, negative control animals that are $hPXR-/-$, or wild type mouse $PXR+/+$ control animals will help identify induction that is specifically human PXR-mediated (Xie et al., 2000; Xie and Evans, 2002).

The following protocol might be used to investigate potential CYP3A4 induction by a test compound in humanized transgenic mice. For generation of transgenic mice, readers are referred to Xie's work (Xie et al., 2000). Details of protocols for specific candidate molecules will depend on, among other factors, chemical stability and solubility, toxicity and side effects of the putative inducer, and the possibility that metabolites, rather than the parent candidate molecule, might mediate induction. Animals are maintained on a 12 h light/dark cycle under controlled temperature (22°C) and humidity (45%) with *ad libitum* access to food and water. Each experimental group consists of four mice; one group is designated as a vehicle control, one group as a positive control (rifampin 5 mg/kg/day), and three groups are exposed to test compound at low, moderate, and high doses. All treatments are administered by gavage, in a dosing volume of 2–5 mL/kg, for 3–7 consecutive days. The frequency of treatment will depend on the rates of clearance of the test compound in its mode of administration. Normally, animals are treated once daily, but this frequency may be adjusted when the pharmacokinetics of the molecule are known. The goal is to maintain blood concentrations of the compound at as constant levels as possible, over the entire treatment period. The mice can also be dosed by intraperitoneal injection, but this route of administration is not recommended for testing drug candidates that are designed to be administered orally in therapeutic situations. At the end of the treatment period, mice are euthanized by IP injection of pentobarbital, and their livers are removed immediately, weighed and placed in a container immersed in an ice bath. The liver tissues are divided to provide samples for mRNA extraction and microsome preparation. Total RNA is extracted using TRI REAGENT™ and 1-bromo-3-chloropropane, and is subsequently used for quantitative real-time RT–PCR (Chomczynski and Sacchi, 1987). Mouse liver microsomes are prepared using standard techniques (Matsuura et al., 1991), and the microsomal pellets are suspended in storage buffer (pH 7.5, K_2HPO_4 5 mM, DTT 0.05 mM, EDTA 0.5 mM, glycerol 10% (v/v), and

0.125 M sucrose), and can be stored at $-80°C$ if they cannot be used immediately. The total protein concentration is determined by the bicinchoninic acid protein assay (BCA-Pierce, Rockford, IL) with bovine serum albumin (BSA) as a standard (Smith et al., 1985).

Quantitative RT–PCR for mouse cyp3a is performed with the ABI PRISM 79000HT Sequence Detection System (Applied Biosystems), using TaqManR Universal PCR Master Mix. Detailed instructions are provided in the kit manual (Prueksaritanont et al., 2005). Fold increase in cyp3a mRNA level (drug treated group compared to vehicle control) can be used independently to identify induction of cyp3a gene expression by the test compound, or to provide confirmatory evidence of induction mediated through enhanced gene transcription, as is normally the case.

Mouse liver CYP3a activity is measured by suspending liver microsomes (0.5 mg/mL) in phosphate buffer (50 mM, pH 7.4), and the probe substrates testosterone (250 μM) or midazolam (50 μM) is added. The reaction is initiated by addition of NADPH at a final concentration of 1 mM, and terminated with an equal volume of acetonitrile after incubation for 10 min (testosterone) or 5 min (midazolam). The rates of generation of the metabolites 6β-OH testosterone or 1′-OH midazolam, which are specifically generated by cyp3a, are determined (Pearce et al., 1996; Wandel et al., 2000).

Western blot analysis (LeCluyse et al., 2000) can also be employed to confirm cyp3a induction. However, this method is time consuming and provides a relatively low throughput.

17.2.2 Assessment of Induction Potential Using *In vitro* Models

The potential of test compounds to induce CYP3A4 can be assessed utilizing cell-based assays such as primary cultures of hepatocytes or the PXR reporter gene assay (Goodwin et al., 1999; Jones et al., 2000; LeCluyse, 2001a; Luo et al., 2002, 2004; Moore and Kliewer, 2000; Zhu et al., 2004). There is a major advantage of employing such assays early in the drug discovery process. They provide the capability to screen a large number of test compounds in a relatively short period of time, thereby detecting the potential for induction by chemotypes in the early stages of drug discovery so that their potential for drug–drug interactions can be timely addressed. More importantly, these *in vitro* screening assays generate results that are more reliably predictive of induction in humans than results obtained from animal studies. In particular, primary cultures of human hepatocytes are considered by academic, industry, and regulatory scientists to be the current "gold standard" for CYP3A4 induction studies. In addition, *in vitro* assays are reproducible and cost-effective. However, the disadvantages of these assays must also be considered. All of the *in vitro* assays are performed under non-physiological conditions, in an over-simplified system, using cell lines. Exposure of cells in culture is not equivalent to human ingestion of test compounds. In addition, each of these assays has its unique disadvantages. For example, the SPA (scintillation proximity assay)-based binding assay is prone to false positive results, while the PXR reporter gene assay only represents the PXR-dependent induction pathway, possibly ignoring other pathways of CYP3A4 induction that exist *in vivo*. Primary cultures of human hepatocytes provide an ideal screening system in many respects. However, these assays are expensive and time consuming. Availability, quality, and variability of donor variations also present potential problems. Finally, induction may be masked by concomitant inhibition of constitutive and induced CYP3A4 by the inducing agent. This "metabolic masking effect" may generate false negative results (Luo et al., 2002, 2003, 2004). In spite of these potential drawbacks, *in vitro* and cell culture models currently provide the best compromise between time- and cost-efficient screening methods, and accurate prediction of induction potential in clinical situations.

17.2.2.1 Primary Cultures of Human Hepatocytes Measurement of CYP3A4 induction in primary human hepatocyte cultures involves three steps: isolation of hepatocytes from autopsy- or biopsy-derived human donor livers; treatment with a test compound for three days following a 2-day recuperation culture period; and measurement of the expression or activity of CYP3A4 (Fig. 17.1).

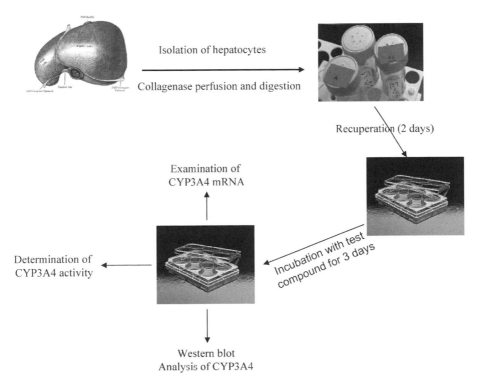

FIGURE 17.1 Isolation and culture of primary human hepatocytes for CYP induction studies.

The most common techniques for isolating human hepatocytes from donor livers involve a two-step collagenase perfusion and digestion (Fabre et al., 1988; LeCluyse et al., 2000; Strom et al., 1982). Encapsulated liver tissue (25–100 g) is perfused with calcium-free buffer containing 5.5 mM glucose and 0.5 mM EGTA for 20 min at a flow rate of 40 mL/min/cannula followed by perfusion with buffer containing 1.5 mM calcium and collagenase (0.4 mg/mL) for 25 min at a flow rate of 35 mL/min/cannula. Hepatocytes are dispersed from the digested liver in Dulbecco's modified Eagle medium (DMEM) containing 5% fetal calf serum, insulin (4 μg/mL), and dexamethasone (1.0 μM) (Supplemented DMEM) and washed by centrifugation at $70 \times g$ for 4 min. The cell pellets are resuspended in 30 mL Supplemented DMEM and 12 mL 90% isotonic Percoll. The cell suspensions are centrifuged at $100 \times g$ for 5 min. The resulting pellets are resuspended in fresh medium and washed once by low speed centrifugation. Hepatocytes are resuspended in Supplemented DMEM and cell viability is determined by trypan blue exclusion (LeCluyse et al., 2000). Cell yields are normally between 5 and 12 million cells per gram of wet liver tissue, and cell viability varies from 75% to 91% (Kostrubsky et al., 1999; LeCluyse et al., 2000).

Hepatocytes are plated in 6-well plates (1×10^6 viable cells/well) and maintained at 37°C, 95% humidity, and 5% CO_2 in Hepatocyte Culture Medium (JRH Bioasciences, Lenexa, KS) supplemented with L-glutamine (0.292 mg/mL), nonessential amino acids (10 μM), insulin (6.25 μg/mL), transferrin (6.25 μg/mL), selenium (6.25 ng/mL), bovine serum albumin (125 mg/mL), linoleic acid (5.35 μg/mL), penicillin (100 U/mL), streptomycin (100 μg/mL), and dexamethasone (0.1 μM) (Prueksaritanont et al., 2005). After a 2-day recuperation period, the cultured hepatocytes are treated for three consecutive days with test compounds dissolved in DMSO. The final DMSO concentration in the medium is 0.1% or less. Cells are normally exposed to drugs at concentrations of 2, 10, and 20 μM. However, some drugs (e.g., phenobarbital) may require 100 μM or higher concentrations to produce induction. Rifampicin (10 μM) is a positive control for CYP3A4 induction, and 0.1% DMSO is used as a negative or vehicle control (Luo et al., 2002).

The extent of CYP3A4 expression and induction is quantitatively assessed by measuring microsomal metabolism of a CYP3A4-specific probe substrate. Testosterone and midazolam are the most well accepted and commonly employed substrates (Bjornsson et al., 2003; Hariparsad et al., 2004; Tucker et al., 2001). The level of CYP3A4 protein expression can be qualitatively and quantitatively determined by Western immunoblotting with CYP3A4- specific antibodies, or by measuring CYP3A4 mRNA by means of RT–PCR, Northern blot, or branched DNA signal amplification technology (bDNA) (Czerwinski et al., 2002; Desai et al., 2002; Luo et al., 2002; Prueksaritanont et al., 2005).

For the measurement of CYP3A4-specific metabolic activity at the end of the drug treatment, the drug-containing medium is removed from the wells, and replaced with drug-free medium for 4 h. Cells are then exposed to testosterone (250 μM) in Williams's E medium (1 mL/well) for 30 min. This medium is then collected, and the concentration of 6β-hydroxytestosterone, a metabolite generated specifically by CYP3A4 is determined (Hariparsad et al., 2004), thereby providing a specific quantitative measurement of CYP3A4-dependent metabolic activity. The direct addition of a CYP3A4-specific substrate to the culture media of whole cell monolayers, followed by measurement of metabolite generation, is generally the most cost effective and time efficient. Possible limitations of this whole-cell method are the potential trapping of metabolites in the monolayers, which prevents their release into the medium, and collateral effects of drug transporters and Phase II enzymes on the overall rates of metabolite formation (LeCluyse, 2001b). Alternatively, CYP3A4-dependent metabolic activity can also be determined by harvesting hepatocytes, preparing microsomes, and measuring CYP3A4 activity in the microsomes (Luo et al., 2002). This method will accurately measure CYP3A4 induction, but is time-consuming and usually not suitable for high throughput screening. For most compounds, the observed fold inductions compared to the vehicle control using intact monolayers and those observed using isolated microsomal fractions are normally quite consistent (LeCluyse, 2001b).

CYP3A4 protein expression can be measured directly by immunodetection (LeCluyse et al., 2000; Luo et al., 2002). Microsomal protein (3 μg) is resolved by SDS–polyacrylamide gel electrophoresis (12% acrylamide). Resolved proteins are transferred to nitrocellulose membranes, which are incubated in 3% bovine serum albumin in phosphate-buffered saline supplemented with Tween 20 (0.1 M, pH 7.4, 0.1% Tween 20) for 45 min to block nonspecific protein binding. Membranes are then treated with primary antiCYP3A4 antibody (Gentest, Woburn, MA), followed by horseradish peroxidase-conjugated antimouse secondary antibody. The antibody-reactive CYP3A4 protein bands are visualized using enhanced chemiluminescence detection and quantitated by photodensitometry (Desai et al., 2002).

For Northern Blot Analysis of CYP3A4 mRNA, total cellular RNA is isolated as described using TRIzol reagent (Invitrogen, Carlsbad, CA) (Desai et al., 2002). The amount of RNA is estimated by the absorbance ratio at 260/280 nm. A 10-μg aliquot is fractionated by electrophoresis in 1% agarose gels containing formaldehyde (2.2 M) and transferred onto a nylon membrane (Millipore Corp., Bedford, MA). The amount of mRNA loaded per lane is verified and normalized by ethidium bromide staining of 18S and 28S rRNA, which is visualized and photographed under UV illumination. The membranes are hybridized with a CYP3A4-specific cDNA probe (780-base pair; Oxford Biomedical Research, Inc., Oxford, MI) labeled with [^{32}P]dCTP (PerkinElmer Life Sciences, Boston, MA) using the random primer method (Desai et al., 2002). CYP3A4 mRNA is quantitated by scanning densitometry of the autoradiographs and normalized against β-actin mRNA (Luo et al., 2002). Alternative assays for quantitation of CYP3A4 mRNA (RT–PCR and bDNA) are described in detail in recent publications (Czerwinski et al., 2002; Prueksaritanont et al., 2005).

The source of hepatocytes, whether via autopsy from accident victims or via biopsy from diseased livers, may affect the induction of CYP3A4. Several studies have indicated that the expression of CYP3A4 is decreased in certain liver diseases such as hepatic cirrhosis, hepatitis B or C, and hepatic tumors (Kirby et al., 1996; Luo et al., 2004; Philip et al., 1994; Sotaniemi et al., 1995; Yang et al., 2003). Therefore, hepatocytes obtained from these patients will probably have lower constitutive or basal CYP3A4 activity compared to hepatocytes from accident victims without liver diseases. These studies also showed that human hepatocytes with lower basal activity exhibit a greater fold induction of CYP3A4 activity (LeCluyse et al., 2000, 2001b; Luo et al., 2002). Because of high interindividual differences in induction potential, experiments should be conducted with hepatocytes obtained from at least three individual donor livers.

Experimental results, including CYP3A4-dependent metabolic activity, mRNA and protein levels, are usually expressed as fold induction, namely (results from test drug treated cells)/(results from vehicle control cells), or as the percentage of positive control, namely [100% × (results from test-drug treated cells)/(results from positive control cells)]. A test compound is considered to be a potential CYP inducer in humans if it produces more than twofold induction, or if its induction is more than 40% of that produced in the positive control. For example, in experiments with human hepatocytes obtained from four donor livers, carbamezepine, clotrimazole, dexamethasone, dexamethasone-t-butylacetate, phenobarbital, phenytoin, rifampin, sulfadimidine, and sulfinpyrazone induced CYP3A4 activity more than twofold, or their induction was more than 40% of positive control (rifampin at 10 μM), while methotrexate, probenecid, ritanovir, taxol, and troleadomycin did not show significant induction in primary human hepatocyte cultures (Luo et al., 2002).

An alternative method of expression of induction potential is the use of EC_{50} or percentage of Y_{max}. EC_{50} (effective concentration of test compound at which 50% maximal induction is achieved) can be used to compare the potency of test compounds, but EC_{50} alone does not provide sufficient information about induction. In addition, the experimental design and the concentration-response function for the inducer can significantly affect the accuracy of the EC_{50} value. The index $\%Y_{max}$ expresses the percentage of maximal induction by a positive control (e.g., rifampin) that a test compound can theoretically produce. However, Y_{max} is most accurate when estimated using concentrations of test compound that correspond to likely therapeutic concentrations.

In general, the estimates of fold induction obtained by measuring mRNA, protein, and enzyme activity are consistent with each other. However, for certain test compounds, results obtained by measuring mRNA or protein directly more accurately represent CYP3A4 induction than those obtained by measuring enzyme activity. Such compounds, including mifepristone, ritonavir, tamoxifen, troleandomycin, and DPC 681, are both inducers and mechanism-based inactivators of CYP3A4, and direct inhibition of drug metabolism will partially mask the induced increase in CYP3A4 expression

(Chan and Delucchi, 2000; Desai et al., 2002; Goodwin et al., 1999; He et al., 1999; Ledirac et al., 2000; Luo et al., 2002, 2003, 2004; Pichard et al., 1999; Zhao et al., 2002). As a result, overall CYP3A4 activity in cells treated with these compounds can actually be lower than activity obtained from vehicle control cells, even though CYP3A4 mRNA levels and protein expression are significantly induced.

In addition to providing good models for prediction of CYP3A4 induction, primary human hepatocytes are also an excellent model for evaluating the potential of a drug to induce other human CYP isoforms, including CYP1A2, 2A6, 2B6, 2C8, 2C9, and 2C19. Positive inducer controls, as well as isoform-specific substrates, are unique for each different CYP isoform. Table 17.1 lists preferred and acceptable inducers and model substrates for each inducible human CYP isoform (Gerbal-Chaloin et al., 2001; Goodwin et al., 2001; LeCluyse et al., 2000; Meunier et al., 2000; Luo et al., 2002).

Studies using primary human hepatocyte cultures are dependent on the availability of donor livers. Recently, scientists from Pfizer have developed the immortalized human hepatocyte cell line Fa2N-4, which may provide a valuable alternative tool for determining the induction of CYP3A4, as well as of CYP1A2, 2C9, UGT1A, and MDR1 (PGP). For example, the levels of induction of CYP3A4 and CYP2C9 by rifampin and phenobarbital, and CYP1A2 by β-naphthoflavone observed in this model, are very close to those obtained using primary human hepatocytes (Mills et al., 2004; Ripp et al., 2006). Cryopreserved human hepatocyte cultures have also been employed for CYP induction studies, but their use is thus far limited to studying CYP3A4, 1A2, and UDP-glucuronosyl-transferase (Kafert-Kasting et al., 2006; Reinach et al., 1999; Roymans et al., 2004). The usefulness of cryopreserved human hepatocytes for induction studies is limited because of the poor attachment of cells to the culture dish and because their basal CYP activities are reduced compared to those of fresh human hepatocytes. In addition, HepG2 cells, derived from human hepatoblastoma, may be useful in combination with RT–PCR to evaluate CYP induction in some limited cases (Matsuda et al., 2002; Sumida et al., 1999).

17.2.2.2 PXR Reporter Gene Assay

17.2.2.2 PXR Reporter Gene Assay   The identification and characterization of members of the human PXR family have permitted the development of cell-based reporter gene assays, which provide an accurate measure of the extent to which putative inducers activate CYP3A4 expression via the PXR signaling pathway. Figure 17.2 illustrates the basic features of the assay, and how the interaction of the components can predict the induction potential of a candidate compound. The basis of the assay is the cotransfection of cells with two DNA vectors, one coding for expression of human PXR, the other containing the reporter gene construct. The reporter gene construct contains PXR responsive elements that activate expression of the reporter gene. A third vector coding for expression of alkaline phosphatase or another control protein can also be cotransfected for normalization of results. The expressed human

TABLE 17.1 CYP isoform-specific inducers and substrates used in *in vitro* models.

CYP	Preferred inducer (μM)	Acceptable inducer (μM)	Substrate	References
1A2	3-Methylcholanthrene (2)	Lansoprazole (10)	7-Ethoxyresorufin	Curi-Pedrosa et al., (1994)
	Omeprazole (25–100)			LeCluyse et al., (2000)
	β-Naphthoflavone (33–50)			Meunier et al., (2000)
2A6	Dexamethasone (50)	Pyrazole (1000)	Coumarin	Meunier et al., (2000)
2B6	Phenobarbital (500–1000)	Phenytoin (50)	S-Mephenytoin	LeCluyse et al., (2000)
2C8	Rifampin (10)	Phenobarbital (500)	Paclitaxel	LeCluyse et al., (2000)
				Gerbal-Chaloin et al., (2001)
2C9	Rifampin (10)	Phenobarbital (100)	S-Warfarin	Gerbal-Chaloin et al., (2001)
2C19	Rifampin (10)		S-Mephenytoin	Gerbal-Chaloin et al., (2001)
3A4	Rifampin (10)	Phenobarbital (100–2000)	Testosterone	Goodwin et al., (1999)
		Dexamethasone (33–250)	Midazolam	LeCluyse et al., (2000)
		Phenytoin (50)		Luo et al., (2002)

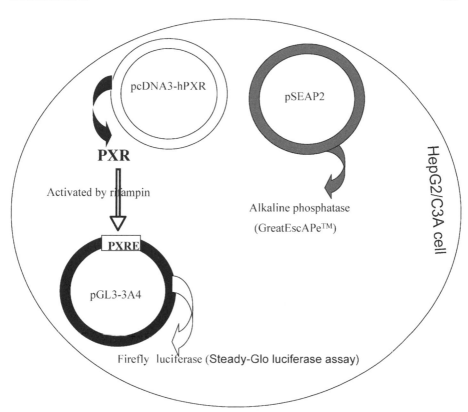

FIGURE 17.2 Illustration of cell-based PXR reporter gene assay. HepG2 cells are cotransfected with three vectors, pcDNA3–hPXR, pGL3–3A4, and pSEAP2. The pcDNA3–hPXR codes for expression of human PXR, whereas pGL3–3A4 is a reporter vector, in which the PXRE is inserted. A CYP3A4 inducer such as rifampin binds and activates the expressed human PXR, which in turn binds to PXRE in the reporter gene construct, and activates the transcription of reporter gene firefly luciferase. The luciferase activity, determined by Steady-*Glo* luciferase assay, is proportional to the activation of PXR by the inducer. pSEAP2 activation results in expression of alkaline phosphatase activity, which is used to normalize the expression of luciferase activity. (Goodwin et al., 1999; Luo et al., 2002).

PXR, in the presence of a CYP3A4 inducer, binds to and activates the responsive elements in a reporter gene construct, leading to higher expression of a reporter gene, which is usually luciferase, chloramphenicol acetyl transferase (CAT), or secretory placental alkaline phosphatase (SPAP). The expression of the reporter gene is determined, and the ratio between the treatment and vehicle control is used to estimate human PXR-dependent induction potential of the test compound.

Variations of the PXR reporter gene assay have been developed in a number of laboratories. In our laboratory, DNA vectors were constructed as follows: For the PXR expression vector, a full-length PXR open reading frame was amplified by RT–PCR from human liver RNA using the gene-specific primers 5′-AACCTGGAGGTGAGACCCAAAGA-3′ and 5′-ATCTCGAGGATCCTCAGCTACCTGTGATGCCGA-3′ (Bertilsson et al., 1998; Lehmann et al., 1998). The resulting amplicon was digested with *Eco*RI and *Xho*I and subcloned into the polylinker site of pcDNA3 for expression, hereafter designated pcDNA3–hPXR. The correct PXR sequence and insert orientation were confirmed by DNA sequence analysis. A chimeric CYP3A4 luciferase reporter vector was prepared following the described method (Goodwin et al., 1999). The resulting construct contains two fragments of the *CYP3A4* 5 flanking region (836 to −7208 and 62 to +53) linked to the luciferase reporter sequence of the pGL3-basic vector (Promega, Madison, WI) and hereafter designated pGL3–3A4. Correct insert orientation was confirmed by DNA sequence analysis.

The PXR reporter gene assay itself is performed in our laboratory as follows: HepG2/C3A cells (American Type Culture Collection, Manassas, VA) are plated in 96-well plates (Packard Bioscience, Meriden, CT) (2×10^4 cells/well) in 100 µL Plating Medium (phenol-red-free Dulbecco Modified Eagle Medium (DMEM) supplemented with 10% fetal bovine serum (charcoal stripped), 2 mM L-glutamine and 1 mM nonessential amino acids), and preincubated overnight. The cells are then transfected by lipofection using Fugene-6 (Roche Molecular Biochemicals, Indianapolis, IN) in the presence of serum per the manufacturer's instructions. Transfection mixes applied to each well contain 2.5 ng of hPXR expression vector pcDNA3–hPXR, 50 ng of luciferase reporter plasmid pGL3–3A4, and 20 ng of alkaline phosphatase expression vector pSEAP2-Control (Clontech, Plao Alto, CA). The media are replaced with 100 µL of fresh Dosing Medium (Plating Medium + test compounds in 0.25% DMSO) in 6 h following transfection, and incubated for an additional 42 h. All test compounds except phenobarbital were tested at final concentrations of 2, 10, 20, and 50 µM. Phenobarbital was tested at 20, 50, 150, and 250 µM. The activities of firefly luciferase and alkaline phosphatase were determined by Steady-Glo™ Luciferase Assay System (Promega, Madison, WI) and GreatEscAPe™ chemiluminescence detection kits (Clontech, Palo Alto, CA), respectively, according to the supplier's specifications.

Although the overall principles of the assay and its basic design are essentially the same in all laboratories, specific details may vary from one laboratory to another. Table 17.2 shows details of how the three key components of the assay (cell line, PXR expression plasmid, and reporter gene construct) may vary among individual laboratories. There are basic requirements for each of the three key components. Cell lines should be immortalized, relatively easy to propagate and maintain in a consistent manner, readily amenable to plasmid transfection, and have little or no background or constitutive expression of PXR. Caco-2 cells, CV-1 (Africa green monkey kidney fibroblast) cells, and HepG2 cells all meet these criteria.

TABLE 17.2 Summary of Human PXR reporter gene assays used in several laboratories.

Author	PXR Construct	Reporter gene construct	Cell line	Maximal induction	References
Bertilsson et al	GAL4–LBD	4USA-Luciferase	Caco-2	~15-fold	Bertilsson et al., (1998)
	pCMV–PXR	CYP3A4(ER)6-Luciferase	Caco-2	~5-fold	
Blumberg et al	GAL–LBD	Tk-USA-Luciferase	CV-1	~12-fold	Blumberg et al., (1998)
	pSG5–PXR	Tk-(ER)6-Luciferase	CV-1	~12-fold	
Lehmann et al	pSG5–PXR	(ER)6-tk-CAT	CV-1	~9-fold	Lehmann et al., (1998)
Goodwin et al	pSG5–PXR	CYP3A4(−7836 to 208)–(−362 to 53)-Luciferase	HepG2	~40-fold	Goodwin et al., (1999)
El-Sankary et al	pSG5–PXR	CYP3A4(−1087 to −57)-SPAP	HepG2	~3-fold	El-Sankary et al., (2000)

A eukaryotic expression vector such as pSG5 (Stratagene) or pCMV (Promega), with a human PXR open reading frame cDNA insert, is employed to express human PXR. The most critical component is the reporter gene construct, in which the PXR responsive elements and reporter gene are the key factors. To represent PXR responsive elements, some laboratories have used the CYP3A4 proximal promoter alone, usually three copies of ER6 (Bertilsson et al., 1998; Blumberg et al., 1998; Lehmann et al., 1998); some have used the entire 1 kb (-1087 to -57) CYP3A4 $5'$ untranslated region (El-Sankary et al., 2000); and some have used both the proximal promoter and distal enhancer elements (Goodwin et al., 1999). These approaches differ in their indication of the maximal fold induction by rifampicin as shown in Table 17.2. In our own laboratory, we have used several different PXR reporter gene constructs, and find that constructs where both the CYP3A4 proximal promoter and distal enhancer regions are present in the reporter gene construct provide the most facile and sensitive assays (Goodwin et al., 1999; Luo et al., 2002). Among the commonly employed reporter gene assays, the luciferase assay is simple and sensitive, whereas the measurement of CAT is much more time-consuming, and there is a considerably higher background. Finally, an alternative reporter gene system has been developed, by creating a construct that inserts the LBD region of PXR into a *GAL4* reporter system. In this case, PXR LBD cDNA was fused with *GAL4* in an expression vector, while the reporter gene construct was either 4USA-luciferase or tk-USA-luciferase (Bertilsson et al., 1998; Blumberg et al., 1998; Lehmann et al., 1998). This assay employs the same principle as other PXR reporter gene assays, except that the PXR DBD is replaced with *GAL4* in the expression vector construct, and a PXR responsive element is substituted for five tandem repeats of the 17-bp *GAL4* binding element in the reporter gene construct. This assay is quite straightforward and the maximal induction is more than 10-fold. We have studied the correlation between the prediction of CYP3A4 induction by the PXR reporter gene assay originally established by Goodwin et al. (Goodwin et al., 1999), and the actual induction and expression of CYP3A4 in primary cultures of human hepatocytes for 14 clinically used drugs (Luo et al., 2002). This study indicated that the PXR reporter gene assay generally predicted the degree of CYP3A4 induction observed with primary cultures of human hepatocytes (Luo et al., 2002). In some cases (e.g., ritonavir and troleandomycin), compounds that induce CYP3A4 are also mechanism-based inactivators of CYP3A4 *in vivo*. For these compounds, the PXR reporter gene assay identified the inducers, even though the net CYP3A4-dependent metabolism of the model substrates was not increased in primary human hepatocyte cultures.

17.2.3   Direct Assessment of CYP3A4 Induction *In vivo* in Humans

As mentioned above, all the methods currently employed for measuring CYP3A4 induction, including humanized mice, PXR reporter gene assays, and primary cultures of human hepatocytes, have limitations. Most importantly, these methods do not directly assess the potential of a compound to induce CYP3A4 in humans under clinical conditions. The prediction of CYP3A4 induction based on results obtained indirectly could be inaccurate. The direct measurement of CYP3A4 activity in humans exposed to drugs or test compounds will, therefore, provide the most directly useful assessment of CYP3A4 induction.

In the late 1980s, Watkins et al. developed the erythromycin breath test (Watkins et al., 1989). Erythromycin is metabolized by human CYP3A4 via N-demethylation. Metabolism of erythromycin that is isotopically labeled at the N-methyl group results in the elimination in expired air of $[^{14}C]CO_2$, while the N-desmethyl erythromycin is excreted into bile. In the standard assay, patients receive $4\,\mu Ci$ of $[^{14}CN$-methyl] erythromycin (equivalent to $0.074\,\mu mol$), dissolved in $2\,mL$ of 5% dextrose immediately before intravenous administration. The breath of patients is collected at timed intervals thereafter through a tube containing $4\,mL$ of hyamine hydroxide and ethanol (1 : 1, v/v), and a trace amount of phenolphthalein. The amount of trapped $[^{14}C]CO_2$ is then determined using scintillation counting. Hepatic CYP3A4 activity, represented by the production of $[^{14}C]CO_2$ in the breath (expressed as percentage of total administrated radiolabel eliminated in breath per minute), was significantly increased in patients treated with the known CYP3A4 inducers dexamethasone or hydrocortisone for 2 days or longer, or with rifampicin (600 mg) for 4 days. CYP3A4-dependent generation of $[^{14}C]CO_2$ was decreased by 80% in patients who received a single oral dose of the CYP3A4-specific inhibitor troleandomycin (500 mg). This method is now widely accepted and utilized. For example, the erythromycin breath test has been used to demonstrate CYP3A4 induction by many xenobiotics, including St. John's wort (Durr et al., 2000), efavirenz (Mouly et al., 2002), progestins (Tsunoda et al., 1998), dexamethasone (McCune et al., 2000), and rifampicin (Floyd et al., 2003).

In the format originally devised, the erythromycin breath test requires intravenous injection of a compound that is radiolabeled, and therefore potentially hazardous. Several nonradiolabeled CYP3A4-specific substrates have been identified that make it possible to more safely assess CYP3A4 induction directly in test subjects. These CYP3A4 selective substrates include some clinically important drugs such as midazolam, buspirone, felodipine, simvastatin, lovastatin, as well as testosterone (Bjornsson et al., 2003; Tucker et al., 2001). For example, the induction by rifampicin of the CYP3A4-dependent systemic clearance of midazolam was measured in 57 healthy subjects (Floyd et al., 2003). The subjects received rifampicin (600 mg PO, once daily) for 16 consecutive days. The systemic clearance of midazolam was measured on day 1, and again on day 15. Five milliliter blood samples were collected into EDTA-containing tubes via an intravenous catheter, before and 5, 15, 30 min, and 1, 2, 3, 4, 5, 6, and 8 h after intravenous administration of 1 mg of midazolam. Plasma concentrations of midazolam and its major metabolite, 1'-OH midazolam, were determined using high-performance liquid chromatography-tandem mass spectrometry (Wandel et al., 2000). The study found that rifampicin significantly enhanced the total body clearance of midazolam, from 5.5 mL/min/kg on day 1 to 10.1 mL/min/kg on day 15. Notably, the twofold induction of CYP3A4-dependent activity measured by the systemic clearance of midazolam was consistent with the twofold increase in erythromycin *N*-demethylation, measured by the erythromycin breath test performed in the same subjects. A similar study by Backman et al., utilizing midazolam clearance as a measure of CYP3A4 induction, demonstrated that rifampicin treatment (600 mg daily for 5 days) reduced C_{max} plasma concentrations of midazolam (15 mg administered PO) from 55 to 3.5 ng/mL, reduced the AUC_{total} from 10.2 to 0.4 μg.min/mL, and reduced the midazolam half-life from 3.1 to 1.3 h (Backman et al., 1996).

Theoretically, the least hazardous method of directly assessing CYP3A4 induction *in vivo* in humans would be to measure the CYP3A4-dependent generation of a metabolite of an endogenous substrate, which is predominantly excreted into urine. This would provide a noninvasive method that would not involve administration of a xenobiotic or a radiolabeled compound. Recently, the clearance of 4β-hydroxycholesterol, an endogenous CYP3A4-dependent metabolite with an apparent half-life of 52 h, was found to be a good indicator of CYP3A4 induction (Bodin et al., 2001, 2002). For example, the plasma levels of 4β-hydroxycholesterol in humans treated with antiepileptic drugs phenobarbital, carbamazepine, or phenytoin were significantly increased by 7- to 20-fold over normal levels (Bodin et al., 2001, 2002). In addition, 6β-hydroxycortisol is a metabolite of cortisol that is believed to be formed exclusively by CYP3A4, and excreted into the urine. Rifampicin treatment (600 mg daily for a week) increased the renal excretion of 6β-hydroxycortisol by more than threefold (Dilger et al., 2005). In addition, 25- or 6β-hydroxylated bile acids, also believed to be endogenous metabolites produced by CYP3A4, may be useful markers for CYP3A4 induction, because they are excreted in urine; and therefore, easily measured (Araya and Wikvall, 1999; Furster and Wikvall, 1999; Handschin et al., 2002; Ourlin et al., 2002).

Thus far, a qualitative correlation between *in vitro* and *in vivo* estimates of CYP3A4 induction has been established, but only in the case of a few, already well-characterized inducers, and quantitative correlations are less clear-cut. Several potent inducers of CYP3A4, including St. John's wort, efavirenz, progestins, dexamethasone, and rifampicin produced a positive induction response in both *in vitro* assay systems, and with the erythromycin breath test *in vivo*, as described above in this section. Also, a number of compounds that induce CYP3A4 *in vitro* also reduce *in vivo* the AUC of coadministered drugs that are CYP3A4 substrates. These CYP3A4 inducers include carbamazepine, phenobarbital, phenytoin, rifampin, St. John's wort, topiramate, and troglitazone (Luo et al., 2004; Ripp et al., 2006). Several factors should be considered when comparing and extrapolating *in vitro* and *in vivo* results. These factors include, but are not limited to, the compound's dosage, route and duration of administration, pharmacokinetic properties, tissue distribution, and potential to produce mechanism-based inactivation of CYP3A4, all of which may not affect the results of the *in vitro* screening assays, but may have a major effect on the effective concentration of the inducer at its site of induction *in vivo*. For other CYP enzymes, the correlation between induction potential predicted by *in vitro* assays, and actual induction seen *in vivo* is thus far less firmly established, primarily because of a more limited number of studies and a smaller available database.

17.3 FINAL COMMENTS

Clinically relevant induction of CYP enzymes occurs during multiple dosing of the inducing drug and is a dose- and time-dependent phenomenon. We have described several short-term methods in this chapter, both *in vivo* and *in vitro*, for determining the potential of a candidate molecule to induce CYP3A4. A number of criteria for the usefulness of data from such methods, especially *in vitro* assays (Bjornsson et al., 2003; Tucker et al., 2001), have been proposed. The use of both positive and negative controls are crucial in these assays. Compounds should be tested at concentrations covering the targeted human C_{max}, and dose–response curves should be generated when possible. Reproducibility of cultured human hepatocyte-based assays is an important issue, and ideally at least three different replicates from three different human liver sources should be used for each treatment group. We have also suggested that a response of at least 40% of that produced by the positive control compounds level be considered as the criterion for a positive induction response in the *in vitro* assays.

In vitro and animal models possess some advantages over the more direct, but also more expensive clinical studies. However, although they can provide a qualitative or semiquantitative assessment of induction, these preclinical methods cannot as yet predict the degree of CYP induction in humans by a candidate molecule with quantitative certainty. A recent study by Kato et al. (2005) compared literature data on *in vitro* and *in vivo* enzyme induction with their data obtained with human hepatocytes *in vitro*, normalized for non-protein bound mean concentrations of inducers. They generally showed that maximum induction ratios produced by inducers in *in vitro* studies were higher than those seen in *in vivo* studies. The authors provide evidence that it is possible to apply pharmacokinetic corrections to accurately predict, on a quantitative basis, human CYP3A induction *in vivo* using data obtained with human hepatocytes *in vitro*. Using immortalized human hepatocyte Fa2N-4 cells and efficacious free human plasma concentrations, Ripp et al. recently have reported good correlations between percentage AUC changes and relative induction scores with midazolam or ethinyl estradiol as CYP3A4 substrates (Ripp et al., 2006).

Finally, in addition to mediating the process of CYP3A4 induction in humans, the PXR receptor is also responsible for the regulation of expression of Phase II conjugating enzymes such as UGT1A1, and it also upregulates transporter proteins such as Pgp, MRP2, and OATP (Eloranta et al., 2005; Klaassen and Slitt, 2005). When CYP3A4 induction occurs, induction of these additional nonCYP drug disposition proteins is also possible. Therefore, any coadministered drugs that are metabolized or otherwise processed by these proteins could potentially be subject to increased clearance, and the potential for decreased plasma levels and subtherapeutic concentrations at normal doses must be considered as a possible consequence of PXR activation.

REFERENCES

Araya Z, Wikvall K. 6alpha-hydroxylation of taurochenodeoxycholic acid and lithocholic acid by CYP3A4 in human liver microsomes. Biochim Biophys Acta 1999;1438:47–54.

Backman JT, Olkkola KT, Neuvonen PJ. Rifampin drastically reduces plasma concentrations and effects of oral midazolam. Clin Pharmacol Ther 1996;59:7–13.

Bertilsson G, Heidrich J, Svensson K, Asman M, Jendeberg L, Sydow-Backman M, Ohlsson R, Postlind H, Blomquist P, Berkenstam A. Identification of a human nuclear receptor defines a new signaling pathway for CYP3A induction. Proc Natl Acad Sci USA 1998;95:12208–12213.

Bjornsson TD, Callaghan JT, Einolf HJ, Fischer V, Gan L, Grimm S, Kao J, King SP, Miwa G, Ni L, Kumar G, McLeod J, Obach SR, Roberts S, Roe A, Shah A, Snikeris F, Sullivan JT, Tweedie D, Vega JM, Walsh J, Wrighton SA. The conduct of *in vitro* and *in vivo* drug–drug interaction studies: a PhRMA perspective. J Clin Pharmacol 2003;43:443–469.

Blumberg B, Sabbagh W, Juguilon H, Bolado J, van Meter CM, Jr, Ong ES, Evans RM. SXR, a novel steroid and xenobiotic-sensing nuclear receptor. Genes Dev 1998;12:3195–3205.

Bodin K, Andersson U, Rystedt E, Ellis E, Norlin M, Pikuleva I, Eggertsen G, Bjorkhem I, Diczfalusy U. Metabolism of 4 beta-hydroxycholesterol in humans. J Biol Chem 2002;277:31534–31540.

Bodin K, Bretillon L, Aden Y, Bertilsson L, Broome U, Einarsson C, Diczfalusy U. Antiepileptic drugs increase plasma levels of 4beta-hydroxycholesterol in humans: evidence for involvement of cytochrome p450 3A4. J Biol Chem 2001;276:38685–38689.

Chan WK, Delucchi AB. Resveratrol, a red wine constituent, is a mechanism-based inactivator of cytochrome P450 3A4. Life Sci 2000;67:3103–3112.

Chomczynski P, Sacchi N. Single-step method of RNA isolation by acid guanidinium thiocyanate-phenol-chloroform extraction. Anal Biochem 1987;162:156–159.

Curi-Pedrosa R, Daujat M, Pichard L, Ourlin JC, Clair P, Gervot L, Lesca P, Domergue J, Joyeux H, Fourtanier G, Maurel P. Omeprazole and lansoprazole are mixed inducers of CYP1A and CYP3A in human hepatocytes in primary culture. J Pharmacol Exp Ther 1994;269:384–392.

Czerwinski M, Opdam P, Madan A, Carroll K, Mudra DR, Gan LL, Luo G, Parkinson A. Analysis of CYP mRNA expression by branched DNA technology. Methods Enzymol 2002;357:170–179.

Desai PB, Nallani SC, Sane RS, Moore LB, Goodwin BJ, Buckley DJ, Buckley AR. Induction of cytochrome P450 3A4 in primary human hepatocytes and activation of the human pregnane X receptor by tamoxifen and 4-hydroxytamoxifen. Drug Metab Dispos 2002;30:608–612.

Dilger K, Denk A, Heeg MH, Beuers U. No relevant effect of ursodeoxycholic acid on cytochrome P450 3A metabolism in primary biliary cirrhosis. Hepatology 2005;41:595–602.

Durr D, Stieger B, Kullak-Ublick GA, Rentsch KM, Steinert HC, Meier PJ, Fattinger K. St John's wort induces intestinal P-glycoprotein/MDR1 and intestinal and hepatic CYP3A4. Clin Pharmacol Ther 2000;68:598–604.

Eloranta JJ, Meier PJ, Kullak-Ublick GA. Coordinate transcriptional regulation of transport and metabolism. Methods Enzymol 2005;400:511–530.

El-Sankary W, Plant NJ, Gibson GG, Moore DJ. Regulation of the CYP3A4 gene by hydrocortisone and xenobiotics: role of the glucocorticoid and pregnane X receptors. Drug Metab Dispos 2000;28:493–496.

Fabre G, Rahmani R, Placidi M, Combalbert J, Covo J, Cano JP, Coulange C, Ducros M, Rampal M. Characterization of midazolam metabolism using human hepatic microsomal fractions and hepatocytes in suspension obtained by perfusing whole human livers. Biochem Pharmacol 1988;37:4389–4397.

Floyd MD, Gervasini G, Masica AL, Mayo G, George AL, Bhat K Jr, Kim RB, Wilkinson GR. Genotype-phenotype associations for common CYP3A4 and CYP3A5 variants in the basal and induced metabolism of midazolam in European- and African-American men and women. Pharmacogenetics 2003;13:595–606.

Furster C, Wikvall K. Identification of CYP3A4 as the major enzyme responsible for 25-hydroxylation of 5beta-cholestane-3alpha,7alpha, 12alpha-triol in human liver microsomes. Biochim Biophys Acta 1999;1437:46–52.

Gerbal-Chaloin S, Pascussi JM, Pichard-Garcia L, Daujat M, Waechter F, Fabre JM, Carrere N, Maurel P. Induction of CYP2C genes in human hepatocytes in primary culture. Drug Metab Dispos 2001;29:242–251.

Goodwin B, Hodgson E, Liddle C. The orphan human pregnane X receptor mediates the transcriptional activation of CYP3A4 by rifampicin through a distal enhancer module. Mol Pharmacol 1999;56:1329–1339.

Goodwin B, Moore LB, Stoltz CM, McKee DD, Kliewer SA. Regulation of the human CYP2B6 gene by the nuclear pregnane X receptor. Mol Pharmacol 2001;60:427–431.

Goodwin B, Redinbo MR, Kliewer SA. Regulation of cyp3a gene transcription by the pregnane x receptor. Annu Rev Pharmacol Toxicol 2002;42:1–23.

Handschin C, Podvinec M, Amherd R, Looser R, Ourlin JC, Meyer UA. Cholesterol and bile acids regulate xenosensor signaling in drug-mediated induction of cytochromes P450. J Biol Chem 2002;277:29561–29567.

Hankinson O. The aryl hydrocarbon receptor complex. Annu Rev Pharmacol Toxicol 1995;35:307–340.

Hariparsad N, Nallani SC, Sane RS, Buckley DJ, Buckley AR, Desai PB. Induction of CYP3A4 by efavirenz in primary human hepatocytes: comparison with rifampin and phenobarbital. J Clin Pharmacol 2004;44:1273–1281.

He K, Woolf TF, Hollenberg PF. Mechanism-based inactivation of cytochrome P450–3A4 by mifepristone (RU486). J Pharmacol Exp Ther 1999;288:791–797.

Jones SA, Moore LB, Shenk JL, Wisely GB, Hamilton GA, McKee DD, Tomkinson NC, LeCluyse EL, Lambert MH, Willson TM, Kliewer SA, Moore JT. The pregnane X receptor: a promiscuous xenobiotic receptor that has diverged during evolution. Mol Endocrinol 2000;14:27–39.

Kafert-Kasting S, Alexandrova K, Barthold M, Laube B, Friedrich G, Arseniev L, Hengstler JG. Enzyme induction in cryopreserved human hepatocyte cultures. Toxicology 2006;220:117–125.

Kato M, Chiba K, Horikawa M, Sugiyama Y. The quantitative prediction of *in vivo* enzyme-induction caused by drug exposure from *in vitro* information on human hepatocytes. Drug Metab Pharmacokinet 2005;20:236–243.

Kirby GM, Batist G, Alpert L, Lamoureux E, Cameron RG, Alaoui-Jamali MA. Overexpression of cytochrome P450 isoforms involved in aflatoxin B1 bioactivation in human liver with cirrhosis and hepatitis. Toxicol Pathol 1996;24:458–467.

Klaassen CD, Slitt AL. Regulation of hepatic transporters by xenobiotic receptors. Curr Drug Metab 2005;6:309–328.

Kliewer SA, Moore JT, Wade L, Staudinger JL, Watson MA, Jones SA, McKee DD, Oliver BB, Willson TM, Zetterstrom RH, Perlmann T, Lehmann JM. An orphan nuclear receptor activated by pregnanes defines a novel steroid signaling pathway. Cell 1998;92:73–82.

Kliewer SA, Willson TM. Regulation of xenobiotic and bile acid metabolism by the nuclear pregnane X receptor. J Lipid Res 2002;43:359–364.

Kocarek TA, Schuetz EG, Strom SC, Fisher RA, Guzelian PS. Comparative analysis of cytochrome P4503A induction in primary cultures of rat, rabbit, and human hepatocytes. Drug Metab Dispos 1995;23:415–421.

Kostrubsky VE, Ramachandran V, Venkataramanan R, Dorko K, Esplen JE, Zhang S, Sinclair JF, Wrighton SA, Strom SC. The use of human hepatocyte cultures to study the induction of cytochrome P450. Drug Metab Dispos 1999;27:887–894.

LeBel M, Masson E, Guilbert E, Colborn D, Paquet F, Allard S, Vallee F, Narang PK. Effects of rifabutin and rifampicin on the pharmacokinetics of ethinylestradiol and norethindrone. J Clin Pharmacol 1998;38:1042–1050.

LeCluyse E, Madan A, Hamilton G, Carroll K, DeHaan R, Parkinson A. Expression and regulation of cytochrome P450 enzymes in primary cultures of human hepatocytes. J Biochem Mol Toxicol 2000;14:177–188.

LeCluyse EL. Pregnane X receptor: molecular basis for species differences in CYP3A induction by xenobiotics. Chem Biol Interact 2001a;134:283–289.

LeCluyse EL. Human hepatocyte culture systems for the *in vitro* evaluation of cytochrome P450 expression and regulation. Eur J Pharm Sci 2001b;13:343–368.

Ledirac N, de Sousa G, Fontaine F, Agouridas C, Gugenheim J, Lorenzon G, Rahmani R. Effects of macrolide antibiotics on CYP3A expression in human and rat hepatocytes: interspecies differences in response to troleandomycin. Drug Metab Dispos 2000;28:1391–1393.

Lehmann JM, McKee DD, Watson MA, Willson TM, Moore JT, Kliewer SA. The human orphan nuclear receptor PXR is activated by compounds that regulate CYP3A4 gene expression and cause drug interactions. J Clin Invest 1998;102:1016–1023.

Lu C, Li AP. Species comparison in P450 induction: effects of dexamethasone, omeprazole, and rifampin on P450 isoforms 1A and 3A in primary cultured hepatocytes from man, Sprague–Dawley rat, minipig, and beagle dog. Chem Biol Interact 2001;134:271–281.

Luo G, Cunningham M, Kim S, Burn T, Lin J, Sinz M, Hamilton G, Rizzo C, Jolley S, Gilbert D, Downey A, Mudra D, Graham R, Carroll K, Xie J, Madan A, Parkinson A, Christ D, Selling B, LeCluyse E, Gan LS. CYP3A4 induction by drugs: correlation between a pregnane X receptor reporter gene assay and CYP3A4 expression in human hepatocytes. Drug Metab Dispos 2002;30:795–804.

Luo G, Guenthner T, Gan LS, Humphreys WG. CYP3A4 induction by xenobiotics: biochemistry, experimental methods and impact on drug discovery and development. Curr Drug Metab 2004;5:483–505.

Luo G, Lin J, Fiske WD, Dai R, Yang TJ, Kim S, Sinz M, LeCluyse E, Solon E, Brennan JM, Benedek IH, Jolley S, Gilbert D, Wang L, Lee FW, Gan LS. Concurrent induction and mechanism-based inactivation of CYP3A4 by an L-valinamide derivative. Drug Metab Dispos 2003;31:1170–1175.

Matsuda H, Kinoshita K, Sumida A, Takahashi K, Fukuen S, Fukuda T, Yamamoto I, Azuma J. Taurine modulates induction of cytochrome P450 3A4 mRNA by rifampicin in the HepG2 cell line. Biochim Biophys Acta 2002;1593:93–98.

Matsuura Y, Kotani E, Iio T, Fukuda T, Tobinaga S, Yoshida T, Kuroiwa Y. Structure-activity relationships in the induction of hepatic microsomal cytochrome P450 by clotrimazole and its structurally related compounds in rats. Biochem Pharmacol 1991;41:1949–1956.

McCune JS, Hawke RL, LeCluyse EL, Gillenwater HH, Hamilton G, Ritchie J, Lindley C. *In vivo* and *in vitro* induction of human cytochrome P4503A4 by dexamethasone. Clin Pharmacol Ther 2000;68:356–366.

Meunier V, Bourrie M, Julian B, Marti E, Guillou F, Berger Y, Fabre G. Expression and induction of CYP1A1/1A2, CYP2A6 and CYP3A4 in primary cultures of human hepatocytes: a 10-year follow-up. Xenobiotica 2000;30:589–607.

Mills JB, Rose KA, Sadagopan N, Sahi J, de Morais SM. Induction of drug metabolism enzymes and MDR1 using a novel human hepatocyte cell line. J Pharmacol Exp Ther 2004;309:303–309.

Moore JT, Kliewer SA. Use of the nuclear receptor PXR to predict drug interactions. Toxicology 2000;153:1–10.

Moore JT, Moore LB, Maglich JM, Kliewer SA. Functional and structural comparison of PXR and CAR. Biochim Biophys Acta 2003;1619:235–238.

Moore LB, Maglich JM, McKee DD, Wisely B, Willson TM, Kliewer SA, Lambert MH, Moore JT. Pregnane X receptor (PXR), constitutive androstane receptor (CAR), and benzoate X receptor (BXR) define three pharmacologically distinct classes of nuclear receptors. Mol Endocrinol 2002;16:977–986.

Mouly S, Lown KS, Kornhauser D, Joseph JL, Fiske WD, Benedck IH, Watkins PB. Hepatic but not intestinal CYP3A4 displays dose-dependent induction by efavirenz in humans. Clin Pharmacol Ther 2002;72:1–9.

Nebert DW, Puga A, Vasiliou V. Role of the Ah receptor and the dioxin-inducible [Ah] gene battery in toxicity, cancer, and signal transduction. Ann N Y Acad Sci 1993;685:624–640.

Okey AB. Enzyme induction in the cytochrome P-450 system. Pharmacol Ther 1990;45:241–298.

Okey AB, Riddick DS, Harper PA. Molecular biology of the aromatic hydrocarbon (dioxin) receptor. Trends Pharmacol Sci 1994;15:226–232.

Ourlin JC, Handschin C, Kaufmann M, Meyer UA. A Link between cholesterol levels and phenobarbital induction of cytochromes P450. Biochem Biophys Res Commun 2002;291:378–384.

Pascussi JM, Gerbal-Chaloin S, Drocourt L, Maurel P, Vilarem MJ. The expression of CYP2B6, CYP2C9 and CYP3A4 genes: a tangle of networks of nuclear and steroid receptors. Biochim Biophys Acta 2003;1619:243–253.

Pearce RE, McIntyre CJ, Madan A, Sanzgiri U, Draper AJ, Bullock PL, Cook DC, Burton LA, Latham J, Nevins C, Parkinson A. Effects of freezing, thawing, and storing human liver microsomes on cytochrome P450 activity. Arch Biochem Biophys 1996;331:145–169.

Philip PA, Kaklamanis L, Ryley N, Stratford I, Wolf R, Harris A, Carmichael J. Expression of xenobiotic-metabolizing enzymes by primary and secondary hepatic tumors in man. Int J Radiat Oncol Biol Phys 1994;29:277–283.

Pichard L, Fabre I, Fabre G, Domergue J, Saint Aubert B, Mourad G, Maurel P. Cyclosporin A drug interactions. Screening for inducers and inhibitors of cytochrome P450 (cyclosporin A oxidase) in primary cultures of human hepatocytes and in liver microsomes. Drug Metab Dispos 1990;18:595–606.

Prueksaritanont T, Richards KM, Qiu Y, Strong-Basalyga K, Miller A, Li C, Eisenhandler R, Carlini EJ. Comparative effects of fibrates on drug metabolizing enzymes in human hepatocytes. Pharm Res 2005;22:71–78.

Reinach B, de Sousa G, Dostert P, Ings R, Gugenheim J, Rahmani R. Comparative effects of rifabutin and rifampicin on cytochromes P450 and UDP-glucuronosyl-transferases expression in fresh and cryopreserved human hepatocytes. Chem Biol Interact 1999;121:37–48.

Ripp SL, Mills JB, Fahmi OA, Trevena KA, Liras JL, Maurer TS, de Morais SM. Use of immortalized human hepatocytes to predict the magnitude of clinical drug–drug interactions caused by CYP3A4 induction. Drug Metab Dispos 2006;34:1742–1748.

Ronis MJ, Ingelman-Sundberg M. Induction of human drug metabolizing enzymes: mechanisms and implications. In: Woolf TF, editor. Handbook of Drug Metabolism. New York: Marcel Dekker; 1999 p 239–262.

Roymans D, van Looveren C, Leone A, Parker JB, McMillian M, Johnson MD, Koganti A, Gilissen R, Silber P, Mannens G, Meuldermans W. Determination of cytochrome P450 1A2 and cytochrome P450 3A4 induction in cryopreserved human hepatocytes. Biochem Pharmacol 2004;67:427–437.

Smith PK, Krohn RI, Hermanson GT, Mallia AK, Gartner FH, Provenzano MD, Fujimoto EK, Goeke NM, Olson BJ, Klenk DC. Measurement of protein using bicinchoninic acid. Anal Biochem 1985;150:76–85.

Sotaniemi EA, Rautio A, Backstrom M, Arvela P, Pelkonen O. CYP3A4 and CYP2A6 activities marked by the metabolism of lignocaine and coumarin in patients with liver and kidney diseases and epileptic patients. Br J Clin Pharmacol 1995;39:71–76.

Strom SC, Jirtle RL, Jones RS, Novicki DL, Rosenberg MR, Novotny A, Irons G, McLain JR, Michalopoulos G. Isolation, culture, and transplantation of human hepatocytes. J Natl Cancer Inst 1982;68:771–778.

Sueyoshi T, Negishi M. Phenobarbital response elements of cytochrome P450 genes and nuclear receptors. Annu Rev Pharmacol Toxicol 2001;41:123–143.

Sumida A, Yamamoto I, Zhou Q, Morisaki T, Azuma J. Evaluation of induction of CYP3A mRNA using the HepG2 cell line and reverse transcription-PCR. Biol Pharm Bull 1999;22:61–65.

Thummel KE, Wilkinson GR. *In vitro* and *in vivo* drug interactions involving human CYP3A. Annu Rev Pharmacol Toxicol 1998;38:389–430.

Tsunoda SM, Harris RZ, Mroczkowski PJ, Benet LZ. Preliminary evaluation of progestins as inducers of cytochrome P450 3A4 activity in postmenopausal women. J Clin Pharmacol 1998;38:1137–1143.

Tucker GT, Houston JB, Huang SM. Optimizing drug development: strategies to assess drug metabolism/transporter interaction potential-toward a consensus. Clin Pharmacol Ther 2001;70:103–114.

Wandel C, Witte JS, Hall JM, Stein CM, Wood AJ, Wilkinson GR. CYP3A activity in African-American and European-American men: population differences and functional effect of the CYP3A4*1B5′-promoter region polymorphism. Clin Pharmacol Ther 2000;68:82–91.

Watkins PB, Murray SA, Winkelman LG, Heuman DM, Wrighton SA, Guzelian PS. Erythromycin breath test as an assay of glucocorticoid-inducible liver cytochromes P450. Studies in rats and patients. J Clin Invest 1989;83:688–697.

Wrighton SA, Stevens JC. The human hepatic cytochromes P450 involved in drug metabolism. Crit Rev Toxicol 1992;22:1–21.

Xie W, Barwick JL, Downes M, Blumberg B, Simon CM, Nelson MC, Neuschwander-Tetri BA, Brunt EM, Guzelian PS, Evans RM. Humanized xenobiotic response in mice expressing nuclear receptor SXR. Nature 2000;406:435–439.

Xie W, Evans RM. Pharmaceutical use of mouse models humanized for the xenobiotic receptor. Drug Discov Today 2002;7:509–515.

Yang LQ, Li SJ, Cao YF, Man XB, Yu WF, Wang HY, Wu MC. Different alterations of cytochrome P450 3A4 isoform and its gene expression in livers of patients with chronic liver diseases. World J Gastroenterol 2003;9:359–363.

Zhang H, LeCulyse E, Liu L, Hu M, Matoney L, Zhu W, Yan B. Rat pregnane X receptor: molecular cloning, tissue distribution, and xenobiotic regulation. Arch Biochem Biophys 1999;368:14–22.

Zhao XJ, Jones DR, Wang YH, Grimm SW, Hall SD. Reversible and irreversible inhibition of CYP3A enzymes by tamoxifen and metabolites. Xenobiotica 2002;32:863–878.

Zhu Z, Kim S, Chen T, Lin JH, Bell A, Bryson J, Dubaquie Y, Yan N, Yanchunas J, Xie D, Stoffel R, Sinz M, Dickinson K. Correlation of high-throughput pregnane X receptor (PXR) transactivation and binding assays. J Biomol Screen 2004;9:533–540.

18

ADME STUDIES IN ANIMALS AND HUMANS: EXPERIMENTAL DESIGN, METABOLITE PROFILING AND IDENTIFICATION, AND DATA PRESENTATION

DONGLU ZHANG AND S. NILGUN COMEZOGLU

18.1 OBJECTIVES, RATIONAL, AND REGULATORY COMPLIANCE

Absorption, distribution, metabolism, and excretion (ADME) studies in discovery and exploratory development stages of drugs are conducted to assess the metabolism and excretion of a radiolabeled drug following a single administration (intravenous or oral) to rodents (rats or mice), nonrodents (dogs or monkeys), special species such as rabbits, and humans. These studies will (1) evaluate the exposures of the parent compound and its metabolites in animals and humans for validation of toxicological species, (2) identify the major metabolic pathways in humans to support drug–drug interaction studies, and (3) establish the rate and route of excretion of a drug candidate, and in addition, (4) provide metabolism data of drugs for regulatory filing.

Nonclinical ADME studies are used to link the animal pharmacology and toxicology studies to humans. Data obtained from the animal metabolism and excretory pathways of a drug may be useful to design a clinical mass balance study. Once in the development stage, ADME data in animals and humans with radiolabeled materials will reveal the major circulating metabolite(s) and

Drug Metabolism in Drug Design and Development, Edited by Donglu Zhang, Mingshe Zhu and W. Griffith Humphreys
Copyright © 2008 John Wiley & Sons, Inc.

major clearance pathways. The ADME data will also define relative concentrations of each metabolite in plasma, urine, bile, and feces. Since one of the main objectives for metabolism studies is to support drug safety evaluation studies, the Metabolite In Safety Testing (MIST) committee suggested that any circulating metabolites and any significant metabolites in excreta of humans should be characterized to understand the formation pathways of metabolites (Baillie et al., 2002). Identification of the major metabolites in human excreta not only aids in defining the major clearance pathways but also designing the drug–drug interaction studies. Major metabolites in animals, which may not be important as human metabolites, should also be identified since they may explain the species-specific toxicities observed in animals. Thus, there is a need to identify quantitatively and qualitatively important radioactive metabolites in studies with radiolabeled materials.

A study protocol including any amendment for ADME studies should be finalized before the experimental start date. The ADME studies are not generally conducted under the Good Laboratory Practice (GLP) Regulations; however, the Standard Operating Procedures (SOP) of the Test Facility should be followed. All animal housing and care conformed the standards recommended by the Guide for the Care and Use of Laboratory Animals. Human ADME studies are performed in accordance with the following codes and guidelines: Title 21, Part 56 CFR (Institutional Review Board Approval); Title 21, Part 50 CFR (Protection of Human Subjects); European guidelines (Lellig, 1996; Nuis, 1997); the principles of the Declaration of Helsinki and its amendments; and Good Clinical Practice. The purposes of these guidelines are for the safe and effective use of the radioactive drugs. Based on the tissue distribution study in rats, the extrapolated radiation dose equivalent for tissues in humans following single 50–100 μCi doses would be lower than the annual allowable exposure limits (3000 mrem) (Appendix A describes a typical rat tissue distribution study and dosimetry calculation). In addition, all subjects were advised of the nature and risks associated with the study and required to give informed and written consent prior to participation in the study. All subjects are acceptable as determined by medical history, physical examination, and clinical laboratory tests conducted prior to study based on inclusion and exclusion criteria. Dosing a radiolabeled material to human subjects is not generally done as the first human study although there is no regulation dictating not to do so. Considering the amount and type of information that a human ADME study can provide, it should be conducted earlier in the program, for example, right after the multiple ascending dosing study or parallel to it.

At the early stage of compound selection and optimization for deciding a clinical candidate, drug metabolism studies are often conducted in vitro with liver fractions and in vivo with rats using nonlabeled compounds to define metabolic stability, soft-spot identification, CYP inhibition/induction potential, bioactivation or toxic metabolite formation, major in vitro metabolic pathways, and the limited in vitro interspecies comparison. Mass balance (or ADME) studies with collection of plasma in animals and humans using a radiolabeled compound are often conducted in preclinical and early clinical stages, especially when a toxic dose and the therapeutically effective dose range have been identified. These studies will define the rate and route of excretion of the compounds. In addition, these studies will provide definite *in vivo* metabolism comparison with subsequent metabolite profiling and identification as well as limited data on absorption and distribution of the compound. Metabolites can also be isolated from *in vitro* and animal samples and these metabolites can be used as reference standards for quantification in the clinical and nonclinical studies. At the same time, complementary tissue distribution in rats and bile collection from bile duct cannulated (BDC) animals will provide tissue distribution and biliary elimination profiles of the drug candidate. Sometimes, metabolism studies may be needed with animals of disease states or to investigate toxic lesion discovered in toxicity studies. Metabolism studies can also validate the animal species used in toxicology studies. Later, complete quantitative tissue distribution studies in pregnant rats will be needed to assess placenta transfer, and milk excretion in lactating rats will also need to be evaluated.

The procedures given in this chapter are general recommendations. Since each test compound and its metabolites have their own unique physiochemical characteristics, discretion, and modifications of the procedures are sometimes necessary when handling different test compounds.

18.2 STUDY DESIGNS

18.2.1 Choice of Radiolabel

Radiolabeled drugs have been widely used in mass balance studies because the radioactivity can easily be detected and quantified using liquid scintillation techniques and disposition of drugs can be assessed. The choice of the radioisotope, the position of the radiolabel in the drug compound, radiochemical purity, and the specific activity are important parameters in designing the ADME studies. These parameters can have an effect on the metabolic, chemical, and radiochemical stability of the drug, metabolite formation and detection, and the recovery of radioactivity. Radioisotopes used in ADME studies have included ^{3}H (Heggie et al., 1987), ^{14}C (Chando, 1998), ^{32}P (Boado et al., 1995), ^{35}S (Lu et al., 1982) and ^{131}I (Wafelman et al., 1997). ^{14}C is the isotope of choice in most of the ADME studies. The labeled part of the drug molecule should not be lost in metabolite formation. Tritium-labeled drugs are also used commonly, but the risk of tritium exchange chemically or metabolically needs to be considered. Data from the preclinical studies performed with the radiolabeled drug can provide information to make the choice of the radioisotope and its position in the drug molecule.

The tritium- or carbon-14-labeled drug at the metabolically stable position should have a radiochemical purity of ≥98%, and in special cases ≥95% may

TABLE 18.1   ADME study designs.

Species	N^a	Gender	Age	Weight	Dose^b (μCi/subject)
Mouse	5/3 $(+12)^c$	M/F	7–12 weeks	25–40 g	5–15
Rat^d	3 $(+12)^c$	M/F	7–12 weeks	150–300 g	25–50
Rabbit	3	F/pregnant	3–5 months	2–4 kg	50–100
Dog^d	3	M/F	6–10 months	8–13 kg	100–200
Monkey^d	3	M/F	2–5 years	2.5–4 kg	100–200
Human^d	6–8	M/F	15–65 years^e	65 kg	100

Note: Dose level (mg/kg), dose volume (mL/kg), dose concentration (mg/mL), dose radioactivity (μCi/kg), vehicle, dosing regimen/route depends on the drug program and should be provided in the study protocol before the start date. Specific activity of the radiolabeled drug may be adjusted (isotopicvally diluted) to yield the desired radioactive dose level.
$^a N$: number of animals or subjects.
^{b}For a ^{14}C-labeled drug.
^{c}The numbers in parenthesis are the additional animals that are dosed for blood collection for the biotransformation studies (at least four time points; three animals/time point).
^{d}Additional studies with bile cannulated animals or humans may be conducted if needed.
^{e}Old population may be needed for special experiments. The normal age range is 18–45 years.

be acceptable. Radiopurity of the radiolabeled drug should be checked before dose preparation and stability (radiopurity) of the radiolabeled drug under conditions of administration should also be checked at predose and postdose.

18.2.2   Preparation of Animals and Human Subjects

Two animal species, one rodent (rat or mouse) and one nonrodent (dog or monkey) are normally chosen for ADME studies. A summary of typical study design is presented in Table 18.1. Selection of animals will be based on the animal species used in the toxicological studies. The strain and source of animals should match those used in the toxicological studies. A list of animals used in toxicity studies is presented in Table 18.2. Although male animals are normally used for metabolism studies, female animals should also be used if PK/metabolism difference is known between males and females since both genders are used in toxicological studies. If the drug is being developed for Japanese registration, both male and female animals should be used in mass balance studies. In addition, the dispositional studies may be conducted in other species such as rabbits used for reproduction, teratogenicity, and carcinogenicity studies.

Dogs and monkeys are normally nonnaive animals. Animals usually completes at least 14 days of quarantine or conditioning before dose administration. In general, a certified canine or primate diet such as LabChow™ or LabDiet™ (Purina Mills Inc) is fed to dogs and monkeys and a certified rodent diet such as LabDiet (Purina Mills Inc) is given to rats and mice *ad libitum* daily. Tap water is provided to animals *ad libitum*. Mice are

TABLE 18.2 Species used in toxicological studies.

Species	Strain and genders	Type of toxicological/pharmacological studies
Mouse	CD-1, male and female	Two-year carcinogenicity study
Rat	Sprague–Dawley male and female, Wistar, Fisher	Short term PK, dose selection, 2-year carcinogenicity studies, teratogenecity studies, 2-week single dose and/or repeated dose toxicology studies, 1-month and/or 3-months and 1-year single and/or repeated dose toxicology studies, embryo-fetal development study, study of fertility and early embryonic development, range finding study in pregnant rats
Rabbit	New Zealand White, SPF, Oryctolagus cuniculus; female	Reproductive studies including range finding studies in pregnant rabbits, embryo-fetal development study, teratogenecity studies
Dog	Purebred beagle, male and/or female	Short term PK, dose selection, 2-week and 1-month single and/or repeated dose toxicology studies, 3-months and 1-year repeated dose toxicology studies
Monkey	Cynomolgus (*Macaca fascicularis*), male and/or female	Short term PK, dose selection, range finding studies, 2-week, 1-month, 1-year toxicological studies (single and/or repeated doses)
Human	NA, male and/or female	Intended species for efficacy and safety

Note: Depending on the drug program other toxicological/pharmacological studies may be conducted in addition to the studies listed above.

fasted for 4 h, and rats, dogs, and monkeys are fasted overnight before dose administration.

In general, a total of six to eight healthy male subjects, aging 15–65 years, participate in human ADME studies. Healthy subjects are usually not used for anticancer drugs. For cytotoxic anticancer drugs (mutagenic, carcinogenic, or teratogenic), patient subjects are needed for the ADME studies. After at least an 8 h overnight fasting in general, each subject receives a single clinically relevant dose of [^{14}C]-drug containing 50–100 μCi of radioactivity. If ^{3}H-drug is administered the higher dose of radioactivity (50 1000 μCi) is needed to detect the low energy β-radiation of ^{3}H (Beumer et al., 2006). All subjects remain in the clinical facility for at least 7 days, which may be longer depending on the PK parameters ($t_{1/2}$, C_{max}) of the drug, and are closely monitored for adverse events throughout the study. In general an oral cathartic dose of milk of magnesia (30 mL) is administered 2–3 days before the patient discharge, and additional doses, if required, to ensure defecation prior to release from the clinical facility. Subjects are discharged from the clinic in the afternoon of the day when the measurement of radioactivity in excreta is $\leq 1\%$ of administered radioactivity.

18.2.3 Dose Selection, Formulation, and Administration

A test article (radiolabeled and nonlabeled) should have a certificate of analysis with a current use date (expiration date) for the animals in the mass balance studies. For selection of dose, in general, the toxicity data is used. The dose with the minimum toxicity ($1/7$th–$1/10$th of STD_{10}, i.e., the single dose that is severely toxic to 10% of the animals) should be used. If it is a high dose drug, which is nontoxic to animals, a well tolerated mid-dose may be chosen and consideration should be given to PK and bioavailability of the drug in the species. A dose that gives sufficient circulating radioactivity and sufficient amounts of metabolites in excreta for metabolite identification is preferred.

A dose vehicle used in toxicity studies is preferred in the mass balance studies. Using polyethylene glycol (PEG) should be avoided if possible, since it suppress ionization of compounds during metabolite identification by mass spectrometry (MS). Dose formulation is usually prepared on the day of dosing. If the drug is stable, dose solution may be prepared a day before dosing and stored at $\leq -20°C$. Target dose level (mg/kg) and volume (mL/kg) and drug concentration in dose solution (mg/mL) should be defined in the study protocol. Concentration of drug ($\mu Ci/mg$) in formulation should be verified by analyzing triplicate aliquots ($\sim 100\ \mu L$; diluted if necessary) from top, middle, and bottom of the dose solution by liquid scintillation counter (LSC) at predose and postdose. The specifications (including identification, manufacturer, physical description, lot number, specific activity for radiolabeled compounds, expiration date, storage temperature) of components of the dose solution should be documented in the study file. The route of administration in animals is usually the one proposed for clinical use (IV or oral). IV dosing is performed by bolus injection or by slow injection or infusion depending on the toxicity of drug. For oral dosing by gavages, toxicity studies are used as a reference. Single doses are normally used for ADME studies although repeated dose studies may be required in special cases.

At predose, if the vehicle contains an ingredient such as cremophor and Tween-80 that causes allergic reaction in animals, a pretreatment dosing regimen may be given to animals to minimize the chances for this reaction occur. At postdose, in the event that an adverse reaction is to occur, additional treatment may be followed as judged to be appropriate.

The volume of the radiolabeled dose formulation to be administered to each animal is calculated based on the body weights taken on the day of dosing. If a syringe is used for dosing, the actual amount of administered dose to each animal is determined by weighing the dosing syringe before and after dose administration. The dose apparatus is flushed with an appropriate amount of dose vehicle. If dose vehicle contains an ingredient that is toxic to animal at higher volumes (such as ethanol, cremophor), that is substituted with water, saline, buffer, etc.

In human ADME studies, generally, a fixed dose selected from a single or multiple ascending dose study containing certain amount of radioactivity (50–100 μCi) is prepared in a formulation currently used in the clinic and dosed to each human subject.

18.2.4 In-Life Studies in Animals and Humans and Sample Collection/Pooling

Plasma, urine, bile, and fecal samples are obtained following a single oral or IV administration of a radiolabeled drug to rats, mice, rabbits, dogs, monkeys, or humans. In general, collection times for blood samples are chosen based on the PK parameters of the drug, such as $t_{1/2}$, C_{max} (at least four time points, e.g., 1, 4, 12, 24 h, for metabolite profiling and multiple time points for PK analysis). Urine and feces are collected at specific intervals from zero hours to the end of the study, usually from 0 to 168 h.

18.2.4.1 Rat or Mouse   Blood samples (at least four time points for metabolite profiling) are obtained from male/female Sprague–Dawley rats and CD-1 mice (three rats or three mice per collection time) by terminal bleeding via cardiac puncture following carbon dioxide anesthesia. Urine (0–12, 12–24, and at 24 h intervals thereafter throughout the study period) and feces (24 h intervals throughout the study period) are collected from three rats or five mice on dry-ice. In a separate study, bile, urine, and feces (0–24 h) are collected from three BDC rats or five BDC mice for 0–24 h following oral or IV administration. For bile collection, a bile salt solution (18 mg/mL cholic acid and 1.1 mg/mL sodium bicarbonate in saline, pH 7.2) is infused via the duodenal cannula at 1 mL/h for BDC rats or 0.3 mL/h for BDC mice. For the PK study, blood is collected from three rats (200 μL) via a jugular vein cannula or from five mice per time point by terminal bleeding via cardiac puncture following carbon dioxide anesthesia at various time points such as 0, 0.25, 0.5, 1, 2, 4, 6, 8, 12, 24, 48, 72, 96, 120, 144, and 168 h postdose.

18.2.4.2 Rabbit, Dog, or Monkey   Blood (5–10 mL) at four or more time points is collected from the indwelling venous catheter from three male and/ or female cynomolgus monkeys, male/female beagle dogs, or three female New Zealand white rabbits. Urine (0–12, 12–24, 24 h interval thereafter throughout the study period) and feces (24 h intervals throughout the study period) are also collected on dry-ice. In a separate study, bile, urine, and feces are collected (0–8, 8–24, 24–48 h for 0–48 h) from three bile duct cannulated male animals. For bile collection, a bile salt solution (18 mg/mL cholic acid and 1.1 mg/mL sodium bicarbonate in saline, pII 7.2) is administered via a distal (flushing) catheter at 1 mL/kg/h for BDC animals. In addition, another series of blood samples (2 mL) are collected at various time points such as

0, 0.25, 0.5, 1, 2, 4, 6, 8, 12, 24, 48, 72, 96, 120, 144, and 168 h postdose for PK analysis.

18.2.4.3 Human Blood (10 mL) for at least four time points is obtained by venipuncture from healthy male and female subjects following a single dose of 50–100 μCi ^{14}C-labeled drug. Urine (0–12, 12–24, and 24 h interval thereafter throughout the study) and feces (24 h intervals 0 h to the end of the study) are obtained. In addition, another series of blood samples (5 mL) are collected at various time points such as 0, 0.25, 0.5, 1, 2, 4, 6, 8, 12, 24, 48, 72, 96, 120, 144, and 168 h postdose for PK analysis. Bile is collected over 3–8 h postdose from, for example, 4–8 subjects if needed (Wang et al., 2006). An intravenous dose of cholecystokinin is used to stimulate gallbladder contraction at 7 h postdose in humans during bile collection.

The pH of the biological samples may need to be adjusted by adding acetic acid (2–5%, v/v), for example, to stabilize metabolites such as acylglucuronide metabolites after sample collection prior to sample freezing at −20°C. The blood is collected in tubes containing K_3EDTA and centrifuged within 30 min of collection to harvest plasma (10 min, 1300 × g at 4°C). Fecal homogenates are prepared by mixing feces with appropriate amounts of water or water:ethanol (50 : 50, v/v) following by homogenization, and aliquots of fecal homogenates are analyzed for total radioactivity determination.

Pooled urine (0 h to end of study) and pooled fecal homogenate (0 h to end of study) samples are prepared by mixing, respectively, urine (2% by weight) and fecal homogenates (1% by weight) obtained from all subjects for each collection interval. A pooled plasma sample is prepared by combining equal volumes of plasma (1–2 mL) from all subjects for each collection time point. Pooled plasma, urine, and feces are analyzed for radioactivity distribution (metabolite profiling) and metabolite identification by high performance liquid chromatography (HPLC) and liquid chromatography/mass spectrometry (LC/MS).

18.3 SAMPLE ANALYSIS

18.3.1 Sample Preparation: Plasma, Urine, Bile, and Feces

In general, depending on the amount of radioactivity in sample, different approaches may be used in processing the sample. If sufficient radioactivity is present, the sample may be diluted before analysis or analyzed (HPLC and/or LC/MS) directly following centrifugation (for urine and bile only). For samples that contain limited amount of radioactivity, concentration of the sample after extraction or protein precipitation (in plasma) is needed. It is important to check recovery of radioactivity after concentration. The procedures used in the muraglitazar study (Wang et al., 2006; Zhang et al., 2006, 2007) are generalized and described below as an example.

18.3.1.1 Plasma   The plasma sample is treated with an organic solvent such as acetonitrile, methanol or methanol/acetonitrile (1 : 1) at a ratio of 3 volumes of solvent to 1 volume of plasma to precipitate proteins. Then the plasma sample is mixed by vortexing and/or sonicating for 5 min, and centrifuged at approximately $3000 \times g$ for 10 min. The supernatant is transferred into a clean tube. The remaining pellets are extracted again with 3–5 mL of the organic solvent. The supernatants are combined and the recovery of radioactivity is determined. In general, the radioactivity recovery of >90% is acceptable. The combined supernatant is dried and reconstituted in an appropriate solvent that is determined by considering the initial HPLC solvent composition and polarity of test compound and its metabolites. The reconstituted sample is mixed by vortexing and centrifuged at $3000 \times g$ for 5 min to remove any solid particles. Aliquots of the reconstituted sample are counted by LSC to determine the final recovery of radioactivity (>80% is desired).

18.3.1.2 Urine and Bile   For direct injection to HPLC, an aliquot of the urine sample (or a diluted sample, usually with acetonitrile to a final composition of ~20–30% acetonitrile, v/v) is centrifuged at $3000 \times g$ for 5 min to remove any solid particles. In the case that the concentration of radioactivity in urine is low, a subsample of urine may be concentrated either by passing through a C18 cartridge (e.g., Oasis™ cartridge, Waters) or by direct evaporation under a steam of nitrogen before HPLC and LC/MS analysis. The aliquots of the reconstituted sample are counted by LSC to determine the final recovery of radioactivity (>90% desired). The supernatant is subjected to HPLC analysis.

18.3.1.3 Feces   The amount of fecal homogenate needed for the desired level of radioactivity per injection is determined. The fecal homogenate is extracted with three volumes of an appropriate solvent (e.g., acetonitrile, acetonitrile/ methanol, acetonitrile/acidic acid). The fecal homogenate/solvent mixture is vortexed and sonicated and centrifuged. The extraction is repeated at least one more time to recover a desirable >85% of radioactivity. The sample is combined, dried, reconstituted in an appropriate solvent (e.g., 20–40% acetonitrile in water), vortexed and centrifuged at ~$3000 \times g$ for 5 min. The aliquots of the reconstituted sample are counted by LSC to determine the final recovery of radioactivity (>80% desired).

18.3.2 Radioactivity Determination

The radioactivity in urine, bile, plasma, HPLC fractions, and extracts is determined by mixing aliquots of the samples with a scintillation cocktail (e.g., Ecolite® cocktail) (5 or 15 mL) and counting with LSC (e.g., Packard 2250CA Tri-Carb Liquid Scintillation counter, Packard Instrument Company, Meriden, CT) for 5–10 min. Radioactivity in feces, blood, and tissues is usually determined by combustion of aliquots by an oxidizer followed by LSC. Another method to determine total radioactivity is to digest the aliquots of feces, blood, and tissues for 24–48 h at room temperature with Soluene-350 (Packard Instrument Co., Meriden, CT). The digested aliquots were then bleached with 20% benzoyl peroxide in toluene, neutralized with a mixture of saturated sodium pyruvate in methanol/glacial acetic acid, mixed with scintillation cocktail and counted using LSC (Krishna et al., 2002). Each sample is homogenized before radioanalysis. In the muraglitazar study, the radioactivity in plasma and excreta (urine, bile, and feces) was determined after combustion of the samples (Zhang et al., 2006). HPLC effluent was collected at 0.26 min intervals after sample injection into 96-deep-well Lumaplates® with a Gilson fraction collector (Gilson Medical Electronics, Middleton, WI). The effluent in the plates was dried with a Speed-Vac® (Savant, Holbrook, NY) and the plates were counted for 10 min per well with a TopCount® scintillation analyzer (Packard Instrument Company, Meriden, CT).

18.3.3 LC/MS/MS Quantification and Pharmacokinetic Analysis

The unchanged drug in plasma and urine samples from mass balance studies can be quantified using LC/MS technique. In muraglitazar study, plasma concentrations of the unchanged drug were determined by a validated protein precipitation and LC/MS/MS method (Wang et al., 2006). The internal standard was dissolved into 0.1% formic acid in acetonitrile, which also served as a protein precipitation reagent. Human plasma samples (0.1 mL) and the internal standard solution (0.3 mL) were added to a 96-well plate. The plate was vortexed for 1 min and centrifuged for 5 min, then the supernatant was directly injected into the LC/MS/MS. Chromatographic separation was achieved isocratically on a Phenomenox Luna C18 column (2 × 50 mm, 5 μ). The mobile phase contained 20% of 1 mM formic acid in water and 80% of 1 mM formic acid in acetonitrile. Detection was by positive ion electrospray tandem mass spectrometry on a Sciex API 4000. The standard curve, which ranged from 1 to 1000 ng/mL, was fitted to a $1/x$ weighted quadratic regression model. All plasma samples were analyzed within a total of four analytical runs. QC samples were analyzed along with the study samples to assess the accuracy and precision of the assay. The acceptance criteria established for the analysis of muraglitazar in plasma specified that the predicted concentrations of at least three fourths of the standards and two thirds of the QC samples be within ±15% of their individual nominal concentration values (±20% for the lowest concentration standard). In addition, at least one QC sample at each concentration had to fall within ±15% of its individual nominal concentration value. Values for the between-run precision and the within-run precision for analytical quality control samples were no greater than 0% and 11.1% coefficient of variation (CV), respectively, with deviations from the nominal concentrations of no more than ±5.2%.

18.3.3.1 *Pharmacokinetic Analysis* The plasma concentration versus time data for radioactivity and unchanged muraglitazar (Wang et al., 2006) were

analyzed by a noncompartmental method (Gibaldi and Perrier, 1982). The peak plasma concentration, C_{max}, and the time to reach peak concentration as the first occurrence, T_{max}, were recorded directly from experimental observations. The area under the plasma concentration versus time curve (AUC) was calculated by a combination of the trapezoidal methods. The AUC was calculated from time zero to the time, T, of last measurable concentration [AUC(0–T)]. The first-order rate constant of decline of radioactivity concentrations and unchanged muraglitazar, expressed as equivalents of muraglitazar, in the terminal phase of each plasma concentration versus time profile, K, was estimated by log-linear regression (using no weighting factor) of at least three data points that yielded a minimum mean square error. The absolute value of K was used to estimate the apparent terminal elimination half-life, $t_{1/2}$.

18.3.4 Metabolite Profiling

Various samples (plasma, urine, bile, fecal extracts) are analyzed by HPLC to determine metabolite profiles. In general, HPLC is performed on a system equipped with two pumps, an auto-injector, and a diode array detector. A reverse-phased HPLC column is normally used for metabolite separation. A gradient of two solvents, A and B, is usually used with Solvent A as aqueous and Solvent B as organic. For metabolite profiling of *in vivo* samples, usually a >60 min HPLC run is needed to ensure good separations of all metabolites. Recovery of radioactivity after HPLC column needs to be checked to ensure that all radioactive material is eluted from the column. For detection of radioactivity, HPLC fractions may be collected and radioactivity is determined. To ensure the accuracy of the results, it is important to check the recovery of radioactivity against the known amount of radioactivity injected. Loss of radioactivity could be due to retaining in the HPLC column, injector misfunction, volatile metabolites, or quenching by matrices. The limit of quantification is ~15 disintegrations per minute (DPM) for microplate scintillation counting (MSC). Alternatively, HPLC is coupled to a flow radiochemical detector (e.g., β-RAM, IN/US systems, Tampa, FL). A liquid cell using β-RAM instrument with 1 mL/min HPLC flow and 2 mL/min cocktail flow (INUS 2:1) with a 500–600 μL cell is recommended (for a balance of peak resolution and counting time). It is necessary to calibrate the cocktail pump and check counting efficiency of the β-RAM. Under these conditions, the counting efficiency is ~80% for ^{14}C and is ~30% for ^{3}H, and the quantitation limit is ~700 counts per minute (CPM) per HPLC peak for ^{14}C. The decision on whether to use online or off-line radioactivity detection should be based on the amount of radioactivity in a sample and the desired detection limit of analytes. For example, in order to detect a metabolite that accounts for 1% of total radioactivity in a sample (^{14}C), it requires approximately 1500 DPM per injection for HPLC–MSC and ~70,000 CPM for HPLC–β-RAM. Biotransformation profiles are prepared by plotting the CPM or DPM values against time-after-injection. Radioactive peaks in the biotransformation profiles are reported as a percentage of the total radioactivity collected during the entire HPLC run. The relative distributions of radioactive metabolites in urine, bile, and feces are calculated from the percent of dose excreted in the matrix (based on DPM by scintillation counting) multiplied by the percent of distribution of metabolites in chromatograms of the matrix (based on CPM counting by Top Count).

18.3.5   Metabolite Identification

The structural identification of metabolites can be carried out at multiple levels. The goal of initial identification of metabolites is often accomplished through LC/UV, LC/MS, and LC/MS/MS analyses to determine the biotransformation pathways involved in the clearance of the compound. Determination of the nature of the metabolites such as oxygenation (hydroxylation or oxidation of a heteroatom), dioxygenation, dealkylation, reduction, and conjugation (glucuronide, sulfate, glutathione, etc.) is often sufficient. The identification of metabolites from biological matrices beyond this initial level is often challenging. Due to low concentrations and interference from endogenous components, identification of metabolites with unusual structures presents additional challenges. Neutral loss and product ion scans are used for real-time, data-dependent acquisition of full MS/MS data of expected metabolites to improve the selectivity and sensitivity.

Following HPLC separation, samples are analyzed by a mass spectrometer with an ESI source such as an LCQ, LTQ, (ThermoFinnigan, San Jose, CA), or Q-TOF Ultima mass spectrometer (Micromass, Beverly, MA). Samples are analyzed in the positive or negative ion mode. The capillary temperature used for the LCQ and LTQ, the desolvation temperature used on the Q-TOF, the nitrogen gas flow rate, spray and cone voltages are adjusted to give maximum sensitivity for the parent compound.

In the muraglitazar study, we have employed accurate mass spectrometry for the analysis of the metabolite profiles in biological matrices such as human feces (Zhang et al., 2007). A Q-TOF Ultima mass spectrometer was used. The Q-TOF was tuned to 18,000 resolution at half peak height using an insulin tuning solution, and was calibrated up to 1500 Da using a polyalanine calibration solution. For accurate mass measurement, the m/z 556.2771 of an infused 20 ng/μL leucine enkephalin solution were used as lock mass. The experimentally obtained masses were within 5 mDa compared to their respective calculated values. The accurate mass measurements not only provided cleaner selective ion chromatograms (ion chromatograms extracted with a mass accuracy to the second decimal places) of metabolites in feces but also provided spectra for easy identification of the molecular ions, which greatly enhanced the metabolite identification. The mass defect filtering (MDF) methodology (Zhang et al., 2003) can be used to aid in the identification molecular ions of metabolites with unusual structures or present at low concentrations as well as for metabolites formed from expected biotransformations.

LC/MS/MS methods will often leave significant ambiguity in the exact structural identification of metabolites. The next level of identification involves detailed structural elucidation of the metabolites by NMR and often requires quantities of materials not available from the *in vivo* samples. In the muraglitazar study, we have demonstrated the use of microbial bioreactors to produce large quantities of major metabolites for isolation and identification (Zhang et al., 2006). The hydroxylation site in metabolites can be determined by direct organic synthesis from the tentatively assignment of partially identified metabolites. The synthetic compound should match the human metabolite both by HPLC retention time and LC/MS analysis. In the muraglitazar study, the synthetic materials and microbial isolates were analyzed on a Jeol ECL-500 MHz spectrometer or a Bruker AVANCE 600 MHz system equipped with a 5 mm *Z* gradient probe, or a 3 mm Nalorac probe. 1D ^{1}H, 2D COSY (correlation spectroscopy), 2D TOCSY (total correlation spectroscopy), HMQC (one bond carbon–proton correlation), HMBC (long range carbon-proton correlation), and edited DEPT (distortion-less enhancement by polarization transfer) experiments were performed. All chemical shifts are reported in ppm relative to tetramethylsilane in CD_3CN. For reference, a complete proton and carbon peak assignment of the NMR spectra was made on the parent drug (Zhang et al., 2006).

18.3.6 Metabolite Isolation from *In vivo* Samples and Generation in Bioreactors

For detailed structural elucidation, a large amounts of metabolites (teens of μg–mg) are needed for NMR analysis. These metabolites can be isolated from urine or bile if they present at sufficient concentrations by extraction and multiple chromatographic separations steps. One good example was illustrated by Chando et al. (1998). Very often, microbes also produce the same metabolites as animals and humans and these metabolites can be isolated from microbial bioreactors (Zhang et al., 2006). The following section describes a general procedure of a microbial bioreactor used in the muraglitazar study (Zhang et al., 2007).

The bacterial strain was started from a 1 mL vial (stored in liquid nitrogen) in 100 mL medium grown for 3 days, then 10 mL of the culture from this flask was used to inoculate 100 mL medium. The filamentous fungus were grown from a 1 mL spore suspension in 100 mL of medium. A drug (5–30 mg slurried in 1 mL methanol) was added to each flask after 24 h growth of the second stage bacterial culture or first stage fungal culture. The incubations were continued for 72 h, then the reactions were quenched with 100 mL of acetonitrile. After storage for several hours at room temperature and 4 days at 4°C, cells were removed by centrifugation at 3000 × *g* for 10 min. A 50 mL portion of each supernatant was extracted twice with ethyl acetate (50 mL for each extraction). The combined ethyl acetate extracts were evaporated to near dryness under a stream of nitrogen. The residues were each reconstituted in 2 mL acetonitrile/water (3:7, v/v). The extracts were analyzed by LC/MS. For metabolite isolation, fractions were collected from multiple injections (100 μL) onto a HPLC column. The microbial isolates were analyzed by LC/MS and NMR.

Alternatively, large scale *in vitro* incubations in microsomes, hepatocytes, or expressed enzymes can be used as bioreactors to generate metabolites.

18.4 DATA PRESENTATION USING METABOLISM OF [^{14}C]MURAGLITAZAR AS AN EXAMPLE

18.4.1 Pharmacokinetic Results and Excretion of Radioactivity

The main objectives of a mass balance study are to determine the amount of the radioactive drug administered in excreta and to show that the administered

TABLE 18.3 Mass balance results in rats, mice, dogs, monkeys, and humans following a single oral administration of [^{14}C]muraglitazar.

Species	No. subjects	Dose[a]	Matrix	Sampling time (h)	Recovery (% of dose)	Total recovery (% of dose)
Rat	3	11 mg/kg	Urine	0–168	1.0	
			Feces	0–168	87.7	88.7
	3 BDC	10 mg/kg	Urine	0–24	0.98	
			Bile	0–24	65.1	
			Feces	0–24	0.55	66.6
Mouse[b]	5 BDC	1 mg/kg	Urine	0–24	8.3	
			Bile	0–24	75	
			Feces	0–24	8.2	91.5
	5 BDC	40 mg/kg	Urine	0–24	16.1	
			Bile	0–24	66	
			Feces	0–24	12	94.1
Dog	3	1.9 mg/kg	Urine	0–168	0.5	
			Feces	0–168	89.0	89.5
Monkey	3	2 mg/kg	Urine	0–168	3.7	
			Feces	0–168	80.4	84.1
	3 BDC	5 mg/kg	Urine	0–48	5.0	
			Bile	0–48	35.7	
			Feces	0–48	15.0	55.7
Human[b]	6	10 mg	Urine	0–240	2.6	
			Feces	0–240	61.6	64.2
	4[c]	20 mg	Urine	0–240	3.7	
			Bile	3–8	39.9	
			Feces	0–240	50.7	94.3

[a]Radiospecific activity of [^{14}C]muraglitazar used was: 8.4 and 5 μCi/mg for rats, 11.2 μCi/mg for mice, 2.8 μCi/mg for dogs, 6.6 and 6 μCi/mg for monkeys, and 10 and 5 μCi/mg for humans.
[b]The human and mouse data with bile collection were presented previously (Wang et al., 2006; Li et al., 2006) and were included here to allow a complete species comparison.
[c]Bile was collected for 3–8 h postdose (Wang et al., 2006).

dose is readily excreted from the body. The amount of radioactivity is usually expressed as the percent of the dose in each matrix (urine, feces, bile, etc). A hundred percent of recovery of the administered radioactive dose is usually not possible. Both in animal and human mass balance studies, a recovery of >85% of the administered radioactive dose would be best although there is no specific number requirement from regulatory agencies. A long elimination half-life of the drug and radioactivity binding to tissues are some of the causes for low recoveries. In addition, the radioactive drug related compounds may be lost through expiration if the drug is volatile.

The recovery values of radioactive doses in urine, bile, and feces of rats, monkeys, and humans administered [^{14}C]muraglitazar are shown in Table 18.3 as an example. Urinary excretion of radioactivity was low (1.0, 0.5, 3.7, and 2.6% of dose for intact rats, dogs, monkeys, and humans, respectively) and the primary route of excretion was feces (87.7, 89.0, 80.4, and 61.6% of dose for intact rats, dogs, monkeys, and humans, respectively). The bile contained the majority of radioactivity from rats, mice, monkeys, and humans, representing the major elimination pathway of [^{14}C]muraglitazar in these four species (Table 18.3). Table 18.4 shows the total concentrations of muraglitazar and its metabolites excreted in urine and bile in rats, dogs, monkeys, and humans.

TABLE 18.4   Approximate concentrations of [^{14}C]muraglitazar and metabolites in urine and bile from rats, mice, dogs, monkeys, and humans following oral administration.

	Rat	Dog	Monkey	Human
Urine				
Dose (mg/kg or mg/subject)	11	2	1.9	10
Body mass (kg)[a]	0.25	10	5	70
Urine collection interval (h)[b]	0–12	0–24	0–12	0–8
Concentration in urine (μg eq/mL)[c]	1.60	0.20	0.57	0.17
% Dose in urine of interval collection	0.51	0.49	1.7	1.3
	Rat	Mouse	Monkey	Human
Bile				
Dose (mg/kg or mg/subject)	10	1, 40	5	20
Body mass (kg)[a]	0.25	0.02	5	70
Dose/subject (mg)	2.5	0.02, 0.8	25	20
Bile Flow (mL/h)[a]	22.5	2	125	350
Bile collection interval (h)	0–24	0–24	0–48	3–8
Concentration in bile (mg eq/mL)[d]	0.289	0.375, 13.2	0.007	0.068
% Dose in bile of interval collection	65.1	75, 66	36	40

[a]Literature values from Davies and Morris (1993).
[b]Closest matched intervals; highest concentrations were excreted in these intervals.
[c]Total concentration of muraglitazar plus all metabolites (muraglitazar equivalents).
[d]Total concentration of muraglitazar plus all metabolites (muraglitazar equivalents). Calculated based on theoretical bile excretion volume in 0–24 or 0–48 h for animals and 0–8 h for humans. The bile concentration in rats was an assumption only since there is no gallbladder in the rat.

In addition to determine the major metabolic pathways of a drug, the radioactivity and the unchanged drug concentration in blood, plasma, urine, and feces obtained from mass balance studies can provide information about the pharmacokinetic behavior of a drug. As summarized by Beumer et al. (2006), the pharmacokinetic parameters for unchanged drug can be determined from the plasma data (obtained from e.g., LC/MS analysis). These parameters include C_{max} (maximum drug concentration), T_{max} (time to reach maximum concentration), AUC (area under the plasma concentration–time curve), V_d (volume of distribution) and $t_{1/2}$ (elimination half-life). These results should be comparable to the results from other pharmocokinetic studies (for both human and animal). These parameters can also be calculated using the radioactivity data. Elimination half-life from plasma will be used to determine if the drug metabolites are accumulated upon multiple dosing. The ratio of the AUC of the unchanged drug to total radioactivity AUC will provide information about how much the metabolites are important (Slatter et al., 2000; Walle et al., 1995). Pharmacokinetic parameters of urine together with those of plasma provide data to determine the renal clearance of the drug.

The muraglitazar pharmacokinetic parameters for the unchanged drug and radioactivity are presented in Table 18.5. The animal and human plasma concentration time profiles for radioactivity and muraglitazar are presented in Fig. 18.1. Following oral administration, the radioactivity and muraglitazar concentrations reached a maximum at 0.5–1 h in rats, dogs, monkeys, and humans.

TABLE 18.5   Mean pharmacokinetic parameters of muraglitazar in rats, dogs, monkeys, and humans following oral administration of [^{14}C]muraglitazar.

Species (n)	Dose	Analyte	C_{max}[a] (μg/mL)	T_{max} (h)	AUC$_{0-t}$[a] (μg h/mL)	$t_{1/2}$[b] (h)
Rat (3)	10 mg/kg	Muraglitazar	12.8 ± 3.0	0.5^c	101 ± 20	10.3 ± 0.4
		Radioactivity	13.5 ± 2.6	0.5^c	105 ± 19	3.8 ± 1.8
Dog (3)	2 mg/kg	Muraglitazar	0.8 ± 0.4	0.5	1.3 ± 0.4	5.7 ± 4.6
		Radioactivity	1.0 ± 0.4	1	1.1 ± 0.4	—[d]
Monkey (3)	1.9 mg/kg	Muraglitazar	6.9 ± 2.5	1^e	41 ± 9	18.0 ± 2.1
		Radioactivity	7.1 ± 2.1	1^e	71 ± 8.3	23.5 ± 11.7
Human (6)	10 mg	Muraglitazar	1.8 ± 0.4	1	12 ± 2.8	35.1 ± 9.7
		Radioactivity	1.4 ± 0.4	1	9.0 ± 3.1	19.8 ± 8.5

[a]Units for C_{max}: muraglitazar (μg/mL) and radioactivity (μg-equivalents/mL); units for AUC: muraglitazar (μg h/mL) and radioactivity (μg-eq h/mL). Last detectable time points for respective parent compound and the radioactivity were 96 and 24 h for rat, 12 and 4 h in dog, 144 and 24 h for monkey, and 192 and 72 h for human.

[b]The $t_{1/2}$ values for radioactivity were lower than the parent for rats, dogs, and monkeys because the detection limit for radioactivity was approximately 15–30 times higher than LC/MS-based parent determination.

[c]Second peaks were observed at 8 h in rats.

[d]Not determined due to limited time points.

[e]Second peaks were observed for at 6 h in monkeys.

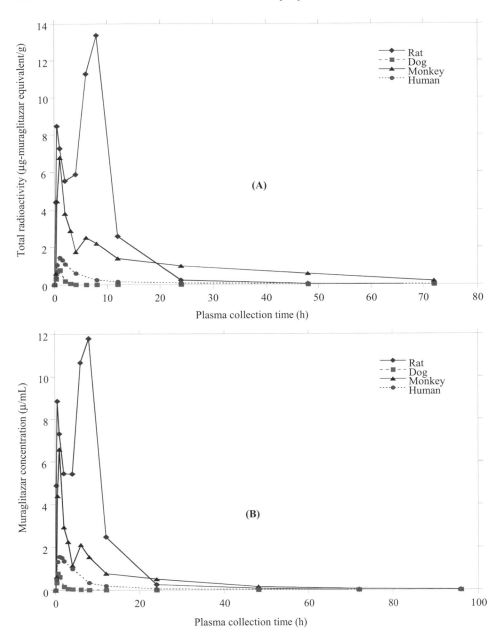

FIGURE 18.1 Total plasma radioactivity and muraglitazar concentration vs. time profiles following a single administration of [^{14}C]muraglitazar in rats (11 mg/kg), dogs (1.9 mg/kg), monkeys (2 mg/kg), and humans (10 mg).

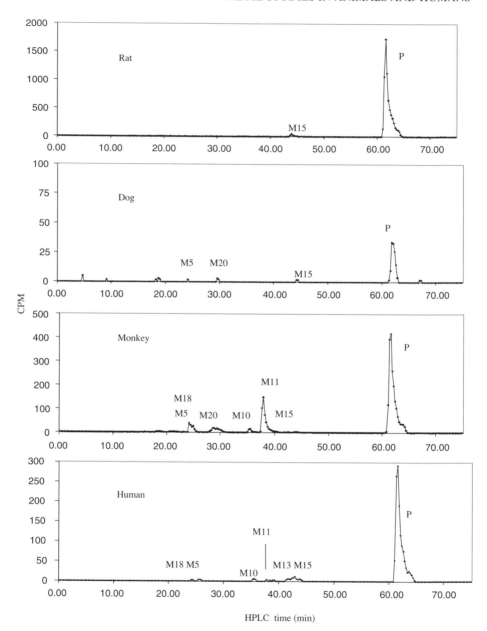

FIGURE 18.2 Radiochromatograms of pooled plasma samples (1 h) after an oral administration of [^{14}C]muraglitazar to rats, dogs, monkeys, and humans (P = parent)

18.4.2   Metabolite Profiles in Plasma, Urine, Bile, and Feces

In order to determine the metabolic fate of a radiolabeled drug, radioactivity in plasma, urine bile and feces is profiled for metabolite distribution. Metabolite profiling in muraglitazar study is described below.

18.4.2.1   Plasma
The extraction recovery of radioactivity from all plasma samples was greater than 98%. The HPLC radiochromatographic profiles of 1-h plasma from rat, dog, monkey, and human are shown in Fig. 18.2. The relative distribution of radioactive metabolites in pooled plasma (rat, monkey and human) is summarized in Table 18.6. The metabolic profiles of plasma were qualitatively similar at different time points within species and across species. Muraglitazar was the major component at 1, 4, and 12 h, accounting for more than 90% of plasma radioactivity at 1 h in all four species. All metabolites (M1, M2, M5, M8, M8a, M10–13, and M18) observed in human plasma were also present in rats, dogs, or monkeys.

TABLE 18.6   Distribution of radioactive metabolites in pooled plasma after an oral dose of [^{14}C]muraglitazar to rats, monkeys, and humans.

Meta	Distribution of Metabolites (% of sample)[a]								
	Rat			Monkey			Human		
	1 h	4 h	12 h[b]	1 h	4 h[b]	12 h	1 h	4 h[b]	12 h
M1	ND[c]	ND	0.05	0.3	0.3	ND	ND	Trace	0.2
M2	ND	ND	0.05	ND	0.1	ND	ND	Trace	NA
M5	ND	ND	Trace[d]	1.5	2.9	2.7	ND	0.3	1.0
M6	ND	ND	Trace	ND	Trace	ND	ND	ND	ND
M8	ND	ND	Trace	1.0	1.5	1.0	ND	Trace	ND
M8a	ND	ND	Trace	0.9	1.3	1.0	ND	Trace	ND
M10	ND	ND	0.2	0.9	1.6	0.4	ND	1.0	1.0
M11	ND	ND	Trace	8.0	16.2	21.4	ND	0.3	ND
M12	ND	ND	Trace	ND	Trace	0.5	ND	0.2	ND
M13	ND	ND	0.05	0.1	0.2	0.1	2.5	4.2	2.5
M15	2.3	2.1	3.5	0.1	0.3	ND	0.2	0.5	0.4
M18	ND	ND	Trace	ND	3.1	4.0	ND	0.3	ND
M20	ND	ND	0.4	0.1	0.2	0.4	ND	ND	ND
P	95.9	95.3	92.2	85.1	67.6	65.9	97.1	92.2	94.0
Other[e]	6.8	10.5	1.6	5.5	1.1	0.6	1.3	0.6	0.2
Total	98.2	97.4	96.6	98.1	92.9	97.1	99.8	99.0	98.1

[a]Radioactive peaks are reported as a percentage of the total radioactivity eluted from the column. Column recoveries were >98%.
[b]LC/MS analysis was performed on this plasma sample.
[c]ND = not detected. Criteria are no discernible radioactive peak, <0.05% radioactivity and not detected by LC/MS.
[d]Trace = <0.05% radioactivity and detected by LC/MS.
[e]Other = not identified peaks.

18.4.2.2 Urine The HPLC radiochromatographic profiles of pooled urine from rats, dogs, monkeys, and humans were also determined (Table 18.7).

18.4.2.3 Feces and Bile Excretion in feces represented more than 95% of the recovered radioactive dose in rats, dogs, monkeys, and humans. Muraglitazar was one of the abundant components present in feces from all species. The HPLC-radiochromatograms of pooled bile samples from rats, monkeys, and humans are shown in Fig. 18.3. The relative distribution of metabolites in urine and feces (as % of dose) is shown in Table 18.7.

TABLE 18.7 Relative distribution (percent of dose) of radioactive metabolites in pooled urine and feces after an oral dose of [^{14}C]muraglitazar to intact rats, dogs, monkeys, and humans.

	Metabolite distribution (% of dose) in urine and feces[b]							
	Rat		Dog		Monkey		Human	
Metab[a]	Urine	Feces	Urine	Feces	Urine	Feces	Urine	Feces
M1	0.22	1.75	0.44	0.89	1.04	2.81	0.06	0.80
M2	0.05	9.65	ND	ND	0.08	1.29	ND	0.56
M3	Trace[c]	ND	ND	ND	ND	2.41	ND	1.36
M4	ND[d]	ND	ND	0.34	ND	0.16	ND	2.83
M5	0.01	2.63	ND	0.38	0.26	1.61	0.16	11.45
M6	0.01	ND	ND	ND	0.01	3.22	ND	1.48
M7	ND	0.43	ND	0.18	ND	0.32	ND	2.21
M8	ND	ND	ND	0.89	0.13	6.43	ND	0.56
M8a	Trace	4.39	ND	0.18	0.12	ND	0.03	0.56
M9	ND	2.63	ND	0.89	ND	0.56	ND	2.46
M10	0.03	0.88	ND	1.10	0.01	7.24	0.03	8.20
M11	Trace	0.88	ND	1.78	0.59	25.73	0.07	5.92
M12	Trace	ND	ND	0.10	0.04	1.61	ND	0.13
M13	Trace	0.88	ND	ND	0.02	0.80	1.00	0.06
M14	Trace	0.88	ND	0.85	0.03	0.56	ND	1.60
M15	0.12	34.20	ND	24.92	0.08	7.24	ND	3.33
M16	ND	0.88	ND	0.18	ND	0.16	ND	3.21
M18	Trace	ND	ND	ND	Trace	ND	0.20	ND
M20	Trace	ND	ND	ND	Trace	ND	ND	ND
M21	0.21	0.88	0.01	ND	0.33	0.40	0.04	0.40
Parent	0.02	13.16	0.01	52.51	0.11	7.24	0.29	9.62
Others[e]	0.3	13.6	0.02	NA	0.79	4.02	0.58	2.40
Total	0.97	74.1	0.48	85.2	3.63	73.49	2.46	59.20
% Excretion	1.0	87.7	0.5	89.0	3.7	80.4	2.56	61.59

[a]Metabolites are assigned based on retention times of radioactive peaks and LC/MS analysis.
[b]Radioactive peaks are reported as a percentage of dose (% of the sample multiplied by % excretion).
[c]Trace = <0.01% radioactivity and detected by LC/MS.
[d]ND = not detected. Criteria are no discernable radioactive peak, <0.01% radioactivity and not detected by LC/MS.
[e]Uncharacterized peaks.

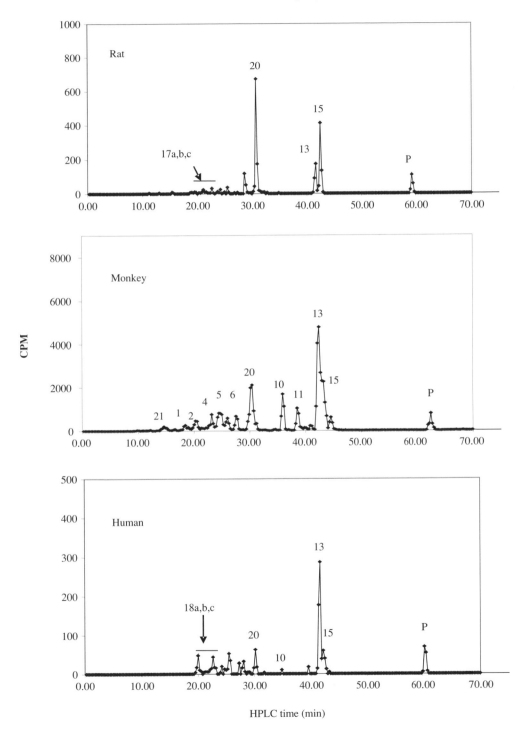

FIGURE 18.3   Radiochromatograms of pooled bile samples after oral administration of [¹⁴C]muraglitazar to rats, monkeys, and humans (1, 2, 3 . . . p = M1, M2, M3 . . . .parent)

18.4.2.4 Summary of Metabolite Profiling Parent compound was the major circulating, drug-related component in rats, dogs, monkeys, and humans following oral administration of [^{14}C]muraglitazar. Fecal excretion of [^{14}C] muraglitazar and its metabolites was the major elimination pathway (>95% of the recovered dose), principally via biliary elimination. Urinary excretion represented a minor elimination pathway (<5% of the recovered dose). Biotransformation profiles of [^{14}C]muraglitazar were qualitatively similar in rats, dogs, monkeys, and humans. Major metabolic pathways of muraglitazar were glucuronidation, alphatic/aryl hydroxylation, O-dealkylation, and a combination of O-demethylation and hydroxylation, and oxazole-ring opening. In addition to those metabolites (M1–M21) identified in other species (rats, dogs, monkeys, and humans), unique metabolites identified in mice included the taurine conjugates of muraglitazar and its metabolites formed from O-demethylation, hydroxylation and dihydroxylation, and glutathione conjugates (Li et al., 2006).

18.4.3 Metabolite Identification by LC/MS/MS

The metabolites appeared in the radiochromatograms of various matrices (plasma, urine, etc.) were analyzed by mass spectrometry. Comparison of the fragment ions of metabolites obtained by LC/MS/MS analysis to the ones from the parent compound provide structural information of the metabolite. Typical mass increases such as +16, +32, +176, +80, +42, +14, −14, −60, and −2 indicate mono-hydroxylation, dihydroxylation, glucuronidation, sulfation, acetylation, methylation, demethylation, deacetylation, and dehydration, respectively.

Following describes metabolite identification process in the muraglitazar study. Fig. 18.4 shows the ion chromatograms of total mass spectral signals versus HPLC run time after mass defect filtering treatment of LC/accurate mass analysis of human, monkey, and rat feces. The relative intensity of the total ion peak of each metabolite and the parent resemble those peaks in the radiochromatograms of these samples. Full scan mass spectra and MS/MS spectra were derived from these ion chromatograms for identification of metabolites. MDF-treated total ion chromatograms were specially useful to identify molecular ions of each metabolite due to the effective removal of endogenous interfering ions (Zhang et al., 2003).

Table 18.8 shows the proposed structures for the muraglitazar metabolites identified in rats, mice, dogs, monkeys, and humans following oral administration of [^{14}C]muraglitazar. Most of the metabolite identification was done by LC/MS analysis. Some metabolite standards were also synthesized and matched the metabolites in human feces by retention times and mass fragmentation patterns. Detailed structural identification of muraglitazar metabolites have been presented previously (Zhang et al., 2007). A typical metabolite identification is described as follows.

Metabolite M15: M15 showed a molecular ion [M + H]$^+$ at *m/z* 503 and major fragment ions at *m/z* 186 and 292 in the LC/MS analysis, consistent with

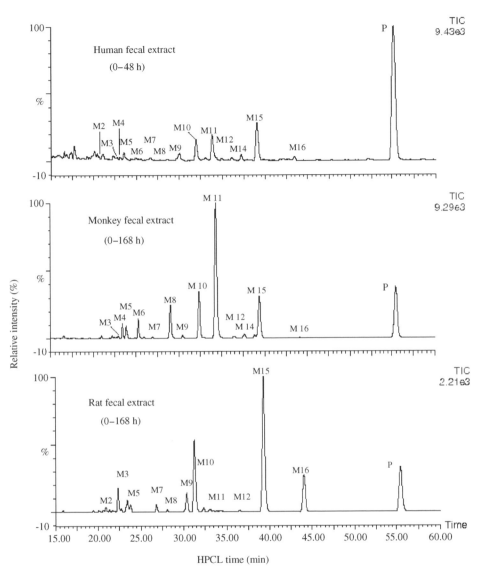

FIGURE 18.4 Comparative ion chromatograms of fecal extracts of humans (0–48 h), monkeys (0–168 h), and rats (0–168 h) following oral doses of [^{14}C]muraglitazar (P = parent) after mass defect filtering analysis of a high resolution spectral set.

O-demethylation. ^{1}H and ^{13}C NMR analyses showed that M15 lacked the OCH$_3$ group (at 3.8 ppm for ^{1}H and 55 ppm for ^{13}C) observed for the parent muraglitazar. The only change was the disappearance of the methoxy resonance of position 23. All other signals were in the same range as in the parent molecule, including the *ortho-* and *meta-* positions in the corresponding aromatic ring. This metabolite was assigned as *O*-demethyl muraglitazar.

TABLE 18.8   Structures of muraglitazar and its metabolites (M1, M2, M4, M4, M5).

Drug or metabolite	Proposed structure[a]	Source
Muraglitazar		Rat: Plasma, urine, feces, bile Dog: Plasma, urine, feces Monkey: Plasma, urine, feces, bile Human: Plasma, urine, feces, bile
M1 (O-Dealkyl muraglitazar)		Rat: Plasma, urine, feces Dog: Plasma, urine, feces Monkey: Plasma, urine, feces, bile Human: Plasma, urine, feces
M2 (8-Hydroxy-O-demethyl muraglitazar)		Rat: Plasma, urine, feces, bile Dog: ND[b] Monkey: Plasma, urine, feces, bile Human: Plasma, urine, feces, bile
M3 (8,12-Dihydroxy muraglitazar)		Rat: Urine Dog: ND Monkey: Feces, bile Human: Plasma, urine, feces
M4 (9,12-Dihydroxy muraglitazar)		Rat: ND Dog: Feces Monkey: Feces, bile Human: Plasma, urine, feces
M5 (12-Hydroxy-O-demethyl muraglitazar)		Rat: Plasma, urine, feces, bile Dog: Plasma, urine, feces Monkey: Plasma, urine, feces, bile Human: Plasma, urine, feces, bile

[a]Structures were proposed based on MS data and comparison with synthetic metabolites.
[b]ND = not detected.

In summary, for comparative biotransformation profiling, both quantitative and qualitative metabolite profiles are important to understand the relative abundance and identities of metabolites in various matrices from different species. LC/MS and LC/MS/MS methods are complementary to quantitative profiling using radioactivity detection when the existence of a metabolite needs to be qualified. MDF-treated ion chromatograms may have provided the first useful LC/MS method for estimation of amounts of metabolites when metabolite standards are not available to develop a quantification method.

18.5   CONCLUSIONS AND PATH FORWARD

Radioactivity profiling in plasma, urine, bile, and feces of appropriate animals and human subjects from mass balance studies enable a quantitative evaluation of the absorption, tissue distribution, and excretion of a drug candidate. The data can be supplemented with tissue distribution information from whole-body analysis and the further defining the sites of absorption and excretion in surgically prepared animals.

Identification of major elimination pathway will help design renal or hepatic impairment clinical study, and identification of major metabolic clearance pathways will help design reaction-phenotyping of major metabolic pathways for potential drug–drug interaction studies. For major circulating metabolites, pharmacological and toxicological activities should be tested and they may need to be monitored in clinical and toxicological studies. Selection of animal species for long-term toxicological studies may need to be reevaluated if the metabolite exposures or major metabolic pathways are significantly different. The ADME study results may help investigate the metabolite and toxicity correlation in animals and humans.

Recently, accelerated mass spectrometry (AMS) have been used in mass balance studies in animals and humans (Garner et al., 2002; Mauthe et al., 1998; Rickert et al., 2005). AMS only requires a small amount of radioactivity (nCi) with a normal dose (not considered as a radioactive study in some countries). However, this technique has limited applications in metabolite identification by connecting LC/MS/MS analysis and radioactivity detection.

ACKNOWLEDGMENTS

We would like to thank Lifei Wang, Mingshe Zhu and W. Griff Humphreys for helpful discussions.

APPENDIX A

Rat Tissue Distribution and Dosimetry Calculation

The objectives of rat tissue distribution studies are to (1) obtain radioactivity distribution in rat tissues after single doses of ^{14}C or ^{3}H labeled compounds, (2) obtain the pharmacokinetic parameter of radioactivity (^{14}C or ^{3}H) in rat major tissues or organs, and (3) calculate dosimetry for human ADME study. Animals used for this study are pigmented male Long–Evan rats, ~250 g body weight, 6–7 weeks. The numbers of rats used for a tissue distribution studies should be based on the pervious PK studies. The total study time should cover at least three to five the plasma half-life ($t_{1/2}$) of the compound. In the traditional method, three rats/time point and 21–30 rats/study will be used (such as at 0.5, 1, 4, 8, 24, 48, 72, 96, 124, and 168 h). In quantitative whole body autoradiography (QWBA) method, 1 rat/time point and 5–6 rats/study will be used (such as at 1, 12, 24, 48 72, and 168 h).

Radioactive dose level should produce no toxicity. The normal radioactivity dose is 15–30 μCi/rat for ^{14}C, and 150–200 μCi/rat for ^{3}H. The route of administration should be that proposed for clinical use. The dose and formulation should be based on pervious toxicity studies.

Rats are sacrificed with an overdose of halothane anesthesia or under CO_2, and following tissues are collected and rinsed with water: adrenal glands, bladder (urinary), blood, bone (femur), bone marrow (from femur), brain, carcass, (residual), cecum, cecum contents/wash, eyes (both), heart, kidneys, large intestine, large intestine contents/wash, liver, lungs, plasma, muscle (pectoral), muscle (thigh), skin (pigmented), small intestine, small intestine contents/wash, spleen, stomach, stomach contents/wash, testes, and thyroid/parathyroid. Radioactivity in all tissue samples are determined by sample oxidation in an oxidizer (Packard model 306). Samples are counted for up to 10 min in a scintillation counter (Packard model 2000CA).

For an QWBA method, the carcasses are immediately frozen in a hexane/dry ice bath for approximately 5 min. The carcasses are drained, blotted dry, and stored at approximately −70°C for at least 2 h. One frozen carcass/time point is embedded in chilled 1.5–3% of carboxymethylcellulose and frozen into a block. Carcasses are sectioned by using a cryomicrotome (such as Leica CM 3600) at a thickness (30–40 μm, and temperature is maintained at approximately −20°C). Appropriate sections are collected on adhesive tape. All major tissues, organs, and biological fluids are represented on five to six sections collected in the sagittal plane. Tissue sections are placed on a holding frame and stand and dried in the chamber of cryomicrotome maintained at approximately −20°C. After freeze-drying, tissue sections are cut away from the holding frame and mounted on a poster board. ^{14}C or ^{3}H-labeled sections are exposed to imaging plates for an appropriate time period (^{14}C: 12–20 h; ^{3}H: 1–2 weeks). WBA images from phosphor imaging plates are acquired (such as by using the Fuji FLA 3000). Tissues, organs, and fluids are sampled, and radioactivity concentrations in tissues are interpolated from the standard curve as nanocuries per gram by using imaging analysis software (such as MCID, Version 4.0, Imaging Research Inc). The tissues that can be analyzed by using the QWBA method are extended from the list for traditional method, including subfraction of a tissue.

For data presentation, tissue radioactivity levels are expressed as nanogram-equivalents per gram tissue (Table 18.A1), and tissue/plasma ratios. The maximum concentration (C_{max}) and the time to reach maximum concentration (T_{max}) are obtained by visual inspection of the raw data. Pharmacokinetic parameters, included half-life ($t_{1/2}$), area under the concentration–time curve

TABLE 18.A1 Concentrations of radioactivity in plasma, blood, and tissues at 1 h following a single oral dose of x mg/kg dose of radiolabeled test article in male Long–Evans rats.

	Nanogram equivalents of drug X/g				
	Animal number				
Tissues	R1	R2	R3	Mean	SD
Adrenal gland					
Bile					
Blood					
Bone					
Bone marrow					
Cecum					
Cecum contents					
Cerebellum					
Cerebrum					
Cerebrospinal fluid					
Diaphragm					
Epididymis					
Esophageal contents					
Esophageal mucosa					
Esophagus					
Exorbital lacrimal gland					
Eye					
Eye (lens)					
Fat (abdominal)					
Fat (brown)					

BLQ Below the limit of quantitation.
SD Standard deviation.

from time 0 to the last time point ($AUC_{0–t}$), and area under the concentration–time curve from 0 to infinity ($AUC_{0–\infty}$), are calculated by using kinetica and presented in Table 18.A2.

For dosimetry calculations, the concentrations of total radioactivity in rat tissues from a tissue distribution study are used as the basis for calculating the estimated radiation exposure in humans following a 100 μCi oral dose of [^{14}C] test article. Pharmacokinetic parameters generated by WinNonLin are transferred into Excel Version 8.0e (Microsoft Corporation) for calculation of human dosimetry parameters.

The dose exposure to each tissue or matrix (μCi/h) is calculated using the following equation.

$$AUC_{0–\infty}(\mu g - eq\ h/g) \times \text{test article specific activity } (\mu Ci/mg)$$
$$\times (\%\text{Tissue weight/Actual dose in rats } \mu Ci/g)$$

TABLE 18.A2 Estimated pharmacokinetic parameters used to calculate radiation absorbed from selected tissues in male Long–Evans rats following a single oral dose of x mg/kg dose of radiolabeled test article.

Tissue	AUC_{0-t} (μg-eq h/g)	$AUC_{0-\infty}$ (μg-eq h/g)	C_{max} (μg-eq/g)	T_{max} (H)	Half-life (H)	xy corr Coefficient (terminal phase)	Dose exposure (μCi H)	Radiation absorbed dose (mRad or mrem)[a]
Adrenal gland								
Bile								
Blood								
Bone								
Bone marrow								
Cecum								
Cecum contents								
Cerebellum								
Cerebrum								
Cerebrospinal fluid								
Diaphragm								
Epididymis								
Esophageal contents								
Esophageal mucosa								
Esophagus								
Exorbital lacrimal gland								
Eye								
Eye (lens)								
Fat (abdominal)								
Fat (brown)								

xy correlation; this coefficient predicts the goodness-of-fit for the concentration–time line (theoretical versus observed).
Note: When $AUC_{0-\infty}$ is not calculated, AUC_{0-t} is used to calculate the absorbed dose.
[a]Calculated for a 100-μCi dose of radiolabeled test article in humans.

$AUC_{0-\infty}$ is obtained from the rat pharmacokinetic data, and the test article specific activity is derived from analysis of the dose formulation administered to the rats.

If a tissue does not have an elimination phase and $AUC_{0-\infty}$ can not be calculated, AUC_{0-t} is used for calculating organ exposure.

The radiation absorbed dose in milliRad (mRad) is calculated as follows:

$$\text{Dose Exposure} \, (\mu Ci/h) \times S - \text{factor} \, (mRad/\mu Ci \, h)$$

where S-factor for $[^{14}C]$ is calculated in the following equation using human tissue mass (Snyder et al., 1975; ICRP Publication 2003):

$$0.105 \, \text{rad} \, g/\mu Ci \, h/(\text{mass of human tissue in grams}) \times 1000 \, mRad/Rad$$

TABLE 18.A3 Effective dose equivalent based on tissue distribution data.

Organ or tissue[a]	Weighting factor	Dose to human organ or tissue	
		Radiation absorbed dose[b] (mRad or mrem)	Weighted dose (mrem)
Bone marrow	0.12		
Large intestine	0.12		
Liver	0.05		
Lung	0.12		
Skin	0.01		
Stomach	0.12		
Testis	0.20		
Thyroid	0.05		
Urinary bladder	0.05		
Remaining five[c,d]			
××	0.06		
××	0.06		
××	0.06		
××	0.06		
××	0.06		

Effective dose equivalent (overall whole-body exposure, mrem) ××[e]

NC Not calculated.

[a]Dose = weighting factor (from × radiation absorbed dose. Organs ICRP (1990) analyzed are critical organs as defined by ICRP 60 (1990).

[b]Calculated for a 100-μCi dose of radiolabeld article in humans.

[c]Dose = weighting factor (from ICRP (1979)) × radiation absorbed dose. Organs analyzed are those five organs (excluding those analyzed per ICRP 60 (1990)) with the highest radiation absorbed dose.

[d]The five organs or tissues receiving the largest absorbed dose (from Table 18.A1).

[e]Value varies depending on the radioactive exposures presented in Table 18.A1.

TABLE 18.A4   Calculated effective dose equivalent and organ doses compared to the applicable FDA limits.

Whole body or organ	Dose limits (mrem)[a]		Radiation absorbed dose[b] (mRad or mrem)	% Limit (Single study)
	Annual	Single study		
Whole body (overall exposure)	5,000	3,000		
Active blood forming organs	5,000	3,000		
Lens of the eye	5,000	3,000		
Gonads	5,000	3,000		
Other organs[c]	15,000	5,000		

NC Not calculated.
[a]FDA limit from US Food and Drug Administration 1999.
[b]Calculated for a 100-μCi dose of ^{14}C- test article in humans.
[c]The other organs with the highest estimated exposure.

The effective dose equivalent in man in millirem (mrem) is calculated using the following equation:

$$\text{Radiation absorbed dose(mRad)} \times Q\,(\text{rem/Rad}) \times \text{Weighting Factor}_{\text{for tissues}}$$

where Q (rem/Rad) for [^{14}C]-labeled compounds is 1(Weber et al., 1989).

Weighting factors for tissues are based on the recommendations of the International Commission of Radiological Protection (ICRP) Publications 60 (ICRP Publication 60, 1990) and 30 (ICRP Publication 30, 1979). ICRP 60 weighting factors are applied to the values for bone marrow, large intestine, liver, lung, skin, stomach, gonads, thyroid, and urinary bladder. ICRP 30 weighting factors are applied to the five remaining tissues with the highest exposure.

According to federal guidelines (U.S. Food and Drug Administration, 1999), the limits for exposure to radioactivity for human volunteers in the course of single dose studies with radioisotopes are 3000 mrem for exposure of the whole body, active blood forming organs, lens of the eye, and gonads, and 5000 mrem for the tissue with the highest effective dose. Annual cumulative limits are 5000 mrem for exposure of whole body, active blood-forming organs, lens of the eye, and gonads, and 15,000 mrem for exposure of other organs.

Based on the pharmacokinetic and dosimetry data, administration of a single oral 100-μCi dose of the radiolabel would not be expected to represent a significant radiation exposure risk in man (Tables 18.A3 and 18.A4).

REFERENCES

Baillie TA, Cayen MN, Fouda H, Gerson RJ, Green JD, Grossman SJ, Klunk LJ, LeBlanc B, Perkins DG, and Shipley LA. Drug metabolites in safety testing. Toxicol Appl Pharmacol 2002;182:188–196.

Beumer JH, Beijnen JH, Schellens JHM. Mass balance studies, with a focus on anticancer drugs. Clin Pharmacokinet 2006;45(1):33–58.

Boado RJ, Kang Y-S, Wu D, Pardridge WM. Rapid plasma clearance and metabolism *in vivo* of a phosphorothiorate oligodeoxynucleotide with a single, internal phosphodiester bond. Drug Metab Dispos 1995;23:1297–1300.

Chando TJ, Everett DW, Kahle AD, Starrett AM, Vachharajani N, Shyu WC, Kripalani KJ, Barbhaiya RH. Biotransformation of irbesartan in man. Drug Metab Dispos 1998;26:408–417.

Garner RC, Goris I, Laenen AAE, Vanhoutte E, Meuldermans W, Gregory S, Garner JV, Leong D, Whattam M, Calam A, Snel CAW. Evaluation of accelerator mass spectrometry in human mass balance and pharmacokinetic study-experience with ^{14}C-labeled (*R*)-6-[amino(4-chlorophenyl)(1-methyl-1*H*-imidazol-5-yl)methyl]-4-(3-chloro phenyl)1-methyl-2(1*H*)-quinolinone (R115777), a farnesyl transferase inhibitor. Drug Metab and Dispos 2002;30:823–830.

Gibaldi M, Perrier D. Pharmacokinetics. 2nd ed.New York: Marcel Dekker, Inc.; 1982.

Heggie GD, Sommadossi J-P, Cross DS, Huster WJ, Diasio RB. Clinical pharmokinetics of 5-fluorouracil and its metabolites in plasma, urine and bile. Cancer Res 1987;47(8):2203–2206.

ICRP Publication 30. Limits for the Intake of Radionuclides by Workers, Part 1. Ann ICRP 1979;2:3–4.

ICRP Publication 60. Recommendations of the International Commission on Radiological Protection. 1990.

ICRP Publication 89. Basic Anatomical and Physiological Data for Use in Radiological Protection: Reference Values. 2003.

Krishna R, Yao M, Srinivas NR, Shah V, Pursley JM, Arnold M, Vacharajani NN. Disposition of radiolabeled BMS-204352 in rats and dogs. Biopharm Drug Dispos 2002;23:41–46.

Li W, Zhang D, Wang L, Zhang H, Cheng P, Zhang DX, Everett D, Humphreys WG. Biotransformation of [14C]muraglitazar in male mice. Interspecies differences in metabolic pathways leading to unique metabolites. Drug Metab Dispos 2006;34:807–820.

Lu K, Benvenuto JA, Bodey GP, Gottlieb JA, Rosenblum MG, Loo TL. Pharmacokinetics and metabolism of β-2′-deoxythioguanosine and 6-thioguanine in man. Cancer Chemother Pharmacol 1982;8(1):119–123.

Lellig H. 1996. European Council Directives. Council Directive 96/29/Euratom: laying down basic standards for the protection of the health of workers and the general public against the dangers arising from ionizing radiation. Available athttp://ec.europa.eu/energy/nuclear/radioprotection/doc/legislation/9629_en.pdf.

Mauthe RJ, Snyderwine EG, Ghoshal A, Freeman SP, Turteltaub KW. Distribution and metabolism of 2-amino-1-methyl-6-phenylimidazo[4,5-b]pyridine (PhIP) in female rats and their pups at dietary doses. Carcinogenesis 1998;19(5):919–924.

Nuis A. 1997. European Council Directives. Council Directive 97/43/Euratom: on health protection of individuals against the dangers of ionizing radiation in relation to medical exposure, and repealing Directive 84/466/Euratom. Available at http://ec.europa.eu/energy/nuclear/radioprotection/doc/legislation/9743_en.pdf.

Owellen RJ, Hartke CA, Hains FO. Pharmocokinetics and metabolism of Vinblastine in humans. Cancer Res 1977;37:2597–2602.

Rickert DE, Dingley K, Ubick E, Dix KJ, Molina L. Determination of the tissue distribution and excretion by accelerator mass spectrometry of the nonadecapeptide ^{14}C-Moli1901 in beagle dogs after inytratracheal installation. Chem-Biol Interact 2005;155:55–61.

Slatter JG, Schaaf LJ, Sams JP, Feenstra KL, Johnson MG, Bombardt PA, Cathcart KS, Verburg MT, Pearson LK, Compton LD, Miller LL, Baker DS, Pesheck CV, Lord III RS. Pharmacokinetics, metabolism and excretion of irinotecan (CPT-11) following I.V. infusion of [14C]CPT-11 in cancer patients. Drug Metab Dispos 2000;28(4):423–433.

Snyder WS. Absorbed Dose Per Unit Cumulated Activity for Selected Radionuclides and Organs, Nm/mird Pamphlet No.11. Oak Ridge: Oak Ridge National Laboratory; 1975.

US Food and Drug Administration.21 CFR 361 April 1,1999.

Wafelman AR, Hoefnagel CA, Maesse HJ. Renal excretion of iodine-131 labeled *meta*-iodobenzylguanidine and metabolites after therapeutic doses in patients suffering from different neural crest-derived tumors. Eur J Nucl Med 1997;24(5):544–552.

Walle T, Walle UK, Kumar GN. Axol metabolism and disposition in cancer patients. Drug Metab Dispos 1995;23:506–512.

Wang L, Zhang H, Swaminathan A, Xue J, Cheng TC, Wu S, Bonacorsi S, Zhu M, Swaminathan A, Hunphreys WG. Glucuronidation as a major metabolic pathway of muraglitazar in humans: metabolite profiles in subjects with and without bile collection. Drug Metab Dispos 2006;34:427–439.

Weber DA, Eckerman KF, Dillman LT, Ryman JC. MIRD Radionuclide Data and Decay Schemes. New York: Society of Nuclear Medicine; 1989.

Zhang HY, Zhang DL, Ray K. A software filter to remove interference ions from drugs metabolites in accurate mass liquid chromatography/mass spectrometric analyses. J Mass Spectrom 2003;38:1110–1112.

Zhang D, Wang L, Raghavan N, Zhang H, Li W, Cheng PT, Yao M, Zhang L, Zhu M, Bonacorsi S, Yeola S, Mitroka J, Hariharan N, Hosagrahara V, Chandrasena G, Shyu WC, Humphreys WG. Comparative metabolism of radiolabeled muraglitazar in animals and humans by quantitative and qualitative metabolite profiling. Drug Metab Dispos 2007;35:150–167.

Zhang D, Zhang H, Aranibar N, Hanson R, Huang Y, Cheng PT, Wu S, Bonacorsi S, Zhu M, Swaminathan A, Humphreys WG. Structural elucidation of human oxidative metabolites of muraglitazar: use of microbial bioreactors in the biosynthesis of metabolite standards. Drug Metab Dispos 2006;34:267–280.

INDEX

Drug Metabolism in Drug Design and Development, Edited by Donglu Zhang, Mingshe Zhu and W. Griffith Humphreys
Copyright © 2008 John Wiley & Sons, Inc.